Fundamentals of Manufacturing Engineering

D. K. Singh

Fundamentals of Manufacturing Engineering

Concepts and Applications

Third Edition

D. K. Singh
Department of Mechanical Engineering
Netaji Subhas University of Technology
New Delhi, India

ISBN 978-981-99-8769-6 ISBN 978-981-99-8767-2 (eBook)
https://doi.org/10.1007/978-981-99-8767-2

This Springer imprint is published by the registered company Springer Nature Singapore Pte Ltd.
The registered company address is: 152 Beach Road, #21-01/04 Gateway East, Singapore 189721, Singapore

Dedicated to my parents,
my wife Alka and my lovely daughters
Shivangi and Shalvi

Preface to the Third Edition

The revised edition of the book tends to eliminate the existing errors noticed in the book from time to time. At the same time, it contains one additional chapter on 'Manufacturing Tools and Workshop Applications', which has enhanced the scope of the book from workshop point of view.

I hope readers will enjoy the revised edition of the book. Any suggestion for the improvement of the book will be highly acknowledged.

New Delhi, India D. K. Singh

Preface to the Second Revised Edition

The revised edition of the book tends to eliminate the existing errors noticed in the book from time to time. At the same time, it contains one additional chapter on 'Manufacturing Tools and Workshop Applications', which has enhanced the scope of the book from workshop point of view.

I hope readers will enjoy the revised edition of the book. Any suggestion for the improvement of the book will be highly acknowledged.

New Delhi, India
January 2011

D. K. Singh

Preface to the First Edition

Manufacturing engineering is an interesting field, and it has vast scope, as it finds extensive applications in many types of industries. The author has made an attempt to sincerely present the subject in an interesting manner using simplified language. This subject forms the core subject for all branches of engineering, which shows its importance. The entire subject is divided into twenty-two chapters, and each chapter contains a number of multiple choice questions along with many short questions to test the knowledge of the readers. The questions will certainly raise curiosity in the minds of the readers.

The author is thankful to his publisher Mr. Sunil Saxena and Mr. Jai Raj Kapoor for their keen interest in the project and Mr. Atin Chatterjee for nice typesetting and the entire unit of the publication division who are directly or indirectly involved in the publication of the book.

Finally, I am also thankful to my wife Alka and daughters Shalu and Sheelu for their constant supports and encouragement without whom the work was impossible to complete.

D. K. Singh

Acknowledgments

I would like to express my sincere thanks to many people who have helped in the publication of this book. First and foremost, my family, who were always supportive of my work and curious about its publication. They always made me in high spirit and confidence. I would like to thank my publisher Springer for publishing this book. I am also thankful to Ms. Priya Vyas, Senior Editor, Springer for taking keen interest in the book. I want to express my sincere gratitude to everyone involved in the production of this book.

I hope this book will meet your expectation. Any suggestion for the improvement of the book will be highly acknowledged.

D. K. Singh

Contents

About the Author

Dr. D. K. Singh is a Professor and the Founder Head of the Department of Mechanical Engineering at Netaji Subhas University of Technology (NSUT), New Delhi, India. He has an extremely rich experience of both teaching and research for nearly three decades. He has contributed over 50 papers to various journals of national and international repute. He has also participated and presented papers at several national and international conferences. He has authored 12 books. His books are published from Springer, CRC Press, and Pearson. Two of his books titled Strength of Materials (Fourth Edition) and Dictionary of Mechanical Engineering, both published by Springer, have received wide acclamation. He is a life member of the Indian Society for Technical Education (ISTE), New Delhi.

Introduction to Materials and Their Properties

1

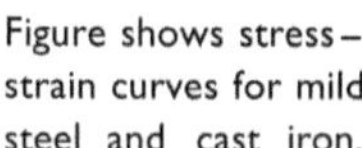
Figure shows stress–strain curves for mild steel and cast iron.

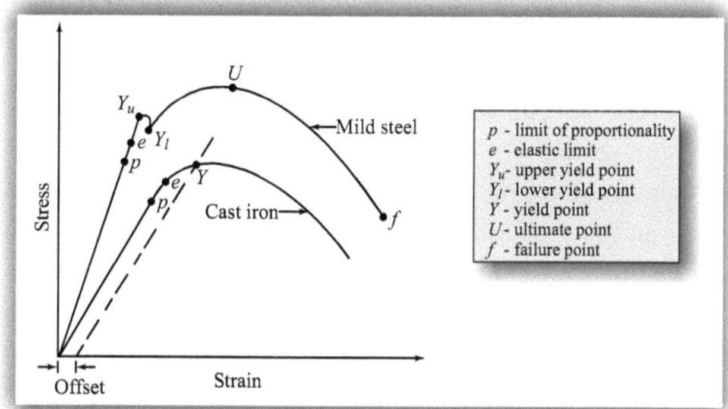

Learning Objectives

After reading this chapter, you will be able to answer some of the following questions:

- What is the basis of classification of materials into different groups?
- Why is oil at a particular temperature more dangerous than water at the same temperature?
- Why are mechanical properties so important for industrial applications?
- Why is true curve for stress–strain different than engineering curve?

D. K. Singh, *Fundamentals of Manufacturing Engineering*,
https://doi.org/10.1007/978-981-99-8767-2_1

1.1 Introduction

Materials are one of the important elements of a manufacturing system. Since the list of materials is large and is constantly increasing with new arrivals, this makes the task of selection of material very difficult. It requires experience to choose a particular material. The selected material should meet the engineering requirements of the designers, the manufacturing requirements of the manufacturers and finally the economics factor of the customer. Engineering materials differ widely in physical properties, mechanical properties, machinability characteristics and manufacturing methods. All these points must be considered during the selection of a suitable material.

1.2 Classification of Materials

Generally materials are classified into two groups, namely metallic and nonmetallic. Some of the elements such as carbon, sulphur, silicon and phosphorous behave as metals and nonmetals both. These elements are referred to as metalloids. Metallic group is further subdivided into ferrous and nonferrous. Nonmetallic materials are subdivided into organic and inorganic groups.

The metallic group is characterized by its lustrous, hard, malleable, heavy and ductile properties and is a conductor of heat and electricity. The important metals include iron, copper, lead, zinc, aluminium, tin, nickel and magnesium. Few of them are used in their pure states e.g. copper and aluminium, but majority of them are used in the alloyed form. An alloy is a combination of two or more metals and its properties are different from its constituent metals. A few examples of alloys include stainless steel, bronze, brass, etc.

Iron is the chief element present in ferrous materials. These materials also contain alloying elements. Important ferrous materials include grey cast iron, malleable iron, white cast iron, wrought iron and steel.

Nonferrous materials do not contain iron and carbon as their constituents. They are not used in their pure state for majority of industrial applications because they lack structural strength. Their strength is considerably increased in the alloyed form. Aluminium, copper, magnesium, nickel, lead, titanium and zinc are some of the important nonferrous metals. Important properties of nonferrous alloys are good resistance to corrosion, high electrical conductivity and good formability. The electrical properties of nonferrous materials are usually superior to iron. Copper has approximately five times and aluminium has approximately three times more electrical conductivity than iron. But their weldability is poor as compared to ferrous materials. They can be cast, formed or machined easily.

Organic materials contain animal or vegetable cells (dead or alive) or organic compounds. Leather and wood are examples. Inorganic materials do not contain animal or vegetable cells or carbon compounds. Organic materials are usually dissolved in organic liquids such as alcohol or carbon tetrachloride, but they are not dissolved in water. Inorganic materials tend to dissolve in water. In general, inorganic materials have better heat resisting characteristics than organic materials.

The scope of nonmetallic materials is constantly increasing. Advanced ceramics, composite materials and polymers are the materials of today and tomorrow.

1.3 Structure of Materials

Materials are composed of three fundamental atomic particles, namely protons, electrons and neutrons. The properties such as electrical, chemical, thermal and mechanical are all ultimately related to the atomic structure of the materials.

In addition to atomic structure factors such as composition of the metal, impurities and vacancies present in the atomic structure, grain size, grain boundaries and the method of forming the materials into useful products greatly influence the material's behaviour.

The smallest group of atoms showing the characteristic lattice structure of a particular metal is known as a unit cell. It is the building block of a crystal, and a single crystal can have many unit cells.

There are four types of atomic arrangement found in metals. They are (a) simple cubic (b) body-centered cubic (bcc) (c) face-centered cubic (fcc) and (d) hexagonal close-packed (hcp). The latter three types of structures, namely (b), (c) and (d) are commonly found in metals at the room temperature.

The simple cubic structure of metals is rarely observed. Only 52% of available space is occupied by the atoms in simple cubic structure, indicating that its atomic packing factor (APF)[1] is 0.52. It contains one atom at each of the eight corners of the cube and each atom is in direct contact with only six neighbour atoms (four in the same plane and two in a vertical plane). Hence, the coordination number for simple cubic structure is 6. Its low APF value and lesser number of nearest neighbour atoms tend to unfavour the formation of metallic bonds.

In bcc structure, there are eight atoms at the corners of the unit cell and one atom at the centre. As every corner atom is surrounded by eight unit cells having eight body-centered atoms, hence the coordination number of the bcc structure is 8. The atomic packing factor (APF) is 0.68 indicating that 68% of the space in the unit cell of bcc structure is occupied by the atoms. This structure is found in iron (α-phase which occurs below 912 °C and δ-phase which occurs above 1394 °C), chromium, manganese, sodium, potassium, tungsten and barium.

In fcc structure, there are eight corner atoms and one atom at the centre of each face. For any corner atom, there will be 4 face-centered atoms of the surrounding unit cells in its own plane, 4 face-centered atoms below this plane and 4 face-centered atoms above this plane. Hence, the coordination number of the fcc structure is 12. The APF is 0.74.

[1] APF is a ratio of volume of atoms to volume of unit cell.

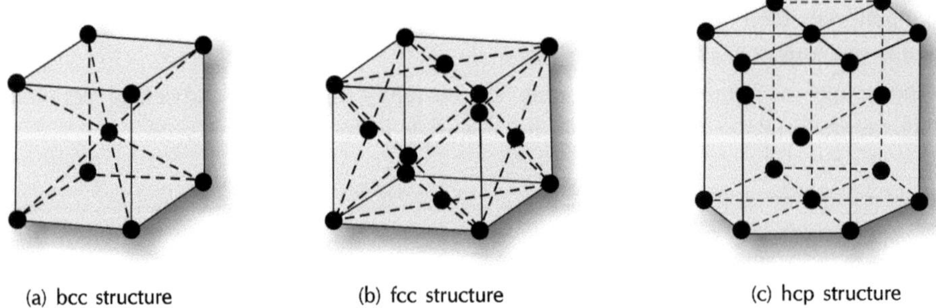

(a) bcc structure (b) fcc structure (c) hcp structure

Fig. 1.1 Atomic arrangements

Materials having fcc structure have very high formability (the ability to be plastically deformed without fracture). Important metals in which fcc structure is found include iron (γ-phase which occurs between 912 and 1394 °C), calcium, aluminium, copper, nickel, silver and lead.

The hcp structure is found in beryllium, cadmium, magnesium and zinc. There is an average of six atoms per unit cell in the hcp metal structure. The coordination number of the hcp structure is 12. The APF is 0.74. Metals having hcp structure have poor formability and often require special processing procedures. The commonly found three structures are shown in Fig. 1.1.

1.4 Physical Properties of Materials

Those characteristics that are employed to describe a material under conditions in which external forces are not concerned are called physical properties. These characteristics are mostly structure-insensitive, and they do not change with change in the structure of the material.

The physical properties of the materials include specific heat, density, thermal conductivity, thermal expansion, melting point, electrical conductivity and magnetic properties.

1.4.1 Specific Heat

The specific heat of a substance is defined as the amount of heat required to raise the temperature of one gram-mole of the substance through 1 K at constant pressure (C_{p}) or constant volume (C_{ν}). In case of solids and liquids, the difference between C_{p} and C_{ν} is negligible on account of very small expansion and therefore they have only one specific heat. On the other hand, in case of gases, there exists a definite difference between C_{p} and C_{ν} expressed as:

$$C_p - C_v = R \tag{1.1}$$

where R is a universal gas constant. This relation was first obtained by Mayer in 1842 but is strictly valid for an ideal gas with $R = 8.31$ J/mol-K and very nearly true for real gases at moderate pressures. Consideration of specific heat is important for processes such as casting, where heat is removed during solidification, and heat treatment, where quantities of materials are heated and cooled.

1.4.2 Density and Specific Gravity

Density of a material is defined as its mass per unit volume, usually expressed at 23 °C, whereas specific gravity is the ratio of density of the material to the density of a substance taken as a standard. In case of solids and liquids, water is taken as standard substance, while air is considered as standard substance for gases. This property is useful to determine the cost of different materials in the design process.

Density is important when weight reduction is an important factor particularly for aircraft and automobile structures and other products where energy consumption and power limitations are major considerations.

1.4.3 Thermal Conductivity

Thermal conductivity indicates the ease with which heat flows within and through the material. Transfer of thermal energy in solids is explained by three mechanisms. The first is by means of free electrons as in the case of metals. The second is by means of molecules as in the case of organic solids, and the third by lattice vibration. In the metals, thermal conduction occurs primarily through electron transfer, but the lattice vibration contributes to the transfer.

The thermal conductivity of metals is lower than their alloys due to scattering of electrons and *phonons*[2] in the presence of alloying (foreign) atoms.

It is essential to dissipate the heat generated during friction or any plastic deformation process to prevent a severe rise in temperature. Low thermal conductivity can produce high thermal gradients and thus cause inhomogeneous deformation during metal working processes.

[2] At temperatures above 0 K, the atoms vibrate randomly about their mean positions. These vibrations can be considered as elastic waves in the crystal and are called phonons. Their random nature destroys the ideal periodicity of a crystal and interferes with the electron motion. Consequently, the mean free path and the conductivity decrease with increasing temperature.

The thermal conductivity of metals is directly proportional to their electrical conductivity. Thus materials such as copper, gold and aluminium that have good electrical conductivity will also be good conductors of thermal energy.

1.4.4 Thermal Expansion

Addition of thermal energy to a substance produces an imbalance between attractive and repulsive forces in its structure. The change in parameters of the structure means a change in dimension, which is evident in the coefficient of thermal expansion.

The potential energy of a pair of atoms in a structure can be expressed as

$$E = -\frac{a}{r^m} + \frac{b}{r^n} \tag{1.2}$$

where a, b, m and n are constants with $n > m$, and r is the atomic distance. If plotted graphically, this relationship takes the form as shown in Fig. 1.2. The shape of the minimum portion of the curve determines the degree of expansion. Covalent as well as ionic solids show a steep curve on each side of the minimum, and these substances have small expansions. Molecular solids have the greatest expansion of all substances. Therefore, by examining the curve of energy versus atomic distance, one can predict about the amount of expansion in a substance.

Study of relative expansion or contraction is important in electronic and computer components, glass-to-metal seals, struts on jet engines, and moving parts in machinery that require certain clearances for proper functioning. Improper selection of materials and assembly can cause thermal stresses, cracking and warping during their service life.

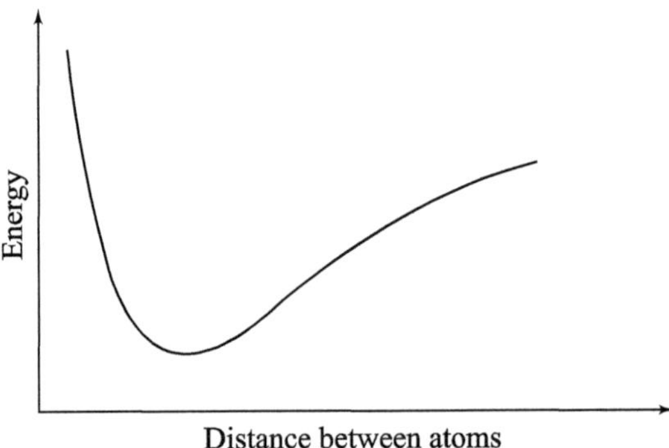

Fig. 1.2 Energy versus interatomic distance curve

1.4.5 Melting Point

Melting point is defined as the temperature at which the solid phase of a substance changes into its liquid phase, and the two phases coexist. It depends upon the types of chemical bond and the amount of heat required to break the bond. Thus materials having stronger bonds have higher melting points. The materials in order of their increasing melting points are molecular, metallic, ionic and covalent bonded respectively. Diamond is an excellent example of the perfect covalent bond, and it has the highest melting point of any solid.

The knowledge of melting points of materials is important for operations such as heat treatment, hot working and casting. Also, melting point consideration is a key factor during the selection of tool and die materials in manufacturing operations. The higher the melting point of the material, the more difficult the operation becomes.

1.4.6 Electrical Conductivity

The electrical conductivity of a material is defined as its ability to conduct electricity. This property is chiefly dependent on the number of free electrons in the materials. Metals are good conductor of electricity because they contain large number of free electrons. Electrical resistivity is the inverse of conductivity and dielectrics or insulators are the materials having high resistivity.

Electrical conductivity or resistivity is often an important design consideration while selecting the materials for electrical appliances. The manufacturing processes where consideration of electrical conductivity of the material is a key factor are magnetic-pulse forming of sheet metals, and electrodischarge machining (EDM) and electrochemical grinding (ECG) of hard and brittle materials.

1.5 Magnetic Properties of Materials

Magnetic properties of solids are connected with the spin of electrons in their atoms and their magnetic moments or dipoles. The electron orbits are always associated with magnetic moments. Magnetic materials can be diamagnetic, paramagnetic, ferromagnetic, antiferromagnetic or ferrimagnetic (ferrite).

In diamagnetic materials, the permanent magnetic moments or dipoles are absent; therefore, they cannot be made magnet. Typical examples are gold, silicon, diamond, ionic solids and molecular solids.

The permanent moments are randomly distributed in paramagnetic materials. They align themselves parallel to the applied magnetic field feebly and become weak magnets. Paramagnetism is found in the oxides, chlorides and sulphates of some metals.

Paramagnetic salts which are rare-earth-based compounds, such as $ErNi_2$, EuS, Gd_5Si_4, $Dy_3Al_5O_{12}$ and $Gd_3Ga_5O_{12}$ also fall in the category of paramagnetic materials.

In ferrimagnetic materials, the moments are aligned systematically (parallel arrangement) and are equal in magnitude. They are very good magnetic materials. Iron, nickel and cobalt are ferromagnetic materials.

Antiferromagnetic and ferrimagnetic materials are the subclasses of ferromagnetic materials. In antiferromagnetic materials, the magnetic moments of equal magnitude are aligned in opposite directions. Oxides of manganese, iron and cobalt and compounds of nickel and manganese fall under this category.

In ferrimagnetic materials, better known as ferrites, the magnetic moments of unequal magnitude are aligned in opposite directions. The magnitude of magnetic moment is more in one direction than in the other. Ferrites are compounds of two metallic oxides of which one is invariably an iron oxide. Magnetite (Fe_3O_4) is a compound of ferrous oxide (FeO) and ferric oxide (Fe_2O_3) and is expressed as (FeO . Fe_2O_3). Other ferrites which are more commonly used are: manganese ferrite ($MnFe_2O_4$), nickel ferrite ($NiFe_2O_4$), magnesium-manganese ferrite, zinc-manganese ferrite and zinc-nickel ferrite. Figure 1.3 shows different types of magnetic materials.

Magnetic materials find extensive applications in electrical equipment such as motors, generators, transformers, alternators, electromagnets and microwave devices.

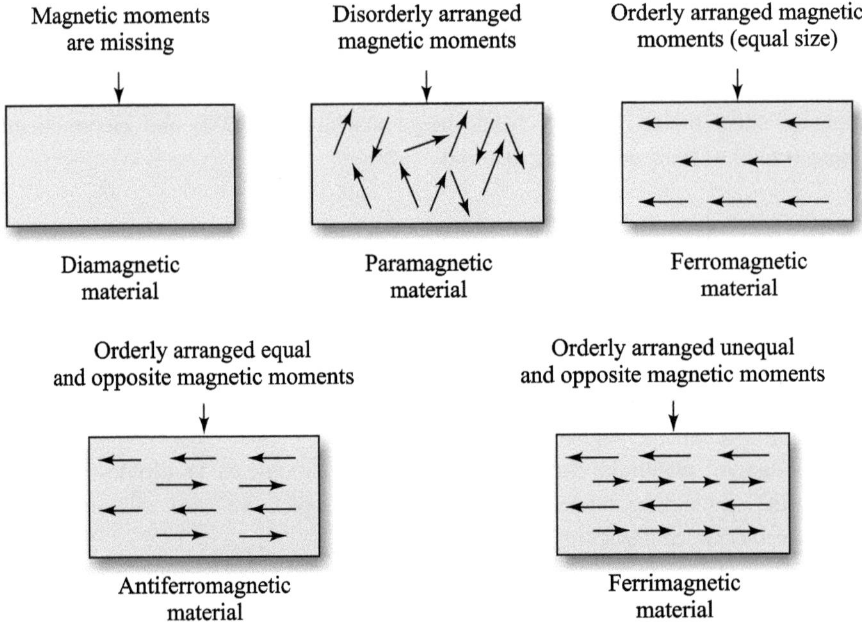

Fig. 1.3 Arrangement of magnetic moments in magnetic materials

1.6 Mechanical Properties of Materials

The reaction of materials in response to applied external force is defined as their mechanical properties. The mechanical properties are structure-sensitive, whereas the physical properties are structure-insensitive. Mechanical properties of materials play a vital role in a manufacturing system in the development of structures, machines, etc. These properties are evaluated by conducting destructive testing. Mechanical properties include tensile strength, ductility, compressive strength, torsional strength, modulus of elasticity, hardness, fatigue and creep.

1.6.1 Tensile Strength

Tensile strength is defined as the ability of a material to support axial load without rupture. A tension test is conducted to know the tensile strength of a material.

In the tension test, a certain length, known as gauge length (usually 50 mm), of the material is subjected to an axial load as shown in Fig. 1.4a. The diameter of the cross section of the material to be tested is generally taken to be 12.5 mm. The behaviour of the material is shown in the form of a stress–strain curve in Fig. 1.5.

As the load on the test specimen increases, the resisting force also increases. Once the load is removed, the material regains its original shape and size. Up to this stage, Hooke's law is strictly valid, i.e. stress (a ratio of the applied load to the original cross-sectional area) is linearly proportional to strain (ratio of change in length to original length). This point is known as limit of proportionality, *p*. Up to this point, the material regains its original conditions very fast once the load is removed. Point, *e* is called elastic limit. Up to *e* material recovers its original conditions on the removal of load but the recovery process is negligibly slow as compared to limit of proportionality. For some materials, these two points, *p* and *e* are almost identical, but in most of the cases, the elastic limit is slightly higher than limit of proportionality.

Elongation beyond the elastic limit becomes unrecoverable and the material remains permanently deformed. This situation is known as plastic deformation. The material does

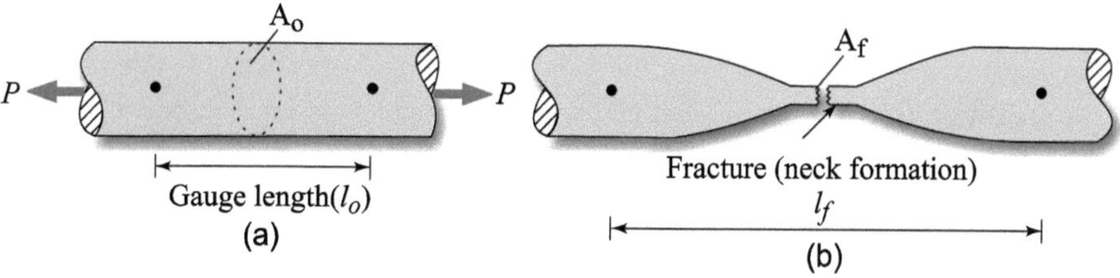

Fig. 1.4 A tension test ($A_f << A_o$ and $l_f > l_o$)

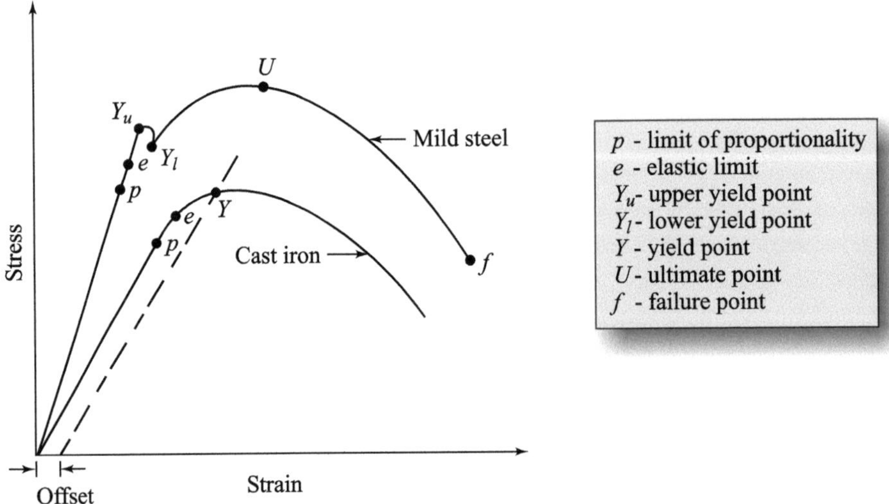

Fig. 1.5 Engineering stress–strain curves

not regain its original conditions on the removal of load. It means the external load has exceeded the resisting force. Plasticity and elasticity are important properties from manufacturing point of view.

When the elastic limit is exceeded, the stress–strain relationship is no longer linear. Rather, increase in strain is much more prominent as compared to the increase in stress. This point is known as yield point, and the corresponding stress is called the yield stress. For low-carbon steels, there are two yield points, the upper and the lower one. But it is the lower yield point which usually appears in the tabulated data. As the load and hence the stress is further increased, it reaches its maximum value and then begins to decrease. The maximum stress is called the ultimate stress or tensile strength or ultimate tensile strength, U of the material.

If the specimen is loaded beyond its tensile strength, neck formation starts. During this period, the cross-sectional area does not remain constant rather it reduces drastically in the necked region. The stress drops further, and the specimen finally fractures in the necked region at the point f as shown in Fig. 1.5. The stress at this point is known as fracture stress.

For some materials, for example, the yield point is not well defined. The elastic-to-plastic transition is not clear in such cases, and the yield point is determined by offset method. In this method, the strain usually known as offset strain is specified at 0.2%. The point of intersection of a line drawn at offset strain with the stress–strain curve gives the position of yield point. If the applied stresses are lower than the yield strength at 0.2% strain, then the plastic behaviour of the material is more deterministic.

The modulus of elasticity, E is obtained as the slope of the liner part of the stress–strain curve.

When a bar is subjected to a tension test, there is an increase in its length in the direction of load applied, but at the same time there is a reduction in the lateral dimension in the transverse direction (direction perpendicular to the direction of load applied). These changes in dimensions are related to each other by Poisson's ratio, v defined as

$$v = -\frac{\text{lateral strain}}{\text{longitudinal strain}}$$

Negative sign indicates that if longitudinal strain increases, lateral strain decreases. For most metals, v is about 0.33, and for cork it is close to zero. In some extreme cases, values like 0.1 (cement concrete) and 0.5 (rubber) have been observed. Its value is the same in tension as well as in compression.

1.6.2 True Stress–Strain Curves

Materials do not obey Hooke's law in plastic region. Figure 1.5 has been drawn considering the original cross-sectional area of the test specimen. But it is not the actual case. The area gradually reduces as the specimen is elongated especially after ultimate point. Thus the engineering stress does not represent the actual or true stress. Similarly the true strain differs from the engineering strain. These quantities are defined as follows:

$$\text{True stress} = \frac{\text{load applied}}{\text{instantaneous area of cross-section}}$$

and

$$\text{True strain} = \ln\left(\frac{l_f}{l_o}\right)$$

where

l_f = instantaneous length
l_o = original length.

True strain is also known as natural or logarithmic strain. For small values of strain, the two strains are approximately equal, but the difference increases for increasing value of strain.

The true stress–strain curve can be represented by the equation

$$\sigma = K.\in^n$$

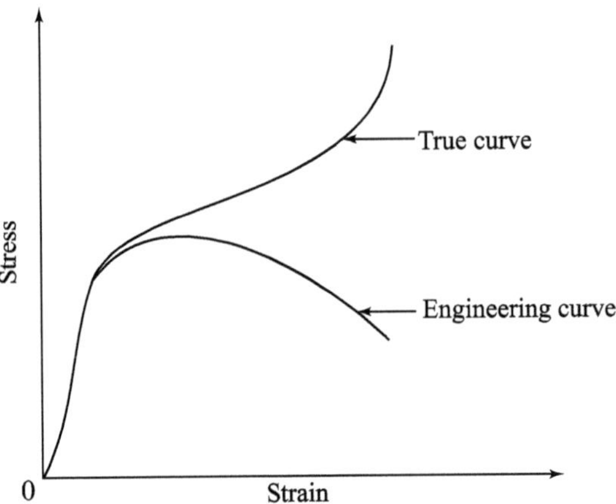

Fig. 1.6 Engineering and true stress–strain curves

where

σ = stress
$\in$ = strain
K = a constant, known as strength coefficient
n = strain-hardening exponent

$\neq 1$ for true curve

$= 1$ for engineering curve, where $K = E$ (modulus of elasticity).

Stress–strain curves in the actual and engineering cases are shown in Fig. 1.6. It is observed that for a certain value of strain, true stress is higher than engineering stress.

The higher value of n is indicative of the relative ease with which a material can be deformed before neck formation. This important aspect is useful in forming operations, where workpiece material is stretched beyond elastic region. The necked region is subjected to three-dimensional tensile stresses which are responsible for increased stress over actual true stress, hence the curve is corrected downward as shown in Fig. 1.7.

1.6.3 Ductility

Ductility of a material is defined as its ability by virtue of which it can be drawn into wire without rupture. The higher ductility of a material is indicative of its ability to sustain higher tensile strain before rupture.

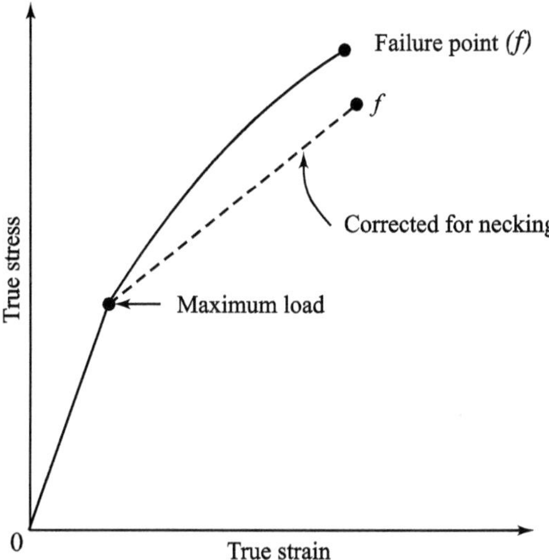

Fig. 1.7 Stress–strain curve under neck formation

Ductility can be measured by two parameters. Percentage elongation, a parameters useful during tension test, is defined as:

$$\text{Percentage elongation} = \frac{l_f - l_o}{l_o} \times 100$$

where

l_f = final length
l_o = original length.

The second parameter is based on change in the cross-sectional area of the test specimen, defined as

$$\text{Percentage reduction of area, } R_A = \frac{A_o - A_f}{A_o} \times 100$$

where A_o and A_f are respectively the original and the final cross-sectional areas.

Values for these two measures run parallel to each other; both are high in a ductile material and are nil in a nonductile material. Chalk has zero ductility because it does not stretch at all. R_A can range from 0% (brittle) to 100% (extremely plastic). An arbitrary strain of 0.05 inch/inch is frequently taken as the dividing line between brittle and ductile materials. For relatively ductile materials, the breaking strength is less than the ultimate tensile strength, and necking precedes fracture. For a brittle material, fracture usually occurs before necking, and possibly before the onset of plastic flow.

1.6.4 Compressive Strength

Compressive strength of a material is determined by conducting a compression test where the test specimen is subjected to a compressive load (Fig. 1.8). Compressive stress–strain curves for composite material, rubber and cork are shown in Fig. 1.9.

The relationship between stress and strain are similar to those for a tension test. But when the engineering stress–strain curve is compared with the true stress–strain curve, some variations are noticed as shown in Fig. 1.10.

Compression test is more difficult to conduct than a standard tension test. Test specimen should have larger cross-sectional areas to resist bending or buckling. Selection of a test is decided by the type of service a material is subjected to. Compressive strength is easily determined for brittle materials that will fracture when a sufficient load is applied,

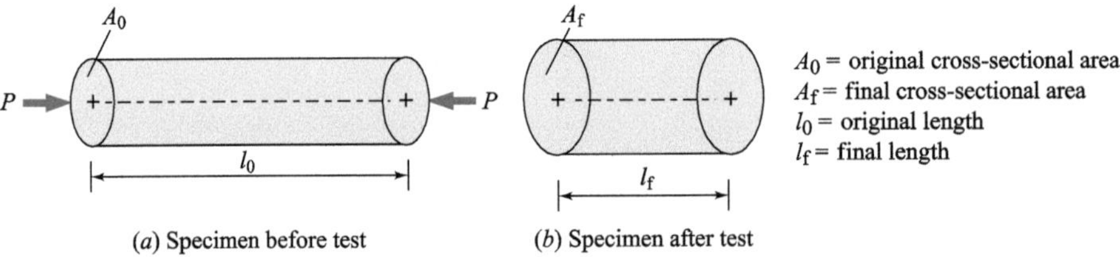

Fig. 1.8 A compression test

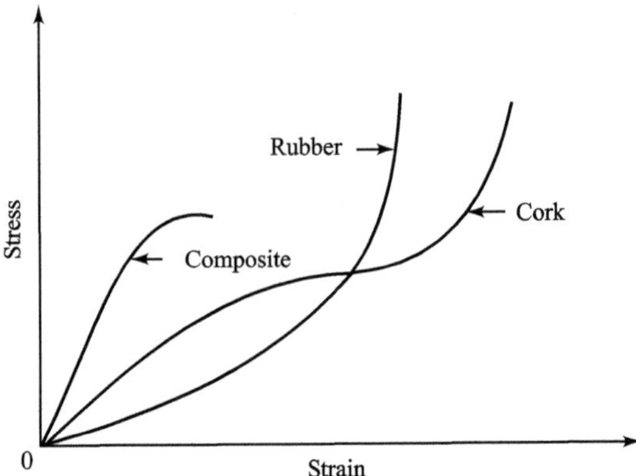

Fig. 1.9 Stress–strain curves in compression

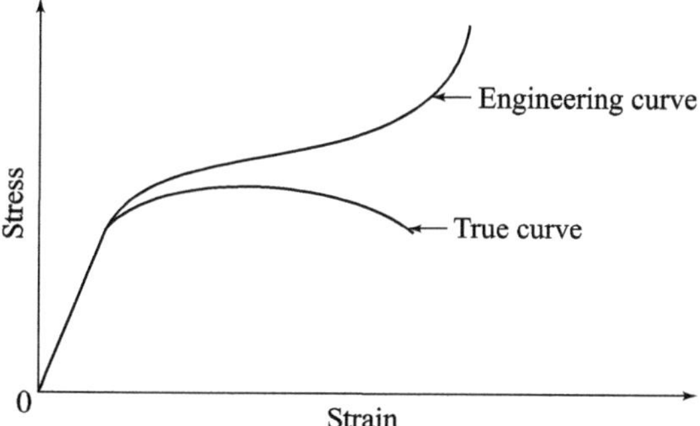

Fig. 1.10 Engineering and true stress–strain curves in compression

but for ductile materials a strength in compression is valid only when the amount of deformation is specified. The compressive strength of cast iron, a relatively brittle material, is three to four times its tensile strength.

1.6.5 Torsional Strength

Torsional strength of the material is measured in terms of shear stress and the test conducted for it is known as torsion test. The test measures the ability of the material to resist twisting moment (torque) and is usually performed on a thin tubular specimen so as to obtain uniform stress and strain distribution along the cross section. Punching of hole in sheet metals and metal cutting are the processes where the material is subjected to shear stress. Figure 1.11 shows the variation of torque with angle of twist, and Fig. 1.12 gives a relationship between shear stress and shear strain. In both figures, the curve is linear up to the point *e*, which means Hooke's law is valid from *O* to *e*. Slope of the curve in Fig. 1.12 gives the value of *G*, the modulus of rigidity or shear modulus.

Shear strain produced either in a hollow circular specimen such as a tube or a solid round bar is given by the expression

$$\phi = \frac{r\theta}{l} \tag{1.3}$$

where

ϕ = shear strain
r = radius of the test specimen
θ = angle of twist (in radian)
l = length of the specimen.

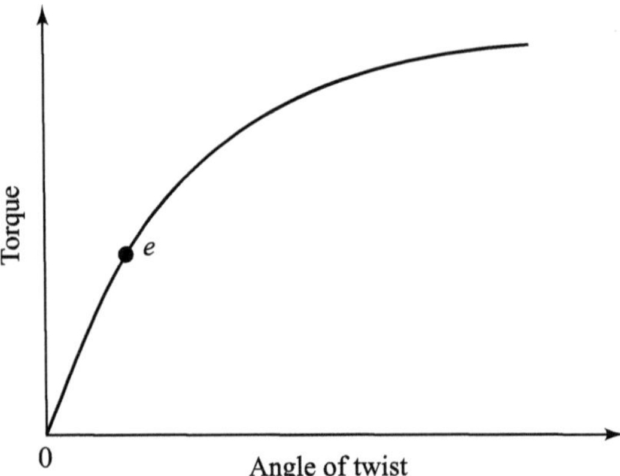

Fig. 1.11 Relationship between torque and angle of twist

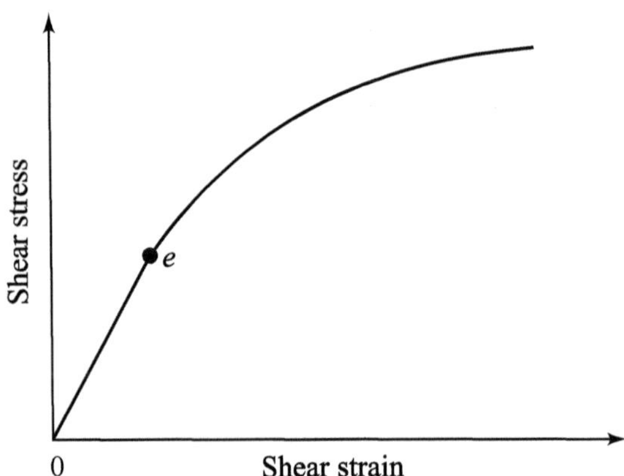

Fig. 1.12 Relationship between shear stress and shear strain

Angle of twist is an important consideration in the forgeability of metals. Higher twist before failure is indicative of greater forgeability of a metal.

The torsional shear strength, τ occurs at the outermost surface of the test specimen. For a solid cylindrical specimen of diameter d subjected to a twisting moment, T

$$\tau_s = \frac{16T}{\pi d^3} \tag{1.4}$$

And τ for a hollow cylindrical specimen is obtained from the expression

$$\tau_h = \frac{16T D_o}{\pi\left(D_o^4 - D_i^4\right)} \tag{1.5}$$

In the above equations, we have

T = twisting moment or torque
d = diameter of the solid section
D_o = outside diameter of hollow section
D_i = inside diameter of hollow section.

1.7 Hardness

Hardness is the resistance of a material to penetration. In other words, we can say that it is the resistance to wear and scratching. There are three important tests to measure the hardness of a material, namely Brinell hardness, Rockwell hardness and Vickers hardness.

1.7.1 Brinell Hardness Test

This test was developed by J. A. Brinell in 1900 and is one of the oldest methods of measuring the hardness of a material. Brinell hardness is determined by applying a predetermined test force of 500 kg, 1500 kg or 3000 kg to a hardened steel or tungsten carbide ball of fixed diameter ($\simeq$ 10 mm), which is held for a predetermined time (5–10 s) till an impression called indentation is left on the material to be tested and then removed. The resulting indentation produced on the test specimen is measured by means of a travelling microscope. The Brinell hardness number (BHN) is then defined as

$$\text{BHN} = \frac{\text{load}}{\text{surface area of indentation}}$$

$$= \frac{2F}{\pi D\left[D - \sqrt{\left(D^2 - d^2\right)}\right]} \tag{1.6}$$

where

F = force (kg)
D = diameter of test ball (mm) = 10 mm
d = diameter of indentation (mm).

The unit of Brinell hardness is kg/mm^2. The smaller the indentation, the harder is the material and hence higher is the BHN. Carbide balls are usually recommended for materials with BHN > 500. The method is extensively used for testing the hardness of iron and steel and on casting and forgings. Two major limitations of the test are:

- It is not suitable for measuring the hardness of very hard and very soft materials.
- Thickness of the test specimen should be at least 10 times the depth of indentation. It puts limitations on the use of thin sections.

1.7.2 Rockwell Hardness Test

Rockwell hardness test was developed by S. P. Rockwell in 1922 and is based on the measurement of depth of penetration rather than diameter of indentation as in case of Brinell hardness test. A smaller indentor, either a small-diameter steel ball or a diamond-tipped cone (included angle of 120°), called a brale, is pressed on the surface of the test specimen, first with a minor load of 10 kg and then with a major load of 60 kg, 100 kg or 150 kg depending on the material to be tested. The difference in the depths of penetration is a measure of the hardness of the material. It is the most widely used method of determining hardness in tool making.

1.7.3 Vickers Hardness Test

This test was developed in 1922 and was earlier known as diamond pyramid hardness (DPH) test. The Vickers hardness test uses a square-based diamond pyramid as the indentor whose apex angle is 136° and is loaded with a force (F) ranging from 1 to 120 kg, depending on the hardness and thickness of the test specimen. The advantage of the test lies in the fact that even a light load can produce plastic deformation, and hence, it can be used to determine microhardness. The indentor is pressed against the surface of the material to get the square impression in contrast to a circular impression as in case of Brinell and Rockwell tests. The Vickers hardness number is defined by the following equation

$$\begin{aligned} \text{VHN} &= \frac{\text{load}}{\text{surface area of indentation}} \\ &= 1.8544\frac{F}{\text{diagonal}^2} \end{aligned} \tag{1.7}$$

1.7.4 Mohs Hardness Test

Mohs test is the oldest one, which was developed by Friedrich Mohs in 1822. This test is based on the capacity of one material to scratch another. Mohs hardness is 1 for talc, more than 1–3 for layered minerals, 4–5 for ionic-bonded materials, 6–8 for ionic-covalent bonded materials, 9–10 for covalent-bonded materials and 10 for diamond. Lead shows its hardness much below unity on Mohs scale.

1.7.5 Knoop's Hardness Test

This test was developed by F. Knoop in 1939. It is regarded as a microhardness test, because light loads ranging from 25 to 3600 g is used in this test. The test is suitable for measuring small or thin specimens and for brittle materials like carbides, ceramics and glass. It can also measure the hardness of individual grains of the materials. The hardness of the material is expressed as Knoop hardness number (KHN), which is simply the ratio of load applied (in kg) to the area of diamond-shaped indentation (in mm^2).

1.7.6 Durometer

Durometer is used to measure the hardness of soft and elastic materials like rubbers and plastics. The indentor of the instrument is pressed against the test specimen, and the hardness is directly read from the dial of instrument which is suitably calibrated. The indentor does not leave any permanent impression on the test specimen, and the impression disappears once the indentor is removed.

1.8 Fatigue

The behaviour of the materials under fluctuating (cyclic or periodic) and reversing loads or stresses is known as fatigue. Figure 1.13 shows the cyclic loading arrangement. The stresses that a material can tolerate under cyclic loadings are much less than under static loading (tensile or compressive), and the failure of the material at this juncture is known as fatigue failure. Majority of the mechanical failures are due to fatigue. Wings of aeroplanes and other aircrafts, leaf springs used in automobiles, connecting rods of I.C. engines, rubber tyres, etc. are the examples which are subjected to rapid fluctuating loads. Figure 1.14 shows the variations of stresses in aeroplane operation.

The maximum stress to which the material can be subjected without fatigue failure, irrespective of the number of cycles, is known as the endurance limit or fatigue limit. This stress is generally 0.3–0.5 times the ultimate strength of materials.

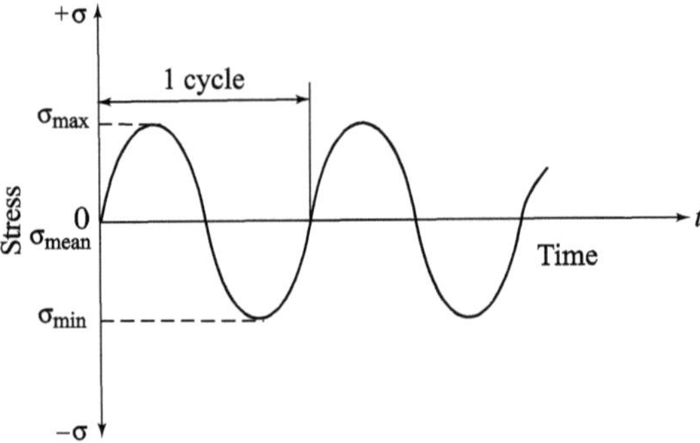

Fig. 1.13 Cyclic loading

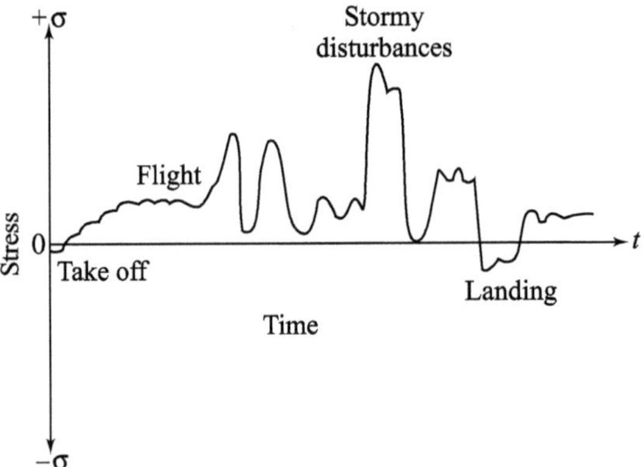

Fig. 1.14 Variation of stresses

A different number of loading cycles are required to determine the endurance limit for different materials. For steels, 10^7 cycles are usually sufficient. Several of the nonferrous metals require 5×10^8 cycles, and some aluminium alloys require such a great number that no endurance limit is apparent under typical test conditions.

If a graph is plotted between the cyclic stress (*S*) and the number of cycles (*N*) of failure, the resulting curve is known as *S–N* curve. The number of cycles *N* that a metal will endure decreases with increased stresses *S*. In designing for unlimited cyclic loading, it is necessary to restrict the stresses to values below the endurance limit of this curve.

Fatigue cracks usually start at the surface where bending or torsion causes the highest stresses to occur and where surface irregularities introduce stress concentrations. As a result, the endurance limit is very sensitive to surface finish. Any design factor that concentrates stresses can lead to premature failure.

1.9 Creep

Creep is a permanent deformation resulting from static loading of members over a long period of time. The thickness of window glass in old houses has been found to be greater at the bottom than at the top of windows, because the glass has undergone creep by its own weight over many years.

Creep occurs at room temperature in many materials such as lead, zinc, white metals, rubber, plastic and leather, etc. It is, however, appreciable at temperatures above 0.4 T_{m}, where T_{m} is the melting point of material (in Kelvin).

Creep is a slow mechanism of strain. The rates range from a few percent per hour at high stress levels, or at high temperatures, down to less than $10^{-4}\%$ per hour ($\simeq 1\%$ per year). Although, the rate of elongation is small, it is of great significance in the design of equipment such as blades of steam or gas turbines, pistons of I.C. engines in power plants and high-temperature pressure vessels such as boilers and steam lines that operate under loads and high temperatures for long periods of time. Creep also occurs in tools and dies that are subjected to high stresses at elevated temperatures during hot-working operations, such as forging and extrusion.

A creep test is conducted at constant tensile load and constant temperature, and change in length is measured over a period of time. A creep curve plotted between elongation (strain) and time is shown in Fig. 1.15. The curve contains three distinct stages.

- a short-lived initial stage, called primary stage. This stage usually occurs at very low temperatures, hence it is also called cold creep.

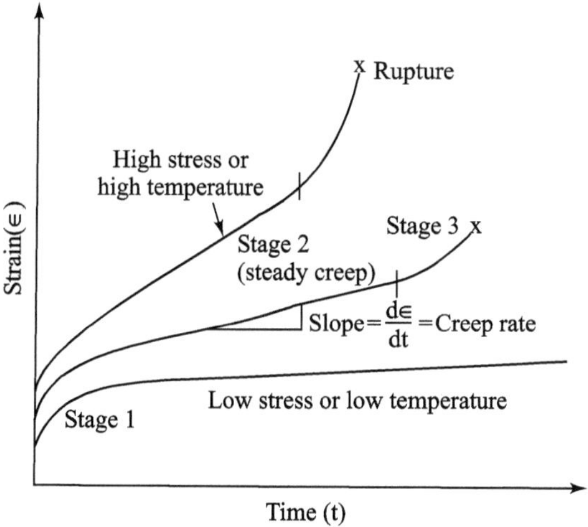

Fig. 1.15 Different stages in creep

- a long second stage, called secondary stage, where the elongation is somewhat linear. This stage dominates at high temperatures, hence also known as hot creep. The steady rate of creep in the second stage determines the useful life of the material, and
- a short-lived third stage, called tertiary stage, which leads to fracture. During this stage, the straining is too fast due to neck formation in the material.

The creep rate increases with increase in temperature and applied load.

Short Answer Questions

1. Why do solids have only one specific heat whereas gases have two?

Ans. The thermal expansion of solids is very low as compared to that of liquids which expand differently upon heating under different pressure conditions.

2. How is elastic deformation of a material different from plastic deformation?

Ans. In elastic deformation, the material regains its original conditions with respect to its shape and size, after the removal of deforming forces. The material follows Hooke's law and the stress is proportional to the strain. In plastic deformation, the material is permanently deformed and the original conditions are not restored back on the removal of deforming forces. The material does not obey Hooke's law.

3. What is the difference between toughness and hardness?

Ans. Toughness of a material is its ability to absorb impact or shock without undergoing permanent deformations, whereas hardness is its ability to resist indentation or scratching.

4. What is ductility? How does it differ from malleability?

Ans. Ductility is a mechanical property of a material which enables it to stretch without failure, that is, a material can be easily drawn into thin wires by applying tension forces. Malleability is also a mechanical property which allows the material to be easily drawn into thin sheets by applying compressive forces.

Multiple Choice Questions

1. Match List I with List II and select the correct answer using the codes given below the lists:

List I (Crystal structure)	List II (Atomic packing factor)
A. Simple cubic	1. 74%
B. Body-centred cubic	2. 74%
C. Face-centred cubic	3. 52%
D. Hexagonal close packed	4. 68%

Codes:	A	B	C	D
(a)	3	4	2	1
(b)	4	3	2	1
(c)	1	2	3	4
(d)	4	3	1	2

2. Atomic packing factor (APF) in case of copper crystal is

(a) 0.52
(b) 0.68
(c) 0.74
(d) 1.633

3. Match List I with List II and select the correct answer using the codes given below the lists:

List I (Materials)	List II (Structure)
A. Charcoal	1. F.C.C.
B. Graphite	2. H.C.P.
C. Chromium	3. Amorphous
D. Copper	4. B.C.C.

Codes:	A	B	C	D
(a)	3	2	1	4
(b)	3	2	4	1
(c)	2	3	4	1
(d)	2	3	1	4

4. Which of the following elements is not a metalloid?

(a) Copper
(b) Carbon
(c) Sulphur
(d) Silicon

5. Consider the following elements:

1. Carbon
2. Phosphorus
3. Sulphur
4. Silicon

Of these elements:

(a) 1 and 4 are metalloid
(b) 2 and 3 are metalloid
(c) 1, 3 and 4 are metalloid
(d) 1, 2, 3 and 4 are metalloid

6. Organic materials contain the following components:

(a) Animal cells
(b) Vegetable cells
(c) Organic compounds
(d) All of the above

7. Of the following nonferrous metals, which one finds application in storage batteries?

(a) Copper
(b) Lead
(c) Nickel
(d) Magnesium

8. Consider the following fundamental particles:

1. Proton
2. Meson
3. Electron
4. Neutron

A material consists of

(a) 1 and 3
(b) 3 and 4
(c) 1, 3 and 4
(d) 1, 2 and 3

9. Consider the following properties of material:

1. Electrical conductivity
2. Thermal conductivity
3. Mechanical properties
4. Formability

Which of the above properties is dominating in ferrous materials?

(a) 1 and 2
(b) 3 alone
(c) 1, 2 and 4
(d) 2 and 3

10. Consider the following chemical bonds:

    1. Covalent bond
    2. Ionic bond
    3. Metallic bond

    Organic compounds consist of

    (a) 1 alone (b) 2 and 3
    (c) 2 alone (d) 3 alone

11. J/kg-K is the SI unit of

    (a) Electrical conductivity (b) Specific heat
    (c) Thermal coefficient (d) Potential energy

12. Gases have

    (a) One specific heat (b) Two specific heats
    (c) Three specific heats (d) Four specific heats

13. The specific heat becomes zero at

    (a) 0 K (b) 273 K
    (c) 273 °C (d) 273 °F

14. Water has its maximum density at

    (a) 4 °F (b) 4 K
    (c) 4 °C (d) 0 °C

15. Consider the following statements:

    1. Density of solids and liquids decreases with rise in temperature
    2. The specific gravity is a dimensionless quantity
    3. The specific gravity of mercury is 13.6

    Of these statements:

    (a) 2 and 3 are true (b) 1, 2 and 3 are true
    (c) 1 and 2 are true (d) 3 alone is true

16. The specific heat of water is

    (a) 4186 J/kg-K (b) 2093 J/kg-K
    (c) 1.00 J/kg-K (d) 0.50 J/kg-K

17. Consider the following statements:

    1. Metals have only one melting point
    2. Alloys have range of melting point
    3. At melting point, a material coexists in two phases
    4. Diamond has lowest melting point of any solid

    Of these statements:

    (a) 1 and 2 are true
    (b) 1, 2 and 4 are true
    (c) 1, 2 and 3 are true
    (d) 1 and 3 are true

18. Arrange the following bonds in order of their increasing melting points:

    1. Covalent, Ionic, Molecular and Metallic
    2. Molecular, Metallic, Ionic and Covalent
    3. Ionic, Molecular, Metallic and Covalent
    4. Metallic, Molecular, Ionic and Covalent

19. Consider the following statements:

    1. Diamagnetic materials have no dipole moments
    2. Diamagnetic materials have orderly arranged dipole moments
    3. Ferromagnetic materials have orderly arranged equal dipole moments
    4. Paramagnetic materials have randomly arranged dipole moments

    Of these statements:

    (a) 1 and 3 are true (b) 1, 3 and 4 are true
    (c) 2, 3 and 4 are true (d) 2 and 3 are true

20. Which of the following materials is not a ferromagnetic material?

    (a) Iron (b) Nickel
    (c) Zinc (d) Cobalt

21. Match List I with List II and select the correct answer using the codes given below the lists:

List I	List II
(Material properties)	(Test)
A. Ductility	1. Impact test
B. Toughness	2. Fatigue test

List I	List II
C. Endurance limit	3. Tension test
D. Resistance to penetration	4. Hardness test

Codes:	A	B	C	D
(a)	3	2	1	4
(b)	4	2	1	3
(c)	3	1	2	4
(d)	4	1	2	3

22. The highest stress that a material can withstand for a specified length of time without excessive deformation is called

(a) Fatigue strength (b) Endurance strength
(c) Creep strength (d) Creep rupture strength

23. Neck formation starts after

(a) Yield point (b) Ultimate point
(c) Elastic limit (d) Failure point

24. Poisson ratio is a ratio of

(a) Longitudinal strain to lateral strain
(b) Lateral strain to longitudinal strain
(c) Lateral strain to shear strain
(d) Lateral strain to volumetric strain

25. The usual value of the offset strain is

(a) 0.3% (b) 0.2%
(c) 0.5% (d) 0.75%

26. Which of the following materials has zero value of Poisson ratio?

(a) Cork (b) Cement concrete
(c) Rubber (d) Wood

27. Tensile load produces

(a) Contraction (b) Elongation
(c) Bending (d) No effect

28. A measure of Rockwell hardness is the

(a) depth of penetration of indentor
(b) surface area of indentation
(c) projected area of indentation
(d) height of rebound

29. Match List I with List II and select the correct answer using the codes given below the lists:

List I	List II
(Mechanical properties)	(Related to)
A. Malleability	1. Wire drawing
B. Hardness	2. Impact loads
C. Resilience	3. Cold rolling
D. Isotropy	4. Indentation
	5. Direction

Codes:	A	B	C	D
(a)	4	2	1	3
(b)	3	4	2	5
(c)	5	4	2	3
(d)	3	2	1	5

30. Match List I with List II and select the correct answer using the codes given below the lists:

List I	List II
(Mechanical properties)	(Testing machines)
A. Tensile strength	1. Rotating bending machine
B. Impact strength	2. Three-point loading machine
C. Bending strength	3. Universal testing machine
D. Fatigue strength	4. Izod testing machine

Codes:	A	B	C	D
(a)	4	3	2	1
(b)	3	2	1	4
(c)	2	1	4	3
(d)	3	4	2	1

Answers

1. (a)	2. (c)	3. (b)	4. (a)	5. (d)	6. (d)
7. (b)	8. (c)	9. (b)	10. (a)	11. (b)	12. (b)
13. (a)	14. (c)	15. (b)	16. (a)	17. (c)	18. (b)
19. (b)	20. (c)	21. (c)	22. (c)	23. (b)	24. (b)
25. (b)	26. (a)	27. (b)	28. (a)	29. (b)	30. (d)

Review Questions and Discussions

Q.1. What are the factors on which selection of a material depends upon?
Q.2. How do metals differ from nonmetals?
Q.3. What are metalloids?
Q.4. What is a unit cell?
Q.5. How does a unit cell differ from a crystal?
Q.6. What is atomic packing factor? What does it indicate?
Q.7. Which metal structure has highest formability? Also state why?
Q.8. Why are ferrous metals extensively used in manufacturing industry?
Q.9. What are the three most common crystal structures found in metals?
Q.10. What is meant by physical properties of a material?
Q.11. How does density differ from specific gravity?
Q.12. Why do metals have high electrical conductivity?
Q.13. What are phonons?
Q.14. Why do diamond has the highest melting point of any solid?
Q.15. What are diamagnetism, paramagnetism and ferromagnetism?
Q.16. Why are diamagnetic materials nonmagnetic?
Q.17. Why is water considered as a standard fluid while defining specific gravity of a material?
Q.18. Gases have two specific heats while solids have only one. Why?
Q.19. What is the significance of specific heat in the manufacturing system?
Q.20. Explain the importance of thermal conductivity in a manufacturing system.
Q.21. What are the three modes of heat transfer? Why are metals good conductor of heat?

Q.22. Why do you mean by structure-sensitive and structure-insensitive properties of metals?
Q.23. How does grain size affect the mechanical properties of metals?
Q.24. Differentiate between engineering stress and true stress.
Q.25. What is ductility, and how is it measured?
Q.26. In the stress–strain curve equation $\sigma = K \,.\, \in^n$, what does the exponent n signify?
Q.27. What is the difference between elastic and plastic behaviour?
Q.28. Is a brittle material necessarily weak?
Q.29. How is toughness of a material defined?
Q.30. How is brittle fracture different from ductile fracture?
Q.31. What is plastic deformation?

Ferrous Materials and Their Heat Treatment 2

Figure shows some products made of ferrous materials.

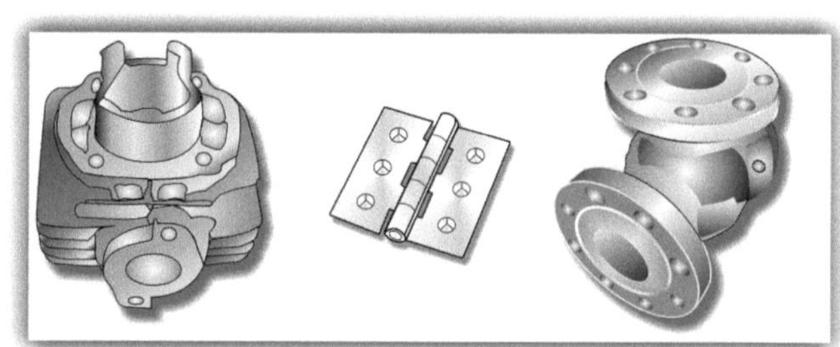

Learning Objectives

After reading this chapter, you will be able to answer some of the following questions:

- Why do ferrous materials find extensive applications in industry?
- What is the purpose of heat treatment?
- Why do some products need only surface hardening?

D. K. Singh, *Fundamentals of Manufacturing Engineering*,
https://doi.org/10.1007/978-981-99-8767-2_2

2.1 Introduction

Ferrous materials are mainly composed of iron and carbon. They may also contain a number of other elements either present in the natural state as impurity elements or are intentionally added alloying elements.

2.2 Iron

Iron is the basic element of ferrous materials. It is rarely found in the metallic state, but occurs in the form of various mineral compounds, known as ores. The chief ore of iron is haematite (Fe_2O_3). Pure iron is a soft metal having a crystalline structure. It has strong magnetic characteristics. It can be readily cold-worked. The strength of the iron can be increased by strain hardening. The magnetic characteristics of pure iron, namely high permeability and low remanence, make its use in direct-current magnetic circuits very desirable. But iron in its pure state has very few practical uses. Obtaining pure iron is a very difficult task, because of the great affinity of iron for certain elements. Therefore iron with 99% purity is regarded as the pure iron from theoretical point of view. The remaining 1% of other substances, however, makes a great difference.

2.3 Wrought Iron

Wrought iron is regarded as the purest form of iron. The chemical analysis shows that it contains as much as 99.9% of iron. It is a mechanical mixture of very pure iron and a silicate slag (about 3%) which is distributed uniformly throughout the metal. It contains very little carbon. Two processes, namely Puddling and Aston, are used for the production of wrought iron.

The ultimate strength of wrought iron can be increased considerably by cold working followed by a period of aging. Wrought iron is famous for its high ductility and hence can be easily shaped in hot forming operations, such as forging. It is also corrosion resistant and has good weldability.

Wrought iron is never cast, because it becomes pasty on strong heating rather than melting. All the shapes are obtained by hammering, pressing or forging.

Wrought iron is used principally in the production of pipe and other products subjected to deterioration by rusting. The iron has the property of being able to withstand sudden and excessive shock loads without permanent injury, because of which, it is used for chains, crane, hooks, railway couplings, etc.

2.4 Pig Iron

Pig iron is the first product in the process of converting iron ore into useful metal by using blast furnace. It has a typical composition of 4% carbon, 1.5% silicon, 1% manganese, 0.04% sulphur and 0.4% phosphorus with the rest being pure iron. The molten metal from the blast furnace was earlier used to be poured into small sand moulds, arranged like a litter of small pigs around a main channel. The solidified metal was called pig, and hence, such iron was named pig iron. Pig iron is used in making iron and steels and may contain element iron up to 98%.

2.5 Steel

Steel is the most common engineering metal, which is mostly made by continuous casting and is also made from cast ingots. It is fundamentally an alloy of iron and carbon, with the carbon content varying up to 1.5%. The higher the carbon content, the higher is the hardenability of the steel and the higher is its strength, hardness and wear resistance. On the other hand, ductility, weldability and toughness are reduced with increasing carbon content. Various alloying elements are added to steel in addition to carbon to impart properties of hardenability, strength, hardness, toughness, wear-resistance, workability, weldability and machinability. The principal alloying elements used in steel are chromium, nickel, manganese, silicon, molybdenum, tungsten and to a lesser extent copper, cobalt, beryllium, boron and silver. Steels are available in the form of plate, sheet, strip, bar, wire, tube, castings and forgings.

2.6 Plain Carbon Steel

Plain carbon steel, also called simply carbon steel, is an alloy of iron and carbon with small amounts of manganese, silicon, sulphur and phosphorous as impurities. The main purpose of these alloying elements is not to modify the mechanical properties of the steel. The carbon content may vary in the range from a trace to 1.7%, although rarely over 1.3%. Plain carbon steels containing more than 1.3% carbon are seldom produced or used. They are used successfully, where strength and other requirements are not severe.

Plain carbon steels are divided into three groups, according to carbon content, as follows:

- ***Low-carbon steel***, also called mild steel, has less than 0.30% carbon. It possesses good formability and weldability but lacks hardness. The structures of such steels are usually ferrite and pearlite. It is used in making bolts, nuts, sheets, plates, tubes and machine components that do not require high strength.

Another very extensive use of low-carbon steel is in beams, plates, channels, angles, etc. for construction purposes. Carbon content in these steels varies between 0.15 and 0.25%, and such steels are known as structural steels. Boiler steels usually come under this range, although the carbon content may be as high as 0.3%.

- ***Medium-carbon steel*** has 0.30–0.60% carbon. It is hard and has higher strength than low-carbon steels. Also, increased carbon content makes them harder. These steels are extremely popular, because the properties of such steels are greatly improved by heat treatment. They are used in making machine parts such as gears, axles, connecting rods, crankshafts, railroad equipment and parts for metal working machinery.
- ***High-carbon steel*** has more than 0.60% carbon. The highest practical limit is 1.3%. They lack toughness and formability but their hardness and wear resistance hardness and wear resistance are high. Therefore, they are generally used for parts such as cutting tools, cable, springs and cutlery.

2.7 Alloy Steel

Alloy steel contains one or more elements in sufficient amounts with a purpose to change the properties substantially. The alloying elements present in the alloy steel not only help to overcome the deficiencies and limitations of plain carbon steels, but may also effect an improvement in some other properties.

The alloying elements tend to

- increase hardenability,
- increase resistance to softening on tempering,
- increase resistance to corrosion and oxidation,
- improve high temperature properties,
- increase resistance to abrasion, and
- strengthen steels that cannot be subjected to quenching.

Alloy steels are classified on the basis of total alloy content into two groups, low alloy and high alloy steel. Low alloy steel contains less than 10% alloy whereas, high alloy steel has more than 10% alloy.

Alloy steels offer many advantages over plain carbon steel as listed below:

- Greater hardenability,
- Less distortion and cracking,
- Less grain growth,
- Higher elastic ratio and endurance strength,

- Greater high temperature strength,
- Better machinability at high hardness, and
- Greater ductility at high strength.

But on the disadvantage side, they are costly, require special handling and can develop temper brittleness in certain grades.

2.8 Stainless Steel

Stainless steels is an iron-based alloy containing 10.5% or more chromium. It was produced by Harry Brerley of Britain for the first time in 1913. Presence of chromium in stainless steel makes it corrosion resistant. It is called stainless, because of the formation of a thin layer of chromium oxide on its surface, which prevents its corrosion. This process is called passivation. Higher carbon content reduces the corrosion resistance of stainless steels. Passivity of the steel is reduced due to the formation of chromium carbide in such cases.

Stainless steels are extensively used in making cutleries, kitchen equipments, health care and surgical equipment. They are also used in chemical, food processing and petroleum industries.

Stainless steels are usually of five types, which are discussed below.

2.8.1 Austenitic Stainless Steel

This steel contains chromium, nickel and manganese. The base element is iron. Such steels are designated as the 200 and 300 series. Steels containing chromium and nickel are of the 300 series, and those containing chromium, nickel and manganese are of the 200 series. Austenitic stainless steels are nonmagnetic and have excellent corrosion resistance against almost all media, except hydrochloric acid and other halide acids and salts. Formability is outstanding due to their fcc crystal structure and they can be hardened by cold working, but not by heat treatment. Type 303 stainless steel is the easiest of all stainless steels to machine and is widely used for machine parts and fasteners. Another popular variety is 304, frequently referred to as 18–8 stainless steel containing 18% Cr and 8% Ni.

2.8.2 Ferritic Stainless Steel

These steels are designated as the 400 series. They cannot be hardened by heat treatment and only moderately by cold working, because their ratio of carbon to chromium is low. They are magnetic and have good corrosion resistance, but have lower ductility than austenitic stainless steels. Poor ductility or formability is because of their bcc crystal structure, but they are readily weldable. They are generally used for kitchen equipment and automotive trim.

2.8.3 Martensitic Stainless Steel

They are placed in 400 series, and type 410 is the most commonly used. They are hardenable because of their high ratio of carbon to chromium. They are magnetic, have high strength and are moderately corrosion resistant. Martensitic stainless steels cost about 1.5 times as much as the ferritic alloys, mainly due to additional heat treatment. These steels are used for cutlery, surgical tools, instruments, valves and springs.

2.8.4 Precipitation-Hardening (PH) Stainless Steel

It is a special type of stainless steel, which contains copper or aluminium in addition to chromium and nickel. Presence of aluminium permits age hardening at relatively low temperatures, as a result of which it attains very high strength. They are costly and find applications mainly in aircraft and aerospace structural components.

2.8.5 Duplex Stainless Steel

The duplex stainless steel is a mixture of austenitic and ferritic stainless steel containing chromium between 21 and 25% and nickel between 5 and 7%. It has higher yield strength and greater resistance to stress-corrosion cracking than the 300 series of austenitic steels. They are mostly used in water-treatment plants and heat-exchanger components.

2.9 High-Speed Steel (HSS)

High speed steel (HSS) is a tool steel consisting of mainly tungsten, molybdenum, chromium, cobalt, and vanadium. It was invented by F. W. Taylor in 1901. It has excellent hardenability, good wear resistance and excellent hot hardness that helps it to retain its hardness at elevated temperatures of around 650 °C. It is used in making cutting tools for various machining operations.

The high-speed steels can be divided into two groups: If the HSS alloy is molybdenum-based, it is known as molybdenum HSS and is identified as *M* series, while tungsten-based HSS is called *T* series. The *M*-series steels contain up to about 10% molybdenum and the T-series steels contain 12–18% tungsten along with other alloying elements mentioned above. The M-series steels are more advantageous as compared to T-series. Certain advantages of M-series steels over T-series are:

- They have higher abrasion resistance.
- They produce less distortion in heat treatment; and
- They are less expensive.

Although there are numerous HSS compositions, they may be grouped in three classes.

18-4-1 HSS: This steel contains 18% tungsten, 4% chromium and 1% vanadium; and the remainder is iron. It is supposed to be one of the best all-purpose tool steels.

Molybdenum HSS: It contains molybdenum as the principal alloying element. One variety of molybdenum HSS is 6-6-4-2, which means it contains 6% tungsten, 6% molybdenum, 4% chromium and 2% vanadium. It has excellent toughness and cutting ability.

Super HSS: It is tungsten-based HSS containing cobalt in higher amount ranging from 2 to 15%. Increased cobalt content increases the cutting efficiency, especially at elevated temperatures. One of the typical HSS contains 20% tungsten, 4% chromium, 2% vanadium and 12% cobalt. The remainder is iron. It is mainly used for heavy cutting operations, involving high pressures and temperatures on the cutting tool. It is very expensive.

2.10 Tool Steel

Tool steels are basically high-carbon steels whose composition and properties are changed through the modification of alloying elements. They have the combined properties of good wear resistance and high toughness and strength. They are used in making cutting tools.

2.11 Cast Iron

Ferrous metal containing 2–4.5% carbon is called cast iron. In addition to carbon, cast iron generally contains small amounts of silicon, phosphorous, sulphur and manganese as impurities. It is the least expensive of all metals because of its abundance occurrence in nature next to aluminium, but the extraction of aluminium is not economical. It requires less refining than steel, has relatively low melting point (= 1200 °C) and is easily melted requiring less fuel. It is extensively used in casting operations as the molten cast iron is easily cast into complex shapes and easily machined to required tolerances. These characteristics makes cast iron a very attractive material. The following are the important types of cast iron:

- Grey cast iron
- White cast iron
- Malleable cast iron
- Ductile iron
- Compacted graphite iron.

2.11.1 Grey Cast Iron

Grey cast iron is the most important variety of cast iron. It looks grayish, because of the large amount of *graphite flakes* present on its surface. It is easily machinable and has a high compressive strength. Its compressive strength varies between three to five times its tensile strength. Presence of graphite flakes in grey iron makes it suitable to absorb vibration. It is the least expensive and most common variety of cast iron. Also, it is the most widely used of all cast metals. Typical applications are engine blocks, pipes and fittings, electric motor housings and machine tools.

2.11.2 White Cast Iron

White cast iron is formed, when the casting is rapidly cooled. It contains carbon in the combined state (iron carbide, Fe_3C). It is also called *chilled cast iron*, as it is produced by rapid cooling of the gray irons castings. On fracture, its structure shows white colour. White cast iron is very hard and brittle due to the presence of Fe_3C and finds applications where high abrasion resistance is required. It is used on hammer mills, crusher jaws, crushing rollers and wear plates.

2.11.3 Malleable Cast Iron

Malleable cast iron is obtained, when white cast iron is subjected to a controlled heat treatment, where the cementite decomposes into iron and graphitic carbon. The graphite exists as clusters and is referred to as *clump* or *popcorn graphite.* It has significantly greater ductility than grey cast iron, because of the presence of soft graphite. It finds extensive applications in making automobile parts, small tools, hardware and pipe fittings.

2.11.4 Ductile Iron or Nodular Iron or Spheroidal Iron

In the ductile iron, carbon is in the form of graphite nodules or spheroids. These spheroids are formed when MgFeSi or MgNi alloy or cesium are added to grey cast iron, which transforms graphite flakes.

Their important properties include good ductility, high strength, excellent castability and machinability and good wear resistance. Typical applications include small as well as large cast products including machinery components and press rolls.

2.12 Alloy Cast Iron

Alloy cast iron contains alloying elements such as nickel, chromium, copper, molybdenum and titanium that are added to cast iron either to improve its mechanical properties or to impart certain special characteristics such as machinability and heat or corrosion resistance. Acicular and spheroidal cast iron are the types of alloy cast iron. The former contains nickel and molybdenum and finds application in making crankshaft by casting process.

2.13 Common Alloying Elements Added to Steel

- **Chromium**: It increases the hardenability of steel significantly along with improving its corrosion resistance.
- **Manganese**: It increases hardenability of steel, if added in significant proportion.
- **Vanadium**: It improves heat-treatability (by quenching), strength and toughness (by reducing the tendency of grain growth at increased temperatures) of steel. It also increases wear resistance of steel by forming carbide.
- **Nickel**: It improves corrosion resistance (if added in large amount), strength and toughness of steel. It also increases hardenability of steel.
- **Molybdenum**: It increases hardenability (significantly), toughness, hot hardness and creep strength of steel. It also improves wear resistance of steel by forming abrasion resisting particles (carbides).
- **Tin**: It increases the corrosion resistance of steel by forming a silvery white protective coating over it.
- **Cobalt**: It increases red hardness by hardening ferrite.
- **Tungsten**: It improves hot strength and hot hardness, if added in significant amount.
- **Silicon**: It improves oxidation resistance of steel. The resulting steel is called silicon steel, which is used for magnetic applications.

2.14 Micro Structures of Steel and Iron

Carbon is found in different forms in iron and steel. When steel is heated above the austenitic temperature followed by cooling under different conditions, the austenite in steel transforms into a number of microstructures.

The microstructures refer to the arrangements of grains and phases within a material. The grains are individual crystals. Each phase of a microstructure has a distinct arrangement of atoms, typically crystalline, but not exclusively so. The microstructures may be polycrystalline too, existing in single phase or multiphase. For the most part, magnification is required to observe a microstructure. The study of microstructures is important in order to understand the iron-carbon equilibrium diagram and time–temperature transformation (TTT) diagram, sometimes known as isothermal transformation diagram or S-curve. The important microstructures are ferrite, cementite, pearlite, ledeburite, austenite, bainite, martensite, troostite and sorbite.

2.14.1 Ferrite or Alpha-Ferrite or Alpha-Iron

It is a solid solution of carbon in alpha-iron and has a maximum solid solubility of 0.022% carbon at a temperature of 727 °C. Ferrite is soft and ductile and is magnetic from room temperature to 770 °C. Low-carbon steel and wrought iron consist of ferrite. It is not responsive to normal heat treatment processes because of its low carbon content and hence has limited use.

2.14.2 Cementite

Iron carbide (Fe_3C) is known as cementite and is the hardest constituent of iron. It has the carbon content of 6.67%, and hence is highly brittle. It is used in making dies and cutting tools and is also used as abrasives.

2.14.3 Pearlite

Pearlite is a mixture of ferrite and cementite and is so called because it resembles mother of pearl at low magnifications. Consequently, the mechanical properties of pearlite are intermediate between ferrite (soft and ductile) and cementite (hard and brittle). It is found in low- and medium-carbon steels.

If ferrite and cementite are closely packed in the pearlitic structure, the microstructure is called fine pearlite and it is produced when the rate of cooling is relatively high. If they are widely spaced, then the resulting microstructure is called coarse pearlite and it is produced when the rate of cooling is slow, as in a furnace.

2.14.4 Ledeburite

It is found in cast iron and consists of austenite and cementite. It possesses lubricating property that imparts good machinability to cast iron. It is brittle, hence cannot be forged or rolled but can be easily cast.

2.14.5 Austenite or Gamma Iron

It is a solid solution of carbon in gamma iron and is nonmagnetic. This structure has a solid solubility of up to 2.11% carbon at 1148 °C. It possesses good formability.

2.14.6 Bainite

It is a mixture of ferrite and cementite and has finer crystals than pearlite. It is visible only through electron microscope and is produced by austempering. It is generally stronger and more ductile than pearlite.

2.14.7 Martensite

Martensite is a supersaturated solid solution of carbon in alpha ferrite, and therefore, is a metastable structure. When heated in the range of 100–700 °C, the excess carbon atoms are rejected from solution and the structure moves toward a mixture of the stable ferrite and cementite phases. This decomposition of martensite into ferrite and cementite is a time and temperature dependent, and diffusion-controlled phenomenon.

Martensite is extremely hard and brittle, lacks toughness and thus has limited use. It can cut other metals and has high wear-resistant characteristics. It has a needle-like structure. Martensite cannot be formed by quenching until the steel is in the austenitic condition, i.e. above the lower critical temperature, the carbon then being in solution.

2.14.8 Troostite

It is a mixture of ferrite and finely divided cementite and differs from pearlite only in the degree of fineness. It is produced on tempering martensite below approximately 450 °C and is the result of decomposition of austenite when cooled at a rate slower than that which will yield a martensitic structure and faster than that which will produce a sorbitic structure.

2.14.9 Sorbite

Sorbite consists of ferrite and finely divided cementite, and is produced on tempering martensite above approximately 450 °C. It is the result of decomposition of austenite, when cooled at a rate slower than that which will yield a troostitic structure and faster than that at which pearlitic structure results.

2.15 Heat Treatment of Ferrous Metals

Heat treatment is the process of heating and cooling a metal or alloy in its solid state to change its metallurgical and mechanical properties. Changing and controlling of these properties are important in manufacturing processes. The usefulness of steel is largely due to the relative ease with which its properties may be changed by properly controlling the manner in which it is heated and cooled. For example, steel can be hardened to resist cutting action and abrasion, or it can be softened to permit machining. Heat treatment removes internal stresses, reduces grain size, increases toughness or produces a hard surface on a ductile interior.

Various heat treatment processes may be classified as follows:

- Tempering: Austempering and martempering
- Annealing: Full annealing, process annealing, stress-relief annealing and cyclic annealing
- Normalizing
- Spheroidizing
- Hardening
- Surface hardening: Carburizing, cyaniding, carbonitriding, nitriding, flame hardening and induction hardening.

2.16 Tempering

The purpose of tempering is to reduce hardness, brittleness and residual stress and improve ductility and toughness. In tempering, the hardened steel is heated below the lower critical temperature and cooled at a definite rate. Martensite, the principal constituent of hardened steel, is extremely hard and brittle and the tool made of it is liable to cracking and chipping. The steel is therefore subjected to tempering, which causes a partial transformation of the martensite back to pearlite, thereby taking away some of the hardness, but making the steel tougher.

The tempering time and the temperature both are important for tempering. Although steel becomes soft within few minutes of heating, there is some additional reduction in hardness, if the temperature is maintained for a prolonged time.

2.16.1 Austempering

It is an isothermal transformation process, that converts austenite to a hard structure called bainite. During the process, the heated steel is quenched rapidly to avoid the formation of ferrite or pearlite. The quenching is performed in a molten salt medium at temperatures ranging from 160 to 750 °C.

Austempering can be used in place of conventional quenching and tempering either to reduce the tendency for cracking and distortion during quenching or to improve ductility and toughness while maintaining hardness. Components of aircraft engines are subjected to austempering.

2.16.2 Martempering

In this process, the steel or cast iron is quenched rapidly from the austenite region in a hot oil or molten salt medium. After holding it at this temperature for a certain time, the steel is allowed to cool gradually in air. During the process, austenite transforms into martensite. The main purpose of martempering is to minimize distortion, cracking and internal stresses that result from quenching in oil or water. Although the resulting product is similar to tempered martensite, a subsequent tempering operation is generally performed. This process is suitable for steels with lower hardenability.

2.17 Annealing

The purpose of annealing is to soften the steel to make it easily machinable and to relieve the internal stresses. It involves heating the metal slowly to the required temperature (usually above the critical temperature), staying it at that temperature for certain time and finally cooling it slowly. On cooling, the steel changes to ferrite and pearlite in case of hypoeutectoid steels, and into pearlite and cementite in case of hypereutectoid steels.

2.17.1 Full Annealing

It is so named because it wipes out all trace of previous structure. Full annealing refines crystalline structure, softens metal, relieves internal stresses and improves ductility. Full annealing is applied on steel castings and ingots. It makes the structure of steel homogeneous and soft. But the process takes much time.

2.17.2 Process Annealing

The process annealing is also known as *intermediate annealing* or *subcritical annealing* or *in-process annealing*. The process involves heating of steel to a temperature close to the lower critical temperature followed by slow cooling. It results in the formation of pearlite. The process is similar to tempering but softness, ductility and grain refinement are poor as compared to that obtained by full annealing. It is suitable for low-carbon steel in the treatment of sheet and wire.

2.17.3 Stress-Relief Annealing

As the name suggests, the process reduces residual stresses induced during various machining and forming processes. Steel casings, welded assemblies and cold-formed products are subjected to this type of annealing. The parts are generally cooled in air after being heated to a particular temperature.

2.17.4 Cyclic Annealing

Cyclic or isothermal annealing reduces the total time required for an annealing operation. Austenite is partially transformed to soft ferrite during quenching of steel and is then completely transformed into pearlite. Part is cooled at desired rate only when transformation is complete. Isothermal annealing results in giving pearlite a more uniform structure than that obtained by other annealing processes.

2.18 Normalizing

It involves heating of steel at about 40 °C above the upper critical temperature, holding it at that temperature for a certain time and then cooling in air to room temperature. It produces higher strength and hardness but lower ductility, when compared to full annealing.

The purpose of normalizing is to refine grain structure, obtain uniform structure (homogenization), decrease residual stresses and improve machinability in low- and medium-carbon steels and alloy steels. Most commercial steels are normalized after being rolled or cast.

There exists a basic difference between full annealing and normalizing. Full annealing produces identical properties in the metal due to uniform cooling in the furnace, whereas normalizing produces different properties due to differential cooling rate.

2.19 Spheroidizing

Spheroidizing is the process of producing a structure in which cementite is in the form of small spheroids surrounded by ferrite. If steel is heated slowly to a temperature just below the critical range and held there for a prolonged duration followed by slow cooling, this structure is obtained. It increases the toughness of steel.

2.20 Hardening

Hardening is the process of heating the steel near to its critical temperature followed by rapid cooling (quenching) in water, oil or brine solution. It imparts hardness to steel by transforming austenite to martensite. Low- and medium-carbon steels are generally quenched in water bath. For high carbon and alloy steel, oil is generally used as the quenching medium. For extreme cooling, brine solution or water spray is most effective. Larger parts are usually quenched in an oil bath. The hardness obtained depends on the quenching rate, carbon content and work size. In alloy steels, the kind and amount of alloying elements influence only the hardenability (the ability of the alloy to be hardened to the desired depth by heat treatment) and does affect the hardness (resistance of a material to indentation or scratching) except in unhardened or partially hardened steels. Low-carbon steel does not respond appreciably to hardening treatment. Medium-carbon and high-carbon steels respond better. Tools and machine parts that are required to undergo heavy duty services are often hardened.

2.21 Surface Hardening

In surface hardening (also called *case hardening*), only the surface properties of a part up to a certain depth (case) are altered, without affecting the core of the alloy. This method is particularly useful for improving resistance to surface indentation, fatigue and wear. Typical applications for case hardening include gear, teeth, cams, shafts and tools.

Carburizing, cyaniding, carbonitriding, nitriding, flame hardening and induction hardening are surface-hardening processes.

2.22 Carburizing

The purpose of carburizing is to increase the carbon content of the surface of low-carbon steel. Mild steel surfaces sometimes need to be hardened varying up to about 2 mm depth. The process involves heating of steel in presence of a carbon-rich substance above critical temperatures for a prolonged time. The carbon is absorbed by the outer surface of steel. The depth of the case depends on the time and temperature of the treatment. Carburizing produces a hard and wear-resistant surface without affecting the inner core.

Carburizing is widely used for the surface hardening of machine parts including cams, piston pins, gears, pump shafts, etc.

There are three general methods of carburizing.

- Solid carburizing
- Gas carburizing
- Liquid carburizing.

2.22.1 Solid Carburizing

Solid or *pack carburizing* consists of placing the parts to be treated in a closed container with some carbonaceous material, such as charcoal or coke. The container is heated at roughly 900 °C. The carbon monoxide (CO) produced as a result of the combustion of carbonaceous material produces carbon on reaction with the metal (iron). The released carbon gets soaked into the surface of the steel. Normally 0.1 mm penetration takes place in an hour. The penetration of carbon into the steel depends on the temperature, the time of heating at a particular temperature and the carburizing agent. It is a very slow and time-consuming process and produces thick cases ranging from 0.76 to 4.06 mm in depth. The demerit of the process is that it suffers from scale formation at the surface of the part.

2.22.2 Gas Carburizing

Gas carburizing uses gaseous medium for heating the parts, which ensures uniform temperature distribution. Here, the part to be treated is heated in the presence of hydrocarbons such as methane, propane, etc. These hydrocarbons produce carbon on being heated to high temperatures which is deposited on the steel surface. The process is fast and more easily controlled.

Gas carburizing is mostly suitable for mass scale production of smaller parts such as ball and roller bearings, pins, axles, etc.

2.22.3 Liquid Carburizing

Liquid carburizing uses molten sodium cyanide (NaCN) bath. The carbon from the cyanide gets absorbed by the steel surface, when the steel part is submerged in the cyanide bath. A small amount of nitrogen is also produced and gets absorbed. The combined effect of carbon and nitrogen enables the process to be carried out at lower temperatures than with carbon alone.

It is a fast and economical method for producing an extremely thin case. It is suitable for case hardening of small- and medium-sized parts.

2.23 Cyaniding

Cyaniding involves heating of steel in a molten sodium cyanide bath. The duration of heating decides the depth of case. The part is then quenched in water or oil to obtain a hard surface. The surface hardness is the result of absorption of carbon and nitrogen both, but nitrogen has more contribution. Liquid carburizing, on the other hand, produces a surface, which contains more carbon than nitrogen.

Cyaniding increases the fatigue limit and surface wear resistance of steel. The process is fast but costly. It is mainly used for the treatment of small parts subjected to light loads.

2.24 Carbonitriding

Carbonitriding is a modification of gas carburizing, but is performed at lower temperature. It involves heating of steel above the critical temperature in a carbon-rich gas containing ammonia. Hence carbon and nitrogen both are absorbed by the surface of steel. The absorption of nitrogen is increased at low temperatures and decreased at high temperatures. It is similar to gas carburizing at higher temperatures. Nitrogen, in addition to increasing the surface hardness of steel, also lowers the lower critical temperature.

2.25 Nitriding

This process involves heating of steel at around 510 °C in ammonia atmosphere and holding there for a certain period of time. Steel absorbs nitrogen from ammonia forming very hard and wear-resistant surface on the steel due to the formation of nitride. No quenching is required in nitriding.

Nitriding develops extreme hardness which is not obtainable by ordinary case hardening processes. Nitrided steels are very stronger, wear and corrosion resistant and more fatigue resistant as compared to carburized steels.

Bearing parts, cylinder liners, aircraft crankshafts and gears are surface hardened by nitriding.

The process is costly, which restricts its use. It is used only when extreme hardness is required, which is not possible even by carburizing. Also, the process is very time consuming and requires close control.

2.26 Flame Hardening

The process consists of heating the surface by using a gas flame above the upper critical temperature followed by rapid spray water cooling. It results in a hard surface with softer and tougher core. The case-depth depends on heating time and flame temperature. It is used in the surface-hardening of pins, gear teeth, etc.

2.27 Induction Hardening

In induction hardening, the steel part to be surface hardened is surrounded by a copper coil carrying high-frequency alternating current, which raises the surface temperature of steel above the upper critical temperature. Subsequently, it is subjected to quenching by spraying water.

The process is very fast and it takes few seconds to produce a skin of about 3 mm. The actual time is decided by the frequency of current, the power input and the depth of hardening required.

Induction hardening is suitable for circular section parts such as camshafts, crankshafts and gears.

Short Answer Questions

1. What is an alloy?

Ans. An alloy consists of two or more elements, out of which one is a metal. For example, steel is an alloy of iron and carbon, and brass is an alloy of copper and zinc.

2. What is the effect of increased carbon content on stainless steel?

Ans. Duralumin is an important alloy of aluminium. It consists of 95% aluminium, 4% copper, 0.5% magnesium and 0.5% manganese. It is extensively used in aircraft and automobile industries because of its lighter weight.

3. What are low and high alloy steels?

Ans. In low allow steel, the total amount of the alloying elements is less than 10%, whereas in high alloy steel, alloying elements exceed 10%.

4. What is case hardening? Name a few case-hardening processes.

Ans. Case hardening, also called Surface hardening, is a process in which a thin layer (called skin) of a mild steel (low-carbon steel) workpiece is hardened by subjecting it to various atmospheres like carbon and nitrogen to make it suitable for further heat treatment processes. Carburizing, cyaniding and nitriding are a few case-hardening processes.

5. What is flame hardening?

Ans. Flame hardening is a very economical and effective surface-hardening method for large ferrous machine parts such as ways of lathe, spindles, shafts, pulleys, and gear teeth.

6. What is soaking?

Ans. Soaking is a process of holding a heated metal for a certain period of time at the elevated temperature in order to improve its mechanical properties.

7. What is the difference between annealing and normalizing?

Ans. In annealing, the heated metal is left in the furnace for slow cooling, whereas in normalizing it is cooled in the air.

8. Name the two principal ores of iron and indicate their iron content.

Ans. Haematite (Fe_2O_3) and magnetite (Fe_3O_4) are the two principal ores of iron. Haematite contains more iron (about 70%) as compared to magnetite. Generally iron content in iron ores varies between 50 and 70%.

9. Why is grey cast iron used in the machine foundation?

Ans. Grey cast iron contains graphite flakes which have vibration dampening capacity, making it suitable for machine foundations.

10. Why is duplex stainless steel so named?

Ans. Because this stainless steel contains two components: austenite and ferrite.

Multiple Choice Questions

1. Match List I with List II and select the correct answer using the codes given below the lists:

List I	List II
(Ferrous materials)	(Carbon content)
A. Mild steel	1. 0.6–1.7%
B. High-carbon steel	2. Less than 0.3%
C. Cast iron	3. 6.7%
D. Cementite	4. 2–4.5%

Codes:	A	B	C	D
(a)	2	4	1	3
(b)	1	2	4	3
(c)	2	1	4	3
(d)	2	1	3	4.

2. 18/8 Stainless steel contains
   (a) 18% nickel, 8% chromium
   (b) 18% chromium, 8% nickel
   (c) 18% tungsten, 8% nickel
   (d) 18% tungsten, 8% chromium
3. Killed steels
   (a) Have minimum impurity level
   (b) Have almost zero percentage of phosphorus and sulphur
   (c) Are produced by LD process
   (d) Are free from oxygen
4. Which of the following irons is the purest one?
   (a) Wrought iron (b) Pig iron
   (c) Electrolytic iron (d) Cast iron

5. Cast iron is used for machine beds, because of its high

(a) Tensile strength
(b) Endurance strength
(c) Damping capacity
(d) Compressive strength

6. Which of the following stainless steels contains maximum carbon content?

(a) Austenitic stainless steel
(b) Martensitic stainless steel
(c) Ferritic stainless steel
(d) Precipitation-hardening stainless steel

7. Addition of silicon to cast iron

1. Promotes graphite module formation
2. Promotes graphite flake formation
3. Increases the fluidity of the molten metal
4. Improves the ductility of cast iron

Of these:

(a) 1 and 4 are true
(b) 2 and 3 are true
(c) 1 and 3 are true
(d) 3 and 4 are true

8. Which of the following pairs regarding the effects of alloying elements in steel are correctly matched?

1. Molybdenum: forms abrasion resisting particles
2. Phosphorus: improves machinability in free cutting steels
3. Cobalt: contributes to red hardness by hardening ferrite
4. Silicon: reduces oxidation resistance

Select the correct answer using the codes given below:

(a) 2, 3 and 4
(b) 1, 3 and 4
(c) 1, 2 and 4
(d) 1, 2 and 3

9. Small amounts of which of the following elements/pairs of elements is added to steel to increase its machinability?

(a) Nickel
(b) Sulphur and phosphorus
(c) Silicon
(d) Manganese and copper

10. Addition of vanadium to steel results in the improvement of

    (a) Heat-treatability by quenching
    (b) Hardenability
    (c) Fatigue strength
    (d) Resistance to oxidation at elevated temperature

11. Consider the following statements about ferrite:

    1. It is magnetic between room temperature and 770 °C
    2. It is soft and ductile
    3. It is found in low-carbon steel and wrought iron

    Of these statements:

    (a) 2 and 3 are true (b) 1 and 2 are true
    (c) 1, 2 and 3 are true (d) 1 and 3 are true

12. Which of the following statements is true about ferrite?

    (a) It is a solid solution of carbon in gamma iron
    (b) It is a solid solution of carbon in α-iron
    (c) It is a supersaturated solution
    (d) It can be easily heat-treated

13. Consider the following statements about cementite:

    1. It is chemically known as iron carbide
    2. It is the hardest constituent of iron
    3. It contains 6.67% of carbon

    Of these statements:

    (a) 1 alone is true (b) 3 alone is true
    (c) 2 and 3 are true (d) 1, 2 and 3 are true

14. Which of the following components is very hard and brittle?

    (a) Ferrite (b) Austenite
    (c) Martensite (d) Cementite

15. Which of the following components has needle-like structure?

(a) Pearlite (b) Austenite
(c) Bainite (d) Martensite

16. The objective of heat treatment is to change

(a) Physical properties of metals (b) Magnetic properties of metals
(c) Electrical properties of metals (d) Mechanical properties of metals

17. Consider the following statements about annealing process:
   1. It reduces hardness and brittleness
   2. It reduces residual stresses induced due to strain hardening
   3. It improves formability

   Of these statements:

(a) 1 alone is true (b) 1 and 2 are true
(c) 1, 2 and 3 are true (d) 2 and 3 are true

18. Quenching involves

(a) Very slow cooling (b) Slow cooling
(c) Rapid cooling (d) No cooling

19. Consider the following parameters:

1. Normalizing 2. Hardening
3. Martempering 4. Cold working

Hardness and tensile strength in austenitic stainless steel can be increased by

(a) 1, 2 and 3 (b) 1 and 3
(c) 2 and 4 (d) 4 alone

20. Pack carburizing uses the following atmosphere:

(a) Sodium carbide (b) Sodium cyanide
(c) Charcoal or coke (d) Hydrocarbons

21. Match List I with List II and select the correct answer using the codes given below the lists:

List I	List II
(Heat treatment)	(Effect on the properties)
A. Annealing	1. Refines grain structure
B. Nitriding	2. Improves the hardness
C. Martempering	3. Increases surface hardness
D. Normalizing	4. Improves ductility

Codes:	A	B	C	D
(a)	4	3	2	1
(b)	1	3	4	2
(c)	4	2	1	3
(d)	2	1	3	4

22. Flame hardening is useful for

(a) Plain carbon and alloy steels (b) Tool steels
(c) Cast irons (d) All of the above

23. Age hardening is employed for

(a) Plain carbon steels (b) Cast irons
(c) Nonferrous metals (d) Tool steels

24. Induction hardening uses

(a) Direct current
(b) Low-frequency alternating current
(c) High-frequency alternating current
(d) High-voltage alternating current.

25. Nitriding uses the following atmosphere:

(a) Sodium carbide (b) Ammonia
(c) Sodium cyanide (d) Sodium phosphide

26. Guideways of lathe beds are hardened by

(a) Carburizing (b) Cyaniding
(c) Nitriding (d) Flame hardening

Answers

1. (c)	2. (b)	3. (d)	4. (c)	5. (c)	6. (b)
7. (b)	8. (d)	9. (b)	10. (a)	11. (c)	12. (b)
13. (d)	14. (d)	15. (c)	16. (d)	17. (c)	18. (c)
19. (d)	20. (c)	21. (b)	22. (d)	23. (c)	24. (c)
25. (b)	26. (d)				

Review Questions and Discussions

Q.1. Why is it easier to weld low-carbon steel than high-carbon steel?
Q.2. Why is malleable iron more ductile than grey cast iron?
Q.3. How is malleable iron produced?
Q.4. What kinds of steels would be used for the following: Knife, chair handle, spring, paper clip?
Q.5. What are the carbon contents of low-carbon, medium-carbon and high-carbon steels?
Q.6. How do stainless steels become stainless?
Q.7. What is a tool steel?
Q.8. What is plain carbon steel?
Q.9. What are the most common alloying elements added to steel?
Q.10. What is the difference between plain carbon steel and alloy steel?
Q.11. Why is pig iron so named?
Q.12. Why is cast iron sometimes alloyed? What are the common alloying elements added to cast iron?
Q.13. Why is cast iron so named?
Q.14. What is the composition of 18-4-1 HSS?
Q.15. What is super HSS? Why is it so named?

Q.16. What are *M* and *T* series of high-speed steel?
Q.17. What is meant by the term 'microstructure'?
Q.18. What is TTT diagram? What is its purpose?
Q.19. Why is ferrite soft and ductile?
Q.20. Cementite is very hard. Why?
Q.21. How is martensite formed?
Q.22. How does troostite differ from sorbite?
Q.23. What is the purpose of heat treatment?
Q.24. How does austempering differ from martempering?
Q.25. Why is annealing performed?
Q.26. What is meant by full annealing?
Q.27. What is the difference between full annealing and normalizing?
Q.28. What is spheroidizing?
Q.29. Why is surface hardening needed?
Q.30. How does carburizing differ from cyaniding?
Q.31. Why is it difficult to heat treat nonferrous metals and alloys?
Q.32. What is age hardening? Which material shows age hardening?
Q.33. How does artificial aging differ from natural aging?
Q.34. What is tempering? Why is it done?
Q.35. What is hardenability?
Q.36. Compare the properties affected by quenching and annealing.

Nonferrous Materials and Their Heat Treatment

3

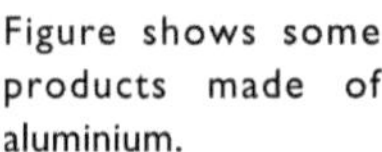

Figure shows some products made of aluminium.

Learning Objectives

After reading this chapter, you will be able to answer some of the following questions:

- Why do non-ferrous materials have wide range of applications?
- How is heat treatment of non-ferrous materials different than that of ferrous materials?

D. K. Singh, *Fundamentals of Manufacturing Engineering*,
https://doi.org/10.1007/978-981-99-8767-2_3

3.1 Introduction

Nonferrous metals and alloys are increasingly used today because of their numerous superior properties over ferrous materials. They constitute lightweight metals and high-strength lightweight alloys. Although they are more expensive as compared to ferrous materials, but they offer properties like better corrosion resistance, higher thermal and electrical conductivity, lower density, relative ease of fabrication and wide colour choices. They are mostly used in the alloyed form. However, a few nonferrous metals such as copper and aluminium are used in their pure states for making electricity conducting wires. Important nonferrous materials include aluminium, copper, lead, tin and zinc, and their alloys.

3.2 Aluminium and Aluminium Alloys

Although aluminium ores are widely distributed in the earth's crust, only bauxite ($Al_2O_3 \cdot 2H_2O$) is used economically in the extraction of aluminium. Due to its light weight, aluminium is used largely for aircraft and automobile components, where the saving of weight is an important consideration. The relative density of aluminium is about 2.68 as compared to 7.8 for steel, thus making aluminium about one-third the weight of steel for an equivalent volume. Other attractive properties of aluminium are its workability, corrosion resistance, and good electrical and thermal properties. In its pure state, aluminium is very weak and soft, but when mixed with small amounts of other alloying elements, its tensile strength, yield strength and hardness are increased. The principal alloying elements include copper, silicon, magnesium, manganese and occasionally zinc, nickel and chromium. It is very ductile and malleable, and can be rolled into thin sheet of 0.02 mm thickness and drawn into wire of 0.1 mm diameter.

The melting point of aluminium is low (649 °C). Thus it can be easily formed into parts by casting. In addition to sand casting, permanent metal moulds and die casting are used extensively. The low density and low melting point combine to practically eliminate most of the problems of sand washes (sand moving out of place) that occur during the pouring of heavier metals in casting.

Machinability of aluminium is not good, but most of the aluminium alloys can be readily formed under cold state. Aluminium requires less energy and power than does the forming of heavier metals. The most serious weakness of aluminium from an engineering point of view is its relatively low modulus of elasticity, about one-third that of steel. Under identical loadings, an aluminium component will deflect three times as much as a steel component of the same design.

In commercially pure form (99% pure aluminium and remaining is iron and silicon), aluminium is used mainly for cooking utensils and chemical equipment. It is available in the form of sheet, plate, tubing, wire, rivets, rod, bar, etc. Aluminium is also used extensively in paint as a paint pigment. Aluminium generally replaces steel or cast iron, where there is a strong need for light weight, corrosion resistance, low maintenance cost or high thermal or electrical conductivity. In modern motor vehicles, aluminium is used in making body parts, engine blocks, manifolds and transmission cases, where the reduced weight is useful to obtain the fuel economy.

3.2.1 Duralumin

Duralumin is an important alloy of aluminium. It consists of 95% aluminium, 4% copper and 0.5% magnesium and manganese each. It is used in making parts of aircrafts and automobiles.

3.2.2 Alclad

The addition of appreciable amounts of copper to aluminium to form precipitation-hardening alloys such as duralumin reduces the corrosion resistance of the aluminium. If a surface layer of pure aluminium or copper-free aluminium alloy is applied to these alloys, then the resulting material is known as alclad. It provides high strength to the core alloy protected by a corrosion-resistant skin. This layer constitutes approximately 5% of the total thickness. Clad products are produced by Alcoa under the general name Alclad. They are available in the form of sheet and plate.

3.2.3 Y-Alloy

It is also called gamma alloy. It contains 3.5–4.5% copper, 1.8–2.3% nickel, 1.2–1.7% magnesium and remaining aluminium. It is also called gamma alloy. It maintains its strength at elevated temperature; therefore, it is used in the casting of engine parts, such as pistons, etc. It is also used largely in the form of sheet and strip.

3.3 Copper and Copper-Based Alloys

Copper is a very important engineering metal not only in the pure state, but also in the alloyed form. Used in its pure state, copper is the backbone of the electrical industry because of its high electrical conductivity. The most common copper alloys are brass and bronze. The presence of a relatively small amount of other elements in the copper greatly reduces its electrical conductivity.

The properties of copper can be altered appreciably by cold working. By cold working, its tensile strength can be raised to over 450 MPa, with a decrease in elongation to about

5%. Its relatively low strength and high ductility make copper a very desirable metal for forming operations.

Electrolytic tough-pitch copper which contains 0.02–0.05% oxygen forming cuprous oxide (Cu_2O) has approximately 101.6% electrical conductivity as compared to pure copper. Because of its high electrical conductivity, copper is used extensively for wire and cable and all parts of electrical appliances such as commutators and others.

Copper is also a good conductor of heat and is highly resistant to corrosion by liquids. These two properties make copper suitable to be used in boiler fire boxes, water heating apparatus, water pipes, and vessels in brewery and chemical plants. The copper alloys also lend themselves nicely to various fabrication processes, including casting, machining and welding. Unfortunately, copper is heavier than iron.

3.3.1 Brass

Brass is an alloy of copper and zinc. It is often referred to by the composition of each component. For example, 70–30 brass means 70% copper and 30% zinc and is more popularly known as cartridge brass. If the zinc content ranges from zero to 38%, the brass is a single-phase solid solution and is known as alpha brass. It is ductile and possesses good formability. Between 38 and 46% zinc, the alloy is a two-phase structure, and is known as beta phase. With more than 50% zinc, another solid solution called gamma is found. The beta and gamma solid solutions are not as ductile as the alpha solid solution; therefore, very few alloys containing more than about 40% zinc are of commercial importance. Most brasses have good corrosion resistance, but brasses with more than 15% zinc often experience season cracking or stress-corrosion cracking.

The commercial brass containing approximately 90% copper and 10% zinc is easily workable and is mainly used for screen wire, hardware, screws, rivets and costume jewellery.

The *cartridge brass* is used in the form of sheets, wire and tubes for all types of drawing and trimming operations. Cartridge brass with tin (about 1%) becomes *admiralty brass* and is used in condenser tubes.

The *high brass* containing about 34% zinc is used extensively for drawing and spinning operations and for springs.

An alloy containing between 50 and 55% copper and the remainder zinc is often used as a filler metal in brazing. It can effectively join steel, cast iron, brasses and copper and produces joints that are nearly as strong as those obtained by welding.

An alloy containing approximately 60% copper and 40% zinc is called *muntz metal*. It is quite resistant to fresh water. For this reason, it is used extensively for condenser tubes, where corrosive actions are not too severe. It is economical as compared to other alloys. Muntz metal with tin is called *naval brass* or *tobin bronze*. Presence of tin (about 1%) in muntz metal improves its corrosion resistance and makes it suitable for use in salt water.

The addition of up to about 4% lead improves the machinability of brass quite appreciably. The resulting brass containing lead is called *leaded brass*. But presence of lead in brass reduces its ductility, and hence, its cold-forming characteristics.

3.3.2 Bronze

Bronze is an alloy of copper and tin. However, some types of bronze contain little or no tin. These materials possess desirable properties of strength, wear resistance and salt water corrosion resistance. They are often used for bearing, gears and fittings that are subjected to heavy compressive loads. Higher tin content (more than 20%) makes the alloy brittle and hard and such alloys are used for making bells mostly in the cast condition.

3.3.3 Gun Metal

Gun metal, an important bronze, contains 88% copper, 10% tin and 2% zinc. Presence of zinc cleans the metal and increases its fluidity. The metal is used chiefly for castings and its typical applications are bearings, steam pipe fittings, hydraulic valves and gears, etc.

3.3.4 Phosphor Bronze

When bronze contains phosphorus, it is called phosphor bronze. Phosphorus acts as a deoxidizer and improves the fluidity and general soundness in casting. Phosphorus is a stronger hardener than tin due to the formation of hard compound Cu_3P. Phosphor bronze can be used in wrought and casting conditions both. Wrought phosphor bronze can be used as rod, sheet and wire. Cast phosphor bronze is used for bearings, worm wheels, gears, nuts, etc.

3.3.5 Aluminium Bronze

Alloys in this system usually contain up to about 10 or 11% aluminium and up to about 4% iron with small amounts of tin. The remaining part is copper. These alloys are best known for their high strength, excellent corrosion resistance, and better wear, impact and fatigue resistance.

They are often considered as cost-effective alternatives to stainless steel and nickel-based alloys. They are used particularly for gears in heavy machinery, bearings, pump parts, valve guides, cold-working dies, etc.

3.3.6 Silicon Bronze

It contains silicon up to 4%, and its strength, formability, machinability and corrosion resistance are very good. They are obtainable in the form of strip, plate, wire, rod, tube, pipe, casting ingots, etc. The silicon bronzes find application in high-strength bolts, bells, rivets, springs, propeller shafts, boiler, tank, etc.

3.4 Nickel-Based Alloys

Those alloys that contain more than 50% nickel are classified as nickel-based alloys.

3.4.1 Monel Metal

Monel metal contains about 67% nickel, 30% copper and small amounts of iron and manganese. Monel metal has better corrosion resistance to acids, alkalis, brines, water and food products than any other commercial alloy. There are five types of monel metal which are classified as Monel, *R*-Monel, *K*-Monel, *H*-Monel and *S*-Monel. Monel is used to represent monel metal.

The characteristics of *R*-Monel are similar to that of Monel but has improved machinability because of presence of little amount (0.35%) of sulphur. *K*-Monel contains about 3% aluminium and is precipitation-hardenable. Its corrosion resistance property is similar to that of Monel. *S*-Monel is used primarily in castings and contains about 4% silicon. It is responsive to precipitation hardening. *H*-Monel is similar to *S*-Monel and contains 3% silicon. It cannot be treated to give high strength and hardness as the *S*-Monel.

3.5 Low-Melting Alloys

Alloys of lead, tin and zinc because of their low melting points are known as low melting alloys.

3.6 Lead

Lead is the heaviest of the common metals, having a relative density of about 11.3 as compared to 7.8 of steel. The chief ore from which lead is extracted is galena (PbS).

It is highly water and acids corrosion resistant by virtue of the stable lead oxide layer formation, which protects the surface. For this reason, it is used for water-pipes, roof covering, the sheathing of electric cables and for containers in chemical plants. Lead is also used for damping sound and vibrations and radiation shielding against X-rays. Pure metal is used in storage batteries. It is used as a base metal in making lead paint. When added in small quantities (alloying element) to steel and brass, lead improves their machinability. Leaded brass and ledloy steel are the suitable examples under this context. Lead, as an alloying element, is also used in solders.

3.7 Tin

Tin has a silvery whiteness and is harder than lead. It is used as a coating on steel sheet to be used as tin cans for food and various other products. Bearing material and solder are the two most important uses of tin. The alloy containing 84% tin, 8% copper and 8% antimony is one of the oldest and best bearing materials and is known as ***genuine or tin babbitt***. Because of the high cost of tin, ***lead babbitt***, composed of 85% lead, 5% tin and 10% antimony and 0.5% copper, is a more widely used bearing material.

Tin is an alloying element for bronze. Tin–lead alloys are common soldering materials.

An important characteristic of tin is the crinkling sound made when a tin bar is bent. This sound is called '*tin cry*' and is a useful method of judging the quality of solder. More tin means louder cry.

3.8 Zinc

The chief ores of zinc are zinc blende (ZnS) and calamine ($ZnCO_3$). Pure zinc is almost as heavy as steel (relative densities of steel and zinc are 7.8 and 7.1 respectively) and is also rather weak and brittle. It has two major uses: one is for galvanizing iron, steel sheet and wire; and the other is as an alloy base for casting (because of its low melting point of 380 °C and high fluidity). Galvanizing accounts for about 35% of all zinc used. In galvanizing, zinc serves as the anode and protects the steel (cathode) from corrosive attack. Zinc also serves as an alloying element. For example, brass is an alloy of copper and zinc.

Zinc-base alloys are used extensively in die casting for making products such as carburettors and fuel pumps, and household components such as vacuum cleaners and washing machines. When rolled into sheets, zinc is used for roof covering and for providing a damp-proof noncorrosive lining to containers, etc.

3.9 Heat Treatment of Nonferrous Metals

Nonferrous alloys cannot be heat treated by the techniques used for ferrous alloys. The reason is that nonferrous alloys do not undergo phase transformations as steels do. The hardening and strengthening mechanisms for these alloys are fundamentally different.

The heat treatment of nonferrous metals and alloys brings the following changes:

- It produces a uniform and homogeneous structure.
- It relieves stresses.
- It induces recrystallization.

The most effective way of increasing strength of nonferrous metals is the precipitation hardening.

3.10 Precipitation Hardening

The strength and hardness of some nonferrous metals and alloys increase with the passage of time at room temperature or slightly increased temperature. It is called age or precipitation hardening. Both temperature and time are important for age hardening. This method is employed for metals not subjected to allotropic transformation. It is based on the precipitation of one element into other because of the difference in their solid solubility. Hence, precipitation hardening can be achieved only with those alloys in which there is a decreasing solubility of one material in another as the temperature is decreased.

Nonferrous metals and alloys such as magnesium and aluminium alloys, copper-beryllium alloys and nickel base alloys are usually hardened by precipitation hardening. Duralumin, an important alloy of aluminium has the property of age hardening. Its hardness at room temperature increases rapidly in the beginning and then slowly. When it is aged, it is too hard for work to be done on it. This is known as natural aging. To make it suitable for a working operation, it is subjected to annealing in which it is heated to about 375 °C followed by slow cooling in air, water or oil.

In artificial aging, the alloy is heated to a higher temperature that accelerates the precipitation of some constituent from the supersaturated solution. The precipitation and hardness both increase with increased temperatures.

The size of the precipitated particles affects the hardness of the material. The hardness is increased with the increase in particle size, but after a critical size is reached, further growth results in brittleness and loss of strength. This phenomenon is called overaging and is undesirable.

Short Answer Questions

1. How are the properties of nonferrous materials superior over ferrous materials?

Ans. Nonferrous materials have corrosion resistance, high thermal and electrical conductivity, low density, ease of fabrication and colour choices, which are missing in ferrous materials.

2. What is duralumin? What is its composition and where is it used?

Ans. Duralumin is an important alloy of aluminium. It consists of 95% aluminium, 4% copper, 0.5% magnesium and 0.5% manganese. It is extensively used in aircraft and automobile industries because of its lighter weight.

3. What is Alclad? In which form is it available?

Ans. Alclad is an alloy of aluminium and consists of a thin layer coating of pure aluminium over duralumin to improve its corrosion resistance. Aluminium coating provides high strength to the core alloy protected by a corrosion-resistant skin. It is available in the form of sheet and plate.

4. What is age hardening? Why is it so named? What it its other name?

Ans. Age hardening is a process of increasing the hardness and strength of nonferrous metals such as aluminium, magnesium, nickel and their alloys. The hardness is obtained with the passage of time, hence called age hardening. It is also called precipitation hardening.

5. What is the difference between natural age hardening and artificial age hardening?

Ans. Natural age hardening occurs at room temperature, whereas artificial age hardening takes place at increased temperatures.

Multiple Choice Questions

1. Which of the following metals is not a nonferrous metal?

(a) Copper (b) Alloy stee
(c) Aluminium (d) Zinc

2. Consider the following properties of metals:

1. High thermal and electrical conductivity
2. High corrosion resistance
3. Low density
4. Better mechanical properties

Of these properties, the nonferrous metals possess the following:

a) 1 and 3 (b) 1, 2 and 3
(c) 1, 2 and 4 (d) 2 and 3

3. Duralumin is an important alloy of

(a) Copper (b) Zinc
(c) Aluminium (d) Magnesium

4. Duralumin has the following composition:

(a) 92% Cu, 4% Al, 2% Mg and 2% Mn
(b) 95% Al, 4% Cu, 0.5% Mg and 0.5% Mn
(c) 85% Al, 10% Cu, 2.5% Mg and 2.5% Zn
(d) 95% Al, 4% Cu, 0.5% Mg and 0.5% Zn

5. White metal is

(a) Copper-based alloy (b) Zinc-based alloy
(c) Lead-based alloy (d) Tin-based alloy

6. Bronze is an alloy of

(a) Copper and Zinc (b) Copper and Tin
(c) Copper and Lead (d) Tin and Lead

7. Gun metal contains the following:

(a) Copper, Tin and Zinc (b) Aluminium, Tin and Zinc
(c) Lead, Tin and Zinc (d) Copper, Tin and Manganese

8. Monel metal is a

(a) Aluminium-based alloy (b) Nickel-based alloy
(c) Copper-based alloy (d) Tin-based alloy

9. Muntz metal finds applications in

(a) Bearings (b) Condenser tubes
(c) Valves (d) Gears

10. An alloy containing approximately 60% of copper and 40% zinc is called

(a) Gun metal (b) Monel metal
(c) Muntz metal (d) Naval brass

11. Match List I with List II and select the correct answer using the codes given below the lists:

List I (Nonferrous metals)	List II (Alloys)
A. Copper	1. Duralumin
B. Aluminium	2. Monel metal
C. Nickel	3. Low melting alloy
D. Lead	4. Brass

Codes:	A	B	C	D
(a)	4	1	3	2
(b)	4	1	2	3
(c)	3	1	2	4
(d)	4	2	3	1

12. Match List I with List II and select the correct answer using the codes given below the list:

List I (Nonferrous metals)	List II (Ores)
A. Copper	1. Galena
B. Aluminium	2. Bauxite
C. Lead	3. Calamine
D. Zinc	4. Pyrite

Codes:	A	B	C	D
(a)	3	2	1	4
(b)	3	2	4	1
(c)	4	2	1	3
(d)	2	3	4	1

13. Match List I with List II and select the correct answer using the codes given below the lists:

List I	List II
(Alloys)	(Applications)
A. Muntz metal	1. Pistons
B. Duralumin	2. Condenser tubes
C. Gun metal	3. Aircrafts
D. Y-alloy	4. Bearings

Codes:	A	B	C	D
(a)	2	3	4	1
(b)	1	3	4	2
(c)	1	3	2	4
(d)	2	4	3	1

Answers

1. (b)	2. (b)	3. (c)	4. (b)	5. (d)	6. (b)
7. (a)	8. (b)	9. (b)	10. (c)	11. (b)	12. (c)
13. (a)					

Review Questions and Discussions

Q.1. Which types of properties do nonferrous metals possess that are not available in the ferrous metals?
Q.2. Why do copper and aluminium have higher electrical conductivity?
Q.3. What is the difference between brass and bronze?
Q.4. What is alclad?
Q.5. Why is duralumin used in making aircraft?
Q.6. What is a solid solution?
Q.7. Lead is used in the manufacture of water-pipes. Why?
Q.8. What is tin cry?

Nonconventional Materials

4

Figure shows chemical structures of polymeric materials such us polyethylene, polyvinyl chloride and polyvinyl benzene.

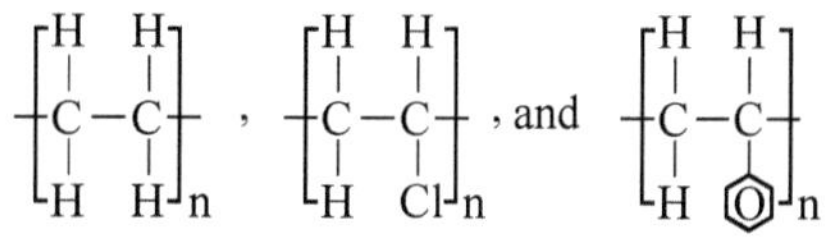

$$\left[\begin{array}{cc} H & H \\ | & | \\ -C- & C- \\ | & | \\ H & H \end{array}\right]_n, \quad \left[\begin{array}{cc} H & H \\ | & | \\ -C- & C- \\ | & | \\ H & Cl \end{array}\right]_n, \text{ and } \left[\begin{array}{cc} H & H \\ | & | \\ -C- & C- \\ | & | \\ H & \text{benzene} \end{array}\right]_n$$

Learning Objectives

After reading this chapter, you will be able to answer some of the following questions:

- Why do composite materials have brightest future?
- Why do polymers offer a wide range of applications?

D. K. Singh, *Fundamentals of Manufacturing Engineering*,
https://doi.org/10.1007/978-981-99-8767-2_4

4.1 Superalloys or High-Temperature Materials

Superalloys are high-temperature materials used for very high-temperature applications in the range of 1100 °C or more. Their important properties include high strength, good resistance to corrosion, better mechanical and thermal fatigue and better creep resistance at elevated temperatures. Superalloys find major applications in jet engines, gas turbines, rocket engines, tools and dies for hot working of metals, and in nuclear, chemical and petrochemical industries.

Superalloys are very difficult to form or machine. They are usually processed by methods such as electro discharge, electro chemical, or ultrasonic machining or they can be made to the final shape by investment castings. Powder metallurgy techniques are also used extensively in the manufacture of superalloys components. Superalloys are very expensive, and this limits their use to small and critical parts.

Iron, nickel or cobalt form the base metal for superalloys. Important alloying elements include chromium, molybdenum, aluminium, tungsten and titanium.

Iron-base superalloys generally contain 32–67% iron, 15–22% chromium and 9–38% nickel. Common alloys in this group are the incoloy series. Cobalt-base superalloys contain 35–65% cobalt, 19–30% chromium and up to 35% nickel. These are not as strong as nickel-base alloys. Stellite or vitallium is an important cobalt-base superalloy. It has been employed extensively in the form of precision investment castings for gas-turbine blades in jet engines and aircraft superchargers.

Nickel-base superalloys are the most common of the superalloys. Such alloys contain nickel in the range of 38–76%, chromium up to 27% and cobalt up to 20%. Common alloys in this group are the hastelloy, inconel, nimonic, rene, udimet, astroloy and waspaloy.

Inconel has been widely used for industrial applications that require corrosion and oxidation resistance at elevated temperatures such as for retorts and fixtures for heat-treating furnaces.

Hastelloys are resistant to attack by a variety of chemical agents. They have good high-temperature strength and ductility.

Nimonic is an age-hardenable turbine blade alloy. Alloys such as rene, udimet and waspaloy possess high strength required in jet-engine turbines. These alloys are also responsive to precipitation hardening, and their vacuum melting produces high-purity turbine blade material with a substantial increase in high-temperature strength and ductility.

4.2 Composite Materials

A composite material is the combination of two or more materials of dissimilar properties.

4.2.1 Composition of Composites

The two major constituents may be metals and ceramics, or metals and polymers, or ceramics and polymers or other combinations. One of the constituents, called reinforcing constituent, may be in particulate form, fibrous form or flake form. Reinforcing fibres include glass, graphite, aramid, boron and others. Glass fibres are the most widely used and least expensive of all fibres. Aramids are among the toughest fibres. It is marketed under the trade name *Kevlar*. *Whiskers* fibres because of their very small size do not contain any imperfections. So their strength remains unaffected and approaches the theoretical strength of the material. Recently, a high-performance polyethylene fibre was introduced under the trade name *spectra*. Whisker-reinforced composites are likely to be the future material.

The other material is called matrix. Matrix materials are usually epoxy, polyester, phenolic, fluorocarbon, polyether-sulphone and silicon. The most commonly used are epoxies (80% of all reinforced plastics) and polyesters, which are less expensive than epoxies.

4.2.2 Reinforcement of Composites

Composites are reinforced to make them stronger. When improved strength is the major goal, the reinforcing component must have a large aspect ratio, that is, its length/diameter ratio must be high, so that the load is transferred across potential points of fracture. Thus, steel rods are placed in concrete structure. Also, glass fibres and polymers are combined for fibre-reinforced plastics.

4.2.3 Examples of Composites

Wood is considered to be a composite of cellulose fibres bonded by a matrix of natural polymers. Concrete can be classified as a ceramic composite in which stones are dispersed among cement. In reinforced concrete, steel rods impart the necessary tensile strength to composite, since concrete is brittle and generally has little or no useful tensile strength.

4.2.4 Properties of Composites

Composites offer several outstanding properties as compared to conventional materials. In composite materials, an attempt is made to increase the stiffness, without the disadvantages of brittleness. Characteristically, composites possess high strength-to-weight and stiffness-to-weight ratios and offer new design flexibilities and improved corrosion and wear resistance.

4.2.5 Advanced Composites

Advanced composite materials, sometimes referred to as fibre-reinforced plastics, are those materials composed of two or more materials, of which one is a fibre (e.g. carbon, fibre glass, kevlar) and the other is a binder or matrix (e.g. epoxy, thermoplastic, polyester). In advanced composites, fibres in the matrix are orderly oriented as compared to random-oriented fibre composites. According to the latest trend, the advanced composite is one that contains a fibre-to-resin ratio of 50% fibre, with the fibres having a modulus of elasticity greater than 16×10^6 psi.

4.2.6 Applications of Composites

Composite materials have wider applications in aircraft, space vehicles, offshore structures, electronics, automobiles, etc.

4.3 Polymers

Polymers, also called plastics, are low-density non-metallic materials, often large molecules of non-metallic elements with many repeatable units (mers). They are one of the most extensively used engineering materials after metals. A few examples of polymers include polyethylene (also called polythene), polyvinyl chloride (PVC) and polyvinyl benzene. Polymers become soft at elevated temperatures, hence to shape them, they are extensively used in various plastic moulding processes such as casting, blow moulding, compression moulding, etc.

4.3.1 Properties and Uses of Polymers

Plastics are low-density materials and are used as insulators, both electrically and thermally. However, the electrical conductivity of some polymers can be increased by doping (introducing certain impurities in the polymer such as metal powder, salts and iodides). Polyacetylene is a conducting polymer. Conducting polymers are used in microelectronic devices, rechargeable batteries and electric power equipment. Plastics are poor reflectors of light and are transparent. They do not crystallize readily because they lack atoms. Some of them are flexible when subjected to deformation, making them suitable for manufacturing. Polyethylene is used in packaging, houseware such as buckets and dust bins, carpet backing and cable insulation. Polyvinyl chloride (PVC) is used in imitation leathers, floor coverings and gramophone records.

4.3.2 Polymerization

The conversion of monomer into polymer is known as polymerization. The polymerization process takes place by either addition or condensation mechanism. Formation of polyethylene and polyvinyl chloride (PVC) is the example of addition polymerization. Here, a number of basic units (monomers) are added together to form a large molecule (polymer).

Polymers which are made from only one kind of monomer are called homopolymers, whereas polymers which are made from two different types of monomer are called copolymers. The formation of copolymers greatly expands the possibilities of creating new types of plastics with improved physical and mechanical properties. Terpolymers consist of three different monomers. Generally, catalysts such as benzoyl peroxide are used to initiate and terminate the polymerization chain.

In contrast to addition polymerization, where all the monomers appear in the product molecule, condensation polymerization takes place by the combination of monomers with the elimination of simple molecules such as water (H_2O) or methyl alcohol (CH_3OH). Heat, pressure and catalyst are often required to drive the reaction. The formation of *Bakelite* as a result of the reaction between phenol and formaldehyde is the example under this category. The structure of condensation polymers can be either linear or a three-dimensional framework in which all atoms are linked by strong primary bonds.

4.3.3 Classification of Polymers

Polymers are broadly classified into two groups:

(a) Thermoplasts and (b) Thermosets

4.3.4 Thermoplastic Polymer

A thermoplastic polymer (also called thermoplast or thermoplastic elastomer[1]) is one which softens on heating and becomes rigid on cooling. On cooling, it returns to its original hardness and strength. In other words, the process is reversible. Thermoplasts have the property of increasing plasticity, that is, increasing ability to deform plastically with increasing temperature. The material can then be cast, injected into a mould, or forced into or through dies to produce a desired shape. Hence, this high degree of plasticity is one of

[1] Elastomer represents a group of flexible materials that can be stretched up to about double their length at room temperature and can return to their original length when stretching load is released.

the technologically attractive properties of plastics. Most addition polymers and some condensation polymers are thermoplastic. Examples are polyethylene, polystyrene (polyvinyl benzene), polyvinyl chloride, nylons and acrylics. Thermoplastic materials are processed by injection or blow moulding, extrusion, thermoforming, calendaring and others.

4.3.5 Thermosetting Polymer

A thermosetting polymer (thermoset) is one which becomes hard on heating. It cannot be softened by heating. Thermosetting plastics are formed to shape with heat and with or without pressure. They are so named because during polymerization the network is completed and the shape of the part is permanently set. Thermosetting plastics, usually have a highly cross-linked or three dimensional framework structure in which all atoms are connected by strong covalent bonds. These materials are generally produced by condensation polymerization, where elevated temperature promotes the irreversible reaction. A typical thermoset is Bakelite (a phenolic resin), which is a product of the reaction between phenol and formaldehyde.

Products such as handles and knobs on cooking pots and pans and components of light switches and outlets are made of thermosetting plastics. The mechanical, chemical and thermal properties; and electrical resistance and dimensional stability of such polymers are superior over thermoplastics. However, their quality deteriorates at increased temperatures. Processes used for thermosetting plastics include compression or transfer moulding, casting, laminating and impregnating.

4.4 Ceramics

Ceramic materials consist of compounds of metallic and nonmetallic elements often in the form of oxides, carbides and nitrides. They are hard and brittle. They tend to be more resistant than either metals or polymers to high temperatures and to severe environments. They are generally insulators, both electrically and thermally. Additionally, ceramics have low thermal expansion, good creep resistance, good chemical stability, high elastic modulus and high compressive strength.

Ceramic materials are characterized by high shear strengths; thus, they are not easily deformed. This leads to high hardnesses and high compressive strengths and low fracture strengths.

Most ceramic phases, like metals, are crystalline but relatively complex. Unlike metals, however, their structures do not contain large numbers of free electrons, and the bonding electrons are generally captive in strong ionic or covalent bonds. This complexity and the greater strength of the bonds holding the atoms together make ceramic reactions sluggish. For example, at normal cooling rates glass does not have time to rearrange itself into a complicated crystalline structure, and therefore at room temperature, it remains as a supercooled liquid for a long time. Due to the absence of free electrons, most ceramic materials are transparent, at least in thin sections, and are poor thermal conductors.

Glass, brick, stone, concrete, abrasives, porcelain enamels, dielectric insulators and high-temperature refractories are some of the examples of ceramics. Carbides of tungsten, titanium, zirconium and silicon, alkali halides and silicon nitride (Si_3N_4) are some important ceramic materials of present time.

4.5 Refractory Materials

Refractory materials are ceramics which can withstand very high temperatures. Most of the refractory materials are the stable oxides of elements such as silicon, aluminium, magnesium, calcium and zirconium, etc. Various carbides, nitrides and borides can also be used in refractory applications. Refractories confine the heat in ovens and furnaces by preventing heat loss to the atmosphere.

One of the most widely used refractories is based on alumina–silica composition, varying from nearly pure silica to nearly pure alumina. Other common refractories are magnesite, forsterite, dolomite, silicon carbide and zircon.

4.5.1 Types of Refractories

Refractories are divided into three groups:

- Acidic refractories
- Basic refractories
- Neutral refractories.

Acidic refractories readily combine with bases and are therefore termed acidic. Silica is their chief constituent. Important acidic refractories include quartz, sand and silica brick.

Magnesium oxide (MgO) is the core material for most basic refractories. They are often required in metal-processing applications to provide compatibility with the metal. The most common basic refractories are magnesite and dolomite.

Neutral refractories do not combine with either acidic or basic refractories. The combination of acidic and basic refractories is very useful. A basic refractory is provided on the surface for chemical reasons and the cheaper, acidic refractory is used beneath to provide strength and insulation. Examples of neutral refractories are silicon carbide, chromite (Cr_2O_3) and carbon.

4.6 Abrasives

An abrasive is a hard, mechanically resistant material used for grinding or cutting. It is commonly made of a ceramic material.

Emery, a mixture of alumina (Al_2O_3) and magnetite (Fe_3O_4), is a natural abrasive used on coated paper and cloth. Corundum (natural Al_2O_3), quartz, sand, garnets and diamonds are other naturally occurring abrasive materials. The last four have commercial importance. Aluminium oxide (Al_2O_3) and cubic boron nitride (CBN) are classified as artificial abrasives, of which aluminium oxide is the most widely used one. CBN and diamond are also known as superabrasives because of their extreme hardnesses.

Hardness is the key property for an abrasive. In addition to hardness, an important characteristics is friability, that is, the ability of abrasive grains to fracture into smaller pieces. It is a useful property of abrasives imparting them self-sharpening characteristics. High friability indicates low strength or low fracture resistance of the abrasive.

Diamond abrasive wheels are used extensively for sharpening carbide and ceramic cutting tools. Diamonds are usually used only when cheaper abrasives will not produce the desired results.

CBN maintains its hardness at elevated temperatures and has several useful properties such as long wheel life, good surface quality, no burn or chatter, and low scrap rate. It can be used successfully in grinding iron, steel, alloys of iron, Ni-based alloys and other materials. It can also solve difficult-to-grind jobs. CBN is popular by its trade name Borazon.

4.7 Coated Abrasives

Coated abrasives are used extensively in finishing flat or curved surfaces of metallic and nonmetallic parts, and in wood working. The precision of surface finish obtained depends primarily on grain size. Important coated abrasives include aluminium oxide, silicon carbide, zirconia alumina and ceramic alluminium oxide. Aluminium oxide and zirconia alumina are most widely used.

Coated abrasives are available as sheets, belts and discs. They are formed by gluing abrasive grains onto a cloth or paper backing by two methods, namely gravity process and electrostatic process. In the gravity process, the abrasive grains are dropped from an overhead hopper into the adhesive-coated cloth. In the electrostatic process, the grains and the backing, both are passed simultaneously through a electrostatic field. The abrasive grains are propelled upward and are embedded in the adhesive on the cloth backing.

Short Answer Questions

1. What is teflon? What is its chemical name and where is it used?

Ans. Teflon is a thermoplastic polymer and its chemical name is polytetra fluoroethylene (PTFE). It is used in making nonlubricating or self-lubricating bearings because of its very low coefficient of friction.

2. What is vulcanization? Why is it carried out?

Ans. Vulcanization is a process in which rubber in its crude form called latex, obtained from the rubber tree, is mixed with small amount of sulphur and then heated. Addition of sulphur improves the mechanical properties of rubber which lack otherwise. Also, before vulcanization filler materials such as carbon black and zinc oxide are added to crude rubber to improve its wear resistance.

3. Give a few examples of thermoplastic and thermosetting polymers.

Ans. A few examples of thermoplastic polymers include polyethylene, polyvinyl chloride (PVC) and Teflon, and examples of thermosetting polymers include epoxies, phenolics and silicones.

4. What is polyethylene? In which form is it used?

Ans. Polyethylene, also called polythene, is a widely used low cost thermoplastic polymer. It is used in the form of sheets, film and wire insulation. Typical polyethylene products include bottles, pipes and housewares.

5. What is the difference between thermoplastic and thermosetting polymers?

Ans. Thermoplastic polymers have linear or branched chain structure, and they soften on heating and become rigid on cooling. Also they do not degrade on repetitive heating and cooling, as they do not cross-link upon heating. On the other hand, thermosetting polymers, because of their cross-link structure, set on heating and become extremely rigid and unsuitable for more than one application.

Multiple Choice Questions

1. Thermosetting plastics are

    1. Formed by addition polymerization
    2. Formed by condensation polymerization
    3. Softened on heating and hardened on cooling

4. Moulded by heating and cooling

Of these:

(a) 1 and 3 are true (b) 2 and 4 are true
(c) 1 and 4 are true (d) 2 and 3 are true.

2. Thermoplastic polymers are

1. Formed by addition polymerization
2. Formed by condensation polymerization
3. Softened on heating and hardened on cooling
4. Moulded by heating and cooling

Of these:

(a) 1 and 3 are true (b) 2 and 4 are true
(c) 1 and 4 are true (d) 2 and 3 are true.

3. Match List I with List II and select the correct answer using the codes given below the lists:

List I	List II
(Materials)	(Nature of products)
A. Polyethylene	1. Adhesive
B. Polyurethane	2. Film
C. Cyano-acrylate	3. Wire
D. Nylon	4. Foam

Codes:	A	B	C	D
(a)	2	4	3	1
(b)	4	2	3	1
(c)	2	4	1	3
(d)	4	2	1	3

4. Which one is weldable plastic?

(a) Thermosets alone
(b) Thermoplastics alone
(c) Both thermosets and thermoplastics
(d) Neither thermosets nor thermoplastics

5. Match List I with List II and select the correct answer using the codes given below the lists:

List I (Materials)	List II (Nature of products)
A. Neoprene	1. Electric switches
B. Bakelite	2. Adhesive
C. Foamed polyurethane	3. Thermal insulation
D. Araldite	4. Oil seal

Codes:	A	B	C	D
(a)	4	1	3	2
(b)	1	4	3	2
(c)	4	1	2	3
(d)	1	2	3	4

6. Which of the following pairs are correctly matched?

   1. Cellulose nitrate—Table tennis ball
   2. Phenol furfural—Brake linings
   3. Epoxies—Jigs and fixtures

   Select the correct answer using the codes given below:

   (a) 1 and 2 (b) 2 and 3
   (c) 1 and 3 (d) 1, 2 and 3

7. Which of the following processes is used for thermosetting materials?

   1. Compression 2. Transfer moulding
   3. Injection moulding 4. Extrusion

   Select the correct answer using the codes given below:

   (a) 1 and 4 (b) 1 and 2
   (c) 2 and 3 (d) 2, 3 and 4.

8. The strength of the fibre-reinforced plastic product

   1. Depends upon the strength of the fibre alone
   2. Depends upon the fibre and plastic

3. Is isotropic
4. Is anisotropic
Which of these statements is correct?

(a) 1 and 3 (b) 1 and 4
(c) 2 and 3 (d) 2 and 4

9. Which of the following materials is used for car tyres as a standard material?

(a) Styrene-butadiene rubber (SBR)
(b) Butyl rubber
(c) Nitrile rubber
(d) None of the above

10. Match List I with List II and select the correct answer using the codes given below the lists:

List I	List II
(Requirements)	(Type)
A. High-temperature service	1. Teflon bearing
B. High load	2. Carbon bearing
C. No lubrication	3. Hydrodynamic bearing
D. Brushing	4. Sleeve bearing

Codes:	A	B	C	D
(a)	1	2	3	4
(b)	4	1	2	3
(c)	2	1	3	4
(d)	2	3	1	4

11. Tin base metals are used where the bearings are subjected to

(a) Large surface wear (b) Elevated temperatures
(c) Light load and pressure (d) High pressure and load

Answers

1. (b) 2. (a) 3. (c) 4. (b) 5. (a) 6. (a)
7. (b) 8. (d) 9. (a) 10. (d) 11. (d)

Review Questions and Discussions

Q.1. What are superalloys? Why are they so named?
Q.2. What methods are used for the manufacture of superalloys?
Q.3. Why is concrete generally not recycled?
Q.4. Plywood has normally an uneven number of wood plies. Why?
Q.5. Paper generally does not tear in a straight line unless it is creased. Why?
Q.6. What are the major differences between plastics and metals?
Q.7. What are polymerization and degree of polymerization? What is the effect of polymerization on the properties of a material?
Q.8. Differentiate between thermoplastic and thermosetting plastics.
Q.9. How are thermoplastic polymer made stronger?
Q.10. Why are thermosetting polymers brittle?
Q.11. How is a composite material defined?
Q.12. Why are most ceramics transparent in thinner sections?

Introduction to Casting 5

Figure shows a cool-ing curve for an alloy, where solidificatior takes place not at a definite temperature as in case of pure metals, but over a definite range of temperatures.

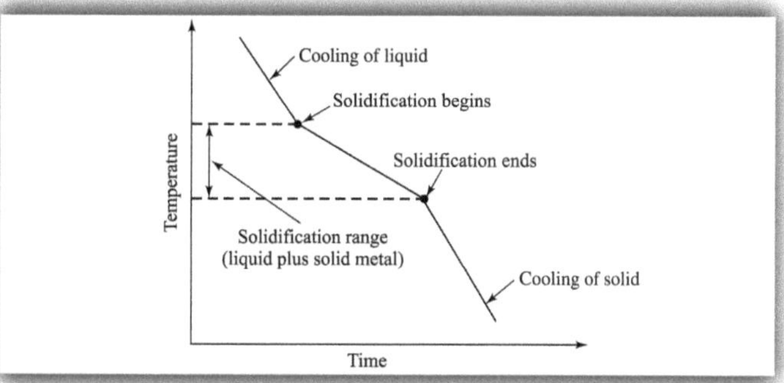

Learning Objectives

After reading this chapter, you will be able to answer some of the following questions:

- Why is casting one of the preferred manufacturing methods?
- Why do casting need heat treatment?
- What are the possible reasons for casting defects?

D. K. Singh, *Fundamentals of Manufacturing Engineering*,
https://doi.org/10.1007/978-981-99-8767-2_5

5.1 Introduction

Casting is one of the oldest methods of manufacturing and even today it finds extensive applications in manufacturing industries.

In casting, metals or alloys are initially heated to melt them. The molten metal is then poured into a mould cavity, where it is allowed to solidify. After solidification, the product is taken out of the mould cavity and subjected to finishing operation as per requirement. The term casting is used to describe both the process and the product. Usually the mould is made of sand and the method of casting using sand mould is called sand casting. Sand casting of cast iron, steel and metals is extensively used in industry, but sand cast parts have usually rough surfaces and are dimensionally less accurate than any other type of casting. However, sand casting is invariably the cheapest method for small batch production of parts.

Casting can produce complex shapes (internal or external), parts with hollow sections or internal cavities and parts having irregular curved surfaces. It can also make parts which are difficult to produce by any other methods. It offers high production rate with improved material properties, produces good surface finish and ensures close dimensional tolerances.

Casting parts have good compressive strength but poor tensile strength. Important casting materials include cast iron, copper alloys, aluminium, zinc, nickel and magnesium.

Pulleys, flywheels, engine blocks, connecting rods, crankshafts, pistons, carburetors, railroad wheels, machine tool beds, gear blanks, turbine housings and turbine blades are some of the important casting products.

There are many casting processes available today and the selection of the best method to produce a particular part depends on several basic factors, such as cost, size, finish (surface finish), production rate, tolerance, section thickness, physical-mechanical properties, intricacy or design, machinability and weldability.

5.2 Casting Terminology

A ***moulding flask*** is a box that contains the moulding aggregate. In a two piece mould, *cope* refers to the top half of the pattern, flask, mould or core and drag, the bottom half of any of these features. The seam between them is called ***parting line*** or ***parting surface*** (Fig. 5.1). When more than two pieces are used, the additional part is called cheek.

A ***pattern*** is the replica of the part to be cast. It is used to mould the sand mixture into the shape of the casting. The moulding material (usually sand) is packed around the pattern and then the pattern is removed to produce a mould cavity exactly of the size of the casting to be produced. A pattern may be made of wood, plastic or metal. The selection of a pattern material depends on the following factors:

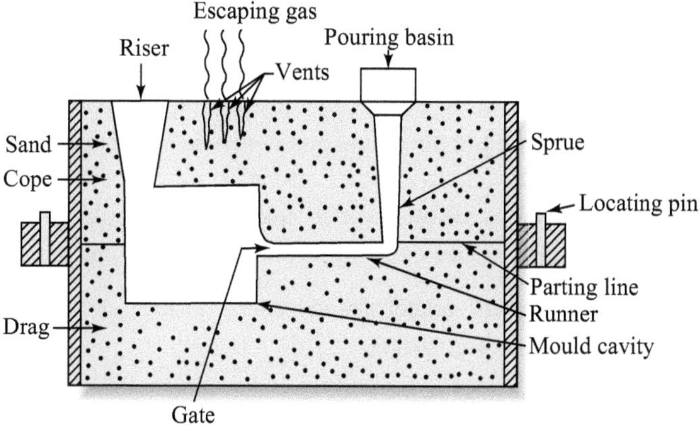

Fig. 5.1 Cross section of a sand mould

- Size and shape of the casting,
- Desired dimensional accuracy,
- Number of castings to be produced, and
- Moulding process to be used.

Since patterns are used repeatedly to make moulds; the strength and durability of the material selected for patterns are very important. Patterns are usually coated with a parting agent to facilitate their removal from the moulds. Pattern may be ***removable*** or ***disposable*** type.

In case of casting using removable pattern, sand is packed around the pattern. Later the pattern is removed and the cavity thus produced is filled with molten metal.

A removable pattern may be made of wood, plastic or metal. Disposable patterns are made from polystyrene (a type of plastic) and, instead of being removed from the sand, are vaporized when the molten metal is poured into the mould.

The following advantages are obtained using a disposable pattern:

- The process is fast since it consumes less time.
- No pattern allowances are required.
- Castings have better surface finish.
- No cores are required.
- Moulding is greatly simplified.

The disadvantages include the following:

- The pattern is lost in the process.
- Patterns are more delicate to handle.

A ***mould*** refers to a void created in a compact sand mass which, when filled with molten metal, will produce a casting. Obviously, it is the impression left behind by a pattern after the withdrawal of the latter from the mould. Mould can also be defined as an assembly of two or more metal blocks and is used to produce castings of good surface finish and detail. The ***mould cavity*** lies within the mould and holds the liquid material. A mould may be made of metal, plaster, ceramics and other refractive substances. The mould made of sand is called sand mould and is the most widely used type of mould in the foundry shop. The process of producing the mould or cavity is known as moulding.

Pouring Basin (Cup) is a small funnel shaped cavity at the top of the mould which contains the molten metal used for casting.

Sprue is a vertical passage connected to the pouring basin through which the molten metal flows downward, hence also called downsprue.

Runner is a horizontal channel in the parting line to carry the molten metal from the sprue to the mould cavity.

Gates are the inlets into the mould cavity.

Gating system is a network of channels used to deliver the molten metal into the mould cavity. It consists of pouring basin, sprue, runner and gate.

Vents are the small openings in the mould to carry off gases when the hot molten metal comes into contact with the sand in the moulds and core. They also exhaust air from the mould cavity as the molten metal flows into the mould.

Core is a sand mass, which is inserted into the mould to produce identical shaped regions such as holes or passages for water cooling or otherwise define the interior surface of the casting. Cores are also used on the outside of the casting to form features such as lettering on the side of a casting or deep external pockets.

They are placed in the mould cavity before casting to form the interior surfaces of the casting and are removed from the finished part during shakeout and further processing. Like moulds, cores must possess strength, permeability, ability to withstand heat and collapsibility. Therefore, cores are made of sand aggregates.

Core Box is a mould or die which is used to produce casting cores.

Core Prints are the projected parts added to the pattern, core or mould and are used to locate and support the core within the mould.

Riser is a reservoir of molten metal, which supplies additional metal to the casting to compensate for shrinkage during solidification of the casting. After filling the mould cavity, the molten metal enters the riser and its level rises up gradually in the riser.

If the riser is contained entirely within the mould, it is known as a ***blind riser***, and if it is open to the atmosphere, it is called an ***open riser*** (conventional riser). The open riser derives feeding pressure from the atmosphere and from the force of gravity on the metal contained in the riser. Since the blind riser is closed from all sides, there is no question of atmospheric pressure. The pressure due to the force of gravity is also reduced due to the formation of vacuum within its body. Due to its permeable nature, air is able to enter the riser and exert some pressure. Sometimes, to create artificial pressure in the blind riser some explosive material is used which explodes while coming in contact with the molten metal, creating high pressure within the riser.

Blind risers are also known as ***side risers*** since they are horizontally located adjacent to the mould cavity usually along the parting line. An ***open riser*** also called ***top riser*** is one that sits on the top of a casting. Because of their location, top risers have shorter feeding distances and occupy less space within the flask.

The ***blind riser*** has the following advantages:

- The hottest metal lies in the riser and the coldest in the casting. This promotes directional solidification.
- A blind riser can be smaller than a comparable open riser.
- Blind risers can be removed more easily from a casting.

Risers may not be always required. For alloys with large freezing ranges, the risers would not be particularly effective, and one generally accepts the fine, dispersed porosity. Die casting, low pressure permanent moulding and centrifugal casting are the processes, where no risers are required and the positive pressures provide the feeding action that is required to compensate for solidification shrinkage.

A common problem with cores is that for certain casting requirements as in the case, where a recess is required, they may lack sufficient structural support in the cavity. To keep the core from shifting, metal supports, known as ***chaplets***, may be used to anchor the core in place.

Chills are metallic objects of high-heat capacity and high thermal conductivity, which are placed in the mould or mould cavity to increase the cooling rate of castings or to provide uniform or to promote directional solidification. When the casting consists of both thick and thin sections, the thinner sections tend to solidify earlier than the thicker ones. This differential cooling rate produces uneven contraction of parts and gives rise to internal strains in the metal. It may even produce cracks, if the cooling of thinner parts is too severe. For rapid solidification of heavy sections and the achievement of directional solidification, chills are commonly used.

External chills are placed in the mould (adjacent to the casting). They can often be used to reduce the number of risers (by retarding the solidification process) required for a casting.

Internal chills are placed within the mould cavity to absorb heat and promote more rapid solidification. Since some of the chill metals will melt during the operation, it will absorb not only the heat capacity energy, but also some heat of fusion. ***Internal chills*** must be made of same material as used for casting because they ultimately become part of the final casting.

Castability of a material is defined as the ease with which a metal can be cast into useful products. Important factors affecting casting include solidification rate, shrinkage, gas porosity and others. Grey cast iron has the best castability among the cast iron group, since it has high fluidity and slow solidification rate.

5.3 Types of Moulds and Moulding Processes

Sand moulds are classified on the basis of types of sand and the methods used to produce them. The following are the important types of moulds and moulding processes.

- The most common mould material is green sand which is nothing but a natural sand and the mould prepared by using green sand is called ***green sand mould***. The sand is called *green* not because of its colour but because of its water content. Green sand is a mixture of silica sand, clay and water and the moulding process using this sand is known as ***green sand moulding,*** which is a least expensive method of making moulds.
- In the ***skin-dried moulding***, the mould surfaces are dried to a depth of about 25 mm or more, either by drying the mould with gas torches and heaters or by storing the mould in air (***Air-dried moulding***). ***Skin-dried moulds*** are generally used for large castings because of their higher strength. Sand moulds are also oven dried, i.e. baked prior to receiving the molten metal (***Dry sand moulding***). Dry sand moulds are stronger than green-sand moulds and impart better dimensional accuracy and surface finish to the casting. However, distortion of the mould is greater, the castings are more susceptible to hot tearing (in the form of cracks developed in the casting due to high residual stresses), because of the lower collapsibility of the mould, and the production rate is slower because of the drying time required.
- In the ***core-box mould*** process, various organic and inorganic *binders* are blended into the sand to bond the grains chemically for greater strength (***Core sand moulding***). These moulds are dimensionally more accurate than green-sand moulds, but are more expensive.
- ***Loam moulds*** are used for extremely large castings. The mould is prepared by using loam sand, which is a mixture of sand and clay, where clay is in higher amount as high as 50%. The shape of the mould is prepared with sweep or skeleton patterns. Objects such as large cylinders, round-bottomed kettles, chemical pans, large gears and other machine components are produced in loam sands. Because of the more production

time, loam sands are seldom used. The process of preparing loam moulds is ***loam moulding***.

- In ***CO_2 mould***, clean sand (containing no oil, resin or clay as the bonding agent) is mixed with 2–6% of sodium silicate solution and the mixture is rammed about a pattern. CO_2 gas is then pressure-fed to the mould for about one minute which hardens the sand mixture. Very smooth and intricate castings are obtained by this method, although the process was originally developed for making cores. Shakeout and core removal are difficult.
- ***Furan mould*** method (furfuryl alcohol binder moulds) is good for making moulds using disposable patterns and cores. Dry sharp sand is mixed with acids such as phosphoric (H_3PO_4) and Sulphuric (H_2SO_4), which act as an accelerator. Resin is then mixed thoroughly. The sand material begins to air-harden almost immediately but the time delay is sufficient to allow moulding. In use with disposable patterns, furan resin sand is employed as a wall or shell around the pattern supported by green or sharp sand or it can be used as the complete moulding material.
- ***Expendable moulds*** are made of sand, plaster, ceramics and similar materials, which are generally mixed with various binders or bonding agents. They have the capability to withstand the high temperatures of molten metals. After the casting has solidified, the mould in these processes is broken up to remove the casting.
- ***Permanent moulds*** are made of metals and are used repeatedly. Metal moulds have better heat conducting capacity than expendable moulds and therefore casting is subjected to a higher rate of cooling. Such moulds produce casting of better surface finish, high dimensional accuracy and offer high rate of production. Also, the usual defects which are noticed in sand castings are eliminated. One of the serious defects of the permanent mould is the *chilling effect*. Different cooling rates and solidification times can produce substantial variation in the structure and properties of the resulting casting. For example, if a casting consists of sections of uneven thickness, the thin sections tend to solidify quicker than the thick ones, resulting in uneven contraction and severe distortion.
- ***Composite moulds*** are made of two or more different materials such as sand, plaster graphite and metal, combining the advantages of each material, and are used in shell moulding and other casting processes. They are generally employed in casting complex shapes, such as impellers for turbines. Composite moulds increase the strength of the mould, improve the dimensional accuracy and surface finish of castings and may help reduce overall costs and processing time.
- ***Bench moulding*** is used for small work to be done on a bench at a suitable height to the moulder.
- In ***Floor moulding***, work is done on the foundry floor. This moulding is used when castings increase in size and are difficult to handle. Medium and large size castings are made by using floor moulding.

- In ***Pit moulding***, extremely large castings are frequently moulded in a pit (a big hole made on the foundry floor) instead of a flask. The pit acts as the drag part of the flask, and a separate cope is used above it. The sides of the pit are brick-lines, and on the bottom is a thick layer of cinders with connecting vent pipes to the floor level.
- In ***machine moulding***, a number of the operations are performed by a machine, which the moulder ordinarily does by hand. Ramming the sand, rolling the mould over, forming the gate and drawing the pattern are performed by machines much better and more efficiently than by hand.

5.4 Types of Patterns

Selection of a pattern principally depends on the number of castings required and the desired complexity of the part.

5.4.1 Solid Pattern

A solid pattern, also called one-piece pattern, as it is made from one piece and does not contain loose piece or joints. It is the simplest of all the patterns. Solid patterns are generally used when the shape is relatively simple and the number of castings required is rather small. It is usually made from wood.

5.4.2 Split Pattern

Sometimes solid pattern is not suitable because of difficulty in its removal from the sand mould. To eliminate this difficulty, patterns are split in two parts so that one half of the pattern is meant for the cope and the other half for the drag part of the mould. The split occurs along the parting line of the mould.

5.4.3 Match-Plate Pattern

A match-plate pattern consists of several patterns mounted on its both sides. One side has the cope impression and the other side has the drag impression. A match plate is made of wood or metal. This pattern is normally used in machine moulding for large scale production of small castings at faster rates. It is used to produce piston rings of internal combustion (I.C.) engines.

5.4.4 Cope and Drag Pattern

In cope and drag pattern, the cope and drag halves of a split pattern are separately mounted on two match plates. A cope pattern plate is used for making copes and a drag pattern plate for making drags. The cope and the drag flasks are made separately and brought together to produce the complete mould. It is used to produce big size castings which as a whole cannot be conveniently handled by one moulder alone.

5.4.5 Gated Pattern

Gated patterns, usually made of metal, are used for mass production of small castings. Here multicavity moulds are prepared; i.e. a single sand mould carries a number of cavities. Patterns are connected to each other through suitable channels, known as *gate formers*, which are used for feeding the molten metal to these cavities. A single runner can be used for feeding all the cavities, thereby saving considerable moulding time.

5.4.6 Sweep Pattern

Sweep patterns are used for producing moulds of large shapes, which are axi-symmetrical or prismatic in nature such as bell shaped or cylindrical. They are made of wood. Here 'sweep' refers to the section that rotates about an edge to yield circular sections.

5.4.7 Skeleton Pattern

Skeleton pattern consists of a simple wooden frame outlining the shape of the casting and is mounted on a metal base. This pattern is generally useful for large size castings required in small numbers. Typical products made from skeleton pattern include turbine casings, water pipes, L-bends, chutes, etc.

5.4.8 Loose Piece Pattern

Loose piece pattern is used when the contour of the part is such that withdrawing the pattern from the mould is not possible. Such patterns are made in loose pieces to facilitate easy withdrawal. After moulding is over, first the main pattern is removed and then the loose pieces are recovered through the gap generated by the main pattern (Fig. 5.2).

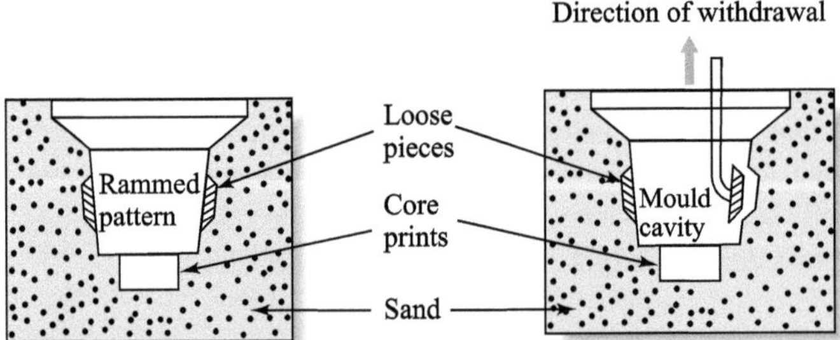

Fig. 5.2 Loose piece pattern

5.5 Pattern Allowances

The size of the pattern is different than the actual size of the casting to be produced and is always made larger to take care of several factors. The excess in dimensions over the actual size of the casting is known as allowance. Shrinkage, draft, finish, distortion and shake are important pattern allowances.

5.5.1 Shrinkage Allowance

Metals and alloys have tendency to shrink on cooling. There are three principal stages of shrinkage: (1) shrinkage of the liquid, (2) solidification shrinkage as the liquid turns to solid and (3) solid metal contraction as the solidified metal cools down to room temperature. During solidification, a casting contracts and this contraction may be as much as 2% or 20.6 mm/m. To take care of this factor, the pattern is made slightly larger than the casting, and this excess material provided on pattern for shrinkage consideration is called shrinkage allowance. The shrinkage allowance depends on the metal being cast. Typical allowances for some materials are given in Table 5.1.

Table 5.1 Shrinkage allowance for some materials

Materials	Shrinkage allowance	
	(%)	(mm/m)
Aluminium	1.0–1.3	10.5–13.1
Brass	1.5	15.7
Cast iron	0.8–1.0	8.2–10.5
Magnesium	1.0–1.3	10.5–13.1
Steel	1.5–2.0	15.7–20.6

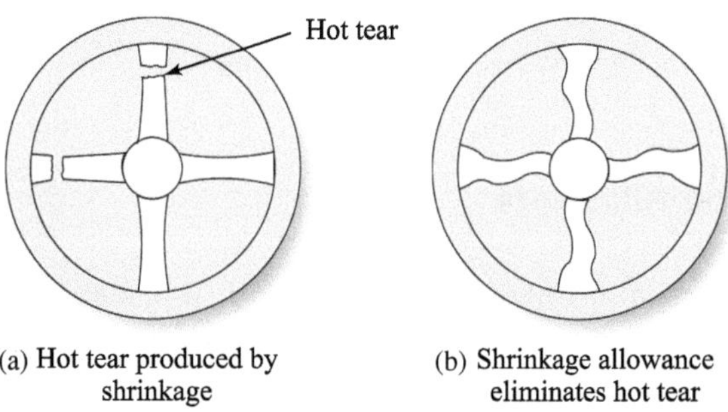

Fig. 5.3 Shrinkage allowance

If shrinkage is not provided, cracks are bound to occur in the casting. Figure 5.3 shows a wheel with spokes. If the spokes are curved, the tensile stress in them resulting from contraction during solidification and hence the tendency for cracking is reduced.

In order to produce a sound casting, a riser is attached to the casting to provide a source of liquid metal.

5.5.2 Draft Allowance

When a pattern is withdrawn from a mould, there are chances that the edges of the mould in contact with the pattern may get damaged. To reduce this happening, the surfaces of the pattern parallel to the direction of its withdrawal are slightly tapered. This tapering of the sides of the pattern is known as draft or draft allowance. It provides a slight clearance for the pattern as it is lifted.

The amount of draft to be provided depends upon the shape and size of the pattern, the depth of the cavity, the pattern and the mould material and the moulding procedure. Draft angles usually vary between 0.5° and 2°. Inner surfaces have twice draft as compared to outer surfaces because the casting shrinks inward towards the core.

5.5.3 Machining Allowance

The dimensional accuracy and the surface finish of the castings produced by sand casting processes are not very good. Their surfaces need machining. The excess in dimensions of the casting and pattern over their final dimensions to take care of machining is called machining allowance. It depends on the type of casting and increases with the size and section thickness. For small castings, machining allowance may vary between 2 and 5 mm and can be more than 25 mm for large castings. Sand castings require more machining allowance. Die castings are sufficiently smooth, and hence very little or no machining is required. Investment castings do not require machining.

5.5.4 Distortion Allowance

Distortion allowance is also called camber allowance. It applies to the castings having different thicknesses at different locations. Due to differential rate of cooling for different parts of the casting, distortion can take place. Distortion occurs due to thermal stresses developed during solidification. U-shaped sections may distort at their arms and base. To eliminate this defect, an opposite distortion of equal amount is provided in the pattern. Distortion depends mainly on the configuration of the casting.

5.5.5 Shake Allowance

The pattern is initially in firm contact with the sand mould. To facilitate easy withdrawal of the pattern from the mould, its vertical surfaces are rapped so that they are detached from the mould. During shaking, the size of the mould cavity may increase. It is the only negative allowance, which is taken care of by decreasing the size of the pattern.

5.6 Melting Furnaces

The following factors are considered during the selection of a melting furnace for casting:

- Economic considerations, such as initial cost, operation cost and maintenance cost
- Control of furnace atmosphere to avoid contamination of the metal
- Composition and melting point of material being cast and ease of controlling metal chemistry
- Capacity and rate of melting required
- Environmental considerations, such as air pollution and noise
- Power supply, its availability and cost of fuels.

The melting furnaces commonly used in foundries are: electric arc, induction, crucible and cupola.

5.6.1 Electric Furnace

Electric furnace is used to produce steels. The source of heat is a continuous electric arc formed between two or three graphite electrodes and the charged metal.

Steel and iron scrap and limestone (which acts as a flux) are dropped in the furnace, which are melted by heat of the arc. The quality of steel produced is better than that of open-hearth or basic oxygen furnace. These furnaces are used extensively in foundries. Advantages of electric furnace include high production rate (due to high melting rate), more environment friendly and its ability to hold molten metal for prolonged time.

5.6.2 Induction Furnace

Induction furnaces are particularly suitable for small foundries. They are basically electric furnaces using alternating current. There are two types of induction furnaces.

- Coreless induction furnace
- Core or Channel furnace.

The *coreless induction furnace* consists of a crucible completely surrounded with a water-cooled copper coil through which high-frequency current passes. Because there is a strong electromagnetic stirring action during induction heating, this type of furnace has excellent mixing characteristics for alloying and adding new charge of metal. The crucible being used here, is a large pot made of refractory materials.

The core or channel furnace uses low frequency (as low as 60 Hz) and has a coil that surrounds only a small portion of the unit. It is commonly used in nonferrous foundries and is particularly suitable for superheating (heating above normal casting temperature to improve fluidity), holding (keeping the molten metal at a constant temperature for a period of time, thus making it suitable for die casting applications) and duplexing (using two furnaces, such as melting the metal in one furnace and transferring it to another). The highest temperature produced in the induction furnaces is 1750 °C.

5.6.3 Crucible Furnace

The furnace is in the form of a crucible, which is heated from outside by an external flame using commercial gases, fuel oil, fossil fuel, as well as electricity. Stirring action, temperature control and chemistry control of such furnaces are poor but offer low capital cost. It is the oldest process for making steel castings, but it is used predominantly in nonferrous foundries.

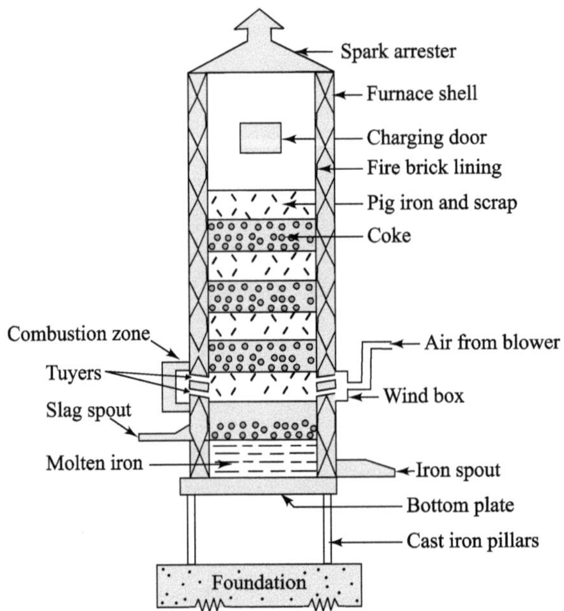

Fig. 5.4 Cross section of a cupola furnace

5.6.4 Cupola Furnace

Cupola furnace is used exclusively for melting cast iron. It converts pig iron into grey cast iron. The furnace is simple, has high melting rate and is available in a wide range of capacities. It can be operated continuously for long hours. It is economical to use, requiring very little maintenance.

The cupola consists of a hollow vertical cylindrical shell made of strong mild steel plates welded at the seams. It has fire bricks lining and rests on a square bed plate supported on four cast iron pillars above the ground. With the help of tuyeres, forced air is introduced inside the furnace. The furnace is provided with a side door called charging door to its top. Spouts are attached at different levels for tapping off the slag and the molten iron (Fig. 5.4).

5.6.5 Charging and Combustion Zone in Cupola

The charge consists of alternate layers of coke, pig iron mixed with scrap castings and a flux usually limestone ($CaCO_3$). It is fed through the charging door. In operation, a coke bed is ignited and alternate charges of coke and pig iron are made in the ratio of 1 part coke to 8 or 10 parts iron on weight basis. The amount of air required to melt one tonne of iron depends upon the quality of coke and coke-iron ratio. Theoretically, 3.19 m^3 of air at 100 KPa and 15.5 °C is required to melt 0.5 kg of carbon. The highest temperature produced in the combustion of the cupola is 1650 °C. The chemical reactions taking place inside the combustion zone are exothermic in nature evolving heat. The representative chemical reactions are:

$$C + O_2 \rightarrow CO_2 + \text{Heat}$$
$$Si + O_2 \rightarrow SiO_2 + \text{Heat}$$
$$2Mn + O_2 \rightarrow 2MnO + \text{Heat}$$
$$4Fe + 3O_2 \rightarrow 2Fe2O_3 + \text{Heat}$$

5.7 Suitability of Casting Process

Casting offers the following advantages over other manufacturing methods:

- Complex shapes with internal cavities or hollow sections can be easily made. Many cast parts do not require subsequent machining.
- Very large size objects can be made, which are otherwise difficult or uneconomical to produce. Typical component weights range from 500 g to 50 kg, but the method can produce weights of several metric tons. Large pump housings, valves and hydro-electric plant parts weighing upto 200 tonnes can easily be made by using casting process.
- Some metals can only be processed by casting and not by other methods because of metallurgical considerations. Cast irons are the best examples under this category.
- The process has high production rate.
- The process is simplified. Production of objects in a single piece is a very attractive feature of casting. Other methods lack this feature.
- Some important engineering properties can only be obtained in cast parts. For example, machinability and vibration damping capacity in cast irons, isotropic properties in cast metals in contrast to nonisotropic properties in wrought metals and good bearing qualities in cast bearing metals.
- They are economical than other manufacturing processes. The necessary tools and equipments are simple and inexpensive.

5.8 Solidification of Casting

The solidification aspect of casting is considered based on the type of material used for casting.

5.8.1 Solidification of Pure Metals

Pure metals have definite and well defined melting and freezing points, and their solidification takes place at constant temperatures. For example, pure aluminium solidifies at 660 °C and iron at 1538 °C.

Figure 5.5 shows the time-temperature curve, also known as cooling curve for a pure metal. During 1–2, temperature of the metal remains constant and solidification takes place due to the release of latent heat of fusion over certain period of time. At the point 2, process of solidification is complete and the solid metal cools to room temperature.

The grain structure of a pure metal observed during the solidification of a casting shows that at the mould walls, where the heat extraction is greatest, the metal cools rapidly due to higher difference in temperature between metal and surrounding. Rapid cooling produces a solidified skin of metal which surrounds the liquids. The thickness of the skin frozen in any given time can be expressed as:

$$D = K\sqrt{t} - C \tag{5.1}$$

where

K and C	= constants
t	= time
D	= thickness of skin.

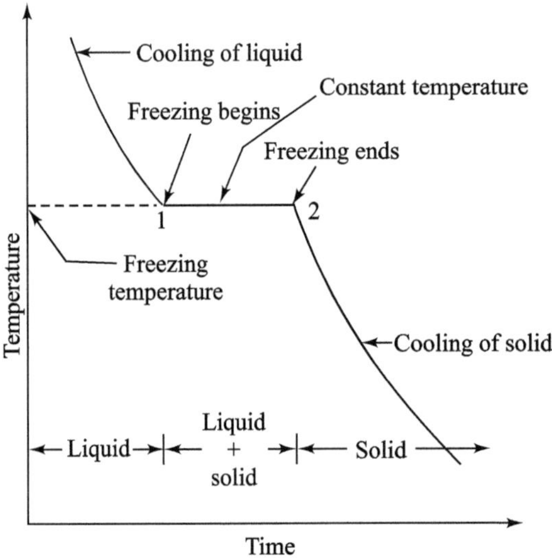

Fig. 5.5 Cooling curve for the solidification of pure metals

The constant K depends upon the size of the casting and the constant C is determined largely by the degree of superheat. The central part of the casting has the reduced rate of cooling due to the formation of this skin.

With sufficient heat extraction through the skin, liquid begins to freeze onto it and the wall thickness increases as we move inward towards the centre of the casting.

5.8.2 Solidification of Alloy

Only a few metals are used commercially in their pure state. For example, copper is used for electrical wiring and zinc for galvanizing. When other elements are added to a pure metal to bring changes in its properties, the resulting material is called an alloy. Brass is an alloy of copper and zinc, bronze and alloy of copper and tin and steel an alloy of iron and carbon.

Although pure metals solidify at a constant temperature, alloys do not, as shown in Fig. 5.6. They solidify over a range of temperatures. This is because of the fact that each element in the alloy has its own melting and freezing point. The solidification of castings of alloys differs from the solidification of castings of pure metals in the sense that the former has mushy zone (a semi-solid state) next to the central part containing molten metal.

A phase diagram, also called an equilibrium diagram, is a graphical representation showing the relationships among temperature, composition and the phase present in a particular alloy system. Equilibrium means that the state of a system remains constant over an infinite period of time. If the alloy system consists of only two elements at atmospheric

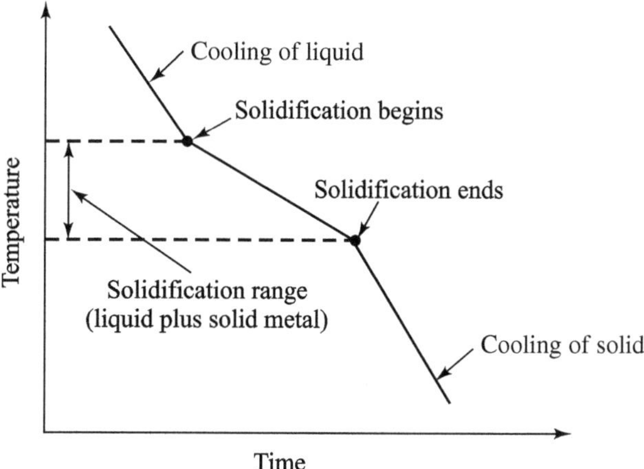

Fig. 5.6 Cooling curve for an alloy

pressures, then it results in a binary phase diagram. Many types of equilibrium diagrams exist depending on the alloys involved.

5.9 Iron-Carbon Equilibrium Diagram

Steels and cast irons are represented by the iron-carbon binary system. Commercially, pure iron contains 99.99% iron along with 0.008% carbon. It is obtained by electrolytic method and hence also known as electrolytic iron. Steel contains carbon starting from very low level upto 2.11%. For cast iron, the practical range of carbon lies between 2.11 and 4.5%, although the higher limit can be upto 6.67%. The diagram shown in Fig. 5.7 is not the full iron-carbon diagram but the iron–iron carbide diagram. The figure is terminated at 6.67% carbon, where cementite (Fe_3C), an intermetallic compound is formed.

An intermetallic compound is chemically produced by the combination of two elements-either two metals or a metal and a nonmetal, such as carbon. However, as shown by the liquidus, liquid iron can dissolve only upto slightly more than 5% of carbon at normal melting temperatures. For this reason, many iron-carbon phase diagrams include only upto about 5% carbon.

There are four single phases in this diagram namely delta ferrite, austenite, alpha ferrite or simply ferrite and cementite. First three of these occur in pure iron. Pure iron melts at 1538 °C. As iron cools, it first forms delta ferrite, then austentite, and finally alpha ferrite. The cooling curve for pure iron is shown in Fig. 5.8.

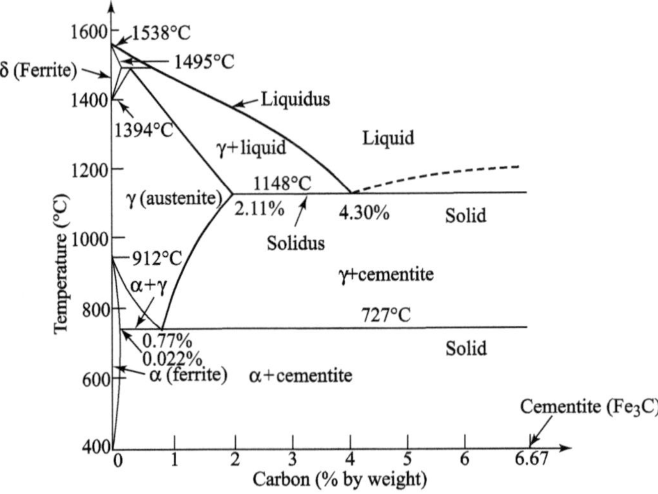

Fig. 5.7 Iron-carbon equilibrium diagram

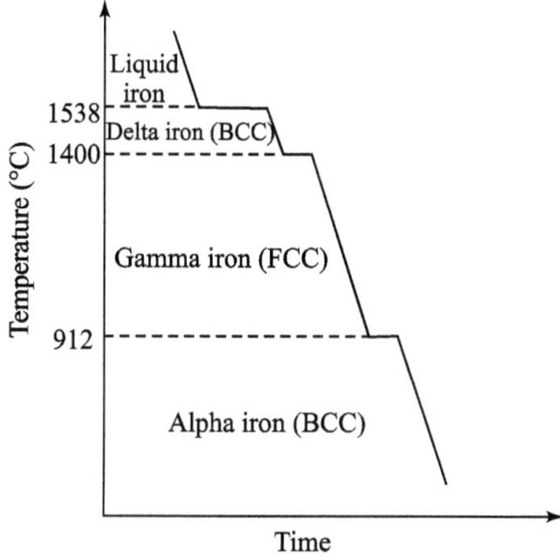

Fig. 5.8 Cooling curve for pure iron

As shown in figure, pure iron changes its crystalline structure at 912 and 1400 °C. Delta ferrite, which has bcc structure, is present only at extreme elevated temperatures very near to the melting point of pure iron and has very limited engineering importance. Between 1394 and 912 °C, iron undergoes a polymorphic transformation from the bcc to fcc structure, forming gamma iron or more commonly austenite. Austenite has high solubility of carbon (upto 2.11% carbon at 1148 °C). It is an important phase in the heat treatment of steels. Hot forming of steel takes advantage of the high ductility and chemical uniformity of austenite. Most of the heat treatments of steel begin with the austenite structure. Steel is nonmagnetic in the austenitic form. Alpha ferrite or simply ferrite has a maximum solid solubility of 0.022% carbon at 727 °C. It has bcc structure. It is relatively soft and ductile and is magnetic in nature from room temperature to 768 °C. Curie point occurs at 770 °C, where nonmagnetic-to-magnetic transition takes place. It is an atomic-level transition and is not associated with any change in phase, therefore, it does not appear on the equilibrium phase diagram. The fourth single phase, comentite, is purely iron carbide. It is very hard and is entirely different from other carbides such as tungsten, titanium or silicon, which are used as dies, cutting tools and abrasives.

In addition to the above-mentioned four single phases, three distinct three-phase reactions take place in the diagram. At 1495 °C, *peritectic* occurs which has very low-carbon content. It has no engineering significance. A *eutectic* is obtained at 1148 °C with 4.3% carbon. All alloys containing more than 2.11% carbon give eutectic reaction and are called cast irons. And finally, the third three-phase reaction is known as *eutectoid,* which occurs at 727 °C with 0.77% carbon. Alloys containing less than 2.11% carbon are known as steels. The basis of separation between steels and cast irons is thus based on the maximum carbon solubility of 2.11% in iron. The structure of eutectoid steel is called *pearlite* because it resembles mother of pearl at low magnifications. The microstructure of pearlite consists of alternating layers of ferrite and cementite. The ferrite and cementite in the pearlite are called eutectoid ferrite and eutectoid cementite respectively. Steels containing less carbon than the eutectoid are called *hypoeutectoid steels* and those with more carbon content are known as *hypereutectoid steels.*

The amount of ferrite, pearlite and cementite can be calculated as follows, by using the lever rule. For a hypoeutectoid steel, the percentage pearlite is given as:

$$\%\mathrm{Pe} = \frac{\%\mathrm{C}}{0.77}$$

where %C is the percentage carbon in the specimen. The reminder is ferrite. For a hypereutectoid steel, the percentage cementite can be expressed as:

$$\%\mathrm{Ce} = 100\left[\frac{1-(6.67-\%\mathrm{C})}{5.87}\right]$$

The remainder is pearlite.

The phase present in an alloy system and its composition can be determined with the help of phase diagram. It can also give an overall picture of the alloy system or determine the transition points for various changes in phase.

5.10 Residual Stresses in Castings

When workpieces are subjected to nonuniform deformation, residual stresses are induced in them. They are present within the material and are independent of any applied load.

Residual stresses are also caused by temperature gradients such as during cooling of a casting or forging due to nonuniform cooling or heating. If all the temperature changes occurred uniformly throughout a part, all of the associated dimensional changes would occur simultaneously and the resultant product would be free of residual stresses.

Sometimes residual stresses are desirable in the material, e.g. compressive residual stresses on the surface of a part. On the other hand, tensile residual stresses on the surface are generally undesirable because they lower the fatigue life and fracture strength of the part.

Residual stresses can be reduced or eliminated by heat treatments, e.g. annealing. Here the metal is reheated below the recrystalline temperature range and cooled very slowly.

5.11 Heat Treatment of Castings

Castings are sometimes heat treated to develop greater strength, refine the grain structure, produce more isotropic properties or produce a more homogeneous structure. A separate heat treatment adds considerably to the cost and hence should be only used when required.

If the material contains less than about 2.0% carbon, it is considered to be a cast steel. Alloys with more than 2% carbon are cast irons.

Cast steels are usually heat treated to produce a quenched-and-tempered structure, and the alloy additions are selected to provide the desired hardenability and balance of properties.

Most cast irons, however, are used in the as-cast condition, with the only heat treatment being a stress relief or annealing.

5.12 Material Selection in Casting

Engineering materials differ widely in physical properties, machinability characteristics, method of forming and possible service life. The designer should consider these factors while selecting an economic material and a process that is best suited to the product. Materials from the casting point of view have been mainly classified under two categories.

- Ferrous casting alloys
- Nonferrous casting alloys

5.12.1 Ferrous Casting Alloys

5.12.2 Cast Iron

Cast iron is a general term applied to a wide range of iron-carbon-silicon alloys containing smaller percentage of several other alloying elements. All cast irons contain iron, carbon, silicon, manganese, phosphorous and sulphur. It has a wide range of properties such as wear resistance, hardness and good machinability due to the presence of alloying elements. Cast irons are the least expensive of all metals. This is because iron is the second most abundantly available metal after aluminium, but the extraction of aluminium is a costly affair. Cast iron requires less refining than steel. They have relatively low melting point ($\simeq$1200 °C) and are easily melted, requiring less fuel and more easily operated furnaces. Pouring temperatures for cast iron are about 1400 °C. Molten cast iron is easily cast into complex shapes and easily machined to required tolerances. These characteristics lead to considerable versatility in product design.

5.12.3 Grey Cast Iron or Grey Iron

Grey cast iron has excellent casting properties, therefore, it is used most widely in castings. Presence of graphite produces excellent fluidity, which makes it possible to produce thin-section castings over a large area. Castings of grey cast iron have relatively few shrinkage cavities and little porosity. It is the least expensive among all other cast irons. Grey cast irons find applications in engine blocks, machine bases, electric motor housings and pipes.

5.12.4 Malleable Cast Iron or Malleable Iron

Malleable cast iron is made from white cast iron. It has high fluidity which allows complicated shapes to be easily cast. Castings made from malleable iron have good surface finish, good dimensional tolerances, better shock resistance and good machinability. They are mainly used by railroad, automobile and agricultural implement industries.

5.12.5 Ductile Iron or Nodular Iron or Spheroidal Iron

Ductile iron and grey cast iron have similar casting characteristics. It has low melting point and good fluidity in the molten state. Therefore, it can be used for intricate shapes and thin sectional parts. Other important characteristics of ductile iron are: excellent machinability, high strength, good toughness and good wear resistance.

5.12.6 White Cast Iron

Its principal constituents are pearlite and cementite. Presence of cementite makes it extremely hard and wear resistant. It is produced by the rapid cooling of cast iron. Hammer mills, crusher jaws and crushing rollers and made of white cast iron.

5.12.7 Cast Steels

Because of the high temperatures required to melt cast steels, upto about 1650 °C, their castings present difficulties in the selection of mould materials particularly in view of the high reactivity of steels with oxygen in melting and pouring the metal.

Steel is not as fluid as cast iron, therefore, complicated shapes, sharp corners and thin sections cannot be cast.

The minimum thickness of steel casting is usually taken as 6 mm. However, in small castings, thickness can be equal to 4.5 mm. All steel castings must be annealed to relieve them of internal stresses. Their strength can be increased by subsequent heat treatment.

Medium-carbon steel castings are the backbone of the steel foundry industries. About 60% of all steel castings are made of medium-carbon steel (contains carbon between 0.30 and 0.70%). This group offers competition to forging. The plain or alloy cast steel is used, where castings of improved properties are required over iron or malleable castings. Steel castings are isotropic than those made by mechanical working processes. However, steel castings are more difficult to produce than iron castings and are more expensive. They are used for machine members of intricate shape that require high strength and impact resistance, such as locomotive frames, large I.C. engine frames, gears, wheels and many small and intricate parts under highly stressed conditions.

5.12.8 Cast Stainless Steels

Casting of stainless steels involves considerations similar to those for steels in general. Stainless steels generally have a long freezing range and high melting temperatures. Cast stainless steels are available in various compositions and can be heat treated and welded. These cast products have high heat and corrosion resistance.

5.12.9 Nonferrous Casting Alloys

The foundry methods for making nonferrous casting in sand slightly differ from that used for iron castings. Moulds are made in the same way and with the same kind of tools and equipment. The moulding sand is usually a fine grain size, because most castings are small, a smooth surface is desired, and the melting temperatures are lower for nonferrous metals. The sand need not be so refractory as for iron and steel castings because of the lower melting temperatures. The addition of an alloy to a pure metal always results in a material with a lower melting point. The pouring temperature is generally about 20% above the melting temperature of the metal but is affected by the type of mould and the section thickness. Crucible furnaces are frequently used for this type of work.

The common elements used in nonferrous casting are copper, aluminium, magnesium, zinc, tin and lead. Many alloys, however, have small amounts of other elements such as antimony, phosphorus, manganese, nickel and silicon.

5.12.10 Aluminium-Base Alloys

Aluminium is considered as one of the most easily cast metals. Aluminium alloys can be cast by any of the commonly used processes: sand casting, plaster moulding, permanent mould and die casting.

Many die castings are made of aluminium alloys, because of their lightweight and resistance to corrosion, but they are more difficult to die cast than zinc (melting points of aluminium and aluminium alloys are 649 °C and 540 °C, respectively, as compared to the melting point of zinc of 380 °C). The combination of low density and low melting point practically eliminates most problems of *sand washes* (sand moving out of place). Consequently, excellent surface finish and dimensional accuracy can be obtained even in large sand castings.

Aluminium has large casting shrinkage ranging from 3.5 to 8.5% by volume, which makes it mandatory that sections should be uniform wherever possible. If section changes are necessary, they should be gradual. The minimum section for sand castings is about 4.8 mm.

Silicon, when added to aluminium, increases its fluidity and allows it to flow further in thin walls of the mould cavity and produces finer detail. It also reduces internal shrinkage and reduces the coefficient of expansion.

Aluminium castings containing more than 8% magnesium will respond to heat treatment, however it makes the metal more difficult to cast.

Copper is one of the principal-hardening elements in aluminium. It increases the strength of the aluminium in both the heat treated and nonheat treated conditions.

Aluminium-base alloys have many applications including architectural and decorative use. Engine blocks of some automobiles are made of aluminium alloy castings. Parts made of aluminium and magnesium base alloys are known as *light-metal castings.*

5.12.11 Copper-Base Alloys

Copper castings are made almost entirely in sand moulds. The technique of casting pure copper having high conductivity is especially difficult, since considerable skill is required in handling the reducing agents. The molten metal does not flow well, making it difficult to fill intricate moulds. However, copper-based alloys are quite easily cast and handled in the foundry.

Die castings of brass and bronze present a problem in pressure casting because of their high casting temperatures. These temperatures range from 870 to 1040 °C and make it necessary to use heat-resisting alloy steel for the dies to reduce their rapid deterioration. Since the high casting temperatures and pressures shorten the die life, the cost of brass die castings is higher than that of other metals. But advantages such as thinner wall sections, the savings in metal cost, coupled with high production rates, help offset the disadvantage of a shortened die life.

5.12.12 Magnesium-Base Alloys

Magnesium alloys are the lightest of all casting metals (specific gravity of magnesium is 1.74 compared to 2.7 for aluminium and 7.9 for iron or steel).

Magnesium alloys due to their low melting point temperatures are suitably used in sand, permanent mould and die castings but die casting dominates over others. Although the magnesium alloys cost about twice as much as aluminium, its hot chamber die casting process is easier, more economical, and 40–50% faster than the cold chamber process generally required for aluminium. Magnesium die castings often replace plastic injection-moulded components when improved stiffness or dimensional stability is required.

The fatigue behaviour of cast magnesium alloys is similar to the aluminium alloys.

5.12.13 Zinc-Base Alloys

It is a low melting point alloy group (melting point of zinc is 380 °C), which has high fluidity making it easily castable. These alloys are mainly used in die castings. More than 75% of die castings are zinc based. Zinc die casting alloys offer a strength greater than that of all other die cast metals, except copper alloys. They can be cast to close dimensional limits with extremely thin sections comparable to the thickness of sheet metal. But because of their low creep strength, castings are not subjected to prolonged high stresses.

Zinc alloys are widely used in automotive industry and in making washing machines, refrigerators, radios and televisions.

5.12.14 Tin-Base Alloys

Tin has the lowest melting point among the casting metals and hence is suitable for casting. But their use is limited because they lack mechanical strength.

5.13 Moulding Sand

Silica sand (SiO_2) is used in various sand casting processes as mould material, because it can withstand high temperature without fusion. Pure silica sand is not suitable for moulding, because it lacks binding qualities. The binding qualities may be obtained by adding 8–15% clay. The three clays commonly used are kaolinite, illite and bentonite. Bentonite is used most often. There are two general types of sand, namely natural and synthetic. A natural moulding sand is one which is ready for use as it is dug from the ground and is naturally bonded since it contains clay and no binder is required to be added to it. *Synthetic sands* are made by mixing together high-purity clay-free refractory grains, usually silica, with about 5% of the most effective clay and other desired ingredients. Additives are also used with synthetic sand which help to improve the quality of casting. Its composition can be controlled more accurately and also it overcomes some of the disadvantages of natural moulding sand, and hence synthetic sand is preferred by most foundries.

The pouring temperature of iron ranges from 1315 to 1538 °C, while for steel it is 1510–1649 °C. These temperatures require a coarser sand for iron, while a highly refractory sand is necessary for steel.

The size of the sand grains depends upon the moulded work. For small and intricate castings, a fine sand with round grains which makes the mould surface smooth, is desirable to allow all the details of the mould to be brought out sharply. As the casting size increases, the sand particles should be coarser to permit the escape of gases. Sharp, irregular-shaped grains are usually preferred because they interlock and add strength to the mould.

Parting Sand is a fine grained, dry silica sand without strength that prevents bonding of sand in the cope with sand in the drag.

Facing Sand, also called fat sand, forms the face (inner surface) of mould; i.e. this sand is rammed around the pattern. It is freshly prepared and well-tempered foundry sand containing small amount of carbonaceous material and is used to give better surface finish to the castings. The remainder of the moulding flask is filled with ***floor sand*** (also known as black or baking or backing sand), which is left on the floor after the castings have been removed from the mould. Backing sand is used to back up facing sand and not used next to the pattern.

Core sand is clay free and has high silica content. Its binding capacity is achieved by using a resin that hardens with the application of heat. **Core sand** is used for making cores.

5.14 Properties of Moulding Sand

5.14.1 Permeability

The permeability, also called porosity, of a moulding sand is its ability to allow the passage of gases through it. Porosity depends on the shape of sand grains, fineness, degree of packing, moisture content and amount of binder. It is measured by the quantity of air that passes through a sample of sand in a prescribed time and under standard conditions. Coarse-grained sands are more permeable. Permeability increases with moisture content upto approximately 5% moisture. This property is also affected by ramming of sand. Soft ramming tends to increase the permeability, while hard ramming will reduce it. Excessive use of binder will adversely affect permeability. If the gas or steam generated in the sand mould are not allowed to escape, they will burst the sand mould affecting the quality of casting.

5.14.2 Refractoriness

Refractoriness is the ability of the sand to withstand high temperatures without fusing. Degree of refractoriness will depend upon the metal of casting. If the sand lacks in this property, it will fuse on coming in contact with hot molten metal and spoil the casting.

5.14.3 Cohesiveness

Sand must be cohesive or it should have strength to the extent that it has sufficient bond. Both water and clay content affect the cohesive properties of sand.

5.14.4 Collapsibility

Collapsibility of the sand makes the sand mould automatically collapsed, when solidification of the casting is over, thereby allowing free contraction of the metal; otherwise some cracks (hot tears) may appear in the castings.

Sand must have a grain size suitable with the surface to be produced and grain shape must be irregular to permit interlocking.

Good moulding sand always represents a compromise between conflicting factors. The size of the sand particles, the amount of bonding agent (such as clay), the moisture content and the organic matters (additives) are all selected to obtain an acceptable compromise of the above-mentioned requirements. The composition of the moulding sand must be carefully controlled to assure satisfactory and consistent results. A typical green-sand mixture contains 88% silica sand, 9% clay and 3% water.

5.15 Testing of Moulding Sand

The nature of moulding sand can be most accurately described by means of the results of various standard sand tests. Complete details of the tests, testing procedure and equipment are mentioned in the American Foundry Society (AFS) handbook.

5.15.1 Test for Moisture Content

Since the moisture content of the sand affects so many properties, this is one of the most frequently performed tests. The usual method for determining the moisture content of a sand is to dry a sample thoroughly at a few degrees above 100 °C and to consider its loss in weight as moisture. A moisture teller is used to conduct this test. The moisture content should vary between 2 and 8%, depending on the type of moulding being done.

5.15.2 Clay Content Test

Clay content can be determined by washing the clay from a 50 g sample of moulding sand in water that contains sodium hydroxide (NaOH, caustic soda). Following several cycles of agitation and washing, the caustic solution, which has absorbed the clay, is siphoned off. The remaining sand is then dried and weighed to determine the amount of clay in the given sample. The loss in weight of the original 50 g sample multiplied by 2 gives the AFS (American Foundrymen's Society) clay percentage in the sand.

5.15.3 Fineness Test

The size and distribution of sand grains in moulding sand are determined with the AFS sieve analysis test. The test is performed on a 50 g clay-free, dried sand sample. The sample is placed on the top of a series of 11 sieves having the numbers as 6, 12, 20, 30, 40, 50, 70, 100, 140, 200 and 270 and shaken for 15 min. After the shaking period is over, the sand retained on each sieve and the bottom pan is weighed and its percentage of the total sample determined. To obtain the AFS fineness number, each percentage is multiplied by a factor, which is the size of the preceding sieve. The fineness number is obtained by adding all the resulting products and dividing the total by the percentage of sand retained in the sieve set and pan.

By definition, the AFS grain fine number is the average grain size, and it corresponds to the sieve number whose openings would just pass all the sand grains if all were of the same size. This number is a convenient means of describing the relative fineness of sand.

5.15.4 Permeability Test

Permeability is a measure of how easily gases can pass through the narrow voids between the sand grains. If these gases are not allowed to escape, they will produce certain defects in the casting, namely porosity or blow holes. Permeability is measured in terms of a number known as *permeability number*, defined as the volume of air in cubic centimetres that will pass per minute under a pressure of 10 g/cm^2 through a sand specimen which is 1 cm^2 in cross section and 1 cm deep. It is given by the following formula:

$$P = \frac{VH}{p \cdot A \cdot T} \tag{5.2}$$

where

P = permeability number (also called AFS number)
V = volume of air = 2000 cm^3
H = height of the sand specimen = 5.08 cm
p = air pressure = 10 gm/cm^2
A = cross-sectional area of the sand specimen = 20.268 cm^2
T = time (in s) for 2000 cm^3 of air to pass through specimen.

Equation (5.2) is finally reduced to

$$P = \frac{3007.2}{T(s)} \tag{5.3}$$

5.15.5 Mould and Core Hardness Test

The hardness achieved by ramming the sand can be measured by a mould-hardness tester. It is an indentation-type test. A spring-loaded steel ball 5 mm in diameter is pressed into the surface of the mould, and the depth of penetration is indicated on the dial in thousandths of an inch. If no penetration occurs, the hardness is arbitrarily assumed as 100. If the ball sinks completely into the sand up to the limiting surface of the tester, the reading is zero; i.e. the sand is very soft. Each scale unit represents 0.025 mm. Common mould hardnesses are 60–90 for machine moulding.

5.15.6 Sand Strength Test

Various types of tests are conducted to measure the strength of moulding sands using a universal type mechanical testing machine. Strength in compression, shear, tension and transverse loading may be performed with this machine.

Most commonly used tests are compression and shear, which involve a cylindrical sample 50 mm high and 50 mm in diameter. During compression test (Fig. 5.9), a compressive load is applied on the sample till it brakes, usually in the range of 0.07–0.2 MPa. Sands containing little moisture and those containing excess moisture both are said to have poor strength. Thus there is a maximum strength and an optimum water content that will vary with the content of other materials in the mix. Tension and transverse tests are used mainly for testing core sands.

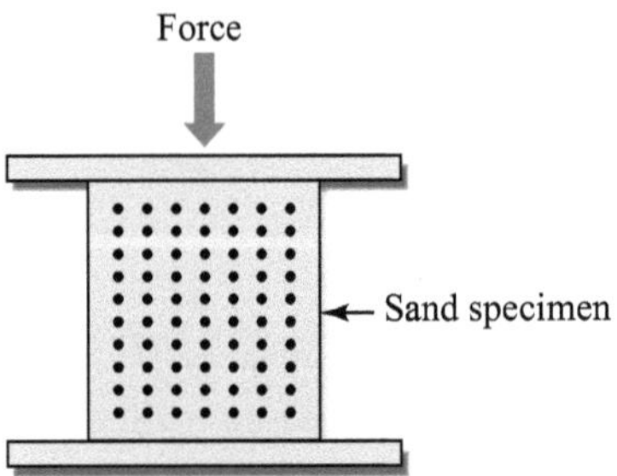

Fig. 5.9 Compression test

5.16 Sand Conditioning and Reconditioning

Making of the moulding sand suitable for mould preparation is known as sand conditioning. Sand conditioning includes the following functions:

- Binder is distributed more uniformly around the sand grains.
- Moisture content is controlled and particle surfaces are moistened.
- Foreign particles are eliminated.

Sand reconditioning consists of recycling the used sand, when the process of casting is over. In mechanized foundries, the sand is automatically returned by belt conveyor for reconditioning. Sand reconditioning includes the following functions:

- Sand is aerated.
- Lumps are broken up.
- Magnetic particles are removed by an electromagnet.
- New clay is added to make up for the 'burn out' or deactivated portion.
- Addition of suitable amount of water.
- Sand is passed through a muller that mechanically fluffs it up and makes it suitable for remoulding.

5.17 Defects in Sand Casting

Several defects can develop in sand castings. These defects may arise due to faulty design of pattern, mould, core and gating system or due to defects in pouring techniques.

The important casting defects occurring in sand castings are discussed below:

Blows (or Blowholes): These are balloon- or irregular shaped holes or gas cavities formed near the casting surface (Fig. 5.10). They are caused due to insufficient venting of the sand mould, possibly that results from hard ramming. To eliminate or minimize this defect, permeability of the moulding sand needs to be improved.

Scar: It is shallow blow, which occurs mainly on the flat surface of a casting (Fig. 5.11).

Blister: It is a variation of scar, where its certain portion is covered by a thin layer of metal (Fig. 5.12).

Scab: This defect occurs on the flat surface of a casting due to thermal expansion of the moulding sand that results in breaking away of small portions of moulding sand (Fig. 5.13). It is caused due to poor ramming of sand mould or because of use of inferior quality of sand in mould.

Wash: This defect occurs on the casting's drag surface near the gate due to high velocity of the molten metal (Fig. 5.14).

Misrun: It is due to the lack of molten metal fluidity. Due to this defect, the molten metal does not reach the desired part of the mould cavity and solidification takes place in the mid-way (Fig. 5.15). This defect occurs when the molten metal has low temperature or the pouring process is too slow.

Cold shut: It is another variation of misrun, which occurs in the centre of a casting having gates at both sides. This defect occurs when two streams of liquid metal from different gates do not fuse together properly because of their low temperatures (Fig. 5.16).

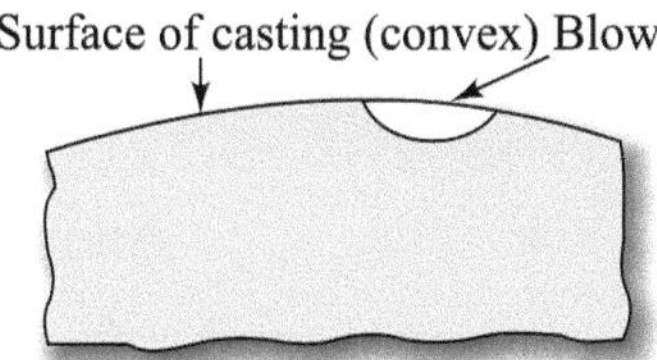

Fig. 5.10 Blow

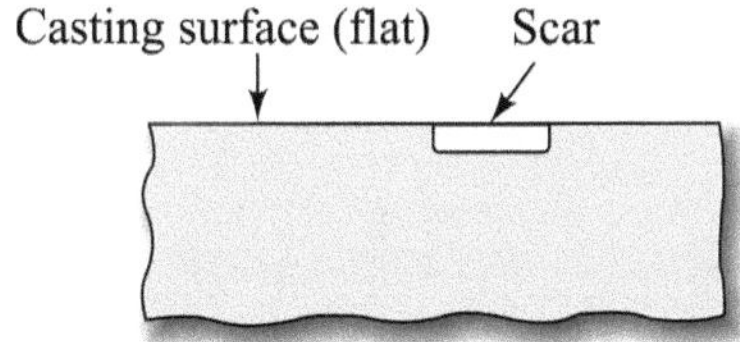

Fig. 5.11 Scar

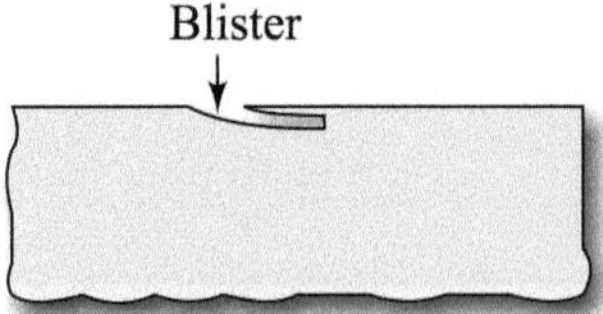

Fig. 5.12 Blister

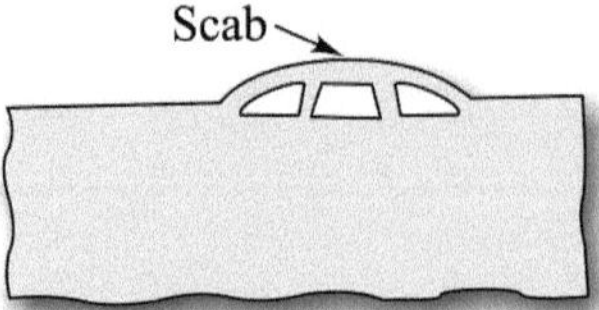

Fig. 5.13 Scab

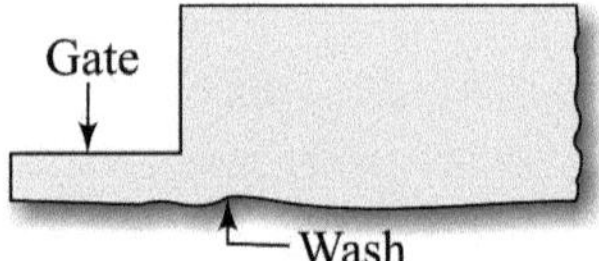

Fig. 5.14 Wash

Fig. 5.15 Misrun

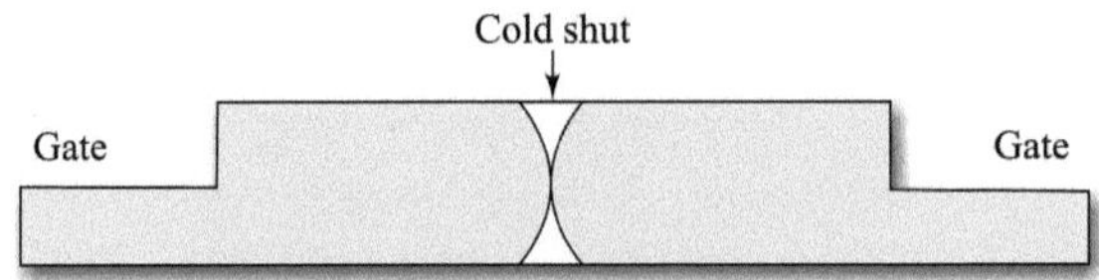

Fig. 5.16 Cold shut

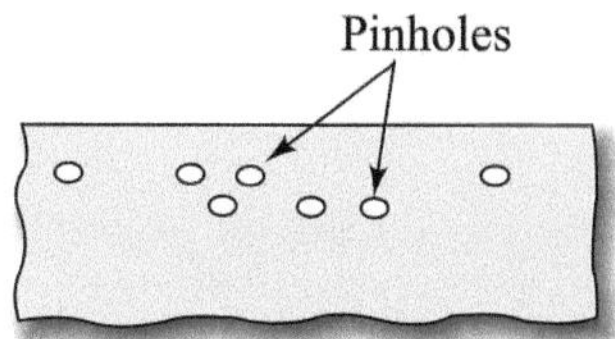

Fig. 5.17 Pinhole

Pin holes: They are in the form of small gas cavities and are produced just below the surface of the casting. Excessive moisture content and poor permeability of moulding sand are responsible for this defect (Fig. 5.17).

Drop: It occurs due to poor sand cohesiveness of moulding sand. Rough handling of the mould can also produce this defect. This defect appears in the form of projection, when a cope part falls on the drag surface (Fig. 5.18).

Dirt: Sometimes sand particles dropping out of the cope get embedded on the top surface of a casting. When removed, these particles leave small and angular holes, known as dirts. Loose sand or easily eroded sand may result in dirt (Fig. 5.19).

Rat tail: This defect looks like a rat tail and is produced mainly on the flat surfaces of a casting (Fig. 5.20). It results due to thermal expansion of the moulding sand.

Buckle: This defect is similar to the rat tail but differs from it in the sense that it occurs in the form of vee-shaped depression on the surface of a casting (Fig. 5.21).

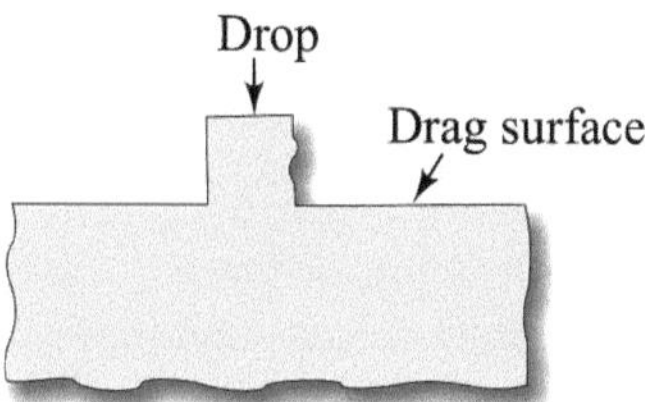

Fig. 5.18 Drop

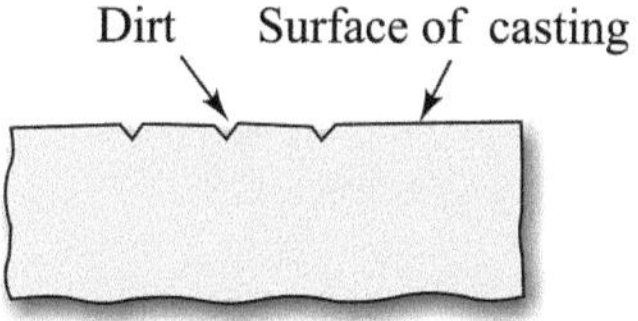

Fig. 5.19 Dirt

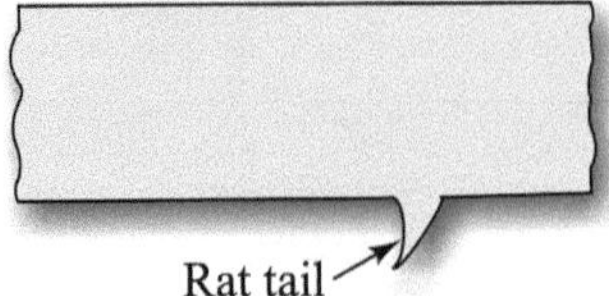

Fig. 5.20 Rat tail

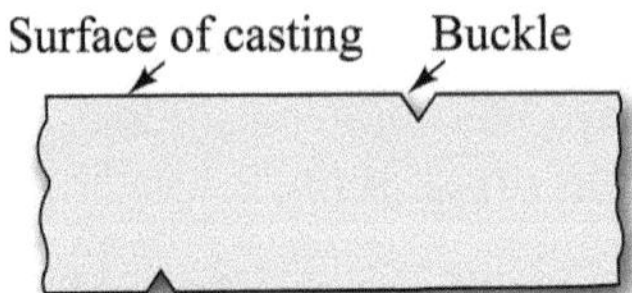

Fig. 5.21 Buckle

Swell: It is caused by the enlargement of the mould cavity after the molten metal has been poured. Hydrostatic pressure of the molten metal or the expansion upon solidification in some cases can push the mould walls outward (Fig. 5.22).

Mould crack: If the mould is not strong enough or not properly supported, it may crack. Metal can run into the crack and appear as a fin on the casting. It is produced by residual stresses or by improper removal of gate and riser.

Core raise: This defect occurs when the core floats and moves up close to the cope. It may even touch the cope surface and seal off the section. A weak core or one which is not properly supported with chaplets may be the cause of this defect. An undersize core not held tightly by the core prints may also raise. It is caused by the buoyancy of the liquid metal.

Penetration: Sand can fuse by the action of the hot metal, producing a mass that may adhere to the casting. Metal can penetrate the sand grains after the surface of the grains has fused (Fig. 5.23).

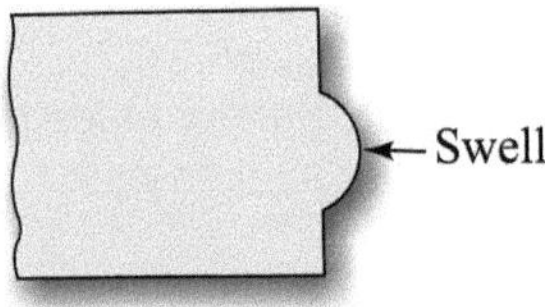

Fig. 5.22 Swell

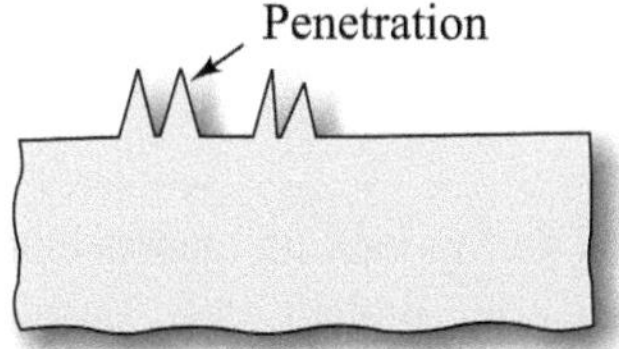

Fig. 5.23 Penetration

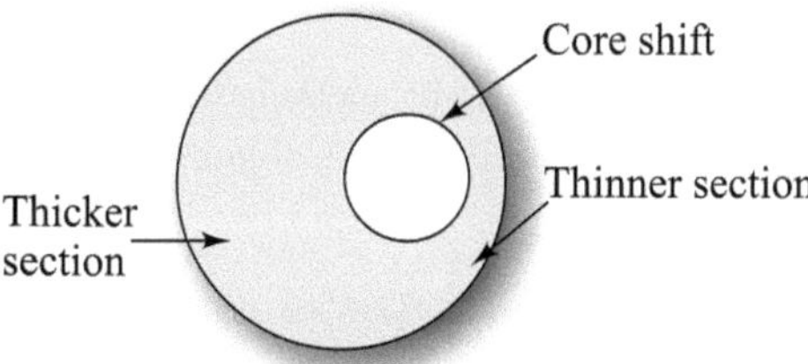

Fig. 5.24 Core shift

Core shift: A core may shift horizontally if it is not securely held by chaplets or it has moved out of position during mould assembly or it has dislodged by the flow of molten metal during pouring. It results in uneven wall thickness in the casting (Fig. 5.24).

Hot tears: These defects are also called cracks. They occur in the form of rupture produced by the tensile stresses on the casting, when it is near the solidus temperature. They are caused due to poor mould design that prevents contraction during cooling or when the mould is too rigid that prevents its automatic collapse after solidification process is over or when the molten metal is too hot when pouring. This defect is most common in malleable iron, steel and nonferrous castings.

Shrinkage: This defect is also called porosity. It occurs in large size castings. It is caused due to impure metal composition containing excess phosphorus that leads to the formation of gases that produce porous regions in castings during solidification because of shrinkage. Another reason of its occurrence is the faulty design of riser (not of the desired size or its wrong positioning in the mould) that can prevent adequate molten metal in castings during their solidification.

Sand inclusions: Sand is entrapped in the mould as a result of crushing or spalling, or as a result of loose sand being in the mould.

Slag inclusions: These result from the presence of slag in the metal.

Veins: Veins on castings are fin-shaped protuberances in cored areas. A crack-shaped opening develops in the core as it becomes heated by the metal. The crack is then filled with molten metal, and a vein appears on the casting, when it is cleaned. This defect is most often encountered in ferrous castings, where a small or thin core is surrounded by a substantial mass of metal.

Fissures: They appear as rough, grainy-looking masses attached directly to cored surfaces of the castings. They occur when the molten metal pushes aside the core sand after its collapse.

Hard spots: They represent areas of the casting surface that have varying level of hardness. They occur because of rapid cooling in localized areas, for example, in thinner sections. Incorrect use of chills can also produce this defect.

Fins: These defects occur in the form of projections of thin slivers of metal around the parting line of the casting either because of poor alignment of its cope and drag parts or due to poorly fitted core prints.

Centre-line shrinkage: It appears in the form of cracks (cavity) in the centre-line of usually large size castings. It occurs due to differential rate of cooling at different sections of the castings. Outside surface of a casting cools faster and solidifies before the central part. As a result, molten metal cannot be fed into centre that has not yet solidified. When the centre begins to solidify, a cavity is formed there because of its contraction. This defect can be reduced or eliminated by the proper use of chills in the affected areas.

5.18 Inspection of Casting

Inspection means to check the dimensions of the object produced and to see whether it complies with the specified dimensional accuracy. It ensures that the specifications of casting quality are being maintained. The earlier the defective casting can be removed from processing, the fewer the operations wasted on them. Unacceptable or defective castings are remelted for reprocessing. Inspection methods may be classified as destructive or nondestructive.

Visual inspection is used to detect surface defects in the castings. Certain casting defects are obvious upon visual examination. These defects are tears, dirt, blow holes, scabs, metal penetration, runouts, swells, etc. Visual inspection is the simplest method of inspection and is conducted with an aim to ensure that the casting looks good. The method of inspection is usually nondestructive in nature.

Dimensional inspection of casting involves the principles of gauging as it is applied to any machine elements. It ensures that if the cast parts are meeting the dimensional requirements. Several gauging instruments are used for this purpose such as callipers, verniers, etc.

Metallurgical inspection includes chemical analysis, mechanical-property tests, evaluation of casting soundness, and product testing of special properties such as electrical conductivity, resistivity, magnetic effects, corrosion resistance, response to heat treatment, strength in assemblies, surface conditions, coatings and surface treatments and others. Some of the tests under metallurgical inspection are destructive in nature while the others are nondestructive.

In *destructive testing*, the test specimen no longer maintains its integrity, original shape, or surface texture. It involves the sawing or breaking of selected parts at places where voids or internal flaws are suspected. This method is used to test for strength, ductility and other mechanical properties, and to determine the presence and location of any defect in the castings. Hardness tests bearing large impressions may be regarded as destructive testing. Destructive tests are not dependable because saw cuts and breaks may easily miss internal flaws. Nondestructive testing methods are discussed below.

5.19 Nondestructive Inspection and Testing

Nondestructive tests (NDT) are conducted to find out defects or flaws in the materials (castings or welds) without damage to the object. The test results are highly reliable because of the involvement of computers and other enhancement techniques.

Some important nondestructive testing techniques are discussed below.

5.19.1 Liquid Penetrant Inspection

It is an effective method of detecting surface defects such as cracks, laps, seams, porosity, folds and inclusions in metals. Internal defects cannot be detected by this technique.

The method is based on the ability of a penetrating liquid to wet the surface opening of a weld defect or casting and to be drawn into it by capillary action. The surface to be inspected is first thoroughly cleaned and dried, and then flooded with the liquid penetrant. After the excess liquid has been removed from the surface, a thin film of developer is applied on it. The developer is an absorbent material capable of drawing the traces of penetrant from the defects back onto the surface, thus magnifying the size of defects. Two common types of liquids, namely fluorescent penetrants which glow under ultraviolet light; and visible penetrants, usually red colour dyes, which appear as bright outlines on the surface, are used as penetrants in the process.

5.19.2 Ultrasonic Inspection

In ultrasonic testing, a high-frequency ultrasonic beam (frequency ranges between 1 and 25 MHz) is directed into the part to be tested. Within the part, the beam is affected by voids, impurities, changes in density, delaminations and other imperfections. The altered ultrasonic signal can be analysed and used to detect the presence and location of defects within the cast material or weld joint.

An ultrasonic inspection process begins with a pulsed oscillator and transducer, a device that transforms electrical signals into ultrasonic sound waves. The pulsed oscillator generating a burst of alternating voltage is applied to a sending transducer, which uses a piezo electric crystal (usually a quartz) to convert the electrical oscillations into sound waves. The transducer is coupled to the test material by a liquid medium, such as water to transmit the ultrasonic waves into the part. Air is not used for the reason that it is a poor transmitter of ultrasonic waves. A receiving transducer converts the transmitted or reflected ultrasonic waves back into corresponding electrical signals. With appropriate instrumentation, the same transducer alternately serves both functions. The received electrical signals are amplified, filtered and processed by the receiving transducer and the time intervals that lapse between the initial pulse and the arrival of the various echoes are displayed on an oscilloscope screen. A flaw is recognized by the relative position and amplitude of the echo.

5.19.3 Radiographic Testing

Radiographic testing employs X-rays, gamma rays, beams or radio isotopes as testing elements. During the test, the rays are allowed to penetrate the weldment or the cast part and the defects or any discontinuities are obtained by the resulting image on a recording film or a viewing screen.

The method is useful to detect internal flaws such as cracks and porosity, and also to check for alignment and operation of assembled parts. Film is positioned behind the part being radiographed. The distance from X-ray source to the part, section thickness and exposure time etc. must be properly selected to give satisfactory results. Since most of the defects transmit the short-wavelength light better than the sound metal does, the film is darkened more where the defects are in the line of the X-ray beam (Fig. 5.25).

5.19.4 Magnetic Particle Inspection

This method of inspection is used to detect surface or near-surface defects or discontinuities in magnetic materials.

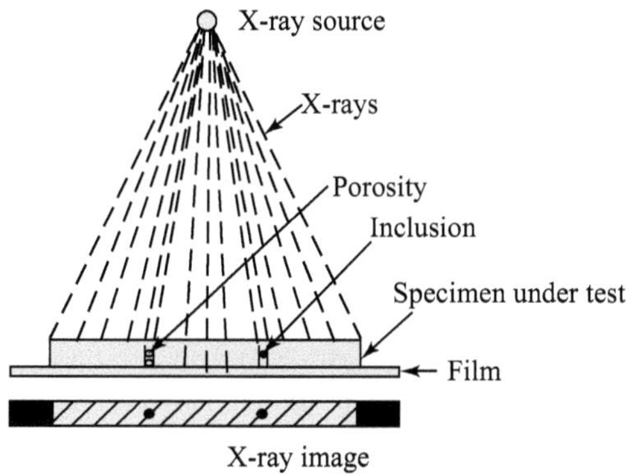

Fig. 5.25 Radiographic test using X-ray

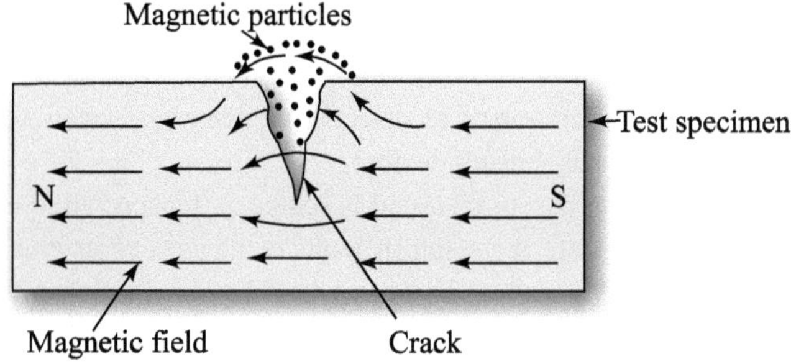

Fig. 5.26 Magnetic-particle inspection of surface defects

In this method, an intense magnetic field is set-up in the part to be inspected. Cracks, voids and material discontinuities cause the lines of magnetic flux to be distorted and they break through the surface (Fig. 5.26). When the part is magnetized with a magnetic field, the ferromagnetic particles placed on the surface of the part build-up at the point, where the discontinuity (defect) occurs. The collected particles generally take the shape and size of the defect. Subsurface defects can be detected best with direct current, although, both alternating and direct current can be used in this process. Subsurface defects can also produce surface-detectable disruptions, if they are sufficiently close to the surface.

5.19.5 Eddy-Current Inspection

The eddy-current testing method can be used to detect surface and near-surface flaws such as cracks, voids, inclusions and seams. Material mix-ups, processing errors, thicknesses or variation in thickness of platings, coatings or even corrosion can be detected or measured by this method.

This method is based on the principle of electromagnetic induction. The part to be tested is placed in the proximity of an electric coil (inspection coil or excitation coil) through which an alternating current of frequencies ranging from 60 Hz to 6 MHz flows. As a result, eddy current starts flowing in the part. Discontinuities in the part change the amplitude and direction of the flow of the induced current in the part. The changes of magnitude and phase difference can be used to defect the flows. The best results are obtained, when the current flow is at right angles to the flaw. This method is not as sensitive to small, open flaws as in liquid penetrant. However, it does not require clean-up operations and is generally faster. While comapring with magnetic-particle testing, eddy-current tests are not as sensitive to small flaws, but work equally well on ferromagnetic and nonferromagnetic materials.

5.19.6 Thermal Testing

Temperature measurement can be used to detect defective components with the help of contact-or noncontact-type heat-sensing devices.

During thermal testing, the part to be tested is heated and inspected to detect abnormal temperature distributions that are the result of faults or flaws such as cracks, poor joints, etc. Thermal anomalies tend to appear in areas where bonding between the components is poor or incomplete.

The contact methods of measurement use thermometers, radiometers, pyrometers or thermocouples. Additionally, this method also uses materials such as heat-sensitive paints and papers, liquid crystals and other coatings that react to temperature changes by altering its colour or appearance.

For noncontact thermal testing, infrared detection systems are most frequently used. Infrared detectors, such as infrared scanning microscopes and cameras, with high response time and sensitivities of 1 °C, produce a thermal picture of a part by representing temperatures as different shades of grey or shades of significant colours.

5.19.7 Advanced Optical Methods

Although visual inspection is often the simplest and least expensive of the nondestructive inspection methods, there are also a number of advanced optical methods. One such method is holography, which records the three-dimensional image of an object on a photographic film. Holographic interferometry is the extension of holography technique, which can detect minute changes in the shape of an object under stress. Acoustic holography can detect internal defects directly from the image of the interior of the part under testing.

Scanning acoustical holography produces the hologram by electronic-phase detection and is more sensitive.

5.19.8 Resistivity Method

The resistivity method is based on the electrical resistivity of a conductive material. By measuring the resistivity, alloy identification, flaw detection or the assurance of proper processing can be done. Micro-ohmmeters are very sensitive instruments that can be used to evaluate the effects of heat-treatment, the amount of cold work, the integrity of weld or the depth of case hardening.

5.19.9 Chemical Analysis and Topography of Surfaces

Chemical analysis is used to determine the chemical and elemental analysis of surface and near-surface material. The techniques included under chemical analysis are auger electron spectroscopy (AES), energy-dispersive X-ray analysis (EDX), electron spectroscopy for chemical analysis (ESCA) and various forms of secondary-ion mass spectroscopy (SIMS). Because of its large depth of focus, the scanning electron microscope is extremely useful for observing the surfaces of materials under testing. Atomic-force microscope and scanning tunnelling microscope give the surface topography with resolution to the atomic scale.

5.19.10 Acoustic-Emission Monitoring

The acoustic-emission method is a monitoring method, which is used to detect a dynamic change in the material taking place during plastic deformation, crack initiation and propagation, phase transformation and sudden reorientation of grain boundaries. It cannot detect an existing defect in a static product.

Engineering materials undergoing stress or plastic deformation emit sound. While these sounds are inaudible to the human ear, they are detectable through the use of sophisticated electronic equipments, which use sensors consisting of piezo electric ceramic elements. The frequencies of the sound waves generated can be as high as 1 MHz. The acoustic emissions can be related to the physical integrity of the material or structure in which they are generated, and the monitoring of these events permits detection and location of flaws as well as prediction of impending failure.

5.19.11 Pressure Testing

Pressure testing is used to locate leaks in a casting or to check the overall strength of a casting in resistance to bursting under hydraulic pressure. Prominent areas of application are pump housing, valve bodies and pipe manifolds. Leakage of fluids through the casting walls may occur as a result of two different defects in the metal, namely the following:

- Highly localized cavities, extending through the leaking section. Gross unfed solidification shrinkage is the usual reason for this defect. An improvement in gating and risering or other foundry practice can eliminate these leakers.
- Dispersed cavities, which when interconnected, permit gradual passage of the fluid through the casting walls. This type of leakage is more difficult to cure. Often it is related to the alloy type and the design of the casting. Dispersed shrinkage is reduced to a minimum in any alloy type by pouring at a minimum temperature (675 °C), making every effort to obtain directional solidification, proper gating and metal flow and use of the best melting practices.

During the pressure testing, air pressure, which ranges from 40 to 90 psi, is applied to the casting and it is immersed in warm soapy water or sprayed with a warm soap solution. If leaks are found, measures may be taken to seal them. In spite of the best of foundry practices, leakers (castings which are not pressure tight) occur. Indeed, sealing is almost a necessity for certain combinations of casting designs and casting alloys to produce leak-proof castings even with the best foundry practices.

The simplest sealing method for localized leakage is peening. The area may be hammered to close up the leak. Impregnation is essential, where dispersed leakage is found. Impregnation consists of causing a liquid to penetrate the shrinkage holes. After being forced into the cavities, the liquid turns into a solid. Commonly used impregnants are the sodium silicate, drying oils, e.g. tung, linseed oil, etc., and various types of synthetic resins.

Short Answer Questions

1. What are the merits and demerits of a wooden pattern?

Ans. A wooden pattern is easy to fabricate because of the favourable properties of wood. But it has the demerit of absorbing the moisture from the sand, and hence its shape may get distorted affecting the size of the casting. Its surfaces are also eroded by the sharp sand particles affecting the surface finish of the casting produced.

2. What is the difference between the match-plate pattern and the cope and drag pattern?

Ans. In the match-plate pattern, there is a single match plate which contains cope and drag parts of the pattern on its opposite sides. In the cope and drag pattern, cope and drag parts of the pattern are separately mounted on two match plates.

3. What are the different stages of shrinkage in casting?

Ans. There are three stages of shrinkage in casting. They include (a) shrinkage of liquid when its temperature drops from the pouring temperature to the freezing temperature (b) shrinkage of metal during change of phase from liquid state to solid state (c) shrinkage of solid when its temperature drops from the freezing temperature to the ambient temperature.

4. What is the function of a riser?

Ans. Being a reservoir of the hot molten metal, its function is to provide extra metal in the mould cavity to take care of the shrinkage of metal due to phase change from its liquid state to solid state during the solidification of the casting.

5. What is a chaplet? Of what material is it made of?

Ans. A chaplet is a casting part which is used to support a core in the mould cavity. As it finally becomes integral part of the casting, hence its material should be the same as that of casting.

6. What is ramming? What is its effect on the permeability of the moulding sand?

Ans. Ramming is the process of distributing the sand mass uniformly all around the mould using a muller. Soft ramming increases the permeability of the moulding sand, but hard ramming reduces the permeability by closing all the smaller holes in the sand mass.

7. What does charging of a cupola mean?

Ans. Charging of a cupola means mixing the different materials in the correct proportion needed to start the combustion in the furnace, which consists of iron (pig iron and scrap), coke and flux.

8. What is core?

Ans. Core is a sand mass which is used to produce hollow cavity in the casting by placing it at the place of cavity desired, which can be destroyed latter on, leaving the vacant space.

9. What is pouring temperature?

Ans. Pouring temperature is the temperature of the hot molten metal when it enters the mould cavity.

10. What is a solid solution? Differentiate between substitutional and interstitial solid solutions.

Ans. A solid solution is a mixture of two or more elements in the solid state. In the substitutional solid solution, the solute atoms replace the solvent atoms. For example, brass is a substitutional solid solution of zinc (solute) and copper (solvent). In the interstitial solid solution, small size solute atoms enter the empty spaces available in the lattice of the solvent atoms. For example, steel is an interstitial solid solution in which small size carbon atoms are dissolved in iron

Multiple Choice Questions

1. Consider the following materials:

1. Wood	2. Plastic
3. Metal	

A pattern is made of

(a) 1 alone	(c) 2 and 3
(b) 2 alone	(d) 1, 2 and 3

2. Consider the following parts of a moulding flask:

1. Cope	3. Drag
2. Cheek	

The lower part of the flask is called

(a) 1	(c) 3
(b) 2	(d) None of these

3. Consider the following casting elements:

1. Pouring basin
2. Runner
3. Sprue
4. Gate

A gating system consists of

(a) 2 and 3
(b) 1, 2, 3 and 4
(c) 1 and 4
(d) 2, 3 and 4

4. Consider the following statements:

   1. Riser is a reservoir of molten metal
   2. Core is used to create void in the casting
   3. Sprue is a horizontal channel in the parting line
   4. A moulding flask has two or three parts

   Of these statements:

   (a) 1 and 2 are true
   (b) 2, 3 and 4 are true
   (c) 1, 2 and 4 are true
   (d) 1, 2 and 3 are true

5. A chaplet is used to

   (a) create void in the casting
   (b) withdraw the pattern from a mould
   (c) act as a reservoir of molten metal
   (d) support the core.

6. The chilling effect is observed in

   (a) plastic moulds
   (b) metallic moulds
   (c) wooden moulds
   (d) CO_2 moulds

7. Consider the following statements:

   1. A cope and drag pattern uses one match plate.
   2. A solid pattern is used for simple casting.
   3. A skeleton pattern is made of wood.
   4. A gated pattern uses a match plate.

   Of these statements:

   (a) 1 and 2 are true
   (b) 1 and 4 are true
   (c) 3 and 4 are true
   (d) 2 and 3 are true

8. A permanent mould has the following advantages:

    1. It has better dimensional accuracy.
    2. It offers better surface finish.
    3. It is used for smaller number of casting.
    4. It suffers from chilling effect.

    Of these statements:

    (a) 1 and 2 are true (b) 12 and 4 are true
    (c) 12 and 3 are true (d) 23 and 4 are ture

9. Consider the following statements:

    1. Grey iron castings are provided with maximum shrinkage allowance.
    2. Rapping allowance is a negative allowance.
    3. Inner surfaces require more draft allowance.
    4. Investment castings require maximum machining allowance.

    Of these statements:

    (a) 1 and 2 are true (b) 1 and 4 are true
    (c) 3 and 4 are true (d) 2 and 3 are true

10. Match List I with List II and select the correct answer using the codes given below the lists:

List I (Pattern Allowances)	List II (Description)
A. Shrinkage allowance	1. Negative allowance
B. Machining allowance	2. Grey iron castings require minimum allowance
C. Draft allowance	3. Die castings require minimum allowance
D. Rapping allowance	4. Facilitates withdrawal of pattern from the mould

Codes:	A	B	C	D
(a)	3	4	2	1
(b)	2	3	1	4
(c)	2	3	4	1
(d)	3	2	4	1

11. The charge in a cupola consists of

(a) coke and pig iron
(b) coke, pig iron and scrap castings
(c) pig iron and scrap castings
(d) coke only

12. Tuyeres in cupola are used to

(a) collect the molten metal
(b) feed the charge
(c) allow the passage of air
(d) arrest the spark

13. A cupola furnace can produce a maximum temperature of around

(a) 500 °C
(c) 1650 °C
(b) 1000 °C
(d) 3250 °C

14. The thickness of skin formation during solidification varies

(a) directly proportional to time
(b) directly proportional to square of time
(c) directly proportional to square root of time
(d) inversely proportional to square root of time.

15. Consider the following statements:

1. Metals solidify at constant temperatures.
2. Phase diagram shows the relationship between temperature, composition and the phase present in an alloy system.
3. Iron has a melting temperature of 1538° F.
4. Pure iron contains 0.008% of carbon.

Of these statements:

(a) 1 and 2 are true
(b) 1, 2 and 3 are true
(c) 1, 2 and 4 are true
(d) 1, 2, 3 and 4 are true

16. The carbon limit for cast iron is

(a) 0 to 0.008%
(c) 2.11 to 4.5%
(b) 0 to 2.11%
(d) 4.5 to 6.67%

17. X contains 6.67% of carbon. What is X?

(a) Mild steel
(b) Tool steel
(c) High-carbon steel
(d) Cementite

18. Hypoeutectoid steel contains

    (a) less than 0.77% of carbon
    (b) greater than 0.77% of carbon
    (c) greater than 2.11% of carbon
    (d) greater than 6.67% carbon.

19. Hypereutectoid steel contains

    (a) less than 0.77% of carbon
    (b) greater than 0.77% of carbon
    (c) greater than 2.11% of carbon
    (d) greater than 6.67% carbon

20. The highest limit of carbon in steel is

    (a) 0.77%
    (b) 2.11%
    (c) 4.5%
    (d) 6.67%

21. Peritectic reaction occurs at

    (a) 1148 °C
    (b) 727 °C
    (c) 1495°F
    (d) 1495 °C

22. Eutectoid contains

    (a) 0.77% of carbon
    (b) 2.11% of carbon
    (c) 4.5% of carbon
    (d) 6.67% of carbon

23. Consider the following statements about ferrite:

    1. It is very hard and brittle.
    2. It is very soft and lacks strength.
    3. It contain maximum carbon of 0.022% at 727 °C.

    Of these statements:

    (a) 1 alone is true
    (a) 1 alone is true
    (c) 3 alone is true
    (d) 2 and 3 are true

24. Pouring temperature for cast iron is

    (a) 727 °C
    (b) 1495 °C
    (c) 1400 °C
    (d) 1538 °C

25. Which of the following cast irons has the least shrinkage?

(a) Grey cast ironv
(b) Nodular iron
(c) White cast iron
(d) Malleable iron

26. Which of the following cast irons is very hard and wear resistant?

(a) Grey cast iron
(b) Nodular iron
(c) White cast iron
(d) Malleable iron

27. The minimum thickness for steel castings is

(a) 2 mm
(b) 3 mm
(c) 4 mm
(d) 6 mm

28. Which of the following elements increases the fluidity of aluminium?

(a) Vanadium
(b) Nickel
(c) Silicon
(d) Chromium

29. Which of the following metals has maximum applications in making die castings?

(a) Copper
(b) Zinc
(c) Tin
d) Magnesium

30. Consider the following statements:

1. Bentonite is the most commonly used clay in moulding sand.
2. Pouring temperature of steel is lower than that of iron.
3. Coarse grains have higher porosity.

Of these statements:

(a) 1 alone is true
(b) 2 and 3 are true
(c) 1 and 3 are true
(d) 1, 2 and 3 are true

31. Consider the following statements:

1. Permeability prevents fusion of moulding sand.
2. Soft ramming increases permeability.
3. A green sand contains moisture.

Of these statements:

(a) 1 and 2 are true
(b) 2 and 3 are true
(c) 1 and 3 are true
(d) 1, 2 and 3 are true

32. The moisture content in the moulding sand varies between

(a) 2 and 8%
(b) 5 and 10%
(c) 10 and 15%
(d) 15 and 20%

33. How much air is allowed to pass through the standard sand sample used in permeability test?

(a) 500 cm^3
(b) 1000 cm^3
(c) 1500 cm^3
(d) 2000 cm^3

34. How much pressure is applied on the sand sample in permeability test?

(a) 5 gm/cm^2
(b) 10 gm/cm^2
(c) 10 Pa
(d) 10 kg/cm^2

35. The cylindrical sand specimen used in strength test of moulding sand has the following dimension (height and diameter in order)

(a) 25 mm, 50 mm
(b) 50 mm, 50 mm
(c) 50 mm, 25 mm
(d) 25 mm, 25 mm

36. Match List I with List II and select the correct answer using the codes given below the lists:

List I (Casting defects)	List II (Reason)
A. Misrun	1. Chaplet is not in position
B. Drop	2. Lack of fluidity of molten metal
C. Scar	3. Sand lacks cohesiveness
D. Core shift	4. Sand lacks permeability

Codes:	A	B	C	D
(a)	2	3	4	1
(b)	3	2	4	1
(c)	2	3	1	4
(d)	4	3	2	1

37. Rat tail occurs due to

(a) lack of molten metal fluidity

(b) poor permeability of moulding sand
(c) thermal expansion of moulding sand
(d) hydrostatic pressure of molten metal.

38. Swell occurs due to

(a) lack of molten metal fluidity
(b) poor permeability of moulding sand
(c) thermal expansion of moulding sand
(d) hydrostatic pressure of molten metal.

39. Cold shut occurs due to

(a) lack of molten metal fluidity
(b) poor permeability of moulding sand
(c) thermal expansion of moulding sand
(d) hydrostatic pressure of molten metal.

40. Pin holes are caused due to

(a) lack of molten metal fluidity
(b) poor permeability of moulding sand
(c) thermal expansion of moulding sand
(d) hydrostatic pressure of molten metal.

41. Consider the following statements:

1. Cold shut is a variation of misrun.
2. Hot tears are common in nonferrous castings.
3. Scar is caused due to thermal expansion of moulding sand.

Of these statements:

(a) 1 alone is true (b) 3 alone is true
(c) 1 and 2 are true (d) 1, 2 and 3 are true

42. Match list I with List II and select the correct answer using the codes given below the lists:

List I	List II
(Casting defects)	(Description)
A. Hot tear	1. Variation of misrun
B. Swell	2. Thermal expansion of moulding sand
C. Cold shut	3. Tensile stress

D. Rat tail	4. Hydrostatic pressure of molten metal

Codes:	A	B	C	D
(a)	2	4	1	3
(b)	3	4	2	1
(c)	3	4	1	2
(d)	4	3	1	2.

43. Consider the following statements about destructive testing methods:

    1. It is mainly used for the testing of mechanical properties.
    2. They are highly reliable.
    3. Ultrasonic testing is a destructive method of testing.

    Of these statements:

    (a) 1 alone is true (b) 2 and 3 are true
    (c) 1 and 2 are true (d) 1, 2 and 3 are true

44. Radiographic testing uses

    1. Red colour dye 2. X-rays
    3. Gamma rays 4. Ultrasonic beam

    Of these:

    (a) 1 alone is true (b) 2 and 3 are true
    (c) 1 and 4 are true (d) 2 alone is true

45. Match List I with List II and select the correct answer using the codes given below the lists:

List I	List II
(Nondestructive tests)	(Description)
A. Radiographic testing	1. Transducer
B. Ultrasonic inspection	2. Peening
C. Magnetic inspection	3. X-rays
D. Pressure testing	4. Surface defects

Codes:	A	B	C	D
(a)	3	1	4	2
(b)	3	4	2	1
(c)	3	4	1	2
(d)	4	3	1	2

46. Which of the following methods is used to find the dynamic defects in materials?

(a) Acoustic-emission monitoring
(b) Radiographic testing
(c) Magnetic inspection
(d) Ultrasonic inspection.

47. Peening is connected with

(a) Magnetic testing (b) Radiographic testing
(c) Pressure testing (d) Ultrasonic testing

Answers

1. (d)	2. (c)	3. (b)	4. (c)	5. (d)	6. (b)
7. (d)	8. (b)	9. (d)	10. (c)	11. (b)	12. (c)
13. (c)	14. (c)	15. (c)	16. (c)	17. (d)	18. (a)
19. (b)	20. (b)	21. (d)	22. (a)	23. (d)	24. (c)
25. (a)	26. (c)	27. (d)	28. (c)	29. (b)	30. (c)
31. (b)	32. (a)	33. (d)	34. (b)	35. (b)	36. (a)
37. (c)	38. (d)	39. (a)	40. (b)	41. (c)	42. (c)
43.(c)	44. (b)	45. (a)	46. (a)	47. (c).	

Review Questions and Discussions

Q.1. How is a casting pattern defined?
Q.2. Why is a core used in casting?
Q.3. What are the differences between expendable and permanent moulds?
Q.4. What are the elements of a gating system?
Q.5. What is the purpose of a parting line or parting surface?
Q.6. How is the core prevented form shifting in sand moulding?

Q.7. What are composite moulds? Discuss their suitability.
Q.8. Pattern allowances are kept as small as possible. Why?
Q.9. Why is pattern made slightly bigger than the actual size of the casting?
Q.10. Why is draft required on a pattern?
Q.11. What are the factors to be considered during the selection of a pattern material?
Q.12. What is the function of a blind riser?
Q.13. Why are blind risers usually smaller than open-top risers?
Q.14. What is the function of a chill?
Q.15. How is castability defined?
Q.16. Which material has the highest castability?
Q.17. Why is a match-plate pattern much easier to use than a standard pattern?
Q.18. What are the three stages of shrinkage when the molten metal is turned into a finished casting?
Q.19. What are the advantages of a split pattern over one-piece pattern?
Q.20. How is a cope-and-drag pattern different from a match-plate pattern?
Q.21. What are the factors to be considered during the selection of a furnace in a casting operation?
Q.22. Why isn't the cupola furnace used for melting steel?
Q.23. For which type of metals a cupola furnace is useful?
Q.24. Why is dry sand moulding not very popular?
Q.25. What are the major types of sand moulds? Describe their characteristics.
Q.26. Name the common mould materials used in permanent mould casting.
Q.27. Why is turbulence of the liquid metal in the gating system and mould cavity is required to be avoided?
Q.28. What are the advantages of casting over other manufacturing processes?
Q.29. For an alloy, solidification takes place not at a definite temperature but over a range of temperatures. Why?
Q.30. Why do metals have unique melting points?
Q.31. How is liquidus different from solidus in a phase diagram?
Q.32. Why do castings sometimes need heat treatment?
Q.33. What is residual stress?
Q.34. What is latent heat? Why is it so called?
Q.35. What is homogeneous nucleation?
Q.36. What is the importance of a phase diagram?
Q.37. What is the importance of a Fe–C equilibrium diagram?
Q.38. What is the highest limit of carbon in cast iron?
Q.39. What is eutectoid?
Q.40. What are hypoeutectoid and hypereutectoid steels?
Q.41. Why is cast iron one of the least expensive materials?
Q.42. Why is it difficult to make large size castings from malleable cast iron?

Q.43. What are the differences among grey cast iron, nodular cast iron and malleable cast iron?
Q.44. Why are majority of the die casting products made from zinc?
Q.45. Why are brass die castings difficult and expensive to produce?
Q.46. Ferrous metals are not suitable for plaster moulds?
Q.47. Why are steels more difficult to cast than cast irons?
Q.48. Why does a grey iron casting require less riser material than a steel casting of the same size?
Q.49. List the important nonferrous metals commonly used for casting.
Q.50. What are the primary requirements of a moulding sand? Discuss them in brief.
Q.51. What is permeability, and why is it important for moulding sands?
Q.52. How does the size and shape of the sand grains affect the properties of moulding sand?
Q.53. What are the limitations of green sand moulds?
Q.54. What is sand reconditioning?
Q.55. If the moulding sand contains excessive moisture, then what can result?
Q.56. What are the possible reasons of casting defects?
Q.57. How does blow differ from scar?
Q.58. What is misrun and how does it result?
Q.59. What is a rat tail?
Q.60. What are the causes of gas porosity in castings?
Q.61. What can happen, if molten metal is not allowed to cool and solidify freely?
Q.62. What is the difference between drop and dirt? How are they produced?
Q.63. How are dissolved gases removed from molten metal?
Q.64. How does hot tearing occur? Explain.
Q.65. What are the possible casting defects arising out of defective design of casting and pattern?
Q.66. Defective gating and risering system can produce a number of casting defects. How?
Q.67. What changes are expected in the design of mould and core to obtain defect-free castings?
Q.68. How does core shift occur?
Q.69. Why is a casting inspected?
Q.70. For what types of defects, visual inspection is useful?
Q.71. How does destructive testing differ from nondestructive testing?
Q.72. Nondestructive testing methods are more accurate and reliable. How?
Q.73. How is ultrasonic inspection performed?
Q.74. X-rays and gamma rays are used for casting inspection. How are they useful?
Q.75. What types of defects are detected by liquid penetrant inspection?
Q.76. Acoustic emission methods can't detect static defects. Why?

Casting Processes

6

Figure shows a vacuum casting process; a process used to produce mechanically superior parts with complex profile.

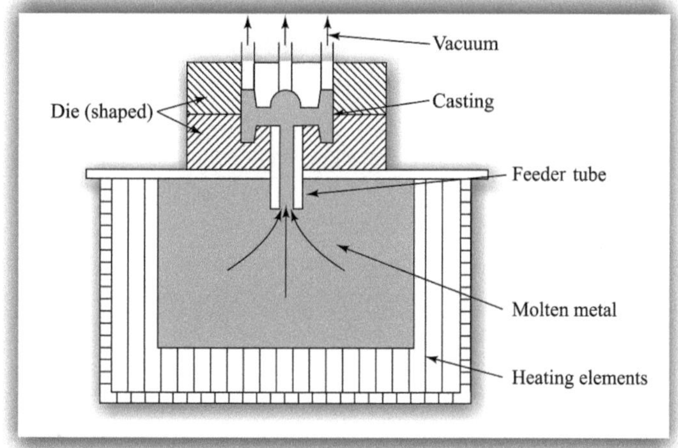

Learning Objectives

After reading this chapter, you will be able to answer some of the following questions:

- Why is investment casting most suitable for small size castings?
- Where are polystyrene patterns used?
- How does hot-chamber die casting differ from cold-chamber die casting?

D. K. Singh, *Fundamentals of Manufacturing Engineering*,
https://doi.org/10.1007/978-981-99-8767-2_6

6.1 Shell Mould Casting

6.1.1 Introduction

Shell mould casting is a variation of sand casting. Here, the mould is made of a mixture of dried silica sand and phenolic resin, formed into thin, half-mould shells, which are clamped together for pouring purposes. It is also known as *Croning* or *C process* after the name of its inventor J. Croning.

6.1.2 Process Details

The clay-free sand is thoroughly mixed with about 5% of either urea or phenol formaldehyde resin and the mixture is put into a dump box. A preheated metal pattern (heated to about 400 °C) is now placed on the top of the dump box. The dump box is then inverted, allowing the sand mixture to fall on the hot pattern.

The heat of the pattern melts the resin and makes the mixture soft enough to form a shell of uniform thickness while sticking to the surface of the heated pattern. After about 10 seconds, the dump box is brought back to its normal position. The actual shell thickness depends upon how long the mixture is allowed to stay on the heated pattern.

The pattern having a thin shell of sand adhering to it is then placed in an oven and the shell is cured for about a minute to make it rigid. Finally, the hardened shell is removed from the pattern by ejector pins. The two half shells, produced as described above, are securely fastened together to form a complete mould. Now, the shell moulds are placed in a flask and are ready for pouring.

6.1.3 Advantages

- Shell mould castings need very little cleaning after separating from the mould.
- It gives better surface finish and higher dimensional accuracy as compared to conventional sand mould casting.
- The process is fast automation of the process is easier.
- It is suitable for mass production.

6.1.4 Limitations

- The use of metallic pattern in the process is a costly affair. The higher cost can only be justified for large production runs.

- The use of resin increases the cost.
- Accurate patterns are required to obtain good quality castings.

6.1.5 Materials to be Cast by This Method

Practically all metals including the high-temperature melting alloy steels can be cast in shell moulds.

6.1.6 Typical Applications

Typical cast products include high-precision mechanical parts, such as gear-housings, cylinder heads and connecting rods; and high-precision moulding cores.

6.2 Plaster Mould Casting

The plaster mould casting is somewhat similar to sand casting in the respect that only one casting is made and then the mould is destroyed.

In this process, the mould material is plaster of Paris (also known as calcium sulphate or gypsum), with the addition of talc or magnesium oxide, which helps to prevent cracking and reduces the setting time; lime or cement helps to control expansion during baking; and glass fibre can be added to improve strength. Sand is often used as a filler. These components are mixed with water and the resulting slurry is poured over a metal pattern. After the pattern sets, it is removed carefully and the mould is dried by proper heating to remove the moisture. The mould halves are then assembled to form the mould cavity and preheated to about 120 °C, before the molten metal is poured in it.

It produces castings of high dimensional accuracy and good surface finish and is therefore known as precision casting.

Typical parts made by this method include gears, valves and ornaments.

6.3 Ceramic Mould Casting

6.3.1 Introduction

This process was developed in England by Clifford and Noel Shaw. It is similar to plaster-mould process, with the exception that it uses refractory moulds which can withstand high temperatures.

6.3.2 Other Names of the Process

- **Ceramic shell process**, because the mould is made of ceramic shell.
- **Shaw Process**, because of the name of its inventor.
- Cope-and-drag investment casting.

6.3.3 Process Details

A creamy ceramic slurry, which is a mixture of fine-grained zircon ($ZrSiO_4$), aluminium oxide (Al_2O_3) and fused silica mixed with bonding agent, is poured over a pattern. The pattern used may be made of wood or metal and is reusable. The slurry hardens on the pattern almost immediately and becomes a strong green ceramic of the consistency of vulcanized rubber. After setting, the mould is removed, dried and burned off using gas torch. During burn-off, most of the volatile matters are removed and a three-dimensional network of microscopic cracks (microcrazing) is formed in the ceramic. The gaps are small enough to prevent metal penetration but large enough to provide venting of air and gases and to accommodate the thermal expansion of the ceramic particles during the pour and subsequent shrinkage of the solidified metal (i.e. provide collapsibility). A subsequent baking operation at 982 °C removes all the remaining volatiles, making the mould hard and rigid. Before pouring, the ceramic moulds are often preheated to ensure proper filling and to control the solidification characteristics of the metal. The ceramic facings are then assembled into a complete mould, ready to be poured. The shell breaks away from the casting as cooling takes place.

6.3.4 Advantages

The Shaw process finds its biggest advantage in being able to mould steels to very precise dimensions. The dimensional accuracy and the surface finish of the casting produced by this method are good. Tolerances of ± 0.13 mm are common.

6.3.5 Limitations

Although the Shaw process is generally quite fast, it may take as long as 5 h to bake out a large ceramic mould. Parting lines may be seen on the casting, where the mould haves have been put back together, and the process is somewhat expensive.

6.3.6 Materials to be Cast by This Method

This process is used for casting ferrous and other high-temperature alloys, stainless steels and tool steels.

6.3.7 Typical Applications

Typical parts made are impellers, cutters for machining and dies for metal working. It is also used in making plastic moulds, glass moulds, stamping dies and extrusion dies.

6.4 Permanent Mould Casting

6.4.1 Introduction

This method uses a metallic mould, made of grey cast iron, steel, bronze or other materials such as graphite and refractories. The moulds can withstand high temperatures and are reusable. That is why the castings produced by such moulds are so named. Among all the listed mould materials, grey cast iron is used most frequently because of its better resistance to thermal fatigue and easy machinability. Graphite is also a good mould material because it can be machined easily, has superior dimensional stability, and has good thermal conductivity, which promotes rapid cooling and thus increasing production rates. The graphite moulds are less expensive and their cost is often only one fifth that of metal moulds. But graphite moulds are restricted to ferrous metal parts due to deterioration caused by the high heat.

6.4.2 Other Names of the Process

Permanent mould casting is also known by three other names:

- **Hard mould casting**, because of the metal mould being used.
- **Near-net shape casting**, because the surface produced is almost final or very little machining is required to produce the finished part.
- **Gravity die casting/Gravity casting**, because the molten metal enters the mould cavity under the action of gravity and no external pressure is required.

6.4.3 Process Details

The two halves of the heated mould are aligned and clamped together before the molten metal is poured by simple gravity. After solidification, the mould is opened and the product is removed. The mould can be reused without preheating for further casting, as the heat of the previous cast keeps the mould heated. The moulds are usually coated with a thin layer of refractory material followed by a layer of lamp black to reduce the chilling effect. It also facilitates the removal of the casting from the mould.

6.4.4 Advantages

- Castings are free from embedded sand.
- Castings have good finish and minute surface details.
- Closer dimensional tolerances (in fraction of a mm) can be maintained.
- The process is economical for large production runs.

6.4.5 Limitations

- The initial cost of the equipment is high because of metal moulds (dies) used in the process.
- The process is generally limited to the lower melting point alloys.
- Because of high pouring temperatures in case of steel and cast iron castings, the life of moulds used for such castings is low.
- There is difficulty in removing the castings from metal moulds which restricts its use for intricate parts. Using semi-permanent-mould casting, which uses collapsable sand cores, intricate details can be made in the cast parts.

6.4.6 Materials to be Cast by This Method

This process can be used for both ferrous and nonferrous castings, but the latter offers less problems because of their lower pouring temperatures. Aluminium, magnesium and copper-based alloys are the best cast metals. Using graphite moulds, iron and steel castings can also be made.

6.4.7 Typical Applications

Typical parts made by permanent mould casting are automobile pistons (made of aluminium), cylinder heads, connecting rods, cooking utensils, refrigerator parts and small gear blanks.

6.5 Slush Casting

Slush casting is a special form of permanent mould casting, which produces hollow casting in metal moulds without the use of cores. In this process, the liquid metal is permitted to remain in the mould till a shell of the desired thickness has formed. The mould is then turned upside down, which causes the remaining liquid metal to run out, leaving a hollow thin-walled casting. This process is frequently used for making ornamental and decorative objects (such as lamp bases and stems), statuettes and toys from low melting point metals such as zinc tin and lead alloys. Slush casts are usually painted or finished to give good looks of expensive metals.

6.6 Vacuum Casting

In *vacuum casting*, vacuum is used to draw the liquid metal into the mould cavity. It is essentially different than vacuum-moulding process. The vacuum reduces the air pressure inside the mould to about two-thirds of atmospheric pressure, which helps to draw the molten metal into the mould cavities through a gate in the bottom of the metal mould (Fig. 6.1). After the mould is filled, it is withdrawn from the molten pool. It is also called **vacuum permanent mould casting**.

This process is an alternative to investment, shell-mould and green-sand casting; and is particularly suitable for thin-walled (0.75 mm) complex shapes with uniform properties. Because of the vacuum being used in the process, the cleanliness of the metal and dissolved gas content are superior. Mechanical properties of vacuum cast parts are 10–15% superior to those of conventional permanent mould products. The process can be automated and the production costs are similar to those for green-sand casting. The process can be used for making parts for gas turbines.

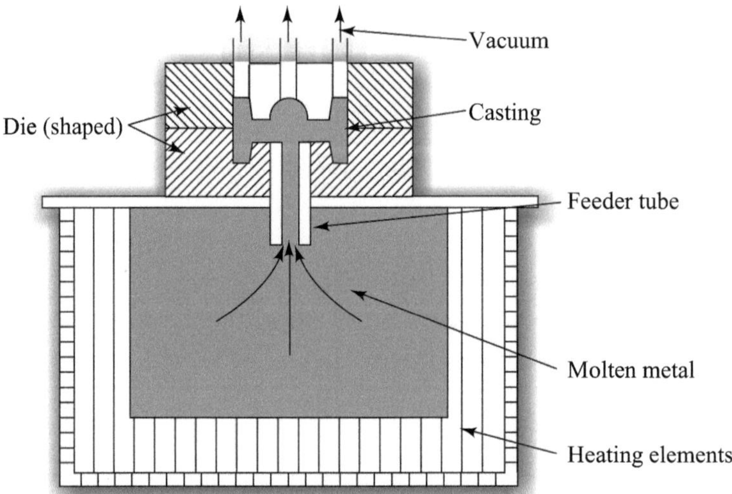

Fig. 6.1 Vacuum casting process

6.7 High-Pressure Die Casting

6.7.1 Introduction

High-pressure die casting may be classified as a permanent mould casting but differs from it in the sense that in the former very high pressure (0.6–275 MPa) is used to force the molten metal in the metal mould, known as a die.

6.7.2 Other Names of the Process

Die casting is also known as **pressure die casting**, because of high pressure involved in the process.

6.7.3 Process Details

The hot molten metal forced in the mould solidifies quickly (within a fraction of a second), because the die is water cooled. Upon solidification, the die is opened and the casting is taken out of the die using ejector pins. Most of the castings have flash, where the two die halves meet.

The dies used may be made single-cavity dies, multiple-cavity dies (with several identical cavities), combination-cavity dies (with several different cavities), or unit dies (simple small dies that can be combined in two or more units in a master holding die).

6.7.4 Advantages

The high-pressure die casting has the following advantages:

- High-pressure die casting can produce parts with good strength and complex shapes.
- Parts have good dimensional accuracy and surface details, thus requiring little or no subsequent machining or finishing operations (**net-shape forming**).
- The production rate is very high, due to metal mould and high pressure involved in the process.
- Because of the high pressures involved, the wall thickness as small as 0.5 mm can be produced, which is not possible by other casting methods.
- No risers are used in high-pressure die casting process, since the high pressures help feed metal from the gating system into the mould cavity.

6.7.5 Limitations

The following are the limitations of the high-pressure die casting process.

- High-pressure die casting have a porosity problem where gases tend to be entrapped. It results in lower strength and leakage. One way to reduce the porosity is to operate the machine under a vacuum.
- The initial cost of the machine is high, so the process is economical only for large production runs. Also, the die costs are high because of the high accuracy with which dies are made and the high polish required on their surfaces.
- There is undesirable chilling of the metal unless high temperatures are sustained before the cast is removed. Metals having a high coefficient of contraction are removed from the mould as soon as possible because of the inability of the mould to contract with the casting.

6.7.6 Materials to be Cast by This Method

High-pressure die castings were once limited to low melting alloys, but with a gradual improvement of heat-resisting metals for dies, their scope broadened and the process is now being used for numerous alloys. Although grey cast iron and low carbon and alloy steels have been produced in dies made of unalloyed sintered molybdenum, die casting is commercially limited to low melting point nonferrous metals and alloys such as zinc, aluminium, magnesium and copper-based alloys, which do not attack the metal die rapidly.

6.7.7 Typical Applications

Typical parts made by high-pressure die casting are carburetors, motors, hand tools and toys. Weight of most of the castings ranges from less than 90 g to about 25 kg. The rapidly solidified surface is usually harder than the interior and is usually sound and suitable for plating or decorative applications.

6.7.8 Types of High-Pressure Die Casting

There are two basic types of high-pressure die casting machines:

- Hot-chamber die casting
- Cold-chamber die casting.

The principal distinction between the two methods is determined by the location of the melting pot.

6.8 Hot-Chamber Die Casting

In ***hot-chamber method***, the melting pot is included within the machine and the injection cylinder (shot cylinder) is immersed in the molten metal all the times. The injection cylinder is actuated by either air or hydraulic pressure, which forces the molten metal into the die cavity (Fig. 6.2). The metal is held under pressure until it solidifies in the die. It is more critical to use the adequate closing pressure between the parts of the die-set to avoid metal leakage along the parting line(s) as it involves very high injection pressure. Dies are generally cooled by circulating water or oil through various passageways in the die block. This improves die life and ensures rapid metal cooling, thereby reducing cycle time.

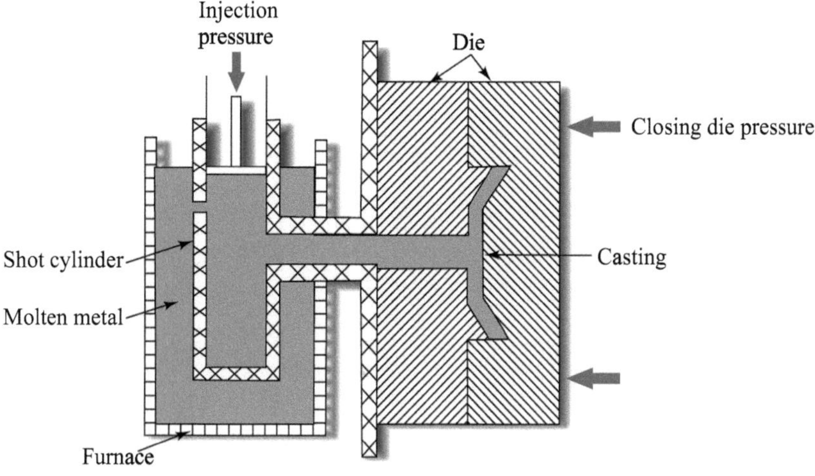

Fig. 6.2 Hot-chamber die casting process

Low melting point alloys such as zinc, tin and lead are commonly cast by this process. Most other materials either have too high a melting point, an affinity for iron or cause other problems that reduce the life of the machine and the die. Hot-chamber castings vary in size from 0.03 to 40 kg, although very small castings are usually cast in multiple-mould dies. It is not suitable for aluminium and its alloys. Molten aluminium and its alloys tend to absorb iron when they come in contact with cold ferrous (iron and steel) surfaces causing erosion of surfaces, thus damaging the die. At the same time, it also contaminates the molten metal.

6.9 Cold-Chamber Die Casting

Machines using cold-chamber process have a separate melting furnace. All other aspects of the cold-chamber die casting are same as the hot-chamber die casting. The molten metal is introduced into the injection cylinder by mechanical means or by hand. Hydraulic pressure then forces the metal into the die. Cold-chamber machines are usually suitable for high melting point alloys of aluminium, magnesium and copper (brass), although other metals (including ferrous metals) can also be cast in this manner.

Molten metal temperatures start at about 600 °C for aluminium and magnesium alloys and increase considerably for copper-base and iron-base alloys. These metals are not melted in a self-contained pot, because the life of the pot would be very short. The usual procedure heats the metal in an auxiliary furnace and ladles it to the plunger cavity next to the dies. The machines may be horizontal or vertical. In vertical cold-chamber machine, the shot cylinder is vertical and the machine is similar to a vertical press.

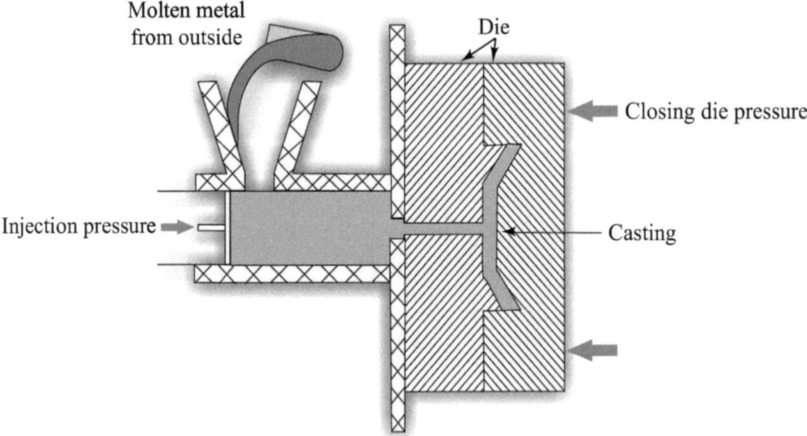

Fig. 6.3 Cold-chamber die casting process

In *horizontal cold-chamber machine*, the shot cylinder remains in the horizontal position (Fig. 6.3). The dies are shown closed and the molten metal is ready to be poured. As soon as the ladle is emptied, the plunger of the shot cylinder moves to the right and forces the metal into the mould. After the metal solidifies, the dies are opened and the casting is ejected from the stationary half. To complete the process of opening, an ejector rod is used to eject the casting from the movable half of the die. These machines are fully hydraulic and semi-automatic. After the metal is poured, the rest of the operations are automatic.

6.10 Low-Pressure Die Casting

6.10.1 Introduction

It is an intermediate process between gravity die casting and high-pressure die casting.

6.10.2 Other Names of the Process

- It is also known as **pressure pouring.**

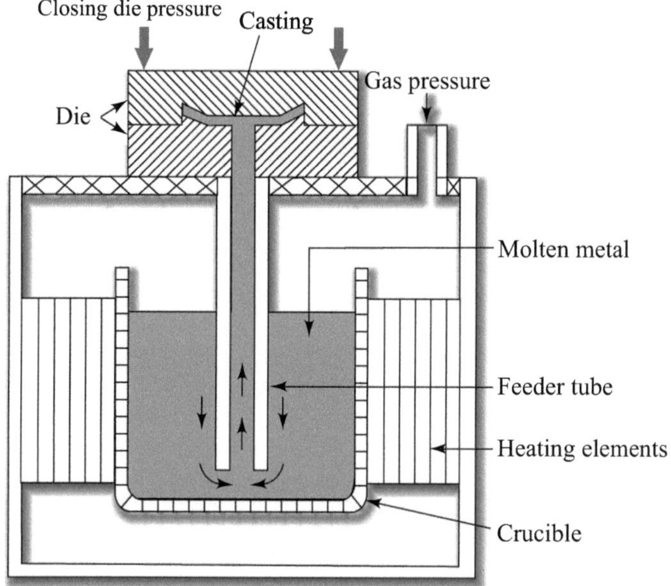

Fig. 6.4 Low-pressure casting process

6.10.3 Process Details

In low-pressure die casting, the molten metal is forced upward by gas (inert) or air pressure into a graphite or metal mould (Fig. 6.4). The mould is located over an induction furnace. It may take half a minute to fill up the mould, so any entrapped gas has time to escape.

Vacuum may be applied to the mould to aid filling and remove entrapped air. The part cools from top to down. The applied pressure continuously feeds molten metal to compensate for shrinkage and the unused metal in the feeder tube simply drops back into crucible after the pressure is released, when the metal has completely solidified in the mould. The upper half of the mould is lifted up with the casting entrapped in it. After tilting or moving laterally, ejector pins drop the casting in a catcher mechanism.

6.10.4 Advantages

The following are the advantage of low-pressure die casting:

- No riser is used, because the pressure is maintained until solidification is complete and the pressurized feed tube itself acts as a riser.

- Because of the pressure system, the minimum wall thickness can be reduced and the transition from heavy to light section can be more easily made.
- High-pressure die castings have a porosity problem that is nonexistent in the low-pressure system.
- The resultant casting is dense, free from inclusions and oxidation, and of good dimensional accuracy.
- The cycle time is reasonably fast, and scrap and remelt is low.
- Because of closely controlled temperature and pressure, mechanical properties of the cast part are improved.
- The low-pressure die casting machines cost about one-third less than high-pressure machines. Similarly the moulds are about one-third less expensive.
- Die life in low-pressure die casting is longer than for high-pressure casting, because of low operating pressure.

6.10.5 Limitations

The following are the limitations of low-pressure die casting:

- Tolerances must be wider for low-pressure casting than for die casting, because a refractory coating is necessary inside the mould.
- Casting modifications are harder to make.

6.10.6 Materials to be Cast by This Method

Aluminium, steel and cast iron.

6.10.7 Typical Applications

Some major automobile manufacturers in both the United States and Japan use low-pressure die casting system to produce aluminium engine blocks. Another example of this process is steel railroad wheels.

6.11 Full Mould Casting

6.11.1 Introduction

In this casting process, the pattern is made of expanded (foamed) polystyrene, which remains in the mould during the pouring of the hot liquid metal. When the molten metal

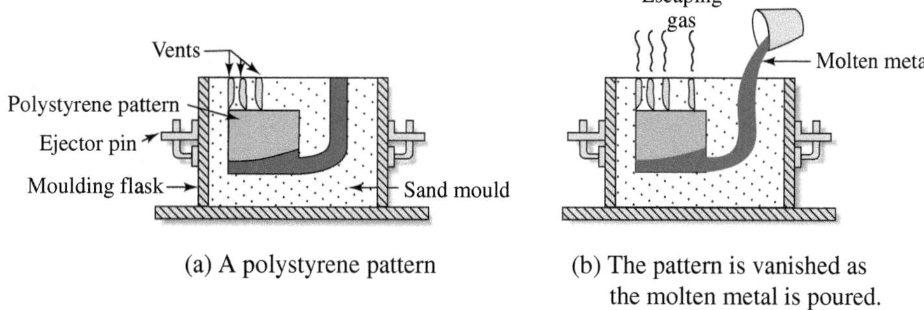

Fig. 6.5 Full mould process

is poured, the polystyrene melts and burns, and the metal fills the space that was occupied by the pattern (Fig. 6.5). Since the pattern is not removed from the mould but is replaced by the molten metal, the term full mould is used. Foamed material in the form of a pouring basin, sprue, runner segments and risers can be glued on to form a complete pattern assembly. A number of patterns can be joined to make one mould, called a **tree**, thus increasing the production rate.

6.11.2 Other Names of the Process

Full mould casting is also known by two other names:

- **Lost-foam casting**, because of the fact that the mould made from foamed polystyrene is finally consumed in the process.
- **Expanded polystyrene (EPS) casting**, because of the fact that the hard beads of polystyrene are first steam expanded and dried. The expanded beads are then injected into a heated metal mould at low pressure, where they further expand to fill up the die.

6.11.3 Process Details

In this process, the polystyrene pattern assembly (tree) is first dipped into a water-based ceramic that wets its surface and forms a coating of about 0.127 mm thick, rigid enough to prevent mould collapse during pouring, but thin and porous enough to permit the escape of the molten and gaseous material. The coated pattern is then positioned in a flask and surrounded by coarse unbounded sand that is vibrated into place. During the pour, the coating contains the metal, isolating it from the loose, unbounded sand.

6.11.4 Advantages

The most obvious benefit of the lost-foam process is the simplified sand handling. The reusable sand, with no additives or binders, reduces capital investment and operating costs. Because of the one-piece flask, there are no parting lines. Since the pattern need not be withdrawn, no draft is required in the design. The elimination of cores means no core sand, core binders or correlated casting defects. Cores are not required as the unbounded sand flows readily into the internal cavities of the pattern. Also, cope and drag problems are missing. Precision and surface finish are sufficiently good so that many machining and finishing operations can be reduced or eliminated.

6.11.5 Limitations

Wall thickness of commercial castings is limited to between 5 and 915 mm due to gas bubbling and turbulence, which affect the solidification of the metal. Steel castings of less than 0.1% carbon should not be made by this process. Patterns have low strength and the process can be costly for low quantities.

6.11.6 Materials to be Cast by This Method

The process can be used for castings of both ferrous and nonferrous materials but latter is most common. Common metals include aluminium, iron, steel, nickel alloys, copper and stainless steel.

6.11.7 Typical Applications

Typical parts made by using this process include cylinder heads, crankshafts, brake components and manifolds for automobiles, and machine bases.

6.12 Investment Casting

6.12.1 Introduction

This is a very old process and has been performed by dentists and jewellers for a number of years. The process uses expendable pattern to form a cavity for the casting. Here, the pattern is used up and is not removed from the mould as it happens in other casting processes, rather it is replaced by the molten metal used for casting and hence it is also called **lost-pattern casting**. It produces very smooth and highly accurate castings from both ferrous and nonferrous alloys. Only die casting can produce intricate parts other than investment casting. The process is useful in casting unmachinable alloys and radioactive metals.

6.12.2 Other Names of the Process

- **Lost-wax process**, because of the fact that the wax pattern used in the process is melted and evaporated, i.e. lost.
- **Precision casting**, because of the high dimensional accuracy of the cast parts.

6.12.3 Process Details

The pattern is made by injecting molten wax or plastic into metal moulds in the shape of the pattern that are usually gated together so that several parts can be made in one go. The pattern is then dipped many times into a slurry of refractory material, such as very fine silica and binders, including water, ethyl silicate and acids to make a coating of slurry over the pattern. This process is called investment process. When the mould becomes hard, it is heated in an oven, melting the wax and at the same time further drying and hardening the mould. The remaining cavity, having all the intricate details of the original wax form, is then filled with metal. Upon cooling, the refractory material invested is broken away, leaving the casting. Wax patterns require careful handling, because they are delicate and weak. But on the merit side, the pattern material is reusable and hence offers economic advantages.

6.12.4 Advantages

Investment casting is particularly advantageous for small precision parts of intricate design that can be made in multiple moulds. Some of the advantages of this casting process are as follows:

- The process is capable of producing intricate shapes with undercuts.
- A very smooth surface is obtained without a parting line.
- Dimensional accuracy is good.
- Certain unmachinable parts can be cast to preplanned shape.
- It may be used to replace die casting, where short runs are involved.

- It overcomes the withdrawal limitations through the use of pattern that can be removed by melting and vaporization.

6.12.5 Limitations

It is an expensive process and is usually limited to small castings but presents some difficulties when cores are involved. Holes cannot be smaller than 1.6 mm and should be no deeper than about 1.5 times the diameter.

6.12.6 Materials to be Cast by This Method

The process is dominantly used for the casting made of aluminium, copper and steel. It is also performed with stainless steel, nickel, magnesium and the precious metals.

6.12.7 Typical Applications

Typical parts made by using this process are gears, cams, valves and ratchets.

6.13 Centrifugal Casting

Centrifugal casting uses a rotating mould (sand, metal or ceramic mould) operating at high speeds. When the molten metal is poured into the mould, it is thrown against the mould wall due to the centrifugal force, where it remains until it cools and solidifies into some form of hollow product (Fig. 6.6). The castings have a dense metal structure, with all the impurities are forced back to the centre, which can be machined out, thus producing a high-grade casting.

6.13.1 Advantages

The centrifugal casting has the following advantages:

- Greater detail on the surface of the casting is obtained.
- The dense metal structure has *superior mechanical properties.* The mechanical properties of centrifugally cast parts lie between those of static castings and forgings.
- Cores in cylindrical shapes and risers or feed heads are eliminated. This makes the process more *economical* than other methods.

- It can produce a wide variety of diameters, lengths and wall thicknesses. Because of the pressure exerted on the metal, the cast parts have *thinner sections* than that can be obtained by using static casting.
- Due to high purity of the outer skin, centrifugally cast pipes have a high resistance to atmospheric corrosion.

6.13.2 Limitations

- If a metal can be melted, it can be cast centrifugally, but for a few alloys, the heavier elements tend to be separated from the base metal. This separation is known as 'gravity segregation'.
- There are also limitations on the size and shape of centrifugally cast parts.

6.13.3 Metals to be Cast by This Method

All castable alloys.

6.13.4 Typical Applications

Centrifugal casting is used for making cast iron pipes, cylinder liners, gun barrels, pressure vessels, brake drums, gears and flywheels. When used in conjunction with investment casting, centrifugal castings of small and detailed complex shapes can be produced.

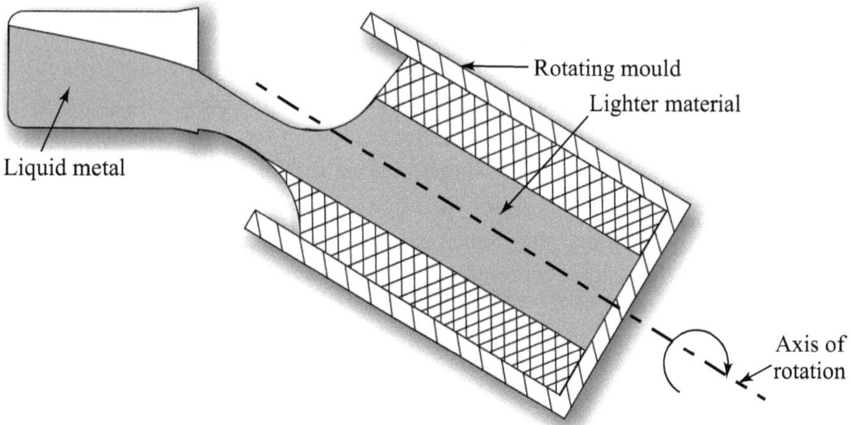

Fig. 6.6 Centrifugal casting process

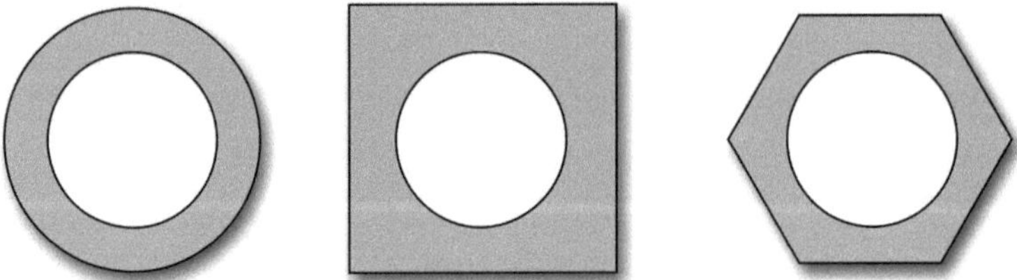

Fig. 6.7 Some cross sections made by centrifugal casting process

Examples include jewellery and dental tooth caps. Figure 6.7 shows some cross sections produced by centrifugal casting.

6.13.5 Types of Centrifugal Casting

There are three types of centrifugal casting:

(a) True centrifugal casting
(b) Semi-centrifugal casting
(c) Centrifuging.

6.14 True Centrifugal Casting

True centrifugal casting is used for pipes, cylinder liners and symmetrical objects by rotating the mould about its horizontal or vertical axis. The metal is held against the wall of the mould by centrifugal force, and no core is required to form a cylindrical cavity on the inside.

In the *horizontal-axis moulds*, thick metal moulds with a thick refractory coating allow the molten metal to solidify faster and the solidification proceeds from the wall of the mould towards the centre of the cast pipe. Such a mould has good solidification, which ensures a more solid casting with any impurities on the inside wall. The wall thickness of the pipe can be controlled by varying the amount of metal poured into the mould. The mould is spinning at the time, the molten metal is introduced, and the spinning action is continued until solidification is complete. A casting produced horizontally is shown in Fig. 6.8.

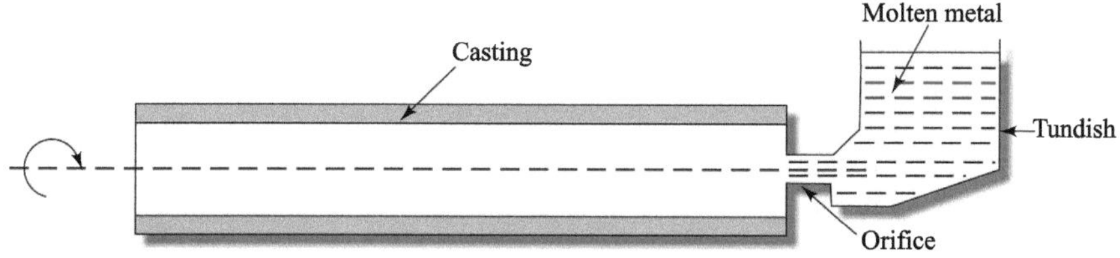

Fig. 6.8 Casting produced by a horizontal-axis mould

Using vertical centrifugal process, paraboloid sections of different sizes can be obtained. Shape and size of the paraboloid will depend upon the rotational speed of the mould. Closer paraboloids are obtained at higher speeds of rotation. To obtain the uniform inside diameter at the top and bottom of a cylinder, rotational speeds are higher in vertical process as compared to horizontal centrifugal process.

6.15 Semi-Centrifugal Casting

In semi-centrifugal casting, the mould is completely full of metal as it is spun about its vertical axis. The central reservoir acts as a riser and cores can be used to increase the complexity of the product (Fig. 6.9).

The centre of the casting is usually solid, but is not as dense, and inclusions and entrapped air are often present. The centre of the casting is removed by machining. Rotational speeds in semi-centrifugal casting are low as compared to true centrifugal one. This method is used to cast parts with rotational symmetry, such as wheel with spokes.

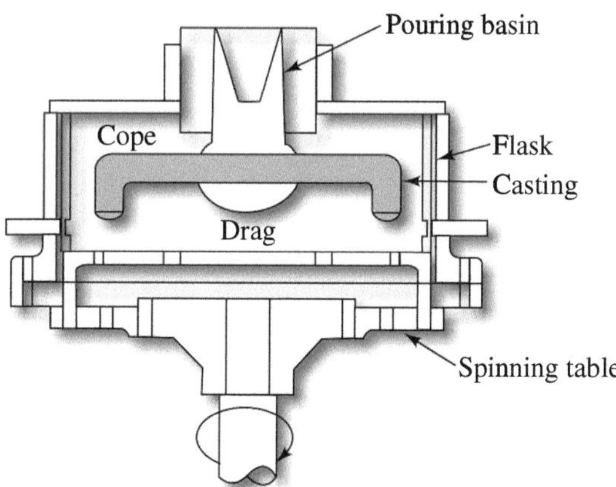

Fig. 6.9 Semi-centrifugal casting process

6.16 Centrifuging

This process is also known as **centrifuge casting**. In this process, several mould cavities are located around the outer portion of a mould, and the metal is fed to these cavities by radial gates from a central pouring reservoir by the action of centrifugal force. Either single or stack moulds (tree structure) can be used. Relatively low rotational speeds are required to produce sound castings with thin walls and intricate shapes. This method is not limited to symmetrical objects but can produce castings of irregular shape such as bearing caps or small brackets. The dental profession uses this process for casting gold inlays. Centrifuging is often used to assist in the pouring of investment casting trees. The centrifused parts have close tolerances, smooth surface and excellent details.

6.17 Electromagnetic Casting

In electromagnetic casting, also called **levitation casting**, an induction coil is used to heat a solid billet. The coil confines the molten metal through magnetic suspension and regularly stirs it. The surfaces of the molten metal are completely exposed and open to the direct impingement of water jets or spray. There is no need for any crucible, which can be a source of contamination with oxide inclusions. The molten metal is then allowed to flow downward into an investment casting mould placed directly below the coil. Investment castings made by this method are free of refractory inclusions and gas porosity. The casting is homogeneous and has fine-grained structure. The exterior surface is quite smooth and the process can be automated and adapted for continuous casting.

6.18 Squeeze Casting

It is also known as **liquid metal forging**, since the process involves casting and forging both. Here, a definite amount of molten metal is poured into the lower part of a die and is allowed to solidify partially. The upper part of the die is then allowed to descend, which applies pressures continuously during the solidification process. The amount of pressure/force as required during the conventional forging process is greatly reduced in squeeze casting as the metal to be given the required shape is still under molten state. Intricate shapes can be produced with reduced pressure/force. Gas and shrinkage porosity are substantially reduced, and mechanical properties are enhanced. This process combines the advantages of casting and forging and can be applied to both ferrous and nonferrous alloys. Typical products made are automotive wheels and mortar bodies (a short-barrelled cannon).

6.19 Continuous Casting

6.19.1 Introduction

The continuouscasting process is used to overcome a number of ingot-related difficulties, such as piping, mould spatter, entrapped slag and structure variation along the length of the product. This process was first developed for continuous casting of nonferrous metal strip, but now it is used for high quality steel production with significant cost reduction. It is a relatively new process, although it was first patented by Sir Henry Bessemer in 1846. It is used to produce blooms, billets, slabs and tubing directly from the molten metal.

6.19.2 Other Names of the Process

Continuous casting is also known as '**Strand casting**' because of the longer continuous length of the casting.

6.19.3 Process Details

In this process, the molten metal is not poured directly into the mould, unlike other moulding methods, rather it flows into a refractory-lined intermediate pouring vessel (tundish), where impurities are skimmed off (Fig. 6.10). From there, the metal travels through a bottomless water-cooled copper mould (in the form of a vertical tube which is open at both ends and has the same cross section as the strand to be cast). The liquid metal begins to solidify over its outer surface as it comes out of the bottom of the mould as a continuous length of metal, which is further cooled by direct water sprays to produce complete solidification.

The cast solid is still hot, which can be bent or fed horizontally by passing it through guide and straightening rolls to make it perfectly straight. Hollow rods or thick-walled tubing are made by placing a graphite core centrally in the mould to a depth below the level at which solidification is complete. The continuously cast metal may be cut into desired lengths (billets and slabs) using an oxy-gas flame cutter or a circular saw. It may also be fed directly into a rolling mill for further reduction in thickness and for shape rolling of products such as channels and I-beams.

6.19.4 Advantages

The continuous casting process has the following advantages:

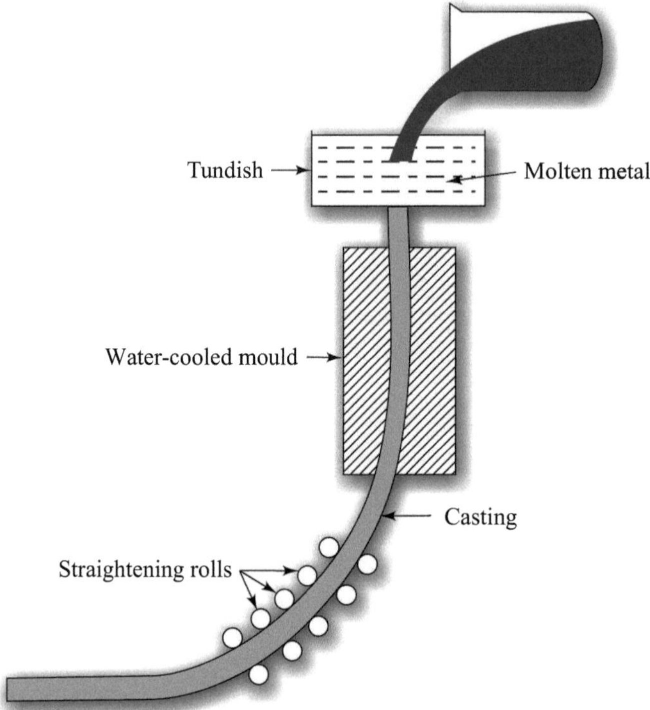

Fig. 6.10 Continuous casting process

- It can produce longer continuous lengths than can be produced otherwise.
- It can substitute some of the hot and cold rolling operations.
- Since each product made by continuous casting is simply a cut-off section of the continuous strand, a single mould is sufficient to produce a large number of pieces. As a result, cost, energy and scrap are all reduced significantly.
- The products have improved surfaces, more uniform chemical composition and there is a little chance of contamination during melting and pouring.

6.19.5 Limitations

The following are the limitations of the continuous casting:

- The high pouring temperature requirement in the preparation of steel offers difficulties in the design of mould.

- The slow solidification rate of steel and the high casting speed make it necessary to have long dies. This increases the chance for bulging of the cast shape due to the deep liquid core.

6.19.6 Materials to be Cast by This Method

This process is applied to copper and copper alloys, aluminium, steel, and grey and alloy-type cast irons.

6.19.7 Typical Applications

Typical parts made by continuous casting are tubes, slabs and gears. These products are obtained by cutting the continuous strand to the required size. Strand can be of rectangular or circular cross sections.

Short Answer Questions

1. How is centrifugal casting different from other casting processes?

Ans. Centrifugal casting uses a rotating mould, whereas other casting processes use a static mould.

2. What are aluminium and magnesium alloys usually not cast by centrifugal casting?

Ans. Aluminium and magnesium alloys have low density, and get segregated during rotation in the rotating mould of the centrifugal casting, thus affecting the composition of the casting produced.

3. Why is investment casting so named? Name a few products made by investment casting.

Ans. The 'word' investment used in investment casting refers to the coating of refractory material on the wax pattern. A few investment cast products include blades of gas turbines, jewellery products and dental fixtures.

4. Why is investment casting not suitable for large size castings?

Ans. The large size wax patterns used for making the large size castings are delicate to handle and can break easily on mishandling. Hence wax patterns are made in small sizes for making small size castings.

5. What is the difference between hot-chamber die casting and cold-chamber die casting?

Ans. The melting furnace is the integral part of the machine in hot-chamber die casting, whereas there is a separate melting furnace in cold-chamber die casting.

6. Which type of metals are usually cast by hot-chamber and cold-chamber die castings?

Ans. Hot-chamber die casting is suitable for low melting point metals such as zinc, tin and lead. On the other hand, cold-chamber die casting is suitable for high melting point metals and alloys such as aluminsium, magnesium and brass.

7. What is meant by 'flash' in die casting?

Ans. Flash indicates the excess metal squeezed out in the space near the parting line or into the clearances around the core due to high pressure of the molten metal used in die casting.

Multiple Choice Questions

1. Which of the following statements is wrong about a shell mould casting process?
   (a) It uses resin which increases the surface finish of the casting obtained.
   (b) It is an improvement over normal sand casting process.
   (c) It is usually suitable for trial production.
   (d) It uses a metal pattern.
2. What does the 'thin shell' mean in the shell mould casting?
   (a) A thin layer of sand mass (b) A thin layer of resin
   (c) A thin layer of metal pattern (d) A thin layer of molten metal
3. Why is magnesium oxide used in plaster mould casting?
   (a) It reduces the setting time of the mould.
   (b) It helps to form a strong slurry.
   (c) It improves the strength of the mould.
   (d) It increases the surface finish of the casting.
4. Consider the following statements about a permanent mould:
   1. It is made of metals.
   2. It is reusable.
   3. It produces precision castings.

Of these statements:

(a) 1 alone is true (b) 1 and 2 are true
(c) 1, 2 and 3 are true (d) 3 alone is true

5. Consider the following statements about a permanent mould casting process:

   1. It is suitable for low melting point materials.
   2. Molten metal flows under the action of simple gravity and no pressure is applied.
   3. Castings require little or no machining.

   Of these statements:

   (a) 1 alone is true (b) 1, 2 and 3 are true
   (b) 2 and 3 are true (d) 1 and 2 are true

6. Why is investment casing so named? Because

   (a) it uses wax pattern
   (b) it uses slurry of refractive material
   (c) the pattern is lost during the process
   (d) it produces intricate parts.

7. Consider the following statements about a disposable pattern:

   1. It produces intricate parts easily.
   2. It requires no draft.
   3. It reduces casting defects.
   4. It requires no machining.

   Of these statements:

   (a) 2 alone is true (d) 1 and 2 are true
   (c) 3 and 4 are true (d) 1, 2, 3 and 4 are true.

8. Investment casting is useful for

   (a) large size castings
   (b) very large size castings
   (c) small size castings
   (d) small size castings having intricate details.

9. Consider the following statements:

1. Centrifugal casting uses a rotating mould.
2. Hot-chamber die casting is useful for high melting point metals.
3. Investment casting is a precision casting process.
4. Slush casting is used for making decorative items.

Of these statements:

(a) 1 and 2 are true (b) 2, 3 and 4 are true

(c) 1, 3 and 4 are true (d) 1, 2, 3 and 4 are true.

10. Match List I with List II and select the correct answer using the codes given below the lists:

List I (Casting process)	List II (Description)
A. Investment casting	1. Rotating mould
B. Cold-chamber die casting	2. Low melting point metals
C. Centrifugal casting	3. wax pattern
D. Hot-chamber die casting	4. High melting point metals

Codes:	A	B	C	D
(a)	3	2	1	4
(b)	3	4	1	2
(c)	1	4	3	2
(d)	1	2	3	4

Answers

1. (c) 2. (a) 3. (a) 4. (c) 5. (b) 6. (b) 8. (d) 9. (c) 10. (b).

Review Questions and Discussions

Q.1. How is shell mould casting different than conventional sand casting?
Q.2. What are the advantages of the shell-mould casting process?
Q.3. Why are the metal moulds expensive?
Q.4. Why are plaster moulds only suitable for metals and alloys having low melting points?

Q.5. What are the advantages of die casting over other casting processes?
Q.6. Why is permanent mould casting sometimes called 'near-net shape' casting?
Q.7. Why is an inert gas used in low-pressure castings?
Q.8. Why are permanent mould castings immediately removed from the mould after solidification?
Q.9. What are the advantages of permanent mould casting?
Q.10. Why is investment casting useful only for small-sized parts?
Q.11. What are the advantages of using disposable patterns in investment casting?
Q.12. Why is investment casting not suitable for large size parts?
Q.13. What materials are used to produce the pattern for investment casting?
Q.14. List those casting processes which do not require any draft.
Q.15. How is full mould casting different from other casting processes? Why is it so named?
Q.16. How does hot-chamber die casting differ from cold-chamber die casting?
Q.17. List the metals and alloys normally cast by hot-chamber and cold-chamber processes?
Q.18. How does the air in the mould cavity escape in the die-casting process?
Q.19. Are risers employed in die casting?
Q.20. How are hollow parts with various cavities made by die casting? Are cores used? If so, how?
Q.21. How is the high-pressure die casting useful for producing the smallest part?
Q.22. Centrifugal castings are of high quality. Why?
Q.23. What is the difference between semi-centrifugal casting and centrifuging?
Q.24. Why is it difficult to make large castings with vertical true centrifugal casting?
Q.25. Name some of the products made by slush casting process?
Q.26. What are the benefits of a continuous casting process?
Q.27. What are the main similarities between the full mould and investment casting processes?
Q.28. What is meant by investment in investment casting process?
Q.29. Why are aluminium and its alloys not cast by hot-chamber die casting?
Q.30. For which type of products, the continuous casting process is useful?

Design Considerations in Casting

7

Draft facilitates the withdrawal of a pattern from sand mould without damaging its edges.

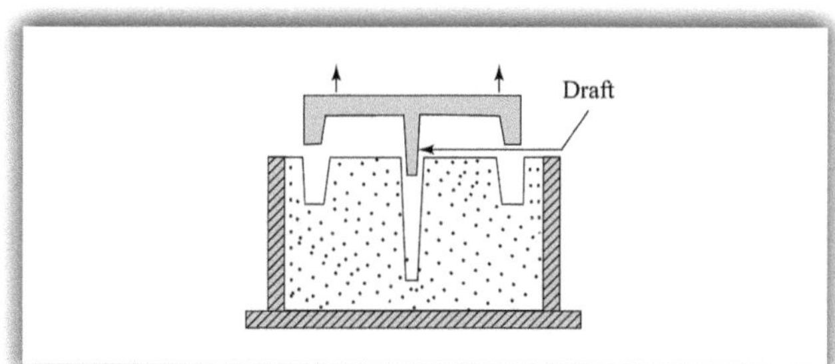

Learning Objectives

After reading this chapter, you will be able to answer some of the following questions:

- Why do gray iron castings do not require shrinkage allowance?
- Why do risers have only cylindrical shape?
- What is the purpose of using draft?

D. K. Singh, *Fundamentals of Manufacturing Engineering*,
https://doi.org/10.1007/978-981-99-8767-2_7

7.1 Design of Section Geometry

Normally a casting should have uniform minimum section thickness everywhere. Table 7.1 gives the recommended minimum section thickness for different materials.

Thinner sections than those listed in Table 7.1 may be cast if they are short in length and are located such that they can be run with hot metal from the ingates. Casting processes differ in the dimension of minimum section with different materials. If large areas of a casting are designed with section thicknesses approaching the limiting ones in Table 7.1, the danger of misrun castings is increased.

There exits a relationship between the minimum section thickness and the length of the section. Thin sections can be obtained without the possibility of having a misrun for a certain limiting distance only. To achieve proper solidification in steel castings, the length of the section should not exceed 4.5 times the minimum thickness. For other metals, length and thickness ratios may be quite different.

In case of nonuniform thickness, change in section should occur gradually. Sharp corners, angles and fillets are the components of bad design and should be definitely avoided because of the following reasons:

- Sharp corners entrap hot liquid metals and are the points of high temperatures as compared to other parts of the casting. Due to differential cooling rates, possibility of occurrence of shrinkage defects is more at such locations.
- These sections are the weakest due to residual thermal stresses and are unsafe.
- They have higher stress concentration leading to cracks in castings.
- Abrupt changes in section prevent directional solidification, making difficult to produce a sound casting.

If fillets are required to be used, then their radii should be properly selected to reduce the stress concentrations and to ensure proper liquid metal flow during the pouring process

Table 7.1 Minimum recommended section thickness

Materials	Section thickness (mm)	Casting processes
Steel	4.76	Sand
Grey iron	3.18	Sand
Malleable iron	3.18	Sand
Aluminium	3.18	Sand
Magnesium	4.76	Sand
Zinc alloys	0.51	Die
Aluminium alloys	1.27	Die
Magnesium alloys	1.27	Die

(Fig. 7.1). Fillet radii usually range from 3 to 25 mm. Fillets of smaller and bigger sizes other than their usual range have limited applications due to reduced rate of cooling. Although use of fillet reduces the stress concentration, it causes the section thickness at the joint to become even larger which increases the possibility of shrinkage defects at that spot (Fig. 7.2). For steel castings, a generally accepted design rule for fillet radii at all types of joints is as follows.

$$\begin{aligned} &\text{For } t < 25\,\text{mm}, && r = t \\ &\text{For } 25 < t < 75\,\text{mm}, && r = 25\,\text{mm} \\ &\text{For } t > 75\,\text{mm}, && r = t/3 \end{aligned}$$

Freezing is delayed at the joint because this is the location of greater mass, i.e., low surface area–volume ratio, and is known as ***hot spot*** which is shown in Fig. 7.2b by a circle larger than the circle at the joint without fillet. These regions could develop shrinkage cavities and porosity as shown in Fig. 7.3(a).

Table 7.2 gives the normal shrinkage allowance for metals commonly sand cast.

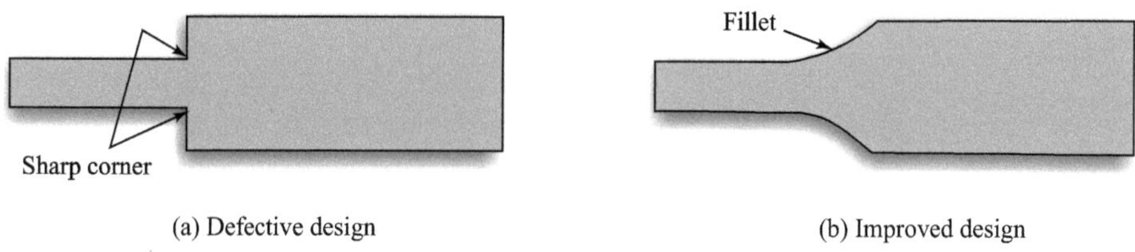

Fig. 7.1 Use of fillet

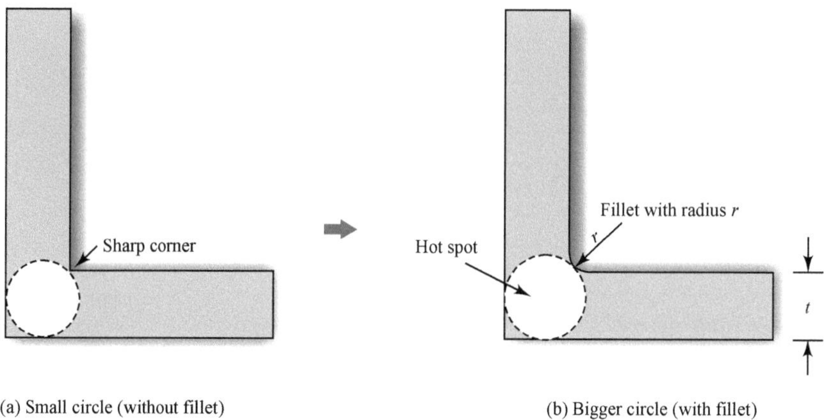

Fig. 7.2 Fillet produces different size circles at the joint

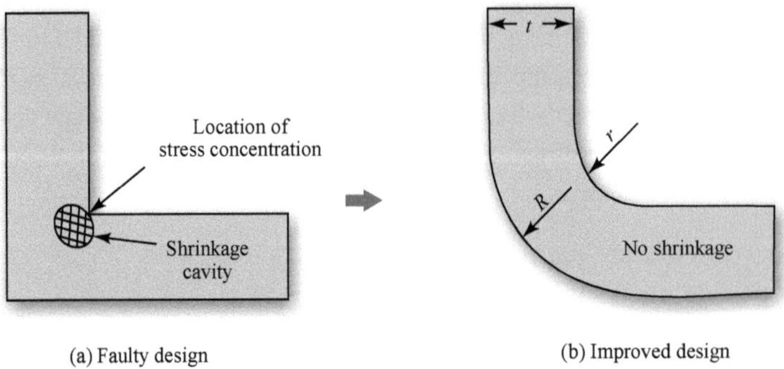

(a) Faulty design (b) Improved design

Fig. 7.3 Shrinkages are eliminated by suitable change in design

Table 7.2 Shrinkage allowances for various materials

Materials	Shrinkage allowance (mm/m)
Cast iron	8.2–10.5
Steel	15.7–20.6
Aluminium	10.5–13.1
Magnesium	10.5 –13.1
Brass	15.7

Because of hot spot problems, it is considered desirable to blend and proportion the sections of castings. Straight uniform sections rather than any change are recommended.

However, by using tapered sections, it may be possible to achieve soundness more readily than with the perfectly uniform sections (Fig. 7.4(b)). Some improvements in the existing design of L and Y sections are shown in Fig. 7.5. Star-marked spots show the locations where modifications are most desired.

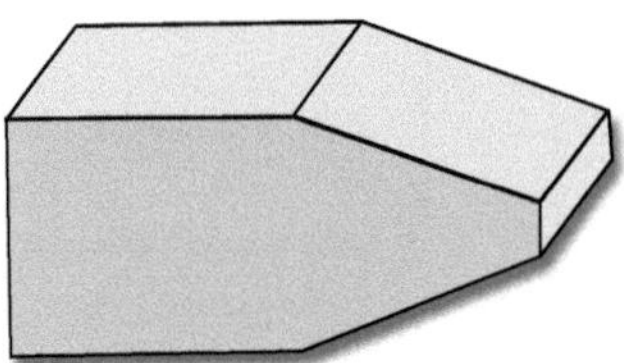

(a) Tapered section is the best option.

(b) Recommended design avoid any change in section.

Fig. 7.4 Recommended change in section

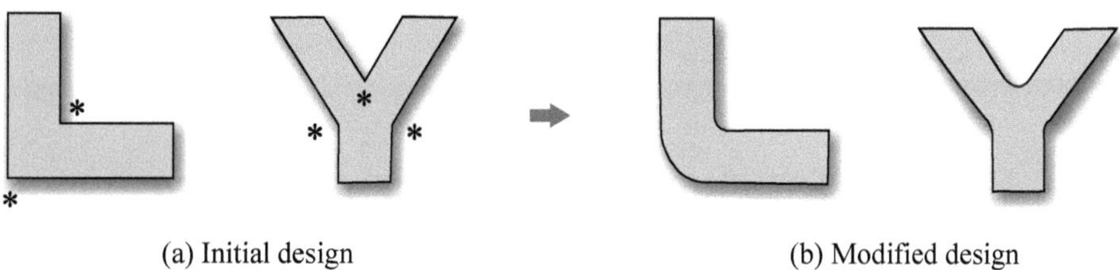

(a) Initial design (b) Modified design

Fig. 7.5 Comparison between initial and modified design

7.2 Design of Gating System

The term gating or gating system refers to all the passageways through which the liquid metal enters a mould cavity. It includes pouring basin, sprue, runner and gates. A good gating design ensures distribution of the metal in the mould cavity at a proper rate without excessive temperature loss, turbulence, and entrapping gases and slags. It also ensures directional solidification in the castings.

The design of a gating system depends on both the metal and mould compositions. Wherever possible, the gating system is made of the same moulding material as that used for the mould cavity. Thus the pouring basins are frequently made of baked sand cores.

There should be minimum turbulence during the flow of metal in the gating system. Many problems such as inclusion of dross or slag, aspiration of air into the metal, erosion of the mould wall and roughening of casting surface may result due to turbulent nature of metal flow. Whether the flow is smooth, i.e., streamlined or turbulent, it is decided by the value of Reynold's number, a nondimensional parameter, which in turn depends upon the velocity of the liquid, the cross section of the flow channel, the density and the viscosity of the liquid. When Reynold's number reaches a certain critical value, turbulent flow prevails. It has been found that steel always flows under turbulent conditions when Reynold's number exceeds 3500.

Flow of metal in the gating system is governed by Bernoulli's theorem. The theorem states that in a flowing liquid stream, without friction, the total energy in the fluid remains constant at any point along its path of flow. The total energy includes potential energy, kinetic energy and pressure energy. For the two points under consideration, the Bernoulli's theorem is mathematically expressed as:

$$\frac{V_1^2}{2g} + \frac{p_1}{W} + h_1 = \frac{V_2^2}{2g} + \frac{p_2}{W} + h_2 = \text{constant} \tag{7.1}$$

where

V_1, V_2 = velocities at the two points

p_1, p_2 = pressures at the two points

h_1, h_2 = height of the two points from a reference level
W = specific weight of liquid

7.2.1 Theoretical Considerations in the Design of Gating System

Gating designs can be mainly classified into three groups, namely vertical gating or parting line gating; bottom gating or side gating, and top gating. In vertical gating system, the liquid metal flows vertically and enters the mould cavity at the parting line. It is the most common and probably the most satisfactory gating system. In bottom gating system, the liquid metal enters the mould cavity at or near the bottom of the mould cavity. It prevents mould erosion and minimizes metal turbulence because metal rises gently in the mould cavity from bottom to top. Vertical and bottom gating systems are shown in Fig. 7.6a and b. By using the principles of Bernoulli's theorem and continuity equation, the following relations have been obtained for the two.

Vertical Gating System [Fig. 7.6(a)]

- Time required to fill up the mould t_m is given as:

$$t_m = \frac{V_m}{A_g \cdot v_g} \tag{7.2}$$

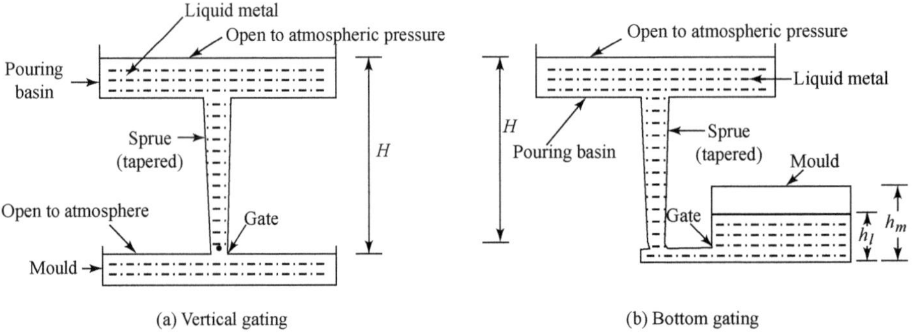

Fig. 7.6 Types of gating systems

where

V_m = volume of the mould
v_g = velocity of the liquid metal at the gate
A_g = cross-sectional area of the gate

Total ingate area in case of multiple-gating system

$$v_g = \sqrt{2gh} \tag{7.3}$$

where

H = height of the liquid in the pouring basin from the gate
g = acceleration due to gravity.

Bottom Gating System [Fig. 7.6(b)**]**

- Velocity of the liquid metal at the gate is given as:

$$v_g = \sqrt{2g(H - h_1)} \tag{7.4}$$

where

$(H - h_1)$ = effective head
h_1 = height of the liquid column in the mould

- Time required to fill up the mould is given as:

$$t_m = \sqrt{\frac{2}{g}}\frac{A_m}{A_g}\left(\sqrt{H} - \sqrt{H - h_m}\right) \tag{7.5}$$

where

A_m = cross-sectional area of the mould
h_m = height of the mould.

In the case of riser being used in the process, the pouring time t_m should also include the time required to fill up the riser.

- Time taken to fill up the riser is given as:

$$t_r = \sqrt{\frac{2}{g}} \frac{A_r}{A_g} \sqrt{H} \tag{7.6}$$

where A_r is the cross-sectional area of riser.

The advantage of top gating system is that all metal enters the casting at the top of the mould cavity. This favours the creation of proper temperature gradients.

7.3 Design of Pouring Basin

A pouring basin may be formed in the moulding sand on top of the cope or it may be made as a dry sand core placed on the cope above the sprue. It is kept filled with the liquid metal during the time of the pouring. The diameters of the cups should be large enough to make it possible to keep the sprue full of metal and to avoid splashing. For active metals such as aluminium and magnesium which react very quickly when exposed to air, it is desirable to have a separate pouring basin made of dry sand core or cast iron.

7.4 Sprue Design

Sprue or downsprue is a vertical channel through which the molten metal flows downward in the mould. It connects the pouring basin to the runner. Sprue size is often selected so that it controls the pouring rate, because the major restriction to flow in the gating system occurs in the sprue.

The application of Bernoulli's theorem and equation of continuity helps to design the actual size of the sprue. Assuming that the pressure at the top of the sprue is equal to the pressure at the bottom and neglecting frictional losses, at any point in the sprue, the relationship between height and cross-sectional area, is given by the following parabolic equation:

$$\frac{A_1}{A_2} = \sqrt{\frac{h_2}{h_1}} \tag{7.7}$$

where A_1 and A_2 are the cross-sectional areas of the liquid stream at the top and bottom of the sprue, respectively, and h_1 and h_2 are the elevations of the two points at the top and bottom of the sprue, respectively, above a certain reference plane.

According to the above equation, on moving downward from the top, the cross-sectional area of the sprue must decrease. Hence, the sprue is tapered downward. This is similar to a situation like a free-falling liquid such as water from a faucet, the cross-sectional area of the stream decreases as it gains velocity downward. By tapering the sprue to proper proportions, the metal does not pull away from the mould as it accelerates downward. Tapering of sprue reduces the formation of vertex on the metal. It also reduces mould erosion and metal turbulence (Fig. 7.6). In case of sprue with constant cross section, the liquid may lose contact with the sprue walls at some places and air will be sucked in or entrapped in the liquid producing porous casting. This is known as ***aspiration effect***. On the other hand, in many systems tapered sprues are now replaced by straight-sided sprues with a choke to allow the metal to flow smoothly.

Sprue may be circular, rectangular or square in cross section. Sprue sizes usually vary from 10 mm^2 for work below 12 kg poured weight to about 50 mm^2 for heavy castings. Sprues larger than 50 mm^2 are seldom used.

7.5 Design of Runners and Gates

Multiple gates and runners are often used to introduce metal to more than one point of a large casting. The runner may be positioned around the casting periphery so as to provide ingates at a number of points. The runner should be streamlined; i.e., it should not have sharp corners or changes of section; otherwise, it will lead to turbulence or gas entrapment. In order to obtain a flow of approximately equal volume through each ingate, the path area of the runner is reduced while moving away from the sprue after each successive ingate by an amount equal to the ingate area. The runner is proportioned in such a way that constant velocity and pressure are maintained in each ingate. The optimum ratio of cross-sectional areas of sprue to runners to total ingates is 1:2:2 to maintain a more uniform distribution of metal in the feeding and hence more constant velocity and pressure conditions.

The gates nearer the sprue will have less metal flowing through them because of higher velocities and lower pressures. In order to satisfy the demands imposed by Bernoulli's theorem, it is necessary that those gates farthest from the sprue be of the smallest cross section so that the volume of metal through these gates is the same as that through those closer to the sprue. If such proportioning is not done, there is a tendency for the pressure to be maximum at the farthest gate and, therefore, for the flow to be greatest at that point.

7.6 Design of Parting Plane

Location of parting plane is particularly useful where segmented or separable moulds are used. In general, it is desirable for the parting line to be along a flat plane, rather than contoured. Whenever possible, the parting line should be at the corners or edges of castings, rather than on flat surfaces in the middle of the casting. If the pattern contains surfaces that are perpendicular to the parting line (parallel to the direction of pattern withdrawal), the friction between the pattern and the mould, or any horizontal movement of the pattern during extraction, would tend the damage the mould. This damage would be particularly severe at the corners where the mould cavity intersects the parting surface. The location of parting plane ensures optimum weight for a casting and makes effective use of the gating system. It also affects the core requirement and method of moulding.

Figure 7.7(b) shows how certain casting defects can be removed if the parting line is oriented along a flat plane, rather than contoure.

Fig. 7.7 Use of straight parting line eliminates possible defects in castings

7.7 Design of Riser

The primary function of a riser is to feed the molten metal to the casting as it solidifies to compensate for the solidification shrinkage. Grey cast iron needs very little feeding since it expands during solidification. Steel, white cast iron and many nonferrous alloys require more feeding because of their higher shrinkages during solidification. The risering of a casting consists of determining locations and sizes of risers. Casting can be subdivided into several sections, and each section can be provided with a riser. A riser should be located close to the heaviest section of simple shape.

It is designed in such a way that it stays molten longer than the casting. The geometry of the casting and of the riser are related, based principally upon Chvorinov's rule. The rule states that the solidification time t is directly proportional to square of the volume of casting and inversely proportional to square of its surface area.

$$t = C_m \left(\frac{V}{A} \right)^2 \tag{7.8}$$

where

V = volume of the casting
A = surface area of the casting
C_m = mould constant.

The mould constant depends on mould material, cast material and pouring temperature.

Thus as long as the ratio shown in Eq. (7.8) for the riser is greater than that for the mould, the riser should feed the casting. The riser takes 25% longer solidification time than the casting, and hence we have

$$t_{\text{riser}} = 1.25 t_{\text{casting}} \tag{7.9}$$

Since both riser and mould receive the same metal and are in the same mould, the mould constant will remain same for both regions.

Using this concept, Eq. (7.10), which determines the criteria for a sound casting, was developed by Caine for steel casting. It expresses the relative freezing time of riser and casting in terms of the relative volume of the riser and casting.

$$X = \frac{0.12}{Y - 0.05} + 1.0 \tag{7.10}$$

where

X = freezing ratio or relative freezing time = $\frac{\text{(Surface area of casting/Volume of casting)}}{\text{(Surface area of riser/Volume of riser)}}$
Y = volume ratio or relative volumes of riser and casting = $\frac{V_r}{V_c}$
V_r = volume of the riser
V_c = volume of the casting

The actual volume of metal needed by most castings for the purpose of preventing shrinkage is relatively small (about 3% for steel). However, the riser as a thermal body must have sufficient heat content and must remain molten longer than the casting. Accordingly, Eq. (7.10) as applied to steel gets modified and takes the form as

$$X = \frac{0.10}{Y - 0.03} + 1.0 \tag{7.11}$$

For an aluminium casting, the above equation is modified to

$$X = \frac{0.10}{Y - 0.06} + 1.08 \tag{7.12}$$

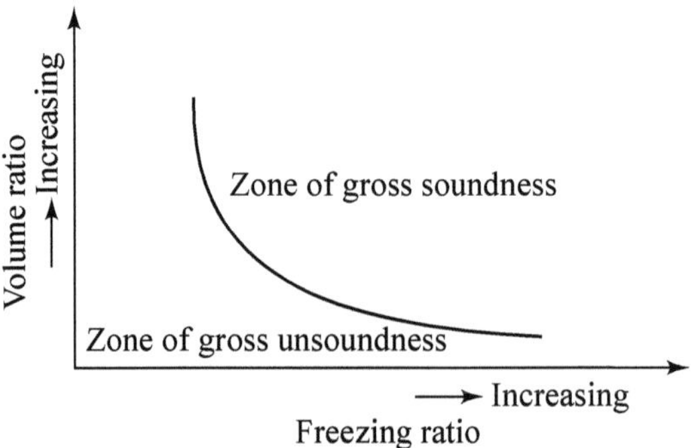

Fig. 7.8 Criteria for a sound aluminium casting

The above equation is plotted in Fig. 7.8.

The figure clearly shows that shrinkages can be avoided by proper use of freezing and volume ratio.

7.7.1 Shrinkage Characteristics of Grey Cast Iron and Riser

The solidification shrinkage (ΔV) is defined by the following relation:

$$\Delta V(\%) = 2 \times (\% \text{ graphitic carbon} - 2.80\%) \tag{7.13}$$

ΔV can be positive or negative. Because of presence of graphitic carbon in the above equation, irons undergoing fast graphitization will have little or no solidification shrinkage.

In grey cast iron, carbon instead of forming cementite precipitates as graphite directly from the melt at the eutectic temperature. The average carbon content in its molten state is around 3.25%. Hence grey cast iron will have virtually no shrinkage or may even expand on solidification. As a result, for ordinary grey irons little or no risering is needed for the purpose of feeding solidification shrinkage. Other cast irons such as white iron have zero percent graphitic carbon and hence have considerable shrinkage, about 5.6%, according to Eq. (7.13).

7.7.2 Riser Shape

Chvorinov's rule is used to decide the shape of the riser. According to this rule, the shape with a long freezing time is preferred for the riser. In other words, the lower the surface area per unit volume, the lower is the solidification rate. Therefore, a riser should be designed to give the least surface area per unit volume. Such a shape would be a sphere. But spherical shapes in most cases would be difficult to mould or to use as risers, because it presents considerable difficulty. Hence, the most popular shape for a riser is a cylinder, where the height–diameter ratio is varied depending upon the nature of the alloy, location of the riser, the size of the flask and other variables. For a cylinder of diameter D and height H, the volume, V and surface area, A can be written as

$$V = \frac{\pi}{4} D^2 H \tag{7.14}$$

and

$$A = \pi DH + 2\left(\frac{\pi}{4} D^2\right) \tag{7.15}$$

7.7.3 Riser Size

The diameter of a conventional riser (top riser) must be larger than that of the casting since it would otherwise solidify before the casting. The height of the riser should be about 1.5 times the diameter with one-third of it in the drag and two-thirds in the cope.

7.7.4 Positioning of Risers

Risers may be attached to the tops or sides of castings, depending upon the sections to be fed and the methods used for removing the risers from the castings. To achieve proper solidification in steel castings, the feeding distance of the riser should not exceed 4.5 times the minimum thickness of the casting. An effective metal chill located at the endmost distance from the riser has been found to add about 50 mm to the riser feeding distance. The feeding distance plays an important role in the proper positioning of riser for making castings to be shrinkage free.

7.8 Draft Consideration in Design

If the surfaces of the pattern parallel to the direction of its withdrawal are slightly tapered, then it can be easily removed from the mould. This tapering of the sides of the pattern is known as draft. It provides a slight clearance for the pattern as it is lifted from the sand mould and prevents its damage (Fig. 7.9).

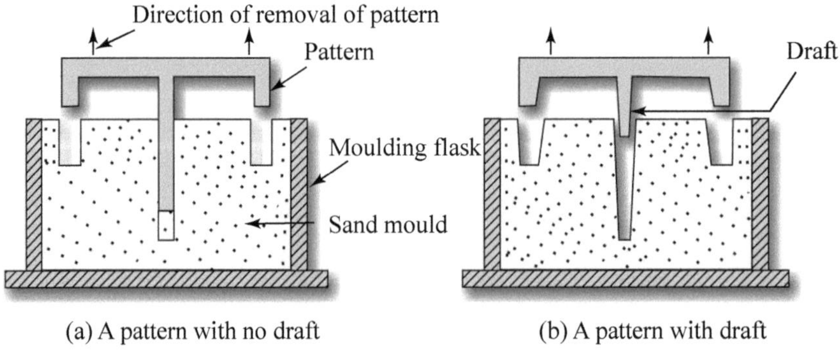

Fig. 7.9 Draft facilitates smooth withdrawal of pattern

Depending on the quality of the pattern, draft angles usually range from 0.5° to 2°. Inside surfaces are provided with more draft (usually twice) as compared to outer surfaces because the casting shrinks inward towards the core. Taper is extremely useful in removing metal damages due to centre-line shrinkage.

The centre-line shrinkage mainly occurs in plate sections in the centre due to columnar solidification of metal. It is actually a shrinkage defect which occurs most frequently in steel sections under 100 mm thickness.

7.9 Design of Core and Cored Hole

When a casting is required to have a cavity or recess, a core which is nothing but a sand mass is introduced into the mould (Fig. 7.10). It is removed from the finished part during shakeout and further processing. Internal or external surfaces of a casting can be formed by a core. The core is formed by being rammed into a core box or by the use of sweps.

A core should be sufficiently strong enough to support itself. Porosity or permeability is an important consideration in making cores. As the hot metal surrounds the core, gases are generated by the heat in contact with binding material. Provision is required to vent these gases. The core must have a smooth surface to ensure a smooth casting. It requires refractoriness to resist the action of the heat until the hot metal has stabilized.

For a casting requiring certain recess, the core needs to be supported properly to provide strength. To keep the core from shifting, chaplets are used. Chaplets are basically small metal shapes made up of low melting point alloys. They are placed between mould walls and cores or between the cores themselves. As the molten metal fills the mould cavity, the chaplets are surrounded by the metal and later should fuse into it. Most common type of chaplet is the double-headed chaplet (Fig. 7.11). The use of chaplets should be limited because of the difficulty in fusing the chaplet with the metal of casing.

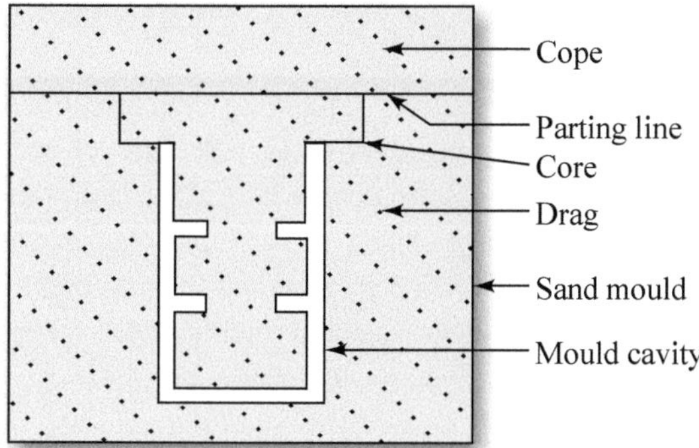

Fig. 7.10 Use of a core

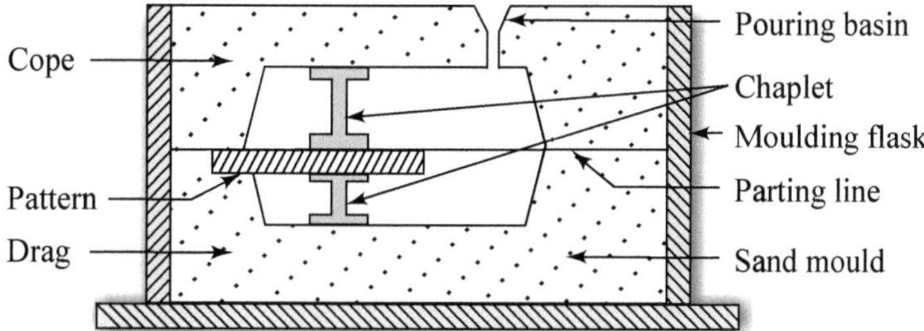

Fig. 7.11 Use of a chaplet

It becomes an integral part of the finished casting and may cause defects or be a location of weakness. Chaplets should be of the same, or at least comparable, composition as the casting material. The use of core should be minimized which ultimately helps in reducing the cost of a casting. The reduction of coring substantially reduces the labour of making the mould for a large casting. By suitable design of parting line in the mould structure, cores can even be eliminated. Cleaning problems are also generally reduced with a decrease in coring, because internal cores, small, long passages, and thin cores relative to the sections surrounding them often pose difficulties in cleaning.

It is always difficult to decide to have cored or drilled holes. It is more economical to core larger holes, when appreciable metal can be saved in the casting and faster machining results. In case of small size hole, the hole is first cored by using casting and then a fast core-drilling operation is followed. It results in less metal removal. The minimum size of cored holes which can be cast depends greatly on the accuracy of core location and tolerance required. If a cored hole is required to be located with extreme accuracy with respect to other surfaces, it is often desirable to drill it rather than core it. In the case of die castings, metal cores may be quite accurately located.

7.10 Machining Allowance Consideration in Casting

Some surfaces of the casting may require machining. Machining allowance is the extra material left on a casting making it oversize to take care of machining. Machining allowances, which are included in pattern dimensions, depend on the type of casting and increase with the size and section thickness of castings. The usual value of machining allowance may be 2–5 mm for small size castings but may be higher for larger castings. Table 7.3 shows the minimum amount of metal normally added for machining aluminium and steel castings. Table 7.3 clearly shows that minimum machining is required if permanent moulds are used with shell cores rather than normal sand casting processes. In general, machining allowance may be a minimum if the surfaces to be machined are entirely in the drag half of the mould, because of minimum possibility of dimensional variation and other defects.

Table 7.3 Minimum recommended machining allowances

Dimension of casting (mm)	Allowances (mm) for sand casting	
	Aluminium	Steel
Up to 150	1.5	4.8
150–300	2.3	4.8
300–450	3.0	4.8
450–600	3.8	6.3
Above 600	According to requirement	

7.11 Design for Dimensional Tolerances

Tolerances represent the permissible variations in the dimensions of a part. It depends upon the size of the casting, the casting process adopted and the type of pattern used. Tolerances are important not only for proper functioning of products, but they also have a major economic impact on manufacturing costs. The smaller we make the tolerance, the higher the production costs. Certain surfaces are needed to be machined. These require a machining allowance as well as allowances for shrinkage. Sand mould castings need more dimensional tolerances because of factors such as changes in the dimensions of the mould cavity due to temperature effect, changes in the mould properties, misalignment of the two mould halves in the flask and others. Die or permanent mould castings have tight tolerances because most of the demerits encountered in sand mould casting are eliminated and the castings produced have better surface finish and good dimensional accuracy. Table 7.4 gives values of tolerances for small- and medium-sized sand castings. Large castings may require different values.

Table 7.4 Tolerances for sand castings

Materials	Tolerance (mm/m)
Aluminium alloys	2.6
Beryllium-copper alloys	5.2
Cast irons	3.9
Copper alloys	7.8
Cast steels	5.2
Magnesium alloys	2.6
Malleable iron	2.6

7.12 Economic Aspects of Casting

The cost of casting includes the cost for raw materials used, time consumed during processing and the effort required to convert the material into the finished form.

Sand casting is relatively cheaper as compared to other casting processes. Processes using dies are expensive because die materials are costlier and involve a great deal of machining and preparation. Melting of the material is itself a costly affair because it requires a particular type of furnace. Finally, costs are involved in heat treating, cleaning and inspecting the castings.

It is important to consider several alternatives such as multiple-cavity casting, fabrication methods, the use of cores, subsequent machining operations and weight reduction while considering the economic aspects of the casting.

Multiple-cavity casting ensures substantial savings which can be realized by incorporating more than one finished part into a single casting. Moulding and handling costs are greatly reduced. They are machined as a single unit and then cut into separate pieces. Combination of casting and welding can also be used to gain economy.

Short Answer Questions

1. Why are the sharp corners in a casting avoided?

Ans. The sharp corners obstruct the smooth flow of molten metal during pouring and tend to entrap it, and are also the points of high temperatures as compared to other parts of casting. It results in differential cooling rate of the casting which increases the possibility of formation of undesirable shrinkage defects at these locations.

2. Why are fillets used in casting?

Ans. Fillets are used to reduce the locations of stress concentrations and allow the smooth flow of molten metal during pouring process.

3. What changes in design are needed to minimize the formation of hot spots in a casting?

Ans. Sharp corners in a casting are designed with smooth curves and are tapered as far as possible. At the same time, uniform sections rather than any abrupt change in sections are recommended.

4. Why is a sprue tapered in the downward direction?

Ans. A tapered sprue reduces mould erosion and tends to prevent metal turbulence which leads to the formation of vortex, by allowing the molten metal to flow smoothly along its surface without getting it separated to fall freely.

5. What is draft? Why is draft provided in a pattern?

Ans. Draft is one of the allowances provided in a wooden pattern which indicates the provision of tapering on the surface of pattern parallel to its withdrawal from the mould. It facilitates easy removal of the pattern from the mould without damaging it.

Multiple Choice Questions

1. Use of fillet

   (a) increases stress concentration
   (b) decreases stress concentration
   (c) improves design of casting
   (d) reduces the cost of casting.

2. Which of the following metals requires the highest shrinkage allowance?

   (a) Aluminium (b) Steel
   (c) Magnesium (d) Cast iron.

3. The Bernoulli's equation is useful in the design of

   1. Sprue
   2. Runner
   3. Moulding flask
   4. Riser

Of these:

(a) 1 alone is true (b) 1 and 2 are true

(c) 2 and 3 are true (d) 1, 2 and 3 are true.

4. The optimum ratio of cross-sectional areas of sprue to runners to total ingates for uniform molten metal feeding is

(a) 1 : 2 : 3 (c) 2 : 1 : 3

(b) 1 : 2 : 2 (d) 3 : 2 : 1

5. Solidification time varies

1. inversely proportional to the surface area of the casting
2. inversely proportional to the square of surface area of the casting
3. directly proportional to the square of volume of casting
4. directly proportional to the volume of casting

Of these:

(a) 1 alone is true (b) 2 and 4 are true

(c) 2 and 3 are true (d) 1 and 4 are true.

6. Molten metal in riser has more solidification time than casting. Approximately it is

(a) 10% (c) 30%

(b) 25% (d) 50%.

7. Caine's formula is

(a) used to find freezing time of riser
(b) connected with the design of sprue
(c) used to find shrinkage allowance
(d) used to find total volume requirement of molten metal.

8. The most feasible shape of riser is

(a) Conical (b) Spherical

(c) Cylindrical (d) All of the above.

9. The usual range of draft is

(a) 5°–10° (c) 0.5°–2°

(b) 10°–20° (d) 2°–5°.

Answers

1. (b) 2. (b) 3. (b) 4. (b) 5. (c) 6. (b)
7. (a) 8. (c) 9. (c).

Review Questions and Discussions

Q.1. Why are sharp sections undesirable?
Q.2. State the functions of a fillet.
Q.3. How is Bernoulli's theorem useful in the design of a riser?
Q.4. Vertical gating system is the best. Why?
Q.5. Differentiate between vertical gating and horizontal gating.
Q.6. What are the elements of a gating system?
Q.7. How is top gating different from bottom gating?
Q.8. Why is a sprue tapered?
Q.9. What is the purpose of a parting line?
Q.10. Why are risers usually cylindrical? Why is spherical riser not feasible?
Q.11. Why are pattern allowances kept as low as possible?
Q.12. Why is shrinkage allowance not required for grey iron castings?

Introduction to Joining Processes 8

Incomplete fusion and penetration are the result of incorrect weld parameters.

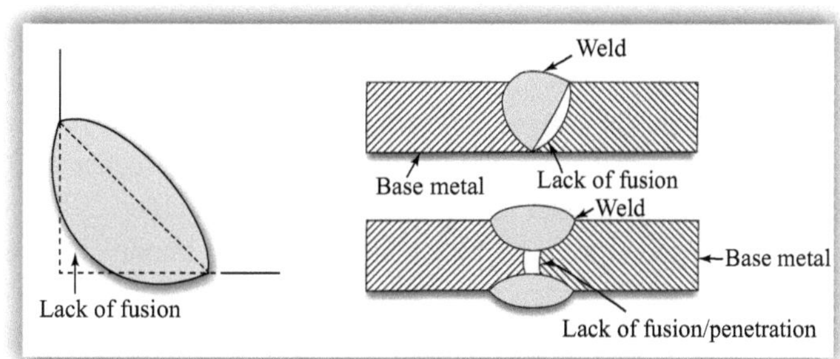

Learning Objectives
After reading this chapter, you will be able to answer some of the following questions:

- How does fusion welding differ from plastic welding?
- Why is oxidation harmful in welding?
- How are welding defects prevented from occuring?

D. K. Singh, *Fundamentals of Manufacturing Engineering*,
https://doi.org/10.1007/978-981-99-8767-2_8

8.1 Introduction

Manufacturing of the part as a single unit is sometimes impossible since the part may be made up of so many intricate shapes. In such situations, different methods are used by the manufacturing systems to join the parts to obtain the complete unit. These parts may be joined together permanently or in such a way that they can be subsequently taken apart again for maintenance or repair purposes. Joining processes include welding, brazing, soldering, adhesive bonding and mechanical fastening. The choice of a particular joining process depends on several important factors such as its application, joint design, the materials involved, and the thickness, size and shape of the components to be joined. Welding and brazing are the most popular techniques to join metals. Adhesive bonding is commonly used to join plastics. Soldering finds extensive applications in electronic manufacturing.

8.2 Welding Process

Welding is a process by which two materials, usually metals, are permanently joined together by means of heat and/or pressure. The coalescence is obtained when atoms of the two parts to be joined come very close to each other so that there is only separation boundary between them. Surface cleaning is very important in welding, because it will lead to oxidation which has detrimental effect. In solid-state welding, contaminated layers are removed by mechanical or chemical cleaning prior to welding. In fusion welding, where a pool of molten metal exists, the contaminants are removed by the use of fluxing agents. Many welding processes are performed in a controlled environment or shielded by an inert atmosphere.

8.3 Classification of Welding Processes

Many welding processes have been developed. They differ not only in the way that heat and pressure are applied but also in the fashion that how they are prevented from possible oxidation or contamination during the welding process. One of the important methods of classification is based on the composition of the joints. On this basis, the joining processes are classified under the following headings:

- Autogeneous joining
- Homogeneous joining
- Heterogeneous joining

In autogeneous joining process, no filler material is used. All types of solid-phase welding and resistance welding come under this category. In homogeneous and heterogeneous methods, filler materials are necessarily used. The composition of the filler material used in homogeneous joining is similar to that of parent materials. Heterogeneous joining uses filler material whose composition is different than that of original parts to be joined. Typical examples of homogeneous joining include arc, gas and thermit welding. Soldering and brazing processes come under heterogeneous class.

Another method of classification is based on the application of pressure. Accordingly, the processes are divided as follows:

- Plastic welding
- Fusion welding

In plastic welding, the metals to be joined are heated to their plastic state and then forged together by applying external pressure. The temperature to which the metals are heated is sufficient for cohesion to take place. This is usually a subfusion temperature. Solid-state weldings fall under this category.

In fusion welding, the metal at the joint is heated to its molten state and then allowed to solidify without applying any pressure. The molten metal flows and fills up the gap between the parts. It is very much similar to a casting process. The process may be either autogeneous or homogeneous. Typical examples include gas welding and arc welding.

8.4 Weldability

Weldability of a material is defined as its ability to be welded. Weldability of a material depends on various factors such as metallurgical changes that occur during the process of welding, its hardness, gas evolution and absorption, extent of oxidation and others. Low-carbon steels ($C \leq 0.12\%$) have the best weldability among metals. As the carbon content increases, weldability decreases.

8.5 Selection Criteria in Welding

Welding has become a dominant joining process in manufacturing, and there is hardly any material which cannot be welded. There are a number of welding processes, but all materials cannot be welded by every process. Therefore, selection of a particular welding process is important. For example, resistance spot welding is extensively used in sheet metal work in automobile industry, electron beam welding is employed for reactive metals and shielded metal arc welding is used for difficult-to-reach sections or for field welding.

The choice of a particular welding process depends on several factors such as its application (automobile industry, aircraft industry, shipbuilding, pressure vessel fabrication, etc.), joint design, materials involved, thickness, size, and shape of the components to be joined. Other considerations are location of the joint (easy accessibility or awkwardly located joints), welding position and economic considerations (equipment and labour cost). Selection of the welding processes based on above-listed factors is given in different Tables 8.1, 8.2, 8.3, 8.4, 8.5, 8.6, 8.7 and 8.8.

Table 8.1 Selection of welding processes on the basis of materials being joined

Materials to be welded	Recommended joining processes
Plain carbon steel and low-alloy steel	Arc welding, Oxyacetylene welding, Resistance welding and Brazing (best results); EBW and Adhesive bonding (commonly performed)
Cast iron	Oxyacetylene welding (best results); Arc welding and Adhesive bonding (commonly performed)
Stainless steel	Arc welding, Resistance welding and Brazing (best result); Oxyacetylene welding, EBW, Soldering and Adhesive bonding (commonly performed)
Aluminium and magnesium	Adhesive bonding (best results); Arc welding, Oxyacetylene welding, EBW, Resistance welding and Brazing (commonly performed)
Copper and copper alloys	Brazing and Soldering (best result); Arc welding, Oxyacetylene welding EBW, Resistance welding and Adhesive bonding (commonly performed)
Thermoplastics and thermosets	Adhesive bonding (commonly performed)
Ceramics	Adhesive bonding (best result); EBW (commonly performed)
Dissimilar metals	Soldering and Adhesive bonding (best result); EBW (commonly performed)

Table 8.2 Selection of welding processes on the basis of material thickness

Material thickness	Recommended joining processes
Very thin category (≤ 3 mm)	Resistance welding, OFW, GMAW, GTAW, FCAW, LBW, Ultrasonic welding and EBW (low power)
Thin category (3–6 mm) and thick category (6–20 mm)	GMAW, SAW, FCAW, LBW low (high power) and EBW (medium power); Multirun weld can be used if necessary
High thick category (20–75 mm) and very high thick category (> 75 mm)	SAW, ESW, EBW (high power) and thermit welding

Table 8.3 Selection of welding processes on the basis of rate of heat input

Rate of heat input	Recommended joining processes
Low	OFW, ESW and Resistance welding (flash)
Moderate	SMAW, FCAW, GMAW, SAW and GTAW
High	PAW, EBW, LBW and Resistance welding (spot, seam and percussion)

Table 8.4 Selection of welding processes on the basis of types of joints

Types of joints	Recommended joining processes
Lap joint	OFW (acetylene), SMAW, GTAW, PAW, GMAW, FCAW, Resistance welding (spot and seam) and EBW
Butt joint (bars)	Resistance welding (butt and flash), Friction welding, EBW, Thermit welding and Diffusion bonding
Butt joint (tubes)	SMAW, GTAW, PAW, GMAW, FCAW, Resistance welding (butt and flash), Friction welding, EBW and Diffusion bonding
T-joint or fillet	SMAW, SAW, GTAW, GMAW, FCAW, ESW, EGW, Resistance welding (projection), EBW, Thermit welding and Diffusion bonding

Table 8.5 Selection of welding processes on the basis of welding positions

Welding positions	Recommended joining processes
Flat or downhand	SAW, GMAW and FCAW
Horizontal (fillet)	SAW, GMAW and FCAW
Horizontal	GMAW and FCAW
Vertical	GMAW, FCAW and ESW
Overhead	GMAW and FCAW

Table 8.6 Selection of welding processes on the basis of deposition rate

Deposition rate	Recommended joining processes
Low	SMAW and GTAW
Medium	GMAW
High	GMAW and FCAW
Very high	SAW and ESW (both using iron powder-coated electrodes)

Table 8.7 Selection of welding processes on the basis of economic consideration

Economic consideration	Recommended joining processes
Low cost	SMAW, ESW GTAW, GMAW, Stud welding, Resistance welding, OFW and Thermit welding
Medium cost	SAW, ESW and Diffusion bonding
High cost	EBW, LBW and Friction welding

Table 8.8 Selection of welding processes on the basis of shape of job

Shape of job	Recommended joining processes
Sheet and plate	OFW (acetylene), SMAW, SAW, GTAW, PAW, GMAW, FCAW, ESW, EGW, Resistance welding (spot and seam) and EBW
Large pipe and cylinder	OFW (acetylene), SMAW, SAW, GTAW, PAW, GMAW, FCAW, Resistance welding (spot and seam) and EBW

8.6 Weld Inspection

The principal objective of weld inspection is to assure the high quality of welded structures through the careful examination of the components at each stage of fabrication. The welds are inspected and tested to locate the presence of defects and flaws. The quality of a welded joint depends on the performance of welding equipment, welding procedure adopted and the skill of the operator.

Weld inspection methods are broadly classified into two groups:

- Destructive testing
- Nondestructive testing (NDT)

8.6.1 Destructive Testing

Destructive tests are mostly mechanical tests and are used for the testing of soundness, strength and toughness of the weld material. As the name suggests, the components after the test can no longer be reused. The following are the important destructive methods:

- Tension test
- Tension-shear test
- Bend test
- Impact test
- Hardness test
- Fatigue test
- Creep test
- Corrosion test
- Nick break test
- Slug and shear test.

8.6.2 Tension Test

Tension tests are conducted to determine the ultimate strength, the yield strength and the ductility (in terms of percentage elongation and percentage reduction of area) of base metal, weld metal and the welded joint. The weld specimen is subjected to the testing machine, and stress–strain curve is plotted, which will give the above-listed parameters. The test may be conducted for both longitudinal and transverse directions.

8.6.3 Tension-Shear Test

In the tension-shear test, the test specimen is subjected to tension and the shear properties of the fillet welds are evaluated. Longitudinal fillet-weld shear test measures the strength of the fillet weld, when the specimen is loaded parallel to the axis of the weld (Fig. 8.1a). During transverse shear test, double-strap fillet joint specimens are used because they are more symmetrical and therefore the stress state under load better approaches pure shear (Fig. 8.1b).

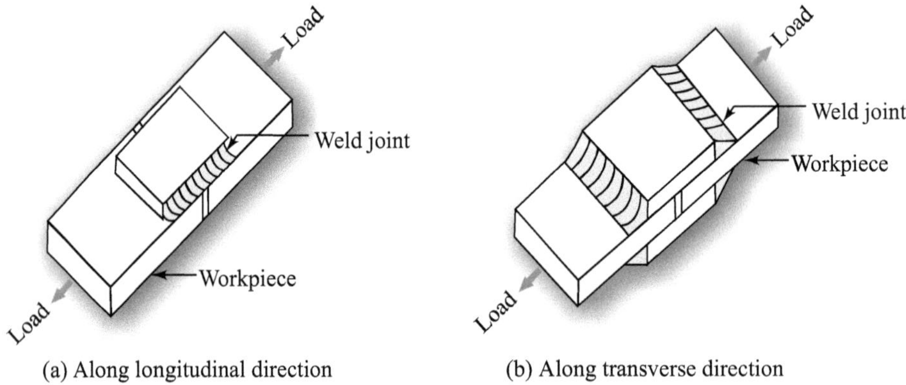

Fig. 8.1 A tension-shear test

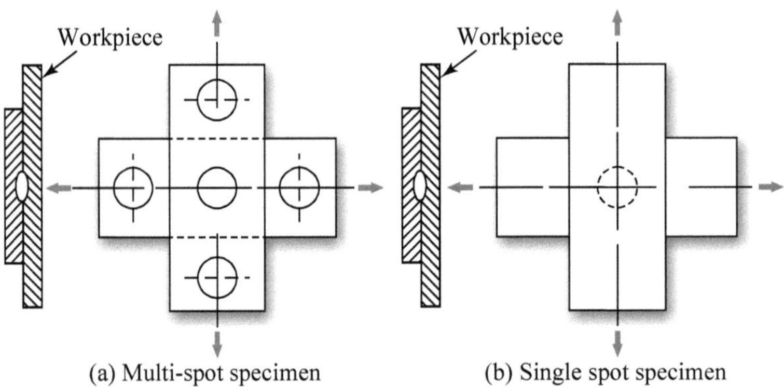

Fig. 8.2 Test specimens in a cross-tension test

8.6.4 Tension Tests for Resistance Welds

The tension-shear test is the most widely used method for determining the strength of resistance spot welds. It is commonly used in fabricating facilities because it is an easy and inexpensive test to perform.

The *cross-tension test* is a tension test, where the test specimens are placed as shown in Fig. 8.2. The test is useful for metal sheet thickness upto about 5 mm and can reveal flaws, cracks and porosity in the weld zone.

The ***slug and shear test*** is used when the sheet thickness is less than 1 mm.

The ***peel test*** is commonly used as a production control test for thin sheets. After bending and peeling the joint, the shape and size of the torn-out weld nugget are observed (Fig. 8.3).

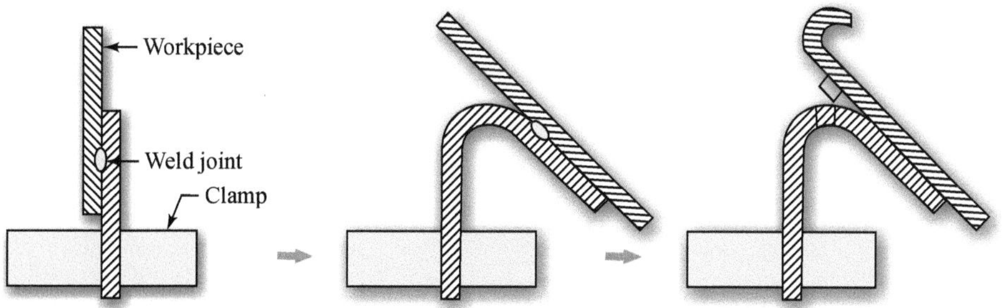

Fig. 8.3 Sequence of operations in a peel test

8.6.5 Bend Test

The bend test is used to determine the ductility and strength of welded joints, and the heat-affected zone (HAZ). The test is further subdivided into three groups:

- Free bend test (Figs. 8.4 and 8.5)
- Guided bend test (Fig. 8.6)
- Controlled bend test (Fig. 8.7).

Free bend test may be conducted on three- or four-point bending.

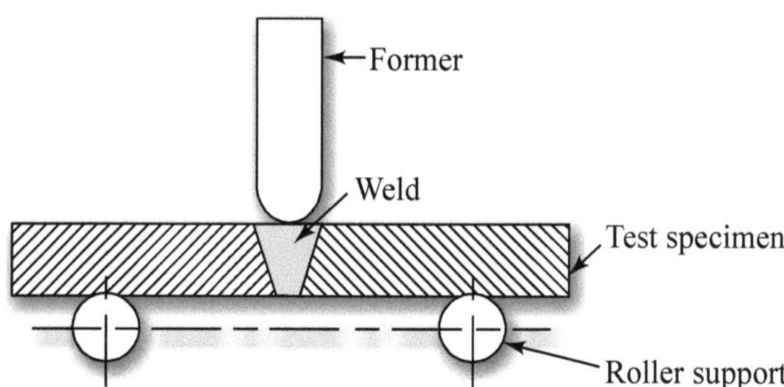

Fig. 8.4 A free bend test (three-point bending)

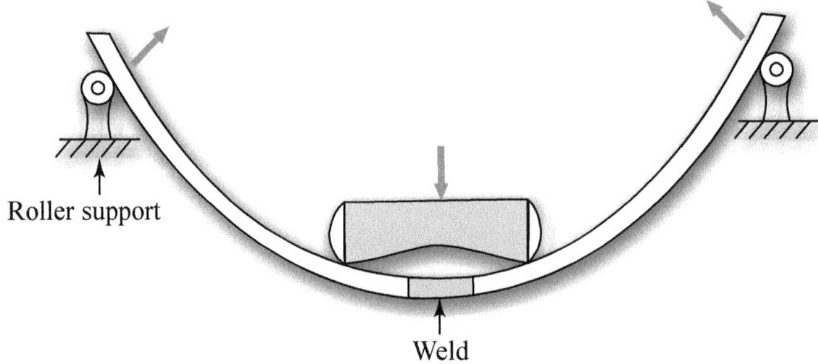

Fig. 8.5 A free bend test (four-point bending)

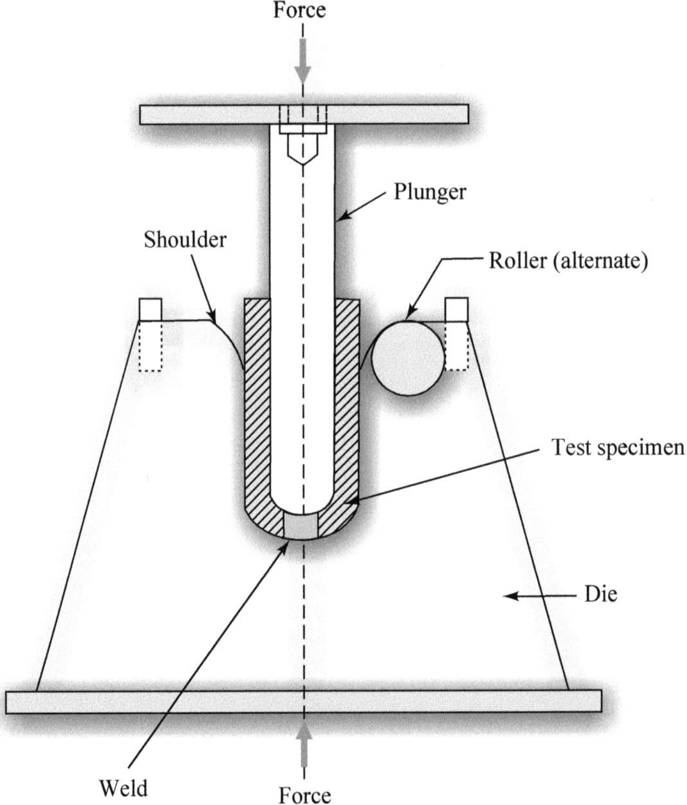

Fig. 8.6 A guided bend test

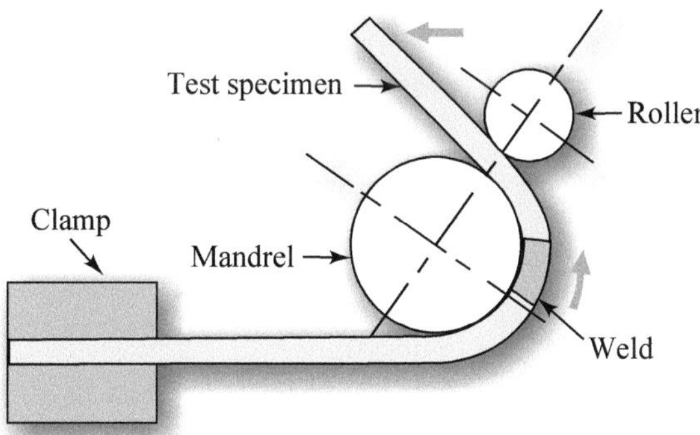

Fig. 8.7 A controlled bend test

8.6.6 Impact Test

The impact test measures the fracture toughness of the welded joint. A number of test methods are available to determine the fracture toughness but the following methods are commonly employed.

- Charpy V-notch test
- Drop weight test
- Dynamic tear (DT) test
- Crack tip opening displacement (CTOD) test.

8.6.7 Hardness Test

The hardness test is the best method of evaluating a weld and heat-affected zone (HAZ) without making a microstructure. The test may be conducted by using Brinell, Vickers and Knoop hardness testing machines.

8.6.8 Fatigue Test

The fatigue test is used to measure the effect of cyclic-stress loading on the welded joint.

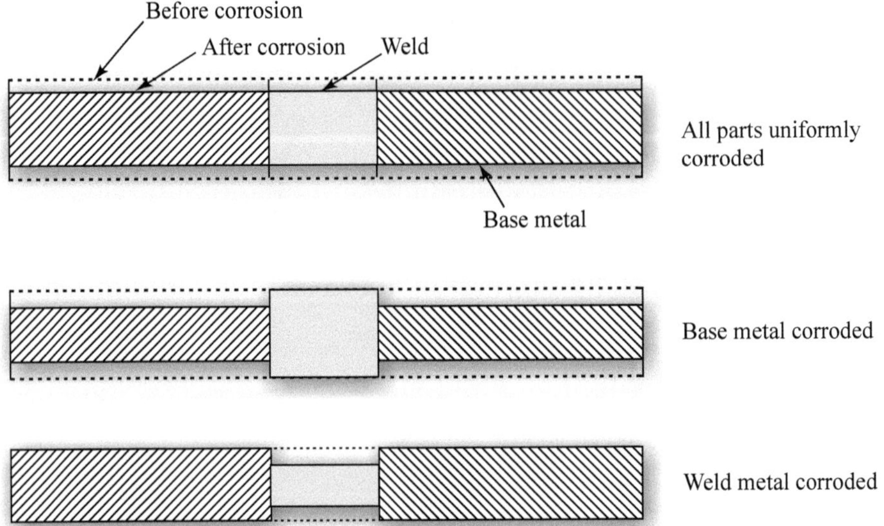

Fig. 8.8 Corrosion conditions in a welded joint

8.6.9 Creep Test

The creep test is conducted to know the behaviour of welded joints at elevated temperatures.

8.6.10 Corrosion Test

A welded joint may corrode uniformly over its entire surface or may be subjected to varying degree of corrosion attack (Fig. 8.8). The most common method of evaluating corrosion resistance is to measure the weight lost during exposure to the corrodent by comparing with a standard formula.

8.6.11 Nick Break Test

The nick break test is used to determine the presence of internal defects such as slag inclusions, gas pockets, poor fusion and oxidized/burnt metal. It is conducted by breaking the test specimen and is usually performed in a butt joint.

Force may be applied by a press or the sharp blow of a hammer (Fig. 8.9). An adaption of the standard nick break test is the fillet nick break test.

The nondestructive testing methods are discussed in Chap. 5.

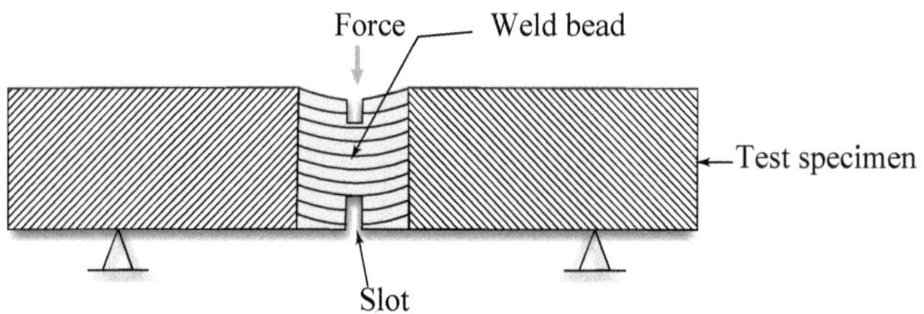

Fig. 8.9 A nick break test

8.7 Weld Defects

The study of defects in a welded joint is necessary to find out the reason of failure because the defective weldment may fail and cause damage during its service life. Defects reduce the strength of the joint and may arise due to the use of substandard materials, defective welding equipment or poor welding skill. The major defects or discontinuities that affect weld quality are discussed below.

8.7.1 Porosity

Porosities are defined as voids, holes or cavities usually of spherical shapes. It is caused by gas entrapped in the weld metal during solidification. Additionally, chemical reactions during welding, contaminants such as dirt, oil, grease, rust and paint on electrode, filler metal and base metal, and insufficient gas shielding may also be the reason for gas entrapment and hence porosity.

Porosity in welds may be of the following types:

- Uniformly scattered porosity
- Cluster porosity
- Starting porosity
- Linear porosity.

All these porosities are shown in Fig. 8.10 and are self-explanatory. Welding starts from leftward and proceeds towards rightward.

The blow holes are voids of large size. Porosity and blow holes are scattered throughout the cross section of a weld. Some types of blow holes reach the surface and appear as pinholes. Some types of porosity are called wormholes when they are long and continuous. Others are called niping, usually long in length and parallel to the root of the weld.

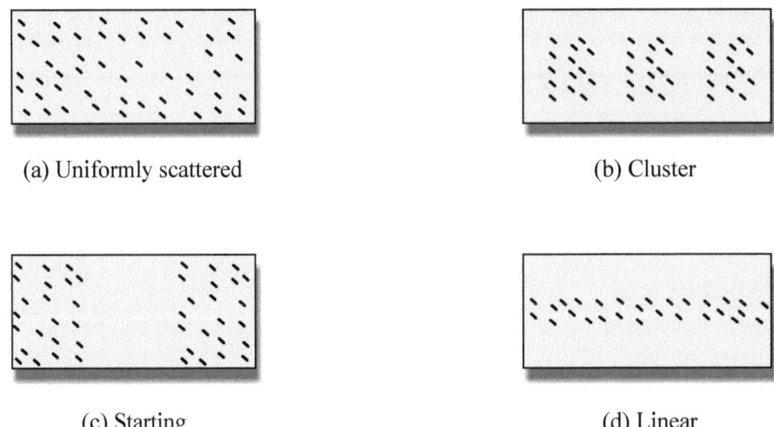

(a) Uniformly scattered (b) Cluster

(c) Starting (d) Linear

Fig. 8.10 Various types of porosity in weld metal

The methods used to reduce porosity include proper selection of electrodes and filler metals, proper cleaning of the above-listed contaminants and preventing them from entering the weld zone and slowing the welding speed to allow sufficient time for gas to escape.

8.7.2 Shrinkage Cavity

Shrinkage cavity normally forms during the welding of thick plates by arc welding processes especially those dealing with large amount of molten metal. This defect occurs at the surface and may extend inside the weld and is formed due to shrinkage of weld metal during its solidification.

8.7.3 Cracks

Cracks are the most dangerous of all weld defects. They may occur in various locations and directions in the weld area such as base metal, base and weld metal interface or weld metal on its surface, under the weld bead or in the crater. Transverse, *longitudinal* and *crater cracks* are defined as surface cracks because they are formed on the surface of the weld. Crater cracks are generated in the crater due to interruption in the welding process and are usually star-shaped. *Toe cracks* are formed in the adjacent base metal. *Underbead cracks* are subsurface cracks (internal cracks), which are usually generated in the high-temperature heat-affected zone (HAZ). Figure 8.11 shows various types of cracks. Cracks may be hot or cold depending upon the temperature conditions.

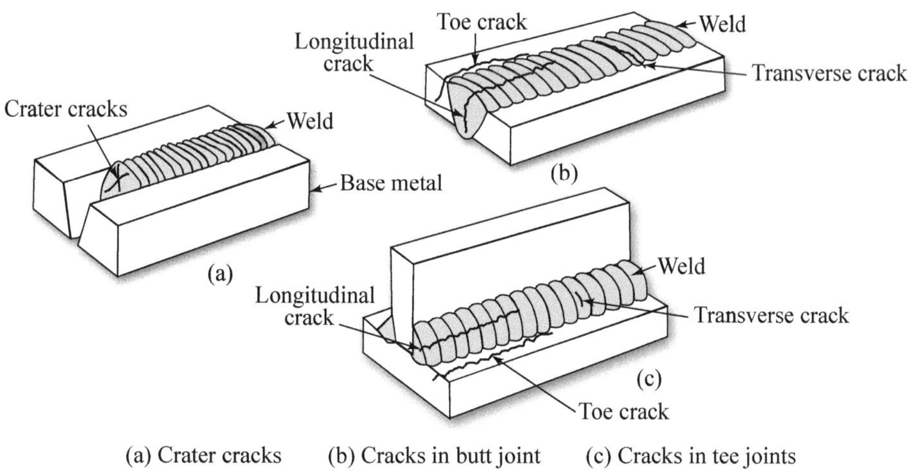

Fig. 8.11 Cracks in welded joints

Formation of cracks may be due to the following reasons:

- Differential thermal stresses generated in the weld zone due to temperature gradient.
- Inability of the weld metal to contract during cooling resulting in the generation of localized stresses.
- High sulphur and carbon content present in the base metal leads to embrittlement.

Cracks formation may be checked by changing the joint design to uniformly distribute the stresses within the weld metal. Additionally, metals to be joined may be preheated and are cooled slowly when the process of welding is over.

8.7.4 Slag Inclusion

Inclusions in the form of slag such as oxides, sulphides, fluxes and other foreign material may get entrapped in the weld metal and base metal and do not get chance to escape and hence float on the surface. Inclusions lower the strength of the joint. The most common cause of slag inclusion is inadequate cleaning of the area near the joint and can be prevented by thoroughly cleaning it.

8.7.5 Distortion

Distortion may result from the differential rate of heating and cooling in the weld zone or adjacent metal leading to the generation of stresses. Parts after joining may not look properly aligned due to distortion. Reducing welding current, using small diameter electrode and proper clamping devices can reduce or even prevent distortion completely.

8.7.6 Incomplete Fusion and Penetration

Incomplete fusion often results from poor joint preparation, wrong joint design or incorrect weld parameters such as welding current and welding speed. However, incomplete or poor penetration results from too little heat input due to decrease in welding current. Both fusion and penetration are related to each other. If fusion is poor, there will be poor penetration. The problem can be overcome by improving the joint design, increasing the heat input and lowering the welding speed. The two defects are shown in Fig. 8.12.

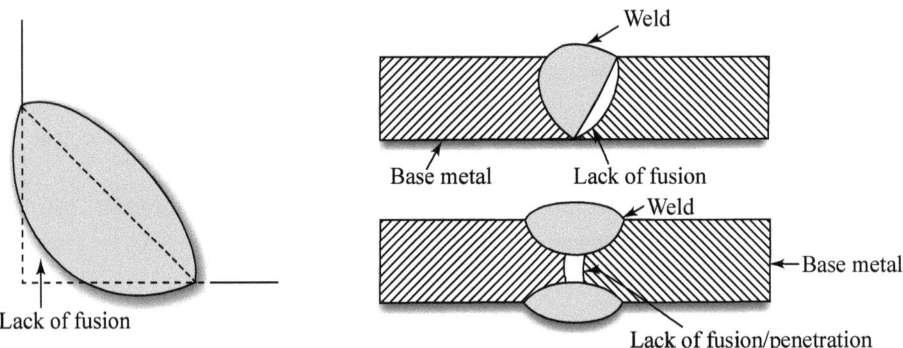

Fig. 8.12 Lack of fusion and poor penetration in fillet and butt welds

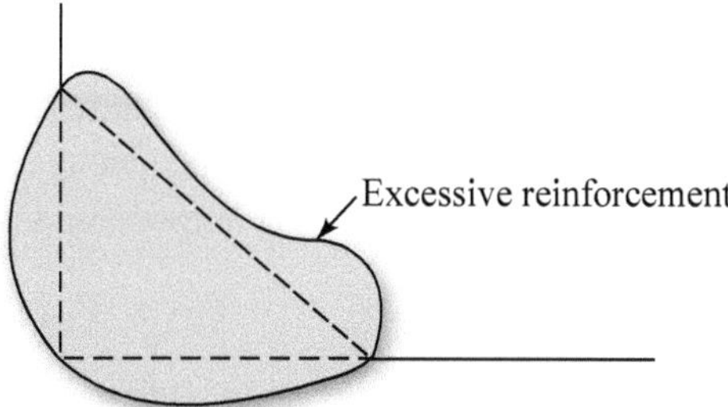

Fig. 8.13 An incorrect weld profile

8.7.7 Incorrect Weld Profile

Wrong bead profile can be a source for lack of strength and bad appearance of the joint. Excessive reinforcement or excessive penetration may be due to bad weld profile and can be a source of stress concentration. Additionally, the excessive reinforcement means the wastage of filler metal, thereby increasing the cost of welding without gaining any worthwhile strength. Skilled workers can reduce this problem to a great extent. The defect is shown in Fig. 8.13.

8.7.8 Undercutting and Overlapping

In undercutting, a groove gets formed in the base metal along the sides of the weld bead. Due to excessive welding current and arc voltage, excessive melting of fusion faces of the base metal takes place and a sharp recess or notch is formed.

Overlapping occurs when excessive molten metal from the filler metal flows to the base metal and gets solidified. This may be caused due to excessive welding current and wrong positioning of the electrode during the welding. Overlapping is just opposite of undercutting (Fig. 8.14). The two defects can be reduced by controlling the welding current and arc voltage and correct positioning of the electrode during the welding operation.

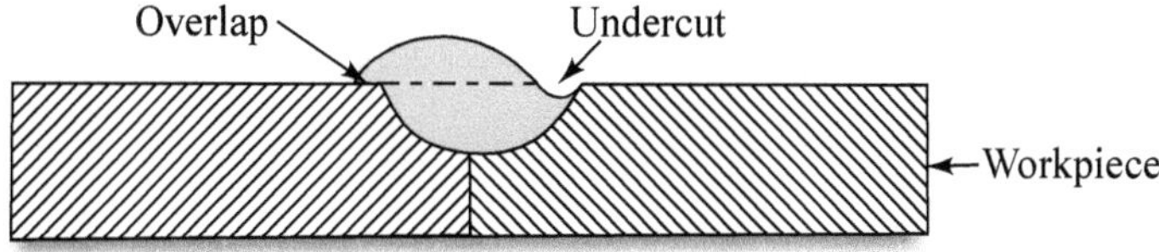

Fig. 8.14 Overlapping and undercutting

8.7.9 Spatter

Spatters are the small metal particles that are thrown randomly in all directions around the arc and are deposited on the base metal. It degrades the quality of the surface around the joint. Excessive spatter occurs due to high welding current, arc blow and wrong selection of electrodes.

8.8 Welding Safety

Safety is a major concern in welding. As in all manufacturing operations, there are certain hazards in welding. The following are the most common hazards in welding.

- Electric shock
- Arc radiation
- Fumes and gases
- Compressed gases
- Fires and explosions
- Noise.

8.8.1 Electric Shock

Electric shock is experienced by the welding operator if he touches a current-carrying part which is at a dangerous voltage. It can also occur when a broken or a naked live wire touches the ground. Even a low current value of 0. 1 A (ac or dc) can prove fatal keeping in mind the fact that human body can have a maximum resistance of 600 Ω for a voltage of 60 V. This hazard is important for welding processes using electric current such as arc welding and resistance welding. The problem can be taken care of by ensuring proper insulation of cables and reliable earthing of welding equipment.

8.8.2 Arc Radiation

During welding, large amount of infrared and ultraviolet radiations are emitted. For example, gas tungsten arc welding (GTAW) performed with argon shielding is the highest emitter of ultraviolet radiation. These radiations are very harmful and may cause damage to eye, face and body. Skin may burn. So proper protection from the arc radiation is essential. Protection against arc radiation can be ensured by using safety equipment and protective clothing. Opaque shields or tinted curtains of polyvinyl chloride plastic sheet can be used as protection against exposure to radiation.

8.8.3 Fumes and Gases

Harmful fumes and gases like ozone, nitrogen dioxide and carbon monoxide can affect respiratory systems and problems such as dryness of throat, coughing, chest tightness and breathing difficulties may arise. To keep the place of welding free from fumes and gases, proper ventilation systems must be installed so that these gases can be thrown out to the surrounding.

8.8.4 Compressed Gases

Compressed gases such as fuel gases, oxygen and shielding gases kept at high pressures in respective cylinders need careful attention, because there is every possibility that these cylinders may burst causing severe injury to the nearby atmosphere. These cylinders should be kept away from combustible materials.

8.8.5 Fires and Explosions

Fires and explosions can cause serious injury and even fatality. Fire-fighting equipments must be installed in the welding workshop areas. The welding operator must wear nonsynthetic clothes.

8.8.6 Noise

Excessive and prolonged noise produced by welding or cutting operations can cause temporary or permanent hearing loss. Noise above 80 dB is considered harmful. Ear protection device should be used.

8.9 Flux

Flux is a substance which is used in joining processes to prevent the formation of oxides and other impurities by forming a protective layer. It melts during joining to mix with the impurities or the by-products and forms slag over the surface of the metal parts being joined and can be removed by chipping or brushing on cooling.

Flux has the following important functions:

- To act as a deoxidiser.
- To act as a slag former.

- To provide a protective atmosphere or a shielding cover for the weld metal pool.
- To stabilize the arc.
- To reduce the spattering of metal during welding.
- To improve the strength of the joint because of the alloying elements in the flux.
- To allow the electrode to carry high current because of flux coating on the electrode.

Fluxes can be used in various forms like powders (granules), pastes and slurries. They are used in bulk quantity in submerged arc welding in the granular form and as a coating over the consumable electrodes in case of shielded metal arc welding (SMAW). It is used in the core of the consumable tubular electrodes in case of flux-cored arc welding.

A good flux has the following properties:

- Low melting temperature to melt it easily.
- Low viscosity to ensure its easy spread (flow) over the surface.
- Low heat transfer rate to ensure the protection of the joint until solidification.
- High wettability for better mixing with impurities.
- Noncorrosive and nonconductive to ensure better quality joint.
- Easily removable upon cooling.

Fluxes are used in joining processes such as welding, brazing, and soldering. The flux used in submerged arc welding consists of calcium or magnesium silicate. Important flux ingredients used in brazing include borax, borates, fluorides and chlorides. Chloride fluxes such as zinc chloride (killed spirits), ammonium chloride and zinc ammonium chloride either alone or in combination of the two are the most efficient fluxes used in soldering, because they clean the surface rapidly and are effective on aluminium, copper, brass, bronze, steel and nickel.

8.10 Automation in Welding

Automation is the process in which predetermined sequence of operations with little or no human labour is followed by using specialized equipment and devices that perform and control manufacturing processes.

The major breakthrough in automation began with numerical control *(NC)* of machine tools in the 1950s. A lot of progress has been made since then. The use of computer has made the process accurate and rapid, and computerized numerical control (CNC), adaptive control (AC), industrial robots, and computer-integrated manufacturing (CIM) systems, including computer-aided design, engineering and manufacturing (CAD/CAE/CAM) are the major components of automation.

The need for improved productivity, the shortage of skilled labour and safety requirements have increased the scope for automation in welding. The extent of automation will depend on the type of welding process to be used and size and geometry of the product. Accordingly, automation may be partial or total. In partial automation, certain functions are performed manually, whereas all functions and steps are performed by the equipment in proper sequence without adjustment by an operator in full automation.

Automatic welding is different from automated welding. Although there are many similarities between them, there is a major difference. Automatic welding involves elaborate fixturing elements with tooling, work holding devices, accurate part location and orientation. It can be used for high-level production so that the high cost of the equipment can be justified. It also reduces manpower requirements and produces high-quality welds.

Automated welding eliminates the elaborate expensive fixtures and a computer programme replaces the complex fixturing and sequencing devices. It can be used for small lot production, even to the welding of single lot production without increasing the cost. Suitable changes in the product can be made during the process without the necessity of redesigning and reworking of the expensive fixtures. Robotic welding, one of the advanced forms of automated welding, uses robot to perform the operations and finds applications in sealing of radioactive materials into metal containers, nuclear installations and some radio chemical processing plants where high corrosive solutions are handled.

Short Answer Questions

1. How does welding differ from other joining processes?

Ans. Welding is used to join two similar metals permanently using heat. On the other hand, other joining processes may not use heat, and also the joint is not permanent and weaker than that produced by welding.

2. What is the principle of fusion welding? Name a few fusion welding processes.

Ans. In fusion welding, the two parts at the joint are heated to their molten states and the molten metal fills up the gap between them, and allowed to solidify without applying any pressure. Examples of fusion welding include gas welding and arc welding.

3. Why is weld inspection essential?

Ans. Weld inspection is required to know the soundness, strength and toughness of the weld material.

4. What is nick break test? For which type of joint it is suitable?

Ans. Nick break test is a destructive method of weld inspection to determine the presence of internal defects such as slag inclusion, gas pockets, poor fusion and burnt metal in the weld material. It is suitable for butt joint.

5. What is heat-affected zone (HAZ)?

Ans. Heat-affected zone (HAZ) indicates the area around the joint of the weld material affected by the application of heat and lies in close proximity to fusion zone and suffers from metallurgical changes.

6. What are the important functions of a flux?

Ans. Flux prevents the formation of oxides, provides a protective atmosphere or a shielding cover for the metal being joined, stabilizes the arc and reduces the spattering.

Multiple Choice Questions

1. Which of the following joining methods finds extensive applications in electronic industry?

(a) Resistance welding (b) Gas welding
(c) Soldering (d) Arc welding

2. Which of the following welding methods is not a fusion process?

(a) Gas welding (b) Arc welding
(c) Resistance welding (d) Thermit welding

3. Which of the following joining methods produces an autogeneous joint?

(a) Soldering (b) Resistance welding
(c) Gas welding (d) Arc welding

4. Match List I with List II and select the correct answer using the codes given below the lists:

List I	List II

(Joining methods)	(Description)
A. Soldering and brazing	1. Autogeneous joint
B. Resistance welding	2. Homogeneous joint
C. Thermit welding	3. Heterogeneous joint
D. Arc welding	4. Mixture of iron oxide and aluminium powder

Codes:	A	B	C	D
(a)	3	1	4	2
(b)	2	1	4	3
(c)	2	3	4	1
(d)	1	2	3	4

5. Fusion process is very much similar to

(a) Spinning (b) Forging
(c) Casting (d) Rolling

6. Which of the following welding methods is more suitable for field operations?

(a) Resistance welding (b) Electron beam welding
(c) Shielded metal arc welding (d) Atomic hydrogen arc welding

7. Resistance spot welding is extensively useful for

(a) Automobile industry (b) Aircraft industry
(c) Reactive metals (d) High melting point metals

8. Thermoplastics and thermosets are commonly joined by

(a) Arc welding (b) Brazing
(c) Gas welding (d) Adhesive bonding

9. Railroad rails are repaired by using which of the following welding methods?

(a) Electron beam welding (b) Thermit welding
(c) Laser welding (d) Resistance welding

10. Heat-affected zone is minimum in which of the following welding processes?

(a) Laser welding (b) Gas welding
(c) Arc welding (d) Thermit welding

11. Consider the following statements:

    1. Wrong bead profile reduces the strength of the weld joint.
    2. Excessive reinforcement is the wastage of filler metal.
    3. Spatter formation increases with increase in welding current.

Of these statements:

(a) 1 alone is true (b) 2 alone is true
(c) 2 and 3 are true (d) 1, 2 and 3 are true

12. Consider the following components. Automation includes

    1. Computer-aided design (CAD)
    2. Computer-aided manufacturing (CAM)
    3. Computer-aided engineering (CAE)

    Of these:

(a) 1 and 2 are true (b) 3 alone is true
(c) 2 and 3 are true (d) 1 and 3 are true.

13. What are the benefits of automation in welding?

    (a) It reduces the scope of welding hazards.
    (b) It increases the welding productivity.
    (c) It reduces the elaborate expensive fixtures.
    (d) All of the above.

14. Light aluminium sheets are joined by

    (a) Tungsten inert gas (TIG) welding
    (b) Metal inert gas (MIG) welding
    (c) Submerged arc welding
    (d) Resistance welding.

15. Which of the following welding processes finds extensive applications in automobile industry?

    (a) Tungsten inert gas (TIG) welding
    (b) Metal inert gas (MIG) welding
    (c) Submerged arc welding

(d) Resistance welding.

16. Which of the following weld defects is inspected by nick break test?

(a) Slag inclusions
(b) Gas pockets
(c) Poor fusion
(d) All of the above

17. Peel test is a

(a) Compression test
(b) Tension test
(c) Impact test
(d) Creep test.

18. Weld spatter occurs due to

1. high voltage current
2. low welding current
3. high welding current
4. arc blow.

Of these:

(a) 2 alone is true
(b) 3 and 4 are true
(c) 1 alone is true
(d) 1 and 2 are true.

19. Presence of sulphur in a welded joint increases its

(a) Ductility
(b) Toughness
(c) Brittleness
(d) Strength.

20. Which of the following decibel (db) is considered safer for human ear?

(a) 100 db
(b) 80 db
(c) 90 db
(d) 120 db

Answers

1. (c)	2. (c)	3. (b)	4. (a)	5. (c)	6. (c)
7. (a)	8. (d)	9. (b)	10. (a)	11. (d)	12. (c)

13. (d) 14. (a) 15. (d) 16. (d) 17. (b) 18. (b)
19. (c) 20. (c)

Review Questions and Discussions

Q.1. What is the importance of joining processes?
Q.2. How does welding differ from other joining processes?
Q.3. What is meant by an autogeneous joint?
Q.4. What is fusion welding?
Q.5. What is the role of pressure in plastic welding?
Q.6. Why is surface cleaning important in welding?
Q.7. What is weldability?
Q.8. Why does weldability decrease with increase in carbon content?
Q.9. How is a particular welding process selected?
Q.10. Why is electron beam welding (EBW) especially useful for reactive metals?
Q.11. Why is weld inspection essential?
Q.12. Why is a tension test performed?
Q.13. How is corrosion test conducted?
Q.14. What causes porosity?
Q.15. How does an incomplete fusion affect a joint?
Q.16. How is the strength of the weld joint affected by incorrect weld profile?
Q.17. What is spatter?
Q.18. What are the different welding hazards?
Q.19. What are the advantages of automation in welding?

Gas Welding Processes 9

Oxyacetylene welding finds extensive applications in sheet-metal fabrication and repair works.

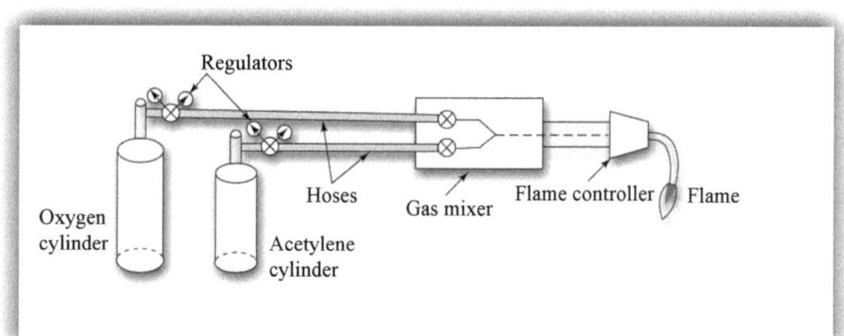

Learning Objectives

After reading this chapter, you will be able to answer some of the following questions:

- Why is oxygen-acetylene combination mostly preferred in gas welding?
- Why do neutral flames find extensive applications?

D. K. Singh, *Fundamentals of Manufacturing Engineering*,
https://doi.org/10.1007/978-981-99-8767-2_9

9.1 Introduction

Gas welding processes use heat of a gas flame by burning gaseous fuels to join the parts. These gases are burnt in the presence of oxygen producing large quantity of heat to be used during welding.

9.2 Oxyfuel Gas Welding (OFW)

Oxyfuel gas welding includes all the processes in which a combination of gases is used to produce a gas flame. This flame is used as the source of heat to melt the metals at the joint. Most common fuel gases include acetylene, propane, propylene, natural gas and hydrogen. But acetylene is still the principal fuel gas employed in the process, since the acetylene–oxygen combination gives the highest temperature as compared to other combinations.

9.3 Oxyacetylene Welding

Oxyacetylene welding uses acetylene–oxygen mixture to produce the required heat. The chemical reaction between acetylene and oxygen is exothermic in nature, and sufficient heat is released at the end of the reaction which has a temperature of around 3250 °C. This heat is used to melt the parts to be joined. The reaction takes place in two stages. In the first stage, acetylene and oxygen react to produce carbon monoxide and hydrogen. Some heat is also liberated during the process.

$$C_2H_2 + O_2 \rightarrow 2CO + H_2 + \text{Heat}$$

In the second stage, the products formed in the first stage further react in presence of excess oxygen producing about two-thirds of the total heat.

$$4CO + 2H_2 + 3O_2 \rightarrow 4CO_2 + 2H_2O + \text{Heat}$$

Acetylene gas used in the process is obtained when calcium carbide (CaC_2) chemically reacts with water.

$$CaC_2 + 2H_2O \rightarrow Ca(OH)_2 + C_2H_2(\uparrow)$$

Acetylene forms explosive mixtures with air and is not safe to store it as a gas at pressures higher than 0.1 MPa. Hence, it is usually dissolved in acetone which has the property of absorbing 25 times its own volume of acetylene for each atmosphere of pressure applied.

9.3.1 Advantages

- The equipment cost is low and requires little maintenance.
- It is portable and can be used anywhere.
- The gas flame is generally more easily controlled.
- The process can also be used for cutting.

9.3.2 Limitations

- Since the heat source is not concentrated, a large area of the metal is heated and distortion is likely to occur.
- In some cases, there is a loss of corrosion resistance.
- The process is very slow. Therefore, it has been largely replaced by arc welding.
- Proper operator training and skill are also essential.

9.3.3 Suitability and Applications

Oxyfuel gas welding is used extensively for sheet metal fabrication and repair works. With proper technique, practically all metals can be welded. Quality of the weld depends upon the way how the surfaces are made clean and prevented from atmospheric contamination.

9.4 Welding Equipment for Oxyacetylene Welding

The equipment for oxyacetylene gas welding consists of a welding torch (blowpipe) which is connected by hoses to high-pressure gas cylinders and equipped with pressure gages and regulators.

Oxygen regulators are usually painted green, and acetylene regulators red. Other equipment includes safety equipments such as goggles with shaded lenses, face shields, gloves and protective clothing. An oxyacetylene welding torch and gas supply is shown in Fig. 9.1. For welding purposes, the oxygen is usually supplied in relatively pure form from the gas cylinders, but in rare cases, air can also be used. Acetylene is compressed in the cylinder under a pressure of 1.7 MPa, and the cylinder can hold upto 300 ft^3 of gas at this pressure.

9.5 Types of Flames used in Oxyacetylene Welding

Three distinct flame variations are produced with an oxyacetylene gas mixture namely, reducing or carburizing, neutral and oxidizing. The correct adjustment of the flame is important for reliable welds.

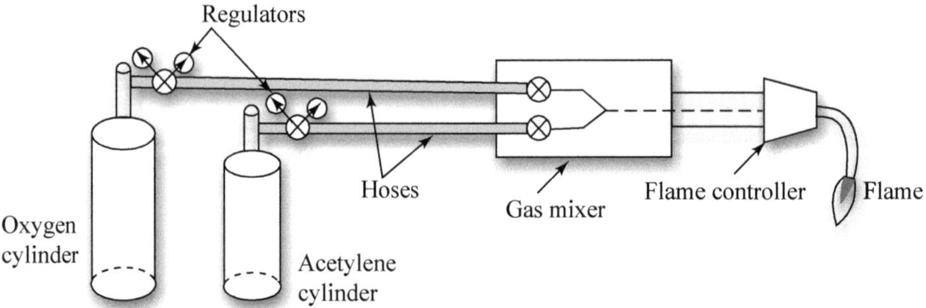

Fig. 9.1 An oxyacetylene welding set-up

If the flame contains excess of acetylene, it is known as ***carburizing flame***. The excess fuel decomposes to carbon and hydrogen in the absence of sufficient oxygen, and the temperature produced by such flame is lower (3050 °C). Hence, it is suitable for applications requiring low heat, such as brazing, soldering and flame hardening. Carburizing flame is used in the welding of monel metal (a nickel-copper alloy), high-carbon steels, nickel, certain alloy steels, and many of the nonferrous, hard-surfacing materials. Steel is not welded by using this flame since it will increase the carbon content in the weld resulting in a hard and brittle weld.

If acetylene and oxygen are present in equal proportions (ratio is 1:1), ***neutral flame*** is produced. For most welding operations, a neutral flame is used, since it has least chemical effect on the heated metal. It is used in the welding of mild steel, alloy steel, grey cast iron and others.

If the flame has high oxygen content (the ratio of oxygen and acetylene is 1.5 :1), then it is known as ***oxidizing flame***. Such flames are used in the welding of copper and copper alloys (brass and bronze). This flame is harmful, especially for steels, because it oxidizes the steel.

9.6 Pressure Gas Welding

Pressure gas welding is used to make butt joints between the ends of objects such as pipe and railroad rail. The ends are heated with oxyacetylene flames to the semi-solid state and then pressure is applied. No filler metal is used. It is also known as ***oxyacetylene pressure welding*** or ***hot pressure welding***. Parts being welded are not melted, and hence it is a solid-state welding process.

Short Answer Questions

1. Why is a filler metal used in welding? What is its composition?

Ans. A filler metal in welding is used to connect thicker parts when more joint strength is required. Its composition is similar to the parts being joined.

2. What is the principle of gas welding? Where is it used?

Ans. Gas welding uses the heat of a gas flame for heating the parts being joined. It is widely used for sheet metal fabrication and repair works.

3. Name the three gas flames used in gas welding and also state their suitability.

Ans. Three gas flames widely used in gas welding include carburizing, oxidizing and neutral. Carburizing flame is used in the welding of monel metal, high-carbon steels, nickel and nonferrous metals. Oxidizing flame is used in the welding of copper and copper alloys such as brass and bronze. Neutral flame is used in the welding of mild steel, alloy steel and grey cast iron.

4. How do three gas flames used in gas welding vary with respect to their compositions?

Ans. Carburizing flame contains excess of acetylene, while oxidizing flame contains excess of oxygen and neutral flame contains equal proportion of acetylene and oxygen.

5. Out of the three gas flames used in gas welding, which flame has extensive industrial applications and why?

Ans. Neutral flame among the three flames is extensively used for industrial applications as it has neutral effect on the joint produced by it.

Multiple Choice Questions

1. Which of the following heat source is used in gas welding?

(a) Resistance heating (c) Arc heating
(c) Gas flame heating (d) Explosion heating

2. Carburizing flame contains excess of

(a) Ethylene (c) Acetylene

(b) Propylene
(d) Methylene

3. The ratio of oxygen and acetylene in oxidizing flame is

(a) 2.0:1.5
(c) 1.0:0.5
(b) 1.5:1.0
(d) 0.8:0.9

4. Neutral flame contains equal amounts of

(a) Acetylene and air
(b) Ethylene and oxygen
(c) Acetylene and hydrogen
(d) Acetylene and oxygen

5. The approximate temperature of the heat produced by a gas flame is

(a) 3200 K
(b) 3200 °F
(c) 3200 °C
(d) 5000 °C

6. Acetylene is produced by pouring heated water on

(a) Sodium cyanide
(b) Sodium carbide
(c) Calcium carbide
(d) Magnesium carbide

7. Acetylene is stored in

(a) Carbon tetrachloride
(b) Ethanol
(c) Acetone
(d) Formaldehyde

8. Consider the following statements about the gas welding:

1. It uses a gas flame.
2. It uses shielding atmosphere.
3. It is a cheaper method of joining.
4. It requires operator's skill.

Of these statements:

(a) 1 and 2 are true
(b) 1, 3 and 4 are true
(c) 3 and 4 are true
(d) 1 alone is true

9. Match List I with List II and select the correct answer using the codes given below the lists:

List I	List II
(Gas flame)	*(Application)*
A. Oxidizing flame	1. Steel
B. Carburizing flame	2. Brass and Bronze
C. Neutral flame	3. Monel metal

Codes	A	B	C
(a)	1	2	3
(b)	3	1	2
(c)	2	3	1

10. If steel is welded by a carburizing flame, what can result?
    (a) The joint becomes stronger in compression.
    (b) The joint becomes hard and brittle.
    (c) The tensile strength of the joint is increased.
    (d) The joint becomes stronger in shear.

Answers

1. (b)	2. (c)	3. (b)	4. (d)	5. (c)	6. (c)
7. (c)	8. (b)	9. (c)	10. (b)		

Review Questions and Discussions

Q.1. What are the possible combination of gases used to produce a gas flame?
Q.2. Why is oxygen–acetylene combination most widely used?
Q.3. How is acetylene prepared?
Q.4. Why is acetylene stored in acetone?
Q.5. Why do the surfaces of the parts to be joined require cleaning before welding?
Q.6. What are the three flames used in gas welding?
Q.7. Differentiate between carburizing and oxidizing flame.
Q.8. Why is neutral flame extensively used for industrial applications?
Q.9. Steel parts are not welded by using carburizing flame. Why?

Arc Welding Processes 10

Electroslag welding process finds extensive applications in joining thick metals because of increased quantity of molten weld metal produced in the process

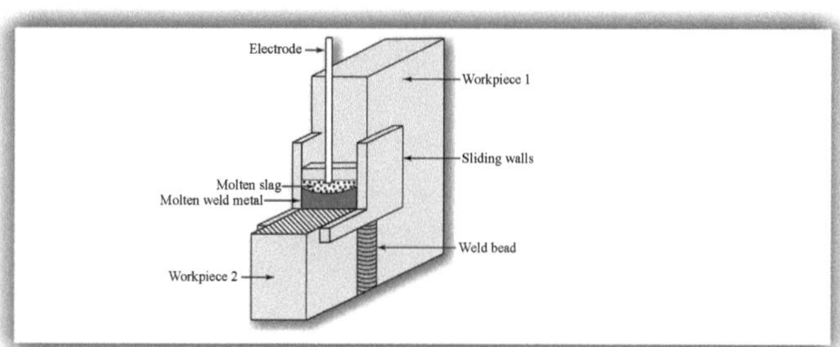

Learning Objectives

After reading this chapter, you will be able to answer some of the following questions:

- Why does shielded metal arc welding find extensive applications?
- How does atomic hydrogen welding differ from other arc welding process?
- Why is carbon arc welding losing its significance?

D. K. Singh, *Fundamentals of Manufacturing Engineering*,
https://doi.org/10.1007/978-981-99-8767-2_10

10.1 Introduction

Arc welding is the most common fusion welding process, where the source of heat is an electric arc. The arc is produced between a metal electrode and the workpiece or between two electrodes and is used to heat the parts to be joined.

The arc acts as a concentrated source of heat. High electrical currents of the order of 1000 amps or more are used, although voltages are low that lie in the range of 30-80 V ac. As a result, temperatures of the order of 6000–7000 °C are produced. A schematic diagram of the arc welding process is shown in Fig. 10.1. Welding may be performed with direct or alternating current. The polarity of the electrode may be positive or negative. If direct current is used and the workpiece is connected to the positive terminal (anode) of the circuit, the condition is known as ***straight polarity***. On the other hand, when the work is made negative and the electrode positive, then the condition is said to be ***reverse polarity***. Approximately 60 to 75% of the heat is liberated near the positive terminal of a DC arc. In case of thicker workpieces, more heat is required to be produced near the workpiece, and hence straight polarity is most effective. Also, it ensures reduced disintegration of electrodes. On the other hand, in overhead welding, where the weld pool should be kept relatively small and a fast freeze helps hold the metal in place, reverse polarity would be best. When it is required to keep the workpiece as cool as possible, as in arc welding of cast iron, reverse polarity is the best choice.

The electrodes may be consumable or nonconsumable in nature. Most of the processes use consumable electrodes. Such electrodes act as filler metal and supply the metal needed to fill the voids in the joint. Nonconsumable tungsten electrodes are used in some arc welding processes, where a separate filler metal is used.

Electrodes are used as bare or in coated form. The coated electrodes consist of the core wire with a covering of coating material. Arc welding processes using coated electrodes are more popular. Shielding gas or fluxing material can be used to protect the molten

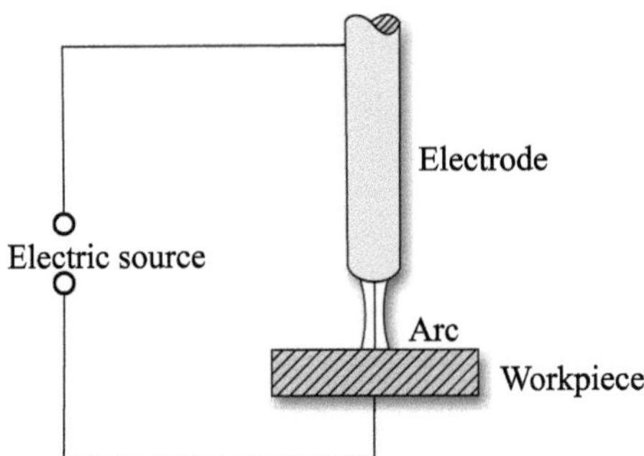

Fig. 10.1 An arc welding process

pool from atmospheric contamination. Shielding also ensures arc stability and affects the metal-transfer characteristics of the arc.

10.2 Shielded Metal Arc Welding (SMAW)

Shielded metal arc welding is the most widely used arc welding process using consumable electrode. It is one of the oldest, simplest and most versatile joining processes. It is also known as *stick welding*, because of stick-like shape of the electrodes used in the process. The heat is produced by an arc between the flux-covered electrode and the workpiece. The process is shown in Fig. 10.2. The heat of the arc melts the tip of the electrode, its coating and the workpiece just below the arc. The electrode coating, on melting, forms shielding gas which protects the molten metal from atmospheric contamination. It also stabilizes the arc. The flux used in the process combines with the impurities present in the molten metal and floats on its surface in the form of slag. It is chipped away from the weld surface on cooling.

10.2.1 Advantages

The SMAW equipment is simple, portable and less expensive compared to other arc welding equipments. Welds can be made in all positions.

10.2.2 Limitations

The process uses finite-length electrode which gets consumed, and it is required to use new electrodes frequently. This makes the process slower. Because some of the heat is

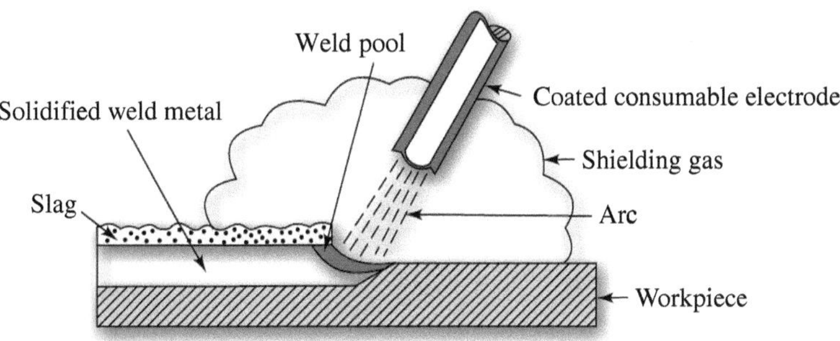

Fig. 10.2 A shielded metal arc welding

taken by the electrode, its quality deteriorates with time. Slag formation is another problem which needs to be taken care of. Slag if left unremoved causes corrosion resulting in the failure of the joint. Low melting point metals such as zinc, lead and tin are not welded by SMAW because of burn-off possibility.

10.2.3 Suitability and Applications

SMAW is used extensively by many of the fabrication industries such as general construction, shipbuilding, pipelines and pressure vessel fabrication. It is also used in repair operations.

Carbon steels, alloy steels, stainless steels and cast irons are commonly welded metals by SMAW.

10.3 Submerged Arc Welding (SAW)

In the submerged arc welding, arc is produced between a consumable metal electrode and the workpiece. Since the arc is submerged (hidden) under a heavy coating of granular flux, hence the name submerged arc welding. The arc travels beneath the surface of flux mass fed from a tube.

The flux consists of calcium or magnesium silicate. A portion of the flux melts and combines with the impurities to form the slag to be finally chipped away on cooling. Slag if left unremoved causes corrosion resulting in the failure of the joint. The remaining unmelted flux provides shielding atmosphere.

Copper-coated steel wire is used as electrode. The larger diameter electrodes are used because of their high current carrying capacity and fast deposition rate. The process can be performed manually or it may be semiautomatic or fully automatic of which the latter is more common today. The process of submerged arc welding is shown in Fig. 10.3.

Both alternating and direct current can be used in this process. Direct current reverse polarity (DCRP) gives deeper penetration. Direct current straight polarity (DCSP) ensures higher deposition rate.

Iron powder can be applied directly into the joint to increase the metal deposition rate which can make the process very fast. The resulting process is called bulk welding.

10.3.1 Advantages

- Since the arc is completely hidden, there is no chance of any flash, spatter or smoke visible from outside.

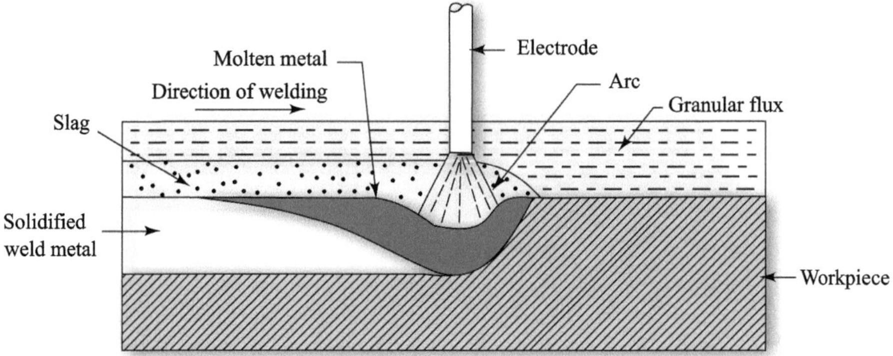

Fig. 10.3 A submerged arc welding process

- Heavy currents can be used. In other arc welding processes, heavier currents are not preferred because of higher infrared and ultraviolet radiations. No such problem arises in this process because the arc is covered by the flux.
- Higher current gives high deposition rate and deep penetration. The deposition rate can be much more than shielded metal arc welding (SMAW) process.
- Welding speed is higher which helps to join thicker plates.
- It produces very high-quality weld because of high degree of cleanliness.

10.3.2 Limitations

- It is largely confined to flat position welding. Overhead welding is not possible because of the possibility of falling down of large molten pool and flux.
- It is not suitable for thinner metals because of burn-off possibility.
- The arc is difficult to guide.
- Due to huge amount of flux being used in the process, it has slow rate of cooling.

10.3.3 Suitability and Applications

Submerged arc welding is most suitable for making flat butt or fillet welds in low-carbon steel. Typical applications include thick plate welding for shipbuilding and pressure vessels.

10.4 Gas Metal Arc Welding (GMAW)

Gas metal arc welding, earlier known as metal inert gas (MIG) welding, was used mainly for welding aluminium and stainless steels with inert gas shielding. In this process, the arc is produced between a consumable electrode and the workpiece. The electrode is in the form of bare wire of small diameter which provides the filler metal, and hence no additional feed is required. Shielding gases such as argon, helium, carbon dioxide and oxygen are used for surrounding the arc in order to protect the molten metal pool from atmospheric contaminants like dirt, dust and metal oxides. These gases can be used both as a single gas and in combination with other gases. When CO_2 alone is used as the shielding gas, the process is termed as CO_2 welding. Originally, argon and helium were used as shielding gases. Presently CO_2 is used extensively, and oxygen and CO_2 are often mixed with inert gases. Generally, argon, helium or mixtures of these gases are used for welding nonferrous metals. CO_2 is widely used for plain carbon and low-alloy steels, and CO_2 with argon and sometimes helium is used for steel and stainless steel. In addition to the use of inert shielding gas, deoxidizers are usually present in the electrode metal itself, in order to prevent oxidation of the molten weld puddle.

A reverse polarity DC arc is generally used because it offers the greatest heat input at the workpieces and thus gives deeper penetration as compared to straight polarity. Also, it produces a stable arc and gives smooth welds. It is generally recommended for aluminium, magnesium, copper and steel. Straight polarity with argon has a high burn-off rate, and the arc is unstable with considerable spatter. It is used only occasionally when a very small penetration is required. Alternating current, being inherently unstable, is seldom used in MIG welding. The welding current can vary in the range of 100–300 A.

Metal is transferred across the arc in the following modes:

- Spray Transfer
- Dip Transfer
- Globular Transfer
- Pulsed Spray Transfer

10.4.1 Spray Transfer

In spray transfer, small droplets of molten metal from the electrode are transferred to the weld area at the rates of several hundred droplets per second. The spray transfer takes place when the current and voltages are very high, and the shielding gas used is argon or argon-rich gas mixtures containing at least 80% argon. The process is ideally suited for fast welding on thicker sections.

10.4.2 Dip Transfer

In dip transfer, the metal is transferred in individual droplets, as the electrode wire dips into the weld pool, causing a short circuit. The arc is immediately extinguished. This causes a rapid rise in current which heats up the electrode wire causing it to melt off and re-establish the arc. This is repeated approximately 100 times per second. It is a low-current and low-voltage process, hence suitable for only thin sheets and sections (<6 mm); otherwise incomplete fusion may occur. It can be used for obtaining welds in all positions; however rate of deposition is low.

10.4.3 Globular Transfer

If the wire feed rate and the voltage are increased, it results in a long arc and the molten metal at the tip of the electrode develops into a big globule which is transferred by free-flight across the arc. Normally the globules are randomly directed across the arc in irregular-fashion, resulting in considerable spatters which are reduced by using CO_2. Globular transfer gives deeper penetration due to high current and high voltages used in the process. Heavier sections are commonly joined by this method.

10.4.4 Pulsed Spray Transfer

It overcomes the difficulties of welding thin metals and out-of-position welds. The high current used in the spray transfer is reduced here by the use of pulsed arcs. It is used for welding thinner sections because the heat input is less but can be used in all welding positions.

10.4.5 Advantages

- The filler metal is transferred through the protected arc, and there is no frequent change of electrodes as in the SMAW process. These two factors help GMAW to produce high-quality welds at high welding speeds. Welding productivity is double that of the SMAW process.
- No flux is required, and no slag forms over the weld. This makes the process cleaner.
- GMAW has excellent penetration, and it produces sound welds at high speeds. This makes the process especially adapted to automatic operation, and the process lends itself readily to robotics and flexible manufacturing systems (FMS).
- Welding can be performed in all positions.

- It is a very versatile process and can be used on both light and heavy-gage structural plates.

10.4.6 Limitations

The cost of equipment and consumables is much higher as compared to SMAW process.

10.4.7 Suitability and Applications

GMAW is perhaps the most widely used process in terms of range of metals. It is suitable for welding a variety of ferrous and nonferrous metals and alloys such as carbon steels, low-alloy steels, stainless steels, aluminium, magnesium, titanium, copper, nickel and zirconium. It is used extensively in the metal fabrication industry and has opened new areas of work in sheet metal industry for which SMAW is found unsuitable; for example, it is found useful in the production of car bodies where freedom from frequent electrode changing and the need to remove flux are important production considerations. Typical applications include structure fabrication, shipbuilding, pressure vessels, tanks and pipes. It is also used successfully for the fabrication of railway coaches and in the automobile industry where long, high-speed welds of fairly heavy sections are used.

10.5 Flux-Cored Arc Welding

This process utilizes a continuous tubular electrode wire (as compared to a finite-length solid wire electrode in SMAW) filled with flux and hence the name flux-cored arc welding. The flux contained in the tubular electrode performs essentially the same functions as the coating on a covered electrode; that is, it acts as a deoxidizer, a slag former, an arc stabilizer and may provide alloying elements as well as the shielding gas. Additional shielding of CO_2 may or may not be provided. The wire diameter normally ranges between 0.5 and 4 mm with flux usually forming 5–25% by weight of the total wire resulting in a deposition efficiency of 85–95%. In other ways, the flux-cored process is similar to GMAW process. The process is shown in Fig. 10.4.

The electrical contact is maintained through the exterior of the electrode wire. A reverse polarity DC power source is used in the process.

10.5.1 Advantages

- The deposition rate is higher than GMAW. It is about twice that of SMAW for a comparable set-up.

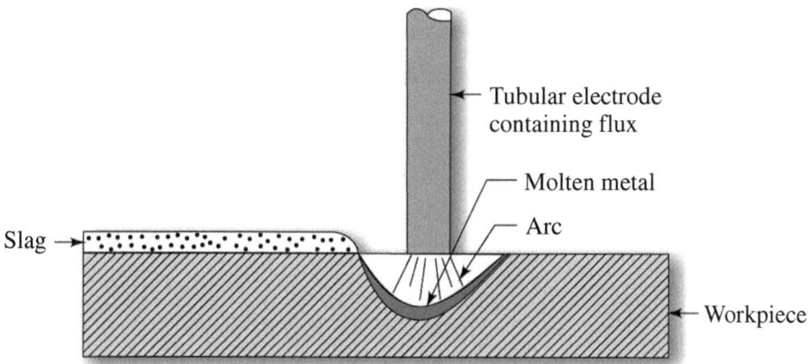

Fig. 10.4 A flux-cored arc welding process

- It can be used for a wide range of metal thicknesses, as thin as 1.57 mm.
- Welds can be made in all positions using smaller diameter wires.
- The process is easy to automate and is readily adaptable to flexible manufacturing systems and robotics.

10.5.2 Limitations

- Large amount of smoke is produced in the operation because of the flux in the electrode wire.
- It is used only to weld ferrous metals, primarily steels.

10.5.3 Suitability and Applications

It is used for welding a variety of joints, mainly with steels, stainless steels, and nickel alloys. Typical applications include bridges, high-rise buildings, shipbuilding and offshore drilling platforms.

10.6 Electroslag Welding

Electroslag welding is a fusion welding process for joining thick materials in a vertical position. Strictly speaking, it is not an arc welding process because arc is not used to obtain the desired heat for joining. The arc simply initiates the process by melting the flux to form the slag. After the molten slag reaches the tip of the electrode, the arc is extinguished. The required heat is generated because of electrical resistance of the molten slag when current flows through it. Resistance heating raises the temperature of the slag to around 1760 °C. The molten slag then melts the edges of the parts to be joined, as well as

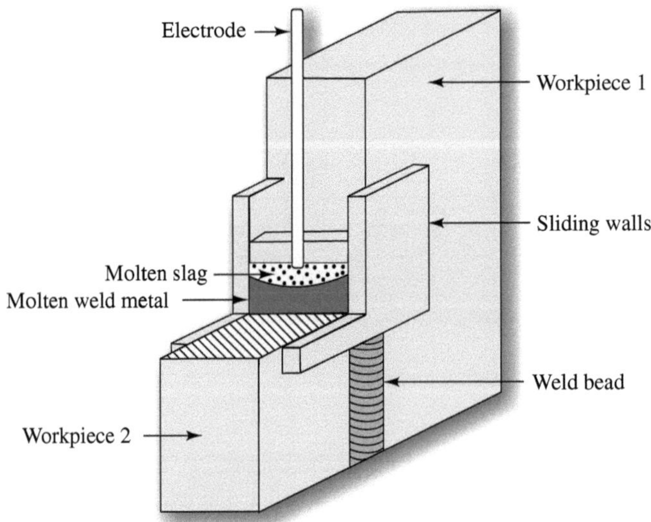

Fig. 10.5 An electroslag welding process

the continuously fed solid or flux-cored electrodes, which supply the filler metal. Multiple electrodes are often used to provide an adequate supply of filler metal and maintain the molten pool. The molten metal is confined between two sliding walls/plates usually made of copper. As the weld metal solidifies, the moulding plates move upward.

The molten metal pool remains shielded by the molten slag which moves along the full cross section of the joint as the welding proceeds. Under the normal operating conditions, the slag layer is 65 mm deep. The process is shown in Fig. 10.5.

10.6.1 Advantages

- It is one of the best methods for joining thick plates. This is due to large amount of molten weld metal involved in the process.
- This process eliminates the problems associated with multirun.
- welds because welding is done in one pass. As a result, weld quality is improved.
- Edge preparation is minimally required.
- It gives extremely high deposition rates.
- Flux consumption is less as compared to submerged arc welding.

10.6.2 Limitations

- Welding is performed in vertical positions only.
- Only thicker plates are welded.

- Controlling of solidification process is essential.
- Post-weld treatment is essential to avoid certain casting defects.

10.6.3 Suitability and Applications

It is most suitable for vertical joints in thick plates. Though cast iron, aluminium, magnesium, copper, titanium, etc., can all be welded by this process, thick steel (carbon steels and low and high alloy steels) parts are most effectively welded. Typical applications include shipbuilding, building construction, machine frames, heavy pressure vessels, turbine casings, and the joining of large castings and forgings.

10.7 Electrogas Welding

It is a variation of electroslag welding. The essential components of equipment for electrogas welding are similar to those for electroslag welding except that there is an arrangement to introduce shielding gas to protect the molten metal. The other point of difference is that in the electroslag welding no arc is used except during the starting stage while the arc is continuously used in electrogas welding. Both are however machine welding processes. The electrogas welding process is shown in Fig. 10.6. The molten weld metal and the molten slag are confined between two water-cooled copper shoes to prevent them from running off. As the molten metal solidifies, the sliding walls move upward with the welding head. Welding also proceeds in the upward direction. Both solid wire electrodes and flux-cored electrodes can be used in the process. Electrodes are fed through welding gun in order to protect them from the heat of the arc. Gases such as CO_2, argon or helium are used for providing shield. Direct current reverse polarity (DCRP) is used in the process.

10.7.1 Advantages

- Welding is normally done in one pass.
- Due to high heat input, preheat is not required even for thick plates.
- Weld metal properties are better as compared to electroslag welding.

10.7.2 Limitations

- Welding is limited to thinner materials.
- Workpieces need to be placed edge to edge (butt joint).
- Welding can be performed in vertical position.

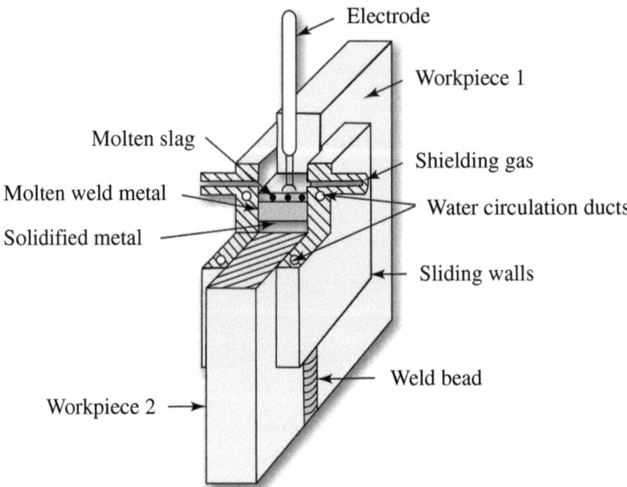

Fig. 10.6 An electrogas welding process

10.7.3 Suitability and Applications

The process is used on thinner materials normally for 15–75 mm thick workpieces. The materials mostly welded by EGW are steels, titanium and aluminium alloys. Typical applications include bridge construction, pressure vessels, thick-walled and large-diameter pipes, storage tanks and ships.

10.8 Carbon Arc Welding

In carbon arc welding, the required heat for joining is obtained from an electric arc between a nonconsumable carbon electrode (pure graphite) and the workpiece. It is the oldest welding process and is no longer used nowadays. Filler metal may or may not be used. Generally filler metals are used to join thick sections. The composition of the filler metal is same as that of the parent metal.

Although carbon electrode is nonconsumable, there is always some loss of carbon from it due to heat of the arc. The lost carbon forms a shielding atmosphere of CO and CO_2 and protects the weld from atmospheric contamination to some extent. Flux is used to neutralize the effect of extra carbon being disintegrating from the electrode.

However, additional flux is used to counteract carbon pick-up from the electrode.

Direct current straight polarity (DCSP) ensures low heat to the carbon electrode and thus restricts electrode disintegration. As a result carbon content of the electrode is not

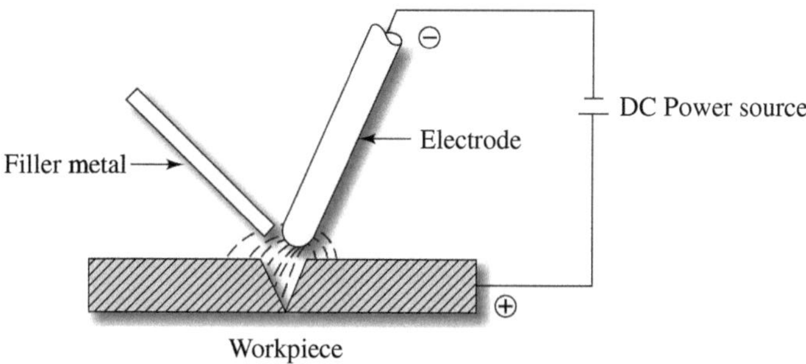

Fig. 10.7 A carbon arc welding process

carried over to the workpiece being welded. The carbon arc welding process is shown in Fig. 10.7.

10.8.1 Advantages

- Equipment is very simple and does not involve too much cost.
- No specialized skill is required on the part of the operator.
- It can be easily automated.

10.8.2 Limitations

- The carbon lost from the electrode can mix with the workpiece making the weld brittle and unsound.
- Blow holes and porosity are frequently occurring welding defects caused due to turbulence in the weld pool by the arc blow.

10.8.3 Suitability and Applications

Steel, copper and its alloys, nickel and monel metal are commonly joined materials by carbon arc welding. It is mainly used for providing heat source for brazing, braze welding and soldering as well as for repairing iron and steel castings.

10.9 Twin Electrode Carbon Arc Welding

It is a variation of the carbon arc welding in which the arc is produced between two nonconsumable carbon electrodes. Workpiece is not a part of the electrical circuit. The process is shown in Fig. 10.8.

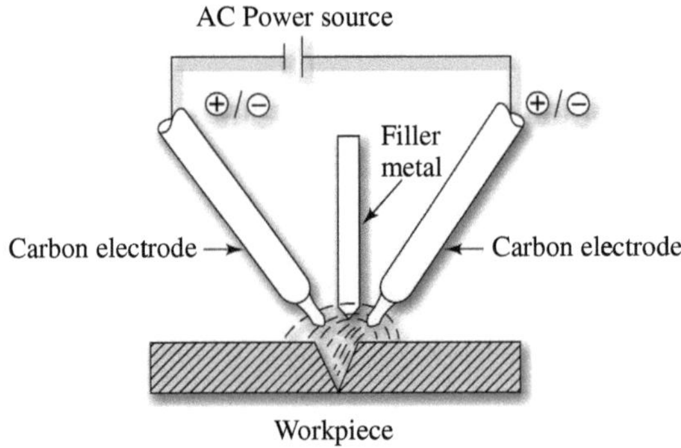

Fig. 10.8 A twin electrode carbon arc welding process

Alternating current (AC) is used to keep the electrodes at the same temperature. This is done by changing the polarity of the electrodes frequently so that the two electrodes are affected equally and present no problem. If connected to a DC set-up, one of the electrodes connected to the positive terminal of the circuit gets heated and consumed at a higher rate because of high heat generation at that side. This makes the arc unstable.

The process is mainly used to join copper alloys and ferrous metals.

10.10 Gas Tungsten Arc Welding (GTAW)

Gas tungsten arc welding (GTAW) was formerly known as tungsten inert gas (TIG) welding. In this process, the arc is produced between a nonconsumable tungsten electrode and the workpiece. Inert gases such as argon, helium or a mixture of the two are employed to provide a protective shield around both the arc and the molten metal pool.

All three welding currents may be used: *DCSP*, *DCRP*, and *AC*. Selection of the current depends upon the type of material to be welded. DCSP gives deep penetration and faster welding on thicker workpieces and is used for welding steel, cast iron, copper alloys and stainless steel. DCRP provides a wide weld pool with shallow penetration and hence is ideal for thin workpieces, but it is rarely used because it tends to melt the tungsten electrode. In general, AC is preferred for aluminium and magnesium because the cleaning action of AC removes oxides and improves weld quality.

A filler metal is generally used when welding thicker pieces with edges prepared. Filler metal has the similar composition as that of the original materials to be welded. In applications where a close fit exists, no filler metal may be needed.

The tungsten electrode has the highest melting temperature (3410 °C) of all metals and is also a strong emitter of electrons, which helps ionize the arc path and thus generates a stable arc. It does not melt in the arc, although a slight loss occurs in the course of time.

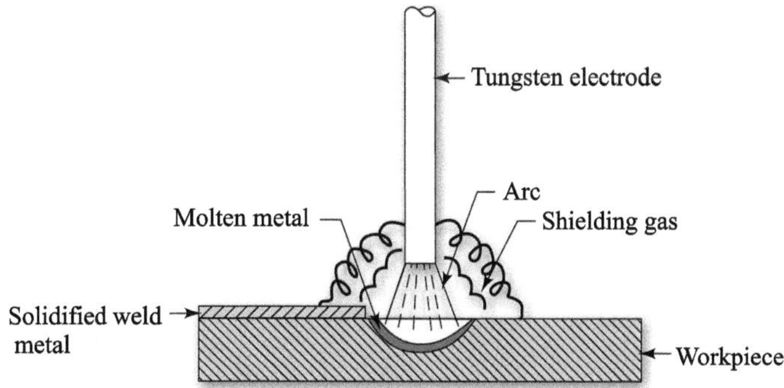

Fig. 10.9 A gas tungsten arc welding process

Thorium or zirconium may be used in the tungsten electrode to increase its melting point and electron emission characteristics and hence stabilize the arc. The GTAW process is shown in Fig. 10.9.

10.10.1 Advantages

- It gives a clean weld joint because of no use of flux in the process.
- There is less chance of weld spatter which makes the joint defect-free.
- It can be easily automated.
- Welding can be done in all positions.

10.10.2 Limitations

- The high cost of the inert gas makes this process more expensive than shielded metal arc welding (SMAW).
- The process is relatively slow as compared to GMAW and hence is not suitable for welding heavier metals.

10.10.3 Suitability and Applications

The process is particularly well suited for welding thin materials where a high-quality finish is desired. All metals and alloys including reactive metals, such as aluminium, magnesium and titanium, as well as the high-temperature refractory metals can be welded. It is extensively used, generally without filler metal, for the welding of longitudinal seams of thin-walled stainless steel and alloy steel pressure pipes and tubing. The process is quite popular for precision welding in aircraft, atomic energy and instrument industries. Aircraft frame, jet engine casing and rocket motor cases are typical examples of its use in aircraft industry.

10.11 Atomic Hydrogen Arc Welding

This welding process differs from other arc welding processes in the sense that the arc is produced between two nonconsumable tungsten electrodes rather than between one electrode and the workpiece. Hence, the workpiece is out of the electrical circuit. Hydrogen gas is passed through the arc. The heat of the arc breaks the hydrogen molecules into atoms. These hydrogen atoms after passing through the arc recombine to form molecules. This is a fusion reaction releasing large amount of heat having temperature of the order of 6100 °C. This heat is utilized for joining the metals. Here heat of the arc is not directly used for welding purposes, rather it is used to break the molecular hydrogen into atomic hydrogen. Normally an alternating current is used because of its better current regulation.

10.11.1 Advantages

- It gives high heat concentration.
- There is no need for additional shield because hydrogen protects the electrodes and molten metal from oxidation.
- Due to high heat capacity, it can be used to join many difficult-to-weld alloys.

10.11.2 Suitability and Applications

It is an older welding method and has limited use today. It is mostly used for the repairing of steel moulds and dies.

10.12 Plasma Arc Welding (PAW)

In plasma arc welding (PAW), the arc is created between a nonconsumable tungsten electrode and the workpiece, as in GTAW process. The electrode is surrounded by a water-cooled copper nozzle with a small orifice which squeezes the arc coming out of it and increases its pressure, temperature and heat intensity. An inert gas, usually argon is forced through the orifice, where it is heated to a high temperature and forms a plasma. A plasma is ionized hot gas, composed of nearly equal numbers of electrons and ions. It conducts electricity. The heat of the plasma which has temperatures as high as 33,000 °C is used to melt the metals to be joined during the welding process. Filler metal may or may not be used depending on the requirements of the joint. Inert gases such as argon,

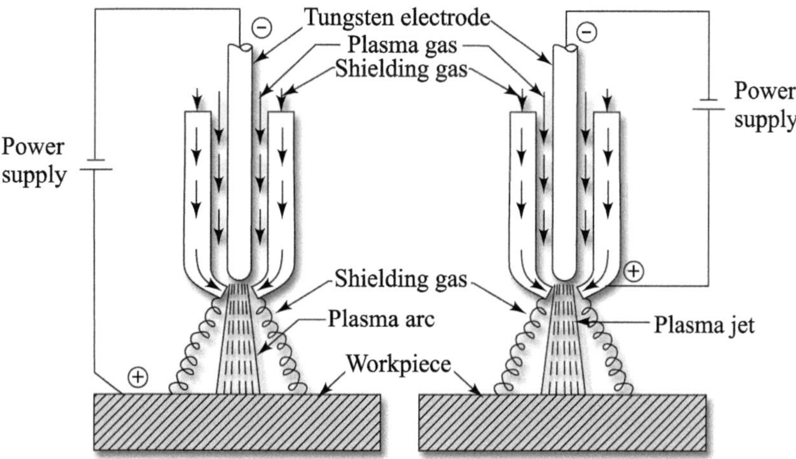

Fig. 10.10 Transferred and nontransferred arc in PAW

helium or mixtures can be supplied through an outer shielding ring to shield the weld pool. PAW is done almost exclusively with DCSP from a constant-current power supply. The welding current ranges between 100 and 300 A.

There are two methods of plasma arc welding, namely *transferred* type and *nontransferred* type. In the transferred type, electrical circuit is between the tungsten electrode and the workpiece. Workpiece acts as anode (+ve electrode) and the tungsten electrode as cathode (−ve electrode). The arc is transferred from the electrode to the workpiece, and hence the term is transferred. Here the arc force is directed away from the plasma torch and into the workpiece, and hence it is capable of heating the workpiece to a higher temperature than with the nontransferred arc. It is more adaptable to melting or cutting metal. In the nontransferred type, the electrical circuit is formed between the electrode (−ve) and the nozzle (+ve), and the heat is carried to the workpiece by the plasma gas. It is completely independent of the workpiece. The two processes are shown in Fig. 10.10.

10.12.1 Advantages

- It gives high heat concentration which results in higher welding speeds.
- It ensures arc stability.
- It produces less thermal distortion.
- Lower width-to-depth ratio of the weld ensures deeper and narrower welds.
- More arc length flexibility is possible without changing the bead width (Fig. 10.11). The higher heat of the plasma arc results in a narrower weld, less heat-affected zone

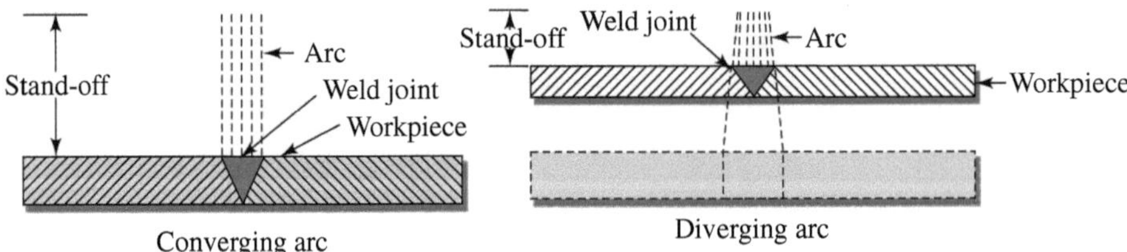

Fig. 10.11 Comparison between PAW and GTAW processes

and deeper penetration. Also, minor variations in the stand-off distance in PAW have little effect on bead width or heat concentration at the workstation.

- With high current plasma process, *Keyhole* effect is observed. The arc creates a hole in the material surrounded by the molten metal. As the torch progresses, the molten metal fills the keyhole to form the bead. If the gas pressure is increased further, the bead is removed from the region and the process can be utilized as plasma cutting.
- The process is readily automated.

10.12.2 Limitations

- The equipment is very expensive. The PAW equipment usually costs two to five times as much as GMAW equipment.
- Intense arc results in excessive ultraviolet and infrared radiations which can harm the skin. This necessitates special protection devices.
- Inert gas consumption rate is high.
- Greater knowledge is required on the part of the welder.
- Noise level is high (about 100 db, which is more than the permissible limit of 80 db for human ears).

10.12.3 Suitability and Applications

The process can be used to weld a variety of metals with part thicknesses generally less than 6 mm. It is often preferred over GTAW for making butt and lap joint because of high heat concentration, better arc stability and higher welding speeds. PAW finds commercial applications in aeronautical industry, precision instrument industry and jet engine manufacturing unit. Typically, the process is used for making piping and tubing made of stainless steels and titanium. PAW is, however, comparatively a new process and not very popular, as yet.

10.13 Electrode

An electrode is a piece of wire or a rod of a metal or alloy having high electrical conductivity which allows the flow of current through them when used in electrical circuit. They help to produce an arc which is a source of welding heat for many welding processes. They are the integral part of a welding system and are usually circular in cross section.

Electrodes may be of limited length (300, 350, 450 mm etc.) or of continuous type. A few welding processes use consumable electrodes which act as a filler metal. Others use nonconsumable electrodes which are usually made of tungsten because of its highest melting point among the metals.

Electrodes are generally of two types:

- Bare electrode
- Coated electrode

Bare electrodes are normally used with straight polarity set-up and are suitable for welding of wrought iron and low- or medium-carbon steel. They are usually solid rods and are used as filler metal in various welding processes. They do not produce sound and satisfactory welds, but still find use where weld strength is not a primary consideration and it is difficult to carry post cleaning of the joint.

Coated electrodes have a light or heavy coatings of slagging, fluxing or some other materials. Electrodes used in manual welding are usually covered with a flux coating that vaporizes in the heat of the arc to form a protective atmosphere (of CO_2) that stabilizes the arc and protects the molten and hot metal from contamination. The flux removes the oxidational impurities and prevents their formation. Some of the electrodes contain flux in their tubular section. Other electrodes have heavier coatings of silicate binders and powdered materials such as oxides, carbonates, fluorides, metal alloys and cellulose (cotton cellulose and wood flour). Generally, electrode coating performs the following functions:

- Provides a protective atmosphere against the formation of oxides, nitrides and other inclusions, thereby protecting the molten weld pool.
- Stabilizes the arc by reducing spatter of weld metal.
- Adds alloying elements to the weld metal to improve the properties of the weld.
- Generates gases (CO_2, CO, H_2 or water vapour) to act as a shield against the surrounding atmosphere.
- Performs metallurgical refining operations.
- Increases metal deposition efficiency. Electrodes with iron powder in the coating can be used to significantly increase the amount of weld metal that can be deposited with a given size electrode wire and current, because the iron in the coating goes into the weld.

Selection of electrodes depends on many factors such as their chemical composition, mechanical properties (especially tensile strength), welding position in which they may be used, preferred type of current and polarity if direct current is used and the type of covering. The type and position of the joint will decide the fill, freeze or follow characteristics of the electrode. Fill electrodes are used primarily for easy-to-weld joints in the flat position. The covering of these electrodes contains as much as 50% iron powder, which increases both deposition rate and current requirements. Fast-freeze electrodes have relatively low deposition rates and are used for vertical and overhead or steeply inclined joints. Coatings of such electrodes contain less than 10% iron powder, and hence less current is required than fill electrodes. Follow electrodes can be used at high welding speed with minimum skips and misses. These electrodes can be used on all type of joints in 10–18 gage (3.42–1.214 mm thick) sheet metal.

10.14 Stud Welding

Stud welding is an arc welding process in which studs (headless threaded bolts) or similar parts are attached to a flat workpiece. The arc is created between the stud and the workpiece by using a DC power supply. The stud gun which is used to hold the stud is positioned over the spot where the stud is to be placed. When the workpiece melts below the arc, the stud is pushed towards it and is allowed to solidify. A disposable ceramic ferrule (ring) is placed around the stud in its front part to concentrate the heat generated, prevent oxidation and retain the molten metal in the weld zone. Sequence of operations in stud welding is shown in Fig. 10.12.

10.14.1 Advantages

- The process is simple and easy to operate.
- The process is fast. It takes less than a minute to connect a stud to a thinner plate.
- No skill is required on the part of the operator.
- The process can be easily automated. Automation makes the process faster.

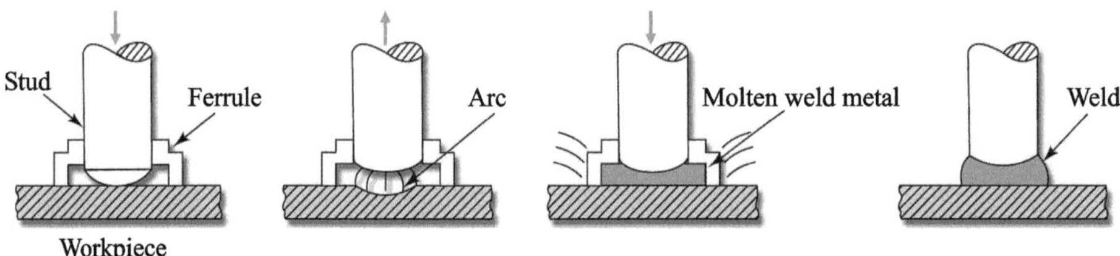

Fig. 10.12 A stud welding process

10.14.2 Limitations

Areas to be welded must be clean and free from impurities such as paint, scale, rust, grease, oil and dirt.

10.14.3 Suitability and Applications

The materials welded by using this method include carbon and alloy steels, brass, bronze and aluminium alloys. It is a quicker and easier way for providing fasteners to plate or round stock. Typically, it can be used for connecting handles to appliances.

10.15 Thermal Cutting

Thermal cutting uses heat of a gas flame, an electric arc, radiant energy or an exothermic reaction to cut a metal into two or more parts. The arc-cutting processes include air carbon arc, shielded metal arc, plasma arc, gas tungsten arc and gas metal arc. Electron beam and laser beam use the radiation energy for cutting. Out of these processes, oxyacetylene, air carbon arc and plasma arc are the three major thermal cutting processes used in the industry.

10.16 Oxyfuel Gas Cutting

It is similar to oxyfuel welding, but here the heat is used to remove a narrow zone from a metal plate or sheet. The workpiece is generally preheated with the fuel gas so that thermal cutting can be achieved at low temperature (870 °C). Cutting occurs mainly by oxidation and burning of the steel, with some melting taking place. The metal which is oxidized (Fe_3O_4) has lower melting point than the melting point of steel, and thus the cut is achieved faster than that by melting. The oxygen is introduced later which also helps in blowing the oxidized metal or slag out of the cut or the kerf (kerf widths range from about 1.5–10 mm), in addition to its primary function of converting the metals into metal oxides. The amount of oxygen required for cutting the metal is given by the following relation.

$$\begin{aligned} V_O &= 1.3 V_s \\ &= 1.3\, lbt \end{aligned}$$

where

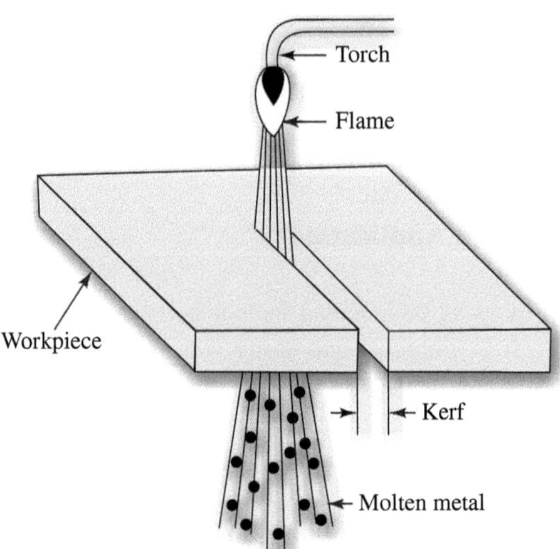

Fig. 10.13 A flame cutting process

V_0 = volume of oxygen required (mm^3)
V_s = volume of steel removed (mm^3)
l = length of cut (mm)
b = width of cut (mm)
t = thickness of steel (mm)

The process of oxyfuel gas cutting is shown in Fig. 10.13. Acetylene is the most widely used fuel gas in oxyfuel gas cutting. Other fuels used in the process include natural gas and propane, but the temperatures produced by such gas flames are lower than acetylene.

10.16.1 Suitability and Applications

Oxyfuel gas cutting works best on metals that oxidize readily. Carbon and low-alloy steels can be readily cut. Cast irons and stainless steels contain oxidation-resistant ingredients and hence require special techniques. Aluminium and copper are difficult to cut. Metal up to 30 inch in thickness can be cut by oxyfuel gas cutting. Underwater cutting can also be performed by oxyfuel gas cutting but with specially designed torches that produce a blanket of compressed air between the flame and the surrounding water. Joint edge preparation for welding is another application area.

10.17 Arc Cutting

The principle of arc cutting is similar to that of arc welding. The intense heat of the arc can cut almost all the metals.

In *air carbon arc cutting,* the metal is melted by the heat of an arc between the carbon electrode and the workpiece, and the molten metal is removed by a high-velocity (compressed) air jet. Metal oxidation is not involved as in case of oxyfuel gas cutting. The process is widely used for gouging, joint edge preparation and removing defective weld metal from a surface. Air carbon arc cutting is particularly effective for cutting cast iron. But the process is noisy and involves safety hazards.

10.17.1 Plasma Arc Cutting

Because of being the highest temperature source (16,500 °C), it can be used for rapid cutting of any material. Compared to flame cutting, it has the following advantages:

- It is more economical.
- It can be used to cut all the metals but is particularly suitable for cutting aluminium and stainless steels.
- Higher speed increases the productivity. The process is about five to eight times faster than flame cutting.
- It produces good surface and narrow kerfs.
- There is no thermal distortion.
- Heat-affected zones are absent.

Electron beam and laser beam are used for very accurate cutting but their use is limited due to high initial cost of equipments.

Short Answer Questions

1. Why is it difficult-to-weld aluminium? Which welding process can be used to join aluminium effectively?

Ans. Welding of aluminium and its alloys are difficult because of formation of its oxide (Al_2O_3) before and during the welding process which prevents its effective joining. Tungsten inert gas (TIG) welding is used successfully as it provides shielding cover of inert gases.

2. How is atomic hydrogen arc welding different from other arc welding processes?

Ans. Welding processes other than atomic hydrogen arc welding use direct heat of the arc for joining the parts. On the other hand, heat of the arc in atomic hydrogen arc welding is not directly used for joining purpose, but is used to break the hydrogen molecules into hydrogen atoms which recombine to form hydrogen molecules after coming out of the arc and this conversion process being exothermic in nature releases a large amount of heat to be used in joining the parts.

3. Why is submerged arc welding not suitable for overhead welding? What is its preferred welding position?

Ans. Overhead welding is not suitable because of the possibility of falling down of large molten pool and flux used in the submerged arc welding. The most preferred welding position for submerged arc welding is flat position.

4. Why is submerged arc welding not suitable for joining thinner metal sheets? Name its few important applications.

Ans. Submerged arc welding is not suitable for joining thinner metal sheets as the sheets may burn because of high temperatures of the molten metal. Its typical applications include thick plate welding for shipbuilding and pressure vessels.

5. Why is tungsten used as nonconsumable electrode?

Ans. Tungsten has the highest melting point temperature (3410 °C) of all metals which prevents its melting in the arc. Also it does not deteriorate even at high temperatures and is a stronger emitter of electrons which helps ionize the arc path and thus produces a stable arc.

Multiple Choice Questions

1. Consider the following statements

   1. An arc is created between the workpiece and the electrode.
   2. An arc is created between two workpieces.
   3. An arc is created between two electrodes.

   Of these statements:

   (a) 1 and 2 are true (b) 1 and 3 are true
   (c) 2 and 3 are true (d) 1 alone is true

2. The temperature of the heat produced by an arc is around

(a) 6500 °F (b) 6500 °C
(c) 3000 °C (d) 10,000 °C

3. Consider the following statements:

   1. In straight polarity, the workpiece is connected to positive terminal and the electrode to negative terminal of the circuit.
   2. In straight polarity, the workpiece is connected to negative terminal and the electrode to positive terminal of the circuit.
   3. In reverse polarity, the workpiece is connected to negative terminal and the electrode to positive terminal of the circuit.
   4. In reverse polarity, the workpiece is connected to positive terminal and the electrode to negative terminal of the circuit.

   Of these statements:

   (a) 2 and 4 are true (b) 2 alone is true
   (c) 1 and 3 are true (d) 3 alone is true

4. Consider the following statements about the submerged arc welding:

   1. It is mainly suitable for flat position welding.
   2. Heavier currents can be used.
   3. It uses huge quantity of fluxing material.
   4. Thinner sections are effectively welded.

   Of these statements:

   (a) 1 and 2 are true (b) 2, 3 and 4 are true
   (c) 1, 2 and 3 are true (d) 1, 3 and 4 are true

5. Thin aluminium sheets can be welded by

   (a) MIG welding (b) TIG welding
   (c) Resistance welding (d) Submerged arc welding

6. Consider the following statements:

   1. MIG welding uses nonconsumable electrode.
   2. Fluxing materials are essentially used in TIG welding.
   3. Approximately 75% of the heat is liberated near the positive terminal of a DC arc.
   4. Thicker sheets are effectively welded by submerged arc welding.

   Of these statements:

(a) 1 and 2 are true (b) 2 and 3 are true
(c) 3 and 4 are true (d) 1 and 4 are true

7. Consider the following statements:

   1. Electroslag welding is used to join thick materials.
   2. Electrogas welding is used to join thinner materials.
   3. Carbon arc welding uses consumable electrodes.
   4. Arc is hidden in submerged arc welding.

   Of these statements:

   (a) 1, 2 and 4 are true
   (b) 1, 3 and 4 are true
   (c) 3 and 4 are true
   (d) 2, 3 and 4 are true.

8. Consider the following statements about the atomic hydrogen arc welding:

   1. It uses tungsten electrodes.
   2. Heat of the arc is not directly used for joining.
   3. It requires inert gas shielding.
   4. The disintegration rate of electrodes is very high.

   Of these statements:

   (a) 2 and 3 are true
   (b) 3 and 4 are true
   (c) 2 and 4 are true
   (d) 1 and 2 are true.

9. Consider the following statements about the plasma arc welding:

   1. Arc is created between a tungsten electrode and the workpiece.
   2. Heat-affected zone is minimal.
   3. Heat concentration is very high.
   4. It does not require shielding atmosphere.

   Of these statements:

   (a) 2, 3 and 4 are true (b) 1, 2 and 3 are true
   (c) 1, 3 and 4 are true (d) 1, 2 and 4 are true

10. Consider the following statements about a coated electrode:

1. It stabilizes the arc.
2. It provides alloying elements.
3. It induces weld spatter.
4. It requires shielding atmosphere.

Of these statements:

(a) 2 and 4 are true (b) 3 and 4 are true
(c) 1 and 2 are true (d) 1 and 4 are true

11. Which of the following processes can be used for thermal cutting?

1. Atomic hydrogen arc 2. Plasma arc
3. Electron beam 4. Laser beam

Of these:

(a) 1, 3 and 4 are true (b) 2, 3 and 4 are true
(c) 1, 2 and 3 are true (d) 1, 2, 3 and 4 are true

Answers

1. (b)	2. (b)	3. (c)	4. (c)	5. (b)	6. (c)
7. (a)	8. (d)	9. (b)	10. (c)	11. (b)	

Review Questions and Discussions

Q.1. How is straight polarity different from reverse polarity?
Q.2. Why is shielded metal arc welding the most widely used welding process?
Q.3. Why is the deposition rate very high in submerged arc welding?
Q.4. What is CO_2 welding? Why is it not used today?
Q.5. What are the different modes of metal transfer in GMAW process?
Q.6. How is electroslag welding different from electrogas welding?
Q.7. What are the different gases used for providing shield?
Q.8. Why is filler metal needed?
Q.9. What is the difference between GMAW and GTAW processes?
Q.10. Why is tungsten used as a nonconsumable electrode in welding processes?
Q.11. Why is GTAW used for reactive metals?
Q.12. How does atomic hydrogen arc welding differ from other welding methods?
Q.13. Why is no extra shielding atmosphere required in atomic hydrogen arc welding?
Q.14. What is plasma?

Q.15. Why is the plasma arc welding not popular today?

Q.16. What are the constituents of a coated electrode?

Q.17. Where is stud welding more suitable?

Q.18. Why do surfaces need to be cleaned before the start of welding?

Q.19. Electron beam welding and laser welding have great potential, but they are not popular. Why?

Solid-State Welding Processes

11

Figure shows joining of a stud to a plate.

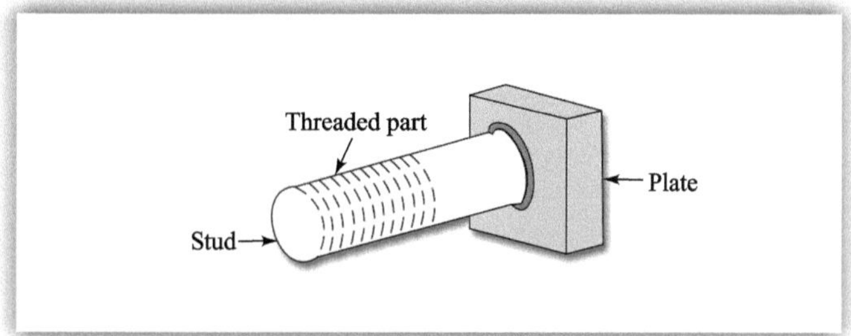

Learning Objectives
After reading this chapter, you will be able to answer some of the following questions:

- What is the significance of cold welding?
- How does inertia welding work?
- Where does explosive welding find applications?

D. K. Singh, *Fundamentals of Manufacturing Engineering*,
https://doi.org/10.1007/978-981-99-8767-2_11

11.1 Introduction

As the name suggests, in solid-state welding processes, parts are joined in the solid state without melting them. The process can be performed both at room temperature and at an elevated temperature. Filler metal is not used in the process. The parts are first cleaned to make them free from any oxide film, oil film or other contaminants. These parts are then brought closer to make a proper contact by applying high pressure.

11.2 Forge Welding

Forge welding is the ancient method of joining two parts. The process consists of heating the part in a forge to a plastic condition and then uniting it by the application of pressure. Village blacksmith uses a charcoal forge for heating purposes but oil or gas furnaces are used to increase the productivity of the process. Borax is used as a fluxing material which cleans the surfaces to be joined by dissolving the oxides. The quality of the weld produced depends on the material temperature, surface cleanliness and the amount of pressure applied. The process is especially used for mild steel and wrought iron parts, but has largely been replaced by other joining processes.

11.3 Cold Welding

Cold welding uses no heating, and the joint is made at room temperature with the application of pressure. The surfaces to be joined are first cleaned and placed in contact. Then the parts are pressed against each other to complete the weld. The joint is the result of intersurface molecular fusion. Pressure is applied either by impact or with a slow squeezing action as produced by hydraulic devices.

Welding may be produced in the shape of a ring, continuous seam or spot. Spot welds are rectangular in shape having an approximate area of $5t \times t$, where t represents the metal thickness.

Usually similar metals are joined by cold welding because of good bond strength. Dissimilar metals produce brittle and weak joint at the interface because of the formation of brittle intermetallic compound.

The process is generally used for joining of small parts made of soft, ductile metals such as aluminium and copper. Cold welding finds application in electrical connections and welding wire stock where welding heat might cause damage. It is particularly useful in explosive areas.

11.4 Friction Welding

Friction welding utilizes the frictional heat generated at the interface of the two parts being joined.

In this process, one of the members remains stationary while the other is rotated at a high constant speed. The stationary member is now brought closer to the moving member with an axial force to make a proper contact. The force is then increased, and contact friction quickly generates enough heat to raise the abutting surfaces to the welding temperature. As soon as this temperature is reached, rotation is stopped, so that the weld is not destroyed by shearing, and the pressure is maintained or even increased to complete the weld. The sequence of operations used in friction welding is shown in Fig. 11.1. Flash produced at the interface can be removed by machining. It is a fast process, and the joint can be made in few seconds. The process is similar to butt and flash resistance welding but differs in the modes of heating. Although friction welding is a solid-state welding process, the contact surfaces reach the plastic state due to frictional heating.

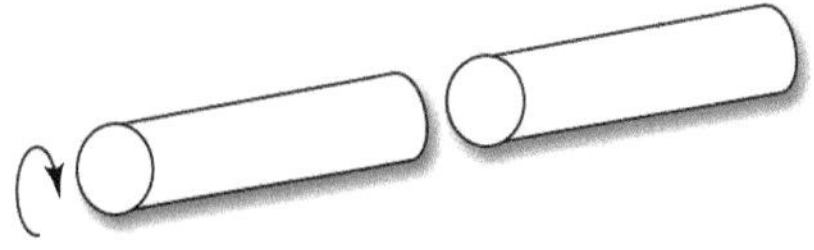

(*a*) Left part is rotated and the right part is held stationary.

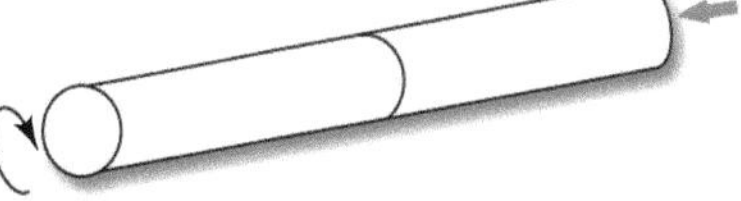

(*b*) Right part is brought closer to left part while the left part is still rotating.

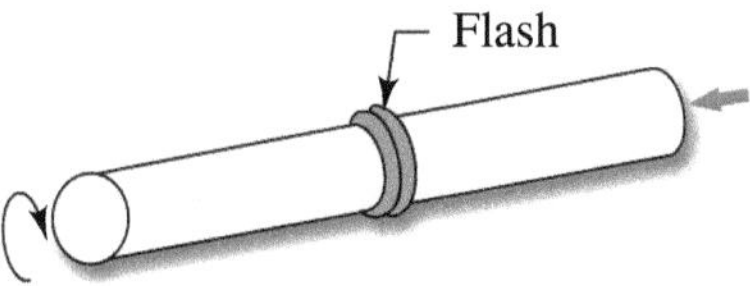

(*c*) Both parts are underpressure and the flash is formed.

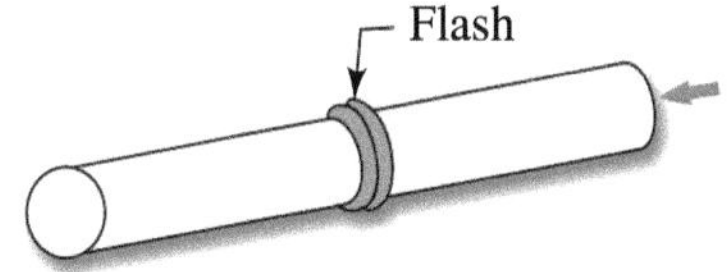

(*d*) The joint is complete with the increased pressure and the left part stops rotating.

Fig. 11.1 Sequence of operations in a friction welding

11.4.1 Advantages

- Friction welding does not require filler metal, flux or shielding gas.
- The process is neat and clean because no smoke, fumes or gases are generated.
- Power requirement is low.
- It is suitable for automation.
- No specialized skill is required on the part of the operator.
- The strength of the joint is equal to that of parent metal.

11.4.2 Limitations

- Only cylindrical parts such as bars, tubes can be joined to flat surfaces.
- The attachment devices should be stronger enough to avoid vibrational problems.
- Flash formation is undesirable, and it needs to be machined for final applications.

11.4.3 Suitability and Applications

Any metal similar or dissimilar can be joined by friction welding. The process finds considerable use in welding plastics. It can suitably replace flash or upset resistance welding, where one of the members to be joined has axial symmetry. Typical applications include production of axle shafts, engine valves, transmission shafts, welding of studs to plates of any thickness, attachment of solid part to a tube and production of cutting tools such as drills, taps and reamers where HSS cutting body is welded to carbon steel shanks. Welding of stud to a plate is shown in Fig. 11.2.

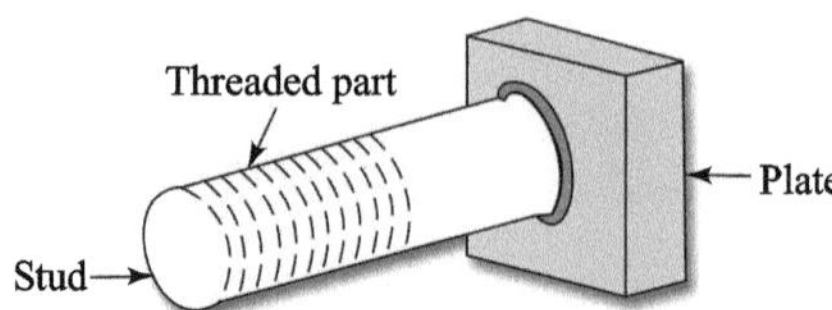

Fig. 11.2 Welding of stud to a plate

11.5 Inertia Welding

Inertia welding is the variation of friction welding. Here, the moving member is attached to a flywheel. The rotating assembly is then pressed against the stationary member. The kinetic energy stored in the flywheel is used to generate the frictional heat for joining. When sufficient temperature is reached, the assembly is further pressed to complete the weld. The process is shown in Fig. 11.3. The main difference between the friction welding and the inertia welding is that in the former, energy is supplied from a conventional drive source such as electric or hydraulic motors, whereas the latter utilizes kinetic energy stored in a rotating flywheel. Inertia welding is a faster method but the strength of the joint is same for both welding processes. Typically a stud can be joined to a flat surface by using this method.

11.6 Ultrasonic Welding

Ultrasonic welding uses the vibration energy for joining the metals. A transducer is used to convert the electrical pulsations into high-frequency mechanical vibrations. No external heat is applied. The vibratory action cleans the oxides and other contaminants on the interface surfaces. After that these surfaces are pressed to make intimate contact resulting in a solid-state bond. At the point of contact, the workpieces are heated by internal friction

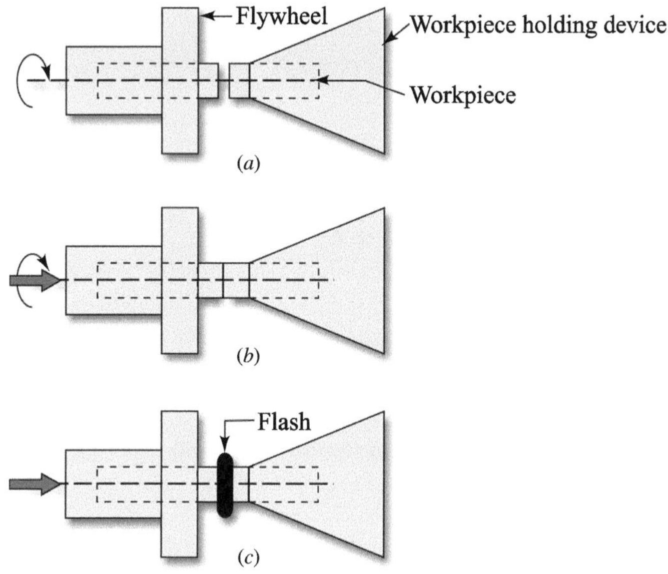

Fig. 11.3 Sequence of operations in inertia welding

which can generate heat having temperatures up to one-half of the melting point of the workpiece material (on absolute scale), making them plastic before the pressure is applied.

11.6.1 Advantages

- Similar and dissimilar metals among themselves or with other materials can be joined. For example, aluminium can be joined to glass.
- Because of no external heating involved and lower-temperature applications, heat-sensitive electronic components can be easily joined by this process.
- It produces joint of adequate strength.
- The process can be easily automated.
- The process is very fast and consumes very little time. Thinner metal sheets can be joined in few seconds.
- The equipment is simple, requiring moderate skill.
- The process is neat and clean.

11.6.2 Limitations

- The process can be performed only in the lap position.
- Thicker workpieces cannot be joined. The maximum thickness can be 3 mm.
- The life of the equipment is short due to fatigue of the horn.

11.6.3 Suitability and Applications

The process is very much suitable for joining thin sheet and foil to itself or to any thickness. Multilayer joining of thinner sheets is another possibility. Aluminium and its alloys, because of their soft nature and low melting points are more easily welded by this method. It is used extensively in joining plastics (melting takes place because of low melting temperatures) and in the automotive and consumer electronic industries for lap welding of sheet, foil and thin wire, and in packaging with foils. Ring-type and continuous seam welds can be used for hermetic sealing.

11.7 Explosive Welding

Explosive welding uses the impact of an explosive material for joining two materials. The impact is so high that it produces pressure of several million pounds per square inch, enough to produce a sound metallurgical bond between the parts being joined.

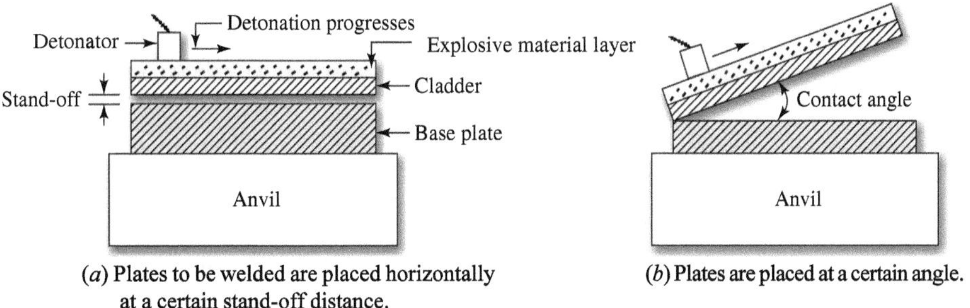

(*a*) Plates to be welded are placed horizontally at a certain stand-off distance.

(*b*) Plates are placed at a certain angle.

Fig. 11.4 An explosion welding process

The two parts to be joined are in the sheet or plate form. One of the parts, the base plate is placed on an anvil. The other part, the cladder (also called clad metal or flyer) is placed above the base plate at a stand-off distance of 3.175 to 6.35 mm. The stand-off distance can be maintained by thin corrugations of stainless steel or even cardboard. An explosive material such as dynamite or nitroguanidine, usually in the form of a sheet is placed on the cladder plate. The explosive detonated progresses in the direction shown by the arrow in Fig. 11.4. As the explosion front moves across the cladder, a moving collision point is formed which generates a jet of air. The jet blows oxides, films or any other contaminants ahead of the collision point leaving clean bonding surfaces. The material placed in the stand-off gap is also blown away by the advancing jet. Detonation speeds are usually 2400–3600 m/s, depending upon the type of explosive, thickness of the explosive layer and its packing density. The impact of the explosive creates a turbulent plastic metal flow, resulting in a characteristic wave or ripple pattern and providing mechanical interlocking of the two surfaces in addition to a fusion bond.

11.7.1 Advantages

- The joint is very strong.
- The composition of the material remains unaffected since there is no heat-affected zone in the joint.
- Similar and dissimilar metals can be joined.
- The process is simple and fast.
- The process can be easily automated.
- There is no limitation on the thickness of materials being joined. Very thin materials can be joined to very thick materials.

11.7.2 Limitations

- Because of explosion involved in the process, the process is dangerous, and hence safety precautions are needed.
- Metals with low melting points and low impact resistance cannot be bonded effectively. Low melting point materials cannot be joined because of burning possibility. Also, those materials which cannot bear large impact are difficult to be welded because of possibility of breaking into pieces.
- It produces loud noise and great vibration.

11.7.3 Suitability and Applications

The explosive welding is particularly suitable for joining large-area sheets. In addition to area welds, seam, spot, lap and edge welds are possible variations. It is a specialized process used for lap joints in difficult-to-weld metals and their combinations. It joins a wide range of cladder materials (such as aluminium, copper, ferrous alloys, nickel alloys, titanium and others) to a variety of base materials, usually carbon and alloy steels.

Cladding of plates is one of the major commercial applications of explosive welding. The most common reason for using clad material is to obtain the corrosion resistance of the cladding alloy. Cladding of cylinders both inside and outside is done by this process. Major areas of use of explosively clad products are heat exchanger tube sheets and pressure vessels. The process can also be used for repairing and building-up particularly both inside and outside of cylindrical components.

11.8 Diffusion Welding

In diffusion welding, the joint is made by the application of heat and pressure, which cause atoms of the two parts being joined to diffuse at the interface. The process creates a metallic bond as a result of the pressure applied causing microdeformation of the surfaces into each other, and diffusion of atoms (because of vacant lattice sites) between the two materials. The process temperature is about 70% of the melting temperature of the base metal on the absolute scale. The parts are usually heated in a furnace or by electrical resistance. Pressure required may be applied by dead weights or a press, but pressures just below the yield stress at the operating temperature are preferred, because diffusion welding is accomplished entirely in the solid state. High-pressure autoclaves are also used for bonding complex parts. The quality of a diffusion weld mainly depends on three factors such as temperature, pressure and time of contact. Other variables that influence the quality of the bond are atmosphere and surface finish. A good vacuum or oxygen-free, inert-gas environments are best suited for diffusion bonding.

11.8.1 Advantages

- The process is versatile. It can be used to join any metal. Plastic materials are best welded.
- A continuous leakproof joint can be made.
- Thinner as well as thicker materials can be joined.
- Because of lesser quantity of heat being used in the process, properties of the materials are remaining unaffected.

11.8.2 Limitations

- Diffusion welding is a slow process because it involves migration of the atoms across the joint between rough surfaces. The time required varies from seconds to 20 h.
- Since the process is slow, it is not suitable for large-scale production.
- Considerable operator skill is required. But the process can be automated for better joint and to reduce the involvement of the operator.
- Production of rough surfaces (surface finish of the order of 0.2–0.4 μm) is a prerequisite requirement for better joint which makes the process costly.
- Surface cleanliness is mandatory.

11.8.3 Suitability and Applications

The process is mainly suitable for joining dissimilar metals and composite materials. However, reactive metals such as titanium, beryllium and zirconium, and the high-temperature refractory metal alloys can also be joined by this process. Generally inert or protective atmospheres are needed to produce a sound joint in such cases. Cutting tools are generally provided coatings of carbide or hard alloys by using diffusion process. Principal users of this process are atomic, aerospace and electronic industries.

Short Answer Questions

1. Why is solid-state welding so named? Name a few solid-state welding processes.

Ans. Solid-state welding is performed in the solid state at room temperature or elevated temperatures with the application of pressure. Examples of solid-state welding include friction welding, diffusion welding, cold welding and ultrasonic welding.

2. What are the important advantages of ultrasonic welding?

Ans. Ultrasonic welding can be used to join similar or dissimilar metals among themselves or with other materials without involving external heating or very high temperatures. As such heat-sensitive electronic components can be easily joined by this process.

3. What is the principle of explosive welding?

Ans. Explosive welding uses the high energy of explosion of a detonator to join the metal sheets or plates in lap position.

4. What are the merits and demerits of friction welding?

Ans. Friction welding does not require filler metal, flux or shielding gas and produces no smoke, fumes or gases, thus ensuring cleanliness of the process. On the demerit side, only cylindrical parts such as bars, tubes can be joined to flat surfaces by this process.

Multiple Choice Questions

1. Consider the following statements about solid-state welding processes.

   1. They can be performed at room temperature or at increased temperature.
   2. Pressure is essentially required.
   3. Filler metal is not needed.

   Of these statements:

   (a) 1 alone is true
   (b) 1 and 3 are true
   (c) 1, 2 and 3 are true
   (d) 2 and 3 are true

2. Consider the following statements about cold welding:

   1. If is performed at room temperature.
   2. Usually similar metals are joined by cold welding.
   3. It is useful for brittle materials.

   Of these statements:

   (a) 1 alone is true
   (b) 1 and 2 are true
   (c) 2 and 3 are true
   (d) 1, 2 and 3 are true

3. Consider the following statements about friction welding:

   1. The process is similar to resistance welding.
   2. Frictional heat is used to join the metals.

3. It is a fusion welding process.
Of these statements:

(a) 2 alone is true
(b) 2 and 3 are true
(c) 1 and 2 are true
(d) 1, 2 and 3 are true

4. Which of the following welding uses a transducer?

(a) Electron beam welding
(b) Ultrasonic welding
(c) Inertia welding
(d) Laser welding

5. Which of the following welding methods finds considerable use in welding plastics?

(a) Electron beam welding
(b) Resistance welding
(c) Shielded metal welding
(d) Friction welding

6. Consider the following statements about ultrasonic welding:

1. It can be used to join plastics.
2. It is only suitable for lap welding position.
3. Thicker workpieces are best welded by this method.

Of these statements:

(a) 1 and 2 are true
(b) 1 and 3 are true
(c) 2 and 3 are true
(d) 1, 2 and 3 are true

7. Consider the following statements about explosion welding:

1. It gives a weaker joint.
2. Thinner and thicker workpieces can be joined by this process.
3. It uses dynamite.

Of these statements:

(a) 1 and 2 are true
(b) 2 and 3 are true
(c) 1 and 3 are true
(d) 1, 2 and 3 are true

8. Consider the following statements about diffusion welding:

1. It uses heat and pressure.
2. It is based on the formation of metallic bond between the joining members.
3. A good vacuum is required.

Of these statements:

(a) 1 alone is true (b) 1 and 2 are true

(c) 2 and 3 are true (d) 1, 2 and 3 are true

Answers

1. (c) 2. (b) 3. (a) 4. (b) 5. (d) 6. (a)
7. (b) 8. (d)

Review Questions and Discussions

Q.1. What is the principle of solid-state welding?
Q.2. What is forge welding?
Q.3. Where is cold welding used?
Q.4. What is the principle of friction welding?
Q.5. What is flash? Is it desirable?
Q.6. How does inertia welding differ from friction welding?
Q.7. What is the range of frequency used in ultrasonic welding?
Q.8. Why is ultrasonic welding popular in electronic industries?
Q.9. Which welding process is widely used for joining plastics?
Q.10. How does explosion welding work?
Q.11. What is the principle of diffusion welding?

12 Resistance Welding Processes

Resistance welding finds extensive applications in automobile industries because of convenience and clean environment.

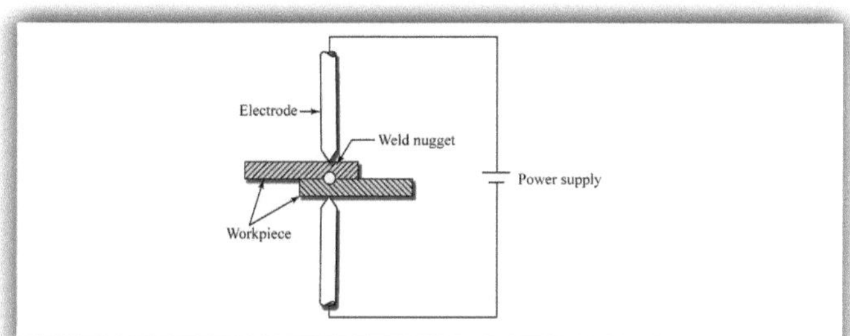

Learning Objectives

After reading this chapter, you will be able to answer some of the following questions:

- How does resistance spot welding differ from projection welding?
- What is meant by high frequency resistance welding?
- How does percussion welding differ from other resistance welding processes?

D. K. Singh, *Fundamentals of Manufacturing Engineering*,
https://doi.org/10.1007/978-981-99-8767-2_12

12.1 Introduction

Resistance welding utilizes the heat generated by the electrical resistance of the workpieces to be joined. Workpieces offer resistance in the flow of electric current which is maximum at the interface in comparison to their other parts. Hence greatest heat is generated at the interface. When the proper temperature is reached, the pressure is applied to complete the weld. Because of application of pressure, the process requires lower temperature as compared to gas or arc welding.

In most of the resistance welding processes, the base metals do not melt, hence they can be included in the category of solid-state welding processes. The resistance welding differs from other welding processes in the sense that it uses no flux and no filler metal and the joints are often of the lap type.

The workpieces are placed between two electrodes and the circuit is completed as shown in Fig. 12.1. Workpieces are part of the electrical circuit. The heat generated during the process is given by the equation

$$H = I^2 R t$$

where

H = heat generated
I = current
R = resistance
t = time of current flow.

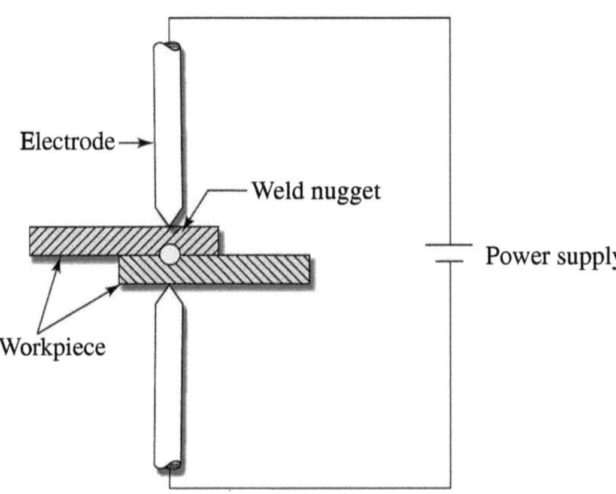

Fig. 12.1 Principle of resistance welding

Because of being in the squared form, the current plays important role in the heat generation.

Both the faying surfaces are heated simultaneously, while remaining material and electrodes are relatively cool. The electrodes are usually water-cooled to keep their temperature low and thereby extend their useful life. Maximum temperature is produced at the point between the faying surfaces where the contact is intimate, i.e., at the point where weld nugget forms. Other areas are relatively at low temperatures and their resistances are partially utilized in the welding process. The process uses low voltage and high current. A suitable amount of pressure is also applied.

12.1.1 Advantages

- The process is fast and suitable for mass production.
- It is economical. Flux and filler metals are not required.
- No specific skill is required.
- Majority of the metals whether similar or dissimilar can be joined.
- Automation of the process is easier.

12.1.2 Limitations

- Only electrical conducting materials can be joined. Metals such as tin, zinc and lead are difficult to be welded because of their low melting points.
- Pressure and current need to be carefully controlled.
- The initial cost of the equipment is high.

12.1.3 Suitability and Applications

Resistance welding is suitable for joining light gage metals and finds applications in manufacturing plants and machine shops.

12.2 Resistance Spot Welding

It is the simplest and most widely used method of resistance welding. A pair of copper electrodes is used to press the sheet metal workpieces kept between them in the lap position as shown in Fig. 12.2. A very small area, known as **spot**, is welded at the point where the workpieces are pressed by the electrodes. The metal at the spot melts and forms nugget. Normally a succession of such spot welds (nuggets) are produced at appropriate

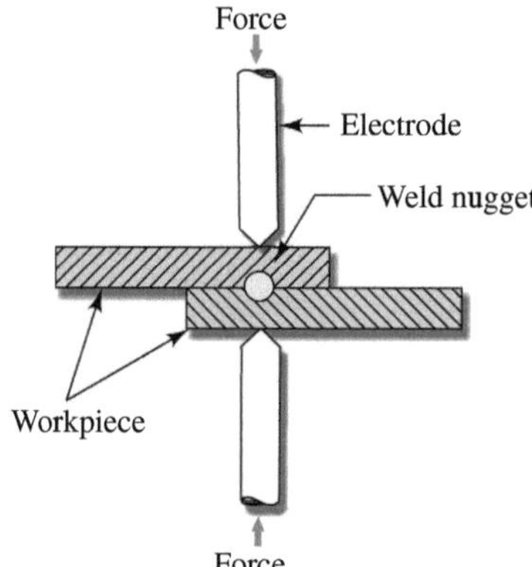

Fig. 12.2 A resistance spot welding process

intervals along the length of the weld, and their size is a function of electrode diameter that varies typically in the range of 2–10 mm. Very high currents are used in the process. Selection of a particular value of current depends on the materials being welded and their thickness. The process uses no filler metal. It produces a clean weld because the joint is fully protected and there is no chance of atmospheric contamination. The welding cycle for the spot welding is divided into four parts.

12.2.1 Squeeze Time

During squeeze time, the workpieces are pressed between the electrodes to make intimate contact. No current flows through the workpieces during this time.

12.2.2 Weld Time

During this period, the current is switched on while maintaining the maximum squeeze pressure. The temperature at the interface starts rising and attains welding temperature to melt the metal at the interface. The pressure is now increased considerably. Pressure plays an important role in resistance welding. The resistance at the interface is inversely proportional to the pressure. The variation is shown in Fig. 12.3. If the pressure is low, the contact resistance will be high which can burn the surface. Application of high pressure will expel the molten metal from between the faying surfaces or few indentation marks will be left by the electrodes on the surfaces.

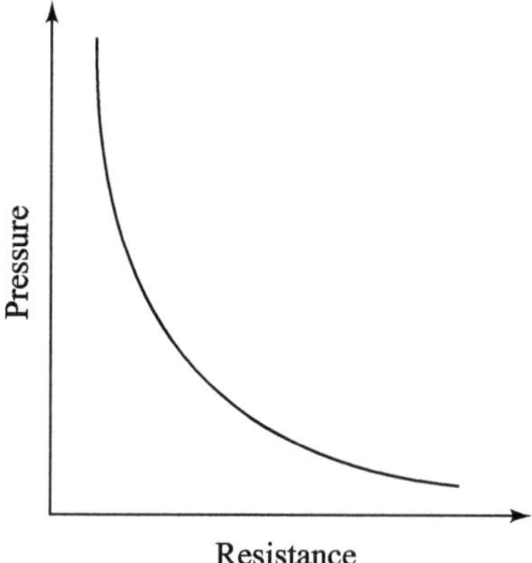

Fig. 12.3 Pressure versus resistance curve

12.2.3 Hold Time

During this time, the increased pressure is maintained on the workpieces. In the beginning of this time, known as forge time, pressure is further increased. The increased pressure is maintained constant afterwards.

12.2.4 Off Time

During this period, pressure is released from the workpieces and the electrodes are removed and positioned for the next position. All the four cycles are shown in Fig. 12.4.

12.2.5 Advantages

- Similar and dissimilar metals can be welded.
- The process is very fast and it takes very little time to complete the weld.
- No edge preparation is required.
- The process can be used for large production run if multiple spot welding machines are used.

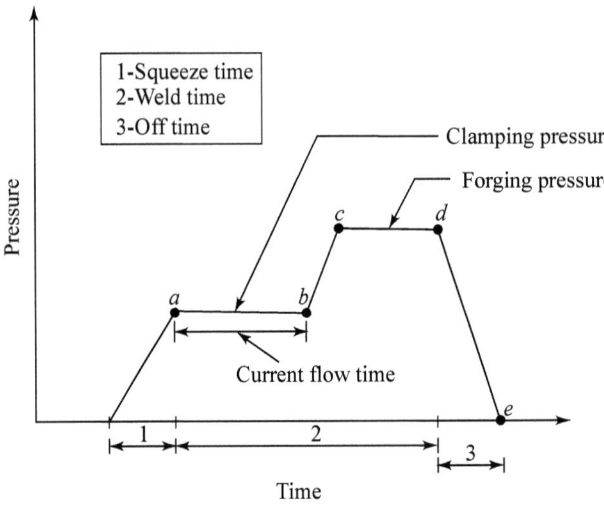

Fig. 12.4 Different cycles in spot welding

12.2.6 Limitations

- Silver and copper are especially difficult to weld because of their high thermal conductivity.
- Welding cannot be performed in all the positions. Only overlapping position is possible.

12.2.7 Suitability and Applications

Spot welding can be used as a substitute for riveting. It gives better joint if the workpiece thickness does not exceed 3 mm. It finds extensive applications in sheet metal industry. Attachment of handles to domestic cookware and spot welding of automobile bodies are some of its examples.

12.3 Resistance Seam Welding

Resistance seam welding can be regarded as a continuous spot welding process. The process uses two copper alloy wheels or rollers that act as electrodes for producing the weld in place of cylindrical electrodes as in case of spot welding. The basic equipment is the same as for spot welding, except for the changes of electrodes. These electrodes also exert pressure as the sufficient heat is generated at the interface due to the electrical resistance of the parts. The resulting weld consists of a series of overlapping spot welds

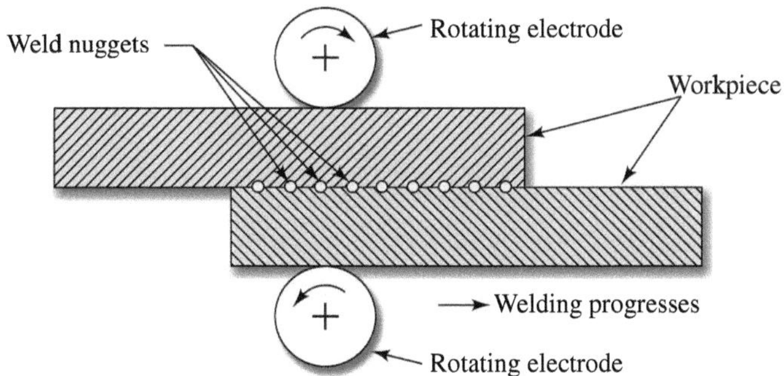

Fig. 12.5 A resistance seam welding process

to provide a pressure-tight joint (Fig. 12.5). If pressure-tight quality of the lap seam weld is not required, the individual nuggets are spaced to give a stitch effect. This process is called roll spot welding.

Seam welding is used to produce a continuous leakproof joint in liquid and gas containers made from thin metal sheet. Typical applications include longitudinal seam of cans for household products, automobile mufflers, gasoline tanks and refrigerator cabinets.

12.4 High-frequency Resistance Welding

It is similar to seam welding, except that a high-frequency current (upto 450 kHz) is used instead of 60 Hz. This process is used to make butt welds between thicker metal plate. A typical application is making butt-welded tubing. In this process, the electrical resistance of the abutting metal is used to generate the heat. The localized high-intensity heating allows the whole edge of the material to reach welding temperature. When they attain the desired temperature, the heated surfaces are pressed together to form the weld. The process is shown in Fig. 12.6. This welding is used to make tube and pipe from metal plates.

12.4.1 Advantages

- The power requirement is very low due to high-frequency current being used in the process.
- Because of low heat-affected zone (HAZ), thermal distortion is minimum.
- Similar and dissimilar metals can be welded. High-conductivity metals such as aluminium and copper can also be joined.
- Welding speed is high which makes it suitable for large-scale production.

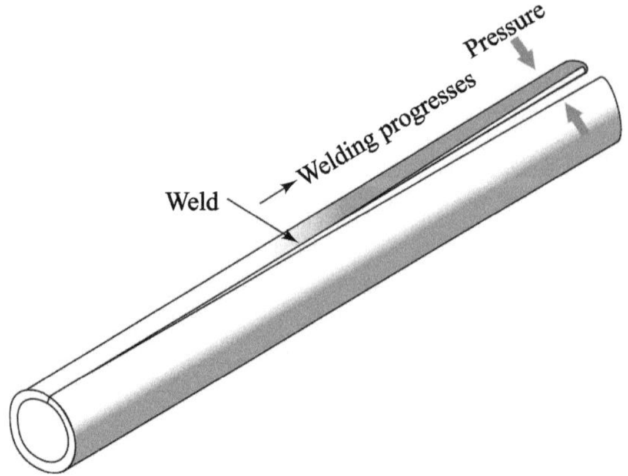

Fig. 12.6 A high-frequency resistance welding process

12.5 Resistance Projection Welding

Projection welding is a variation of spot welding. Small projections (dimples) are produced on one of the surfaces to be welded with a punch and die at the location where a weld is desired. The workpieces are then placed between large-area electrodes, and the current is switched on. High localized temperatures are generated at the projections. During the welding process, the projections collapse, owing to heat and pressure, and the workpieces are brought in close contact. Usually two or three projections are welded at a time. Projections may be round, oval or circular to suit the design conditions. If circular projections are used, their diameter can be equal to the workpiece thickness and extend upto about 60% of the thickness. The weld cycle time for projection welding is the same as for the spot welding. The finished weld is similar to spot weld but has small indentation mark left by the projection. The process is shown in Fig. 12.7.

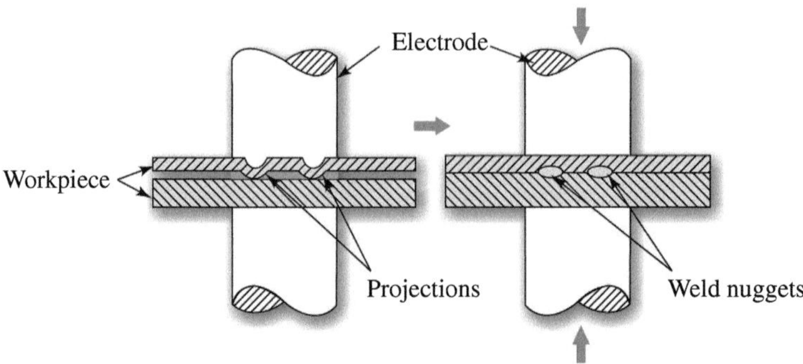

Fig. 12.7 A resistance projection welding process

12.5.1 Advantages

- The process is fast because a number of welds can be made simultaneously.
- Small amount of current is required because of greater current density at the projections.
- The pressure to be applied by the electrode is low.
- Thermal shrinkage and distortion are minimum because of low heat-affected zone.
- The electrode has longer life. Also, it requires little maintenance.

12.5.2 Limitations

Selection of correct current and application of optimum pressure are important.

12.5.3 Suitability and Applications

Projection welding can be used to weld metals of different thicknesses. It finds extensive applications in joining small attachments such as nuts, bolts, studs, and similar other parts to sheet structures. It is used in the production of automobile bodies, domestic equipments, office furnitures and machine parts.

12.6 Flash Welding

Flash welding is also known as flash butt welding. The basic equipment for flash welding is similar to that of butt welding but it differs in the mode of heating the metal. Here the heat required for the joint is obtained by means of an arc rather than the simple resistance heating. So it is classified as arc welding. A high voltage in the circuit produces a flash (arc) between the parts placed end-to-end. The arc produces enough heat to soften the metal at the joint. The weld is completed by applying suitable amount of forging pressure. The joint may later be machined to remove the upsetting. The process is shown in Fig. 12.8.

12.6.1 Advantages

- It consumes less welding current.
- Larger area can be welded.
- The process is fast because of high heat generation.
- Edge preparation is not required.

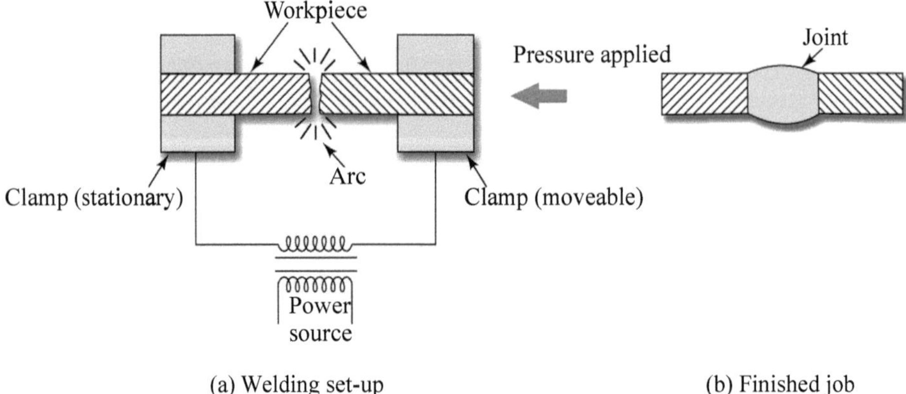

Fig. 12.8 A flash welding

- The joint is stronger.

12.6.2 Limitations

- Metals such as lead, zinc, tin, copper and their alloys are not welded by this method because of their burn-off possibility by the arc.
- Little upsetting can occur at the joint which needs to the machined.
- The welding machine is costly and takes more floor area.

12.6.3 Suitability and Applications

The flash welding is suitable for end-to-end or edge-to-edge joining of similar and dissimilar metal sheets, tubings, bars and rods. Major users are automobile and aircraft industries. Typical applications include joining of pipe and tubular shapes for metal furniture and windows.

12.7 Resistance Butt Welding

Resistance butt welding is also known as upset welding. In this process, the pieces to be joined are placed end-to-end between two clamps which can exert pressure as per requirement. The high resistance of the joint generates heat and causes fusion to take place at the interface. The joint is now compressed by the clamp forces to complete the

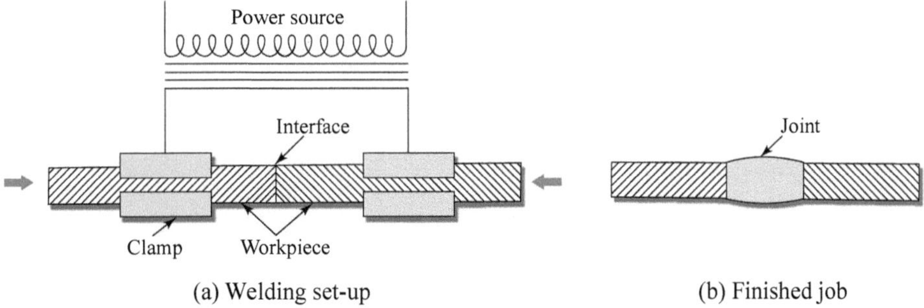

Fig. 12.9 A resistance butt welding

weld. The process is shown in Fig. 12.9a. Because of the high pressure at the interface, the ends slightly upset which can be removed by machining, if required (Fig. 12.9b).

12.7.1 Limitations

- Edge preparation is mandatory.
- It requires very high-density current.

12.7.2 Suitability and Applications

The resistance butt welding is especially suitable for joining rods, pipes and other parts of uniform cross sections.

12.8 Percussion Welding

Percussion welding uses heat of the arc produced by the rapid discharge of stored electrical energy in a capacitor.

The parts to be joined are placed in position similar to flash welding. The intense heat of the arc melts the parts. The heated parts are then pressed together to complete the weld. The process is very fast due to intense heat of the arc. The arc duration is very short.

12.8.1 Advantages

- Due to rapid action of discharge, heat-affected zone (HAZ) is very minimum and thermal distortion is almost absent.

- Parts with different thermal conductivity can be easily joined because heat is concentrated only at the two surfaces.
- Little or no upsetting occurs at the joint.

12.8.2 Limitations

- Only small areas can be welded.
- Parts should have regular section.

12.8.3 Suitability and Applications

The process is very much suitable for butt welding of bar or tube which are heat prone. Typically, the electronic components can be joined by percussion welding.

Short Answer Questions

1. What is the principle of resistance welding?

Ans. Resistance welding uses the electrical resistance of the parts being joined for producing the heat to be used in joining process according to the equation $Q = I^2Rt$, where Q = heat produced (J), I = current (A), R = electrical resistance of the parts being joined (ohm), and t = time (s).

2. Where does resistance welding find extensive applications and why?

Ans. Resistance welding is widely used in automobile industries to join thin metal sheets usually in lap position. As it uses the pressure for completing the joint and eliminates the use of flux and filler metal, thus offering clean atmosphere around the joint and making it suitable for such applications.

3. What are the merits and demerits of resistance welding?

Ans. Resistance welding offers faster operation, eliminates the use of flux and filler metal, can be used to join similar or dissimilar metals, and provides clean joint atmosphere. On the demerit side, only electrical conducting materials can be joined by resistance welding.

4. How does resistance seam welding differ from resistance spot welding?

Ans. Resistance spot welding uses one or two spots for joining two parts and can be used as a substitute for riveting, but the joint is weak. On the other hand, resistance seam welding is a continuous spot welding where the gap between the spots is negligibly small to be treated as a continuous joint with more strength.

Multiple Choice Questions

1. Consider the following statements about resistance welding:

    1. It is based on electrical resistance heating.
    2. It can be used for overhead welding position.
    3. It is extensively used in automotive industry.

    Of these statements:

    (a) 1 and 2 are true (b) 2 and 3 are true
    (c) 1 and 3 are true (d) 1, 2 and 3 are true

2. Which of the following parameters need to be carefully controlled in resistance welding?

    (a) Pressure (c) Heat
    (b) Current (d) Resistance
    Of these:

    (a) 1 and 2 are true (b) 1 and 3 are true
    (c) 2 and 4 are true (d) 1, 2 and 3 are true

3. Which of the following resistance welding is used for making leakproof joint?

    (a) Spot welding (b) Projection welding
    (c) Seam welding (d) Percussion welding

4. Which of the following welding processes uses capacitor?

    (a) Spot welding (b) Projection welding
    (c) Seam welding (d) Percussion welding

5. Which of the following is true about the resistance welding?

    (a) High voltage and low current
    (b) Low voltage and high current
    (c) Low voltage and low current
    (d) High voltage and high current.

6. Consider the followings statements:
   1. Flash welding requires edge preparation.
   2. Spot welding and projection welding are same.
   3. Seam welding is a continuous series of spot welding.

   Of these statements:

   (a) 1 alone is true (b) 3 alone is true
   (c) 2 and 3 are true (d) 1 and 3 are true

Answers

1. (c) 2. (b) 3. (c) 4. (d) 5. (b) 6. (b).

Review Questions and Discussions

Q.1. What is the principle of resistance welding?
Q.2. Why is the control of current very important in resistance welding?
Q.3. What are the different cycles used in spot welding?
Q.4. What is high-frequency resistance welding?
Q.5. Why is HAZ very small in high-frequency resistance welding?
Q.6. How does projection welding differ from spot welding?
Q.7. In what respect, the flash welding is different from butt welding?
Q.8. State the principle of percussion welding.
Q.9. Why is edge preparation not required in flash welding?
Q.10. Some metals such as tin, zinc and lead offer difficulty during resistance welding. Why?

Modern Welding Processes

13

Electron beam welding produces a high quality joint with low heat affected zones.

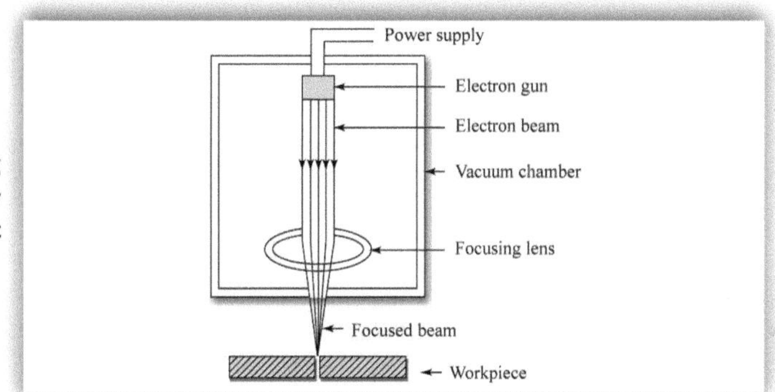

Learning Objectives

After reading this chapter, you will be able to answer some of the following questions:

- Hoe does laser welding differ from electron beam welding?
- Why is vacuum an essential requirement in electron beam welding?
- Why does thermit welding find extensive applications in connecting railroads?

D. K. Singh, *Fundamentals of Manufacturing Engineering*,
https://doi.org/10.1007/978-981-99-8767-2_13

13.1 Introduction

Modern welding processes are used to obtain excellent and precise welds having high depth-to-width ratio, especially in case of electron beam and laser welding. Good surface finish and tight tolerances are characteristic features of such processes. They are used to join super alloys that are otherwise impossible to weld.

13.2 Electron Beam Welding

Electron beam welding (Fig. 13.1) is a fusion welding process and has a lot of potentials. It uses a 0.5–1 mm narrow diameter, high-velocity electron beam which is emitted from a high-voltage source (up to 150 kV) gun that strikes at the point of joining to fuse the parts being joined. The kinetic energy of the electrons is converted into heat as they strike the workpiece. The metal is joined by melting the edges of the workpiece or by penetrating the material. It produces a very concentrated heat that takes very little time to complete the process.

The welding operation can be performed in three situations, namely hard vacuum, partial vacuum and atmospheric condition (air).

In hard vacuum, welding is carried out at high vacuum in a vacuum chamber. Pressure in the vacuum chamber varies between 10^{-3} and 10^{-4} mm of Hg. It gives deep penetration, but the production rate is low because of more time required to produce very low pressure in the vacuum chamber.

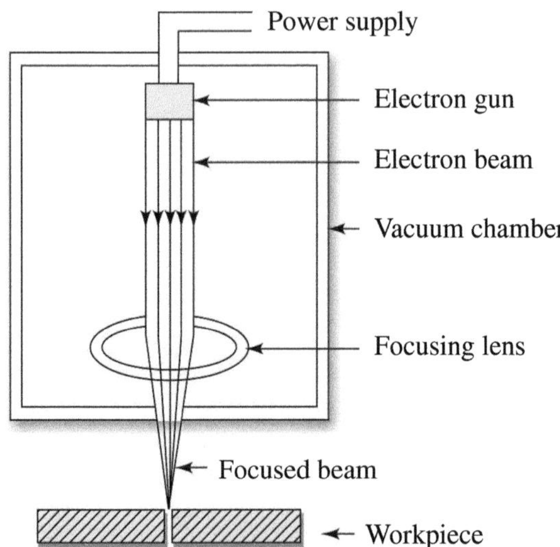

Fig. 13.1 An electron beam welding process

In partial vacuum, welding is performed at low vacuum.

In the atmospheric condition welding, the workpiece is placed outside the vacuum chamber. Production rate is increased considerably in this case because of absence of vacuum. Also, the weld quality is slightly poor, but the penetration is greatly reduced.

13.2.1 Advantages

- It produces very high-quality welds.
- Depth-to-width ratio is very high (up to 25:1) because of heat confinement to a narrow area. The intense heat energy can also produce holes in the workpiece (keyhole effect).
- There is no requirement of shielding gas, flux or filler metal.
- Higher welding speed is possible. It also ensures low-heat-affected zones.
- Distortion and shrinkage in the weld area are minimal.
- Welding can be done in all positions.
- Heat-sensitive materials can be easily welded because of low HAZ.
- Similar as well as dissimilar metals can be joined with equal effectiveness.

13.2.2 Limitations

- The equipment cost is very high.
- It is not suitable for large gap joining because of low quantity of molten metal.
- Vacuum is essential to prevent the scattering of electron beam.
- The vacuum requirement reduces the production rate and increases the operational cost.
- Joint preparation is essential to get a better quality weld.

13.2.3 Suitability and Applications

The process is best employed where extremely high-quality welds are required. Almost any metal can be welded by this process. The process is used to join common metals, refractory metals, highly oxidizable metals and various super alloys.

Thinner as well as thicker workpieces can be joined. Workpiece can be a foil or a plate.

Typical applications include welding of aircraft, missile, nuclear, and electronic components and gears and shafts in the automotive industry.

13.3 Laser Beam Welding

Laser beam welding uses laser as heat-source. The laser is a highly directional, monochromatic and coherent beam. It can be focussed to a very small area giving a very high energy density, and hence has deep penetration capacity. Laser beam can produce very high temperature to melt any known material.

There are three basic types of lasers, namely solid-state laser, gas laser and semiconductor laser. Solid-state lasers include ruby, sapphire and neodymium-doped yttrium aluminium garnet (Nd-YAG). CO_2 laser is the important gas laser. Both the CO_2 and YAG lasers can be used for welding. YAG laser is ideal for producing spot and seam welds because of its pulsed mode. CO_2 lasers can produce deeper welds at higher speed than those of YAG lasers. While the gas laser is the most popular one in laser cutting, over 70% of laser welding is now Nd-YAG based. Laser beam welding process is shown in Fig. 13.2.

13.3.1 Advantages

- It gives deep penetration without affecting the surrounding metal because of coherent nature of laser beam.
- The process is very fast because of very high temperature produced.
- Vacuum is not required. It is a mandatory requirement in electron beam welding.

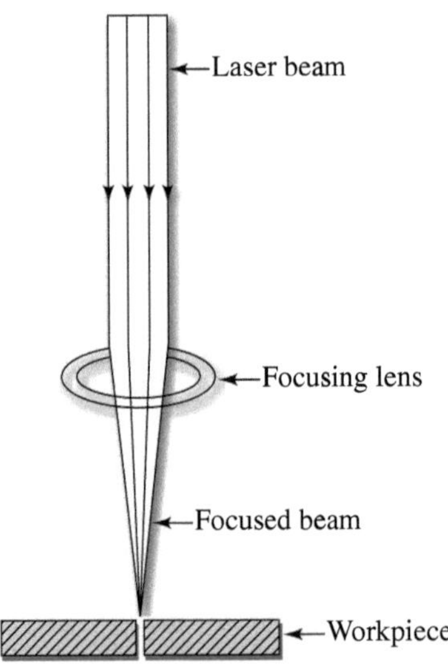

Fig. 13.2 A laser welding process

- Welding can be done in inaccessible locations.
- The quality of the weld is better because thermal distortion is minimum.
- The laser beams do not generate X-rays as in case of electron beam welding. Hence, no protection is needed against X-rays.
- The process can easily be automated.

13.3.2 Limitations

- Better design for the weld joint and edge preparation are absolutely essential to produce the narrow gap.
- The cost of equipment is very high, although not as high as electron beam welding.
- The process is limited to very thinner materials.
- Laser beams are hazardous causing injury to eyes and skin. Solid-state (YAG) lasers are particularly more dangerous.

13.3.3 Suitability and Applications

A wide variety of similar and dissimilar metals can be joined by laser welding. Thin workpieces are most effectively welded. The process is particularly suitable for welding deep and narrow joints. Typical metals and alloys welded include aluminium, titanium, ferrous metals, copper, superalloys and refractory metals.

Because of sharp focus of laser beam, the process finds extensive applications in electronic industries to connect wire leads to small electronic components. Wires can be welded without removing their polyurethane insulation because the laser evaporates the insulation and completes the weld with the internal wire.

13.4 Thermit Welding

In this process, a mixture of finely divided aluminium and iron oxide (Fe_3O_4), called thermit (mixed in the ratio of 1:3), is ignited at a temperature of about 1500 °C by a magnesium fuse to produce the required quantity of molten metal (iron) and heat according to the following chemical reaction.

$$8\,Al + 3\,Fe_3O_4 \rightarrow 9\,Fe + 4\,Al_2O_3 + \text{Heat}$$

This is an exothermic reaction which produces a temperature of the order of 2750 °C in about 30 s. The heat of the reaction is utilized in welding. It superheats the molten iron, which then flows into the prepared joint. Upon solidification it results in a weld joint. It is thus a casting cum welding process. There is no need of additional filler metal because the molten iron itself acts as a filler metal. This is the only welding process in which a chemical reaction is used to produce a high temperature.

13.4.1 Advantages

- It produces high-quality welds.
- There is no limit to the size of welds that can be made by thermit welding.

13.4.2 Limitations

- It is an old process and is now replaced by flash butt welding and electroslag welding.

13.4.3 Suitability and Applications

Thermit welding is suitable for welding and repairing large forgings and castings that would be difficult to weld by other processes, on accounts of their positions particularly in remote locations or where more-sophisticated welding equipment is not available. It can also be used to weld thick steel structural sections, railroad rails, shipbuilding, large size crankshafts, propeller shafts and pipes.

Short Answer Questions

1. What are the salient features of modern welding processes? Name a few modern welding processes.

Ans. Modern welding processes produce excellent and precise welds with good surface finish and tight tolerances, and offer clean joint atmosphere.

2. How does electron beam welding differ from laser welding?

Ans. Electron beam welding requires vacuum atmosphere to prevent the scattering of electrons and increasing their penetrations for effective joining of parts, which increases its operational costs. Laser welding, on the other hand, does not require vacuum atmosphere during welding.

3. What is the principle of thermit welding? Where is it used?

Ans. Thermit welding uses the heat produced by an exothermic reaction between aluminium powder and iron oxide (Fe_3O_4) to produce molten iron to be poured in the gap between the parts being joined, which gets solidified after sometime. It is used for welding and repairing large forgings and castings which are otherwise difficult to weld because of remote locations or unavailability of more sophisticated equipment, for example railroad rails, ship building, etc.

Multiple Choice Questions

1. Which of the following welding processes requires vacuum?

   (a) Thermit welding (b) Laser welding
   (c) Electron beam welding (d) Electroslag welding

2. Consider the following statements about the electron beam welding:

   1. It produces low heat affected zone.
   2. There is no need of shielding gas, flux or filler metal.
   3. Thermal distortion is minimum.
   4. It produces accurate joint.

   Of these statements:

   (a) 1 and 3 are true (b) 2 alone is true
   (c) 1, 2, 3 and 4 are true (d) 2 and 4 are true

3. What will happen, if vacuum is not provided in electron beam welding?

   1. The production rate is increased.
   2. The rate of collisions between electron and air molecules is decreased.
   3. The electron beam is scattered.

   Of these statements:

   (a) 1 alone is true (b) 1 and 2 are true
   (c) 3 alone is true (d) 1, 2 and 3 are true

4. A thermit mixture consists of

   (a) Aluminium oxide and iron (b) Iron oxide and aluminium
   (c) Calcium carbide and iron (d) Iron carbide and calcium

5. LASER stands for

   (a) Light Amplification by Simulated Emission of Radiation
   (b) Light Amplification by Stimulated Emission of Radiation
   (c) Light Amplitude by Simulated Emission of Radiation
   (d) Light Amplitude by Stimulated Emission of Radiation.

Answers

1.	(c)	2.	(c)	3.	(c)	4.	(b)	5.	(b)

Review Questions and Discussions

Q.1. Why is vacuum essential in electron beam welding?
Q.2. Why is electron beam welding useful in electronics industry?
Q.3. Although electron beam welding has several advantages, still the method is not very popular. Why?
Q.4. Differentiate between laser and electron beam welding.
Q.5. What is a thermit mixture?
Q.6. Where is thermit welding used?

Soldering, Brazing and Adhesive Bonding 14

Adhesive bonding resulting due to molecular attractions between adhesive and workpiece offers may advantages over conventional methods of joining.

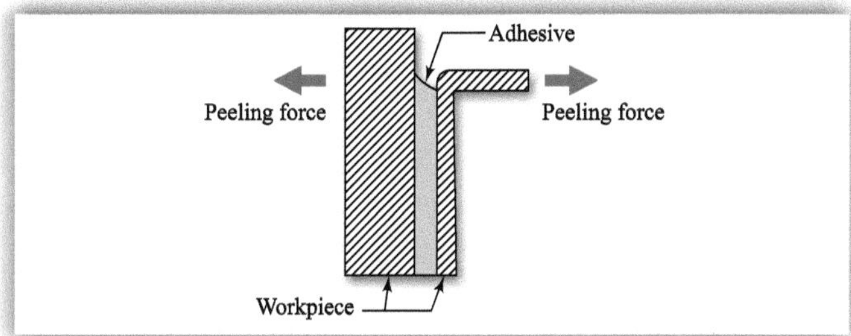

Learning Objectives

After reading this chapter, you will be able to answer some of the following questions:

- How does soldering differ from brazing?
- What is the principle of adhesive bonding?
- How does braze welding differ from brazing?

D. K. Singh, *Fundamentals of Manufacturing Engineering*,
https://doi.org/10.1007/978-981-99-8767-2_14

14.1 Soldering

Soldering or soft soldering as it is often called, is a quick method of making joints in light articles made from steel, copper and brass without melting the material by means of a fusible alloy, called solder. The solder which melts below 450 °C acts as a filler metal and reaches the gap between the two parts to be joined by capillary action. It is extensively used in electrical and electronic works to join wires.

Solders are usually tin–lead alloys. They have good strength at low temperatures. A 60–40 solder in which the first number denotes lead and the second number tin, both expressed on percentage basis, has the lowest melting point. Other solder compositions may be tin-zinc, lead-silver, cadmium-silver, zinc-aluminium and indium-tin. Both tin-zinc and zinc-aluminium solders are used mainly for soldering aluminium. A 50% indium and 50% tin solder is most suitable for joining cryogenic parts and for joining glass and metal.

The strength of a soldered joint is principally due to the metallic bond formed although adhesion and mechanical attachment also contribute. A good joint is characterized by a small amount of solder and perfect adhesion, rather than by large amount of solder.

To produce an effective bond, the joining surfaces need to be cleaned of any impurity such as oxide and others. Fluxes are used for this purpose, but it is essential that all dirt, oil and grease be removed before the flux is applied. A soldering flux may be a solid, a liquid or a gaseous product. Chloride fluxes (zinc chloride, better known as killed spirits, ammonium chloride and zinc ammonium chloride, a combination of the two) are the most efficient one because they clean the surface rapidly and are effective on aluminium, copper, brass, bronze, steel and nickel. These fluxes not only protect the surface, but play an active role in cleaning it chemically. But they leave residues which should be removed by washing thoroughly with water to avoid corrosion. Noncorrosive resin-based fluxes are used in electrical applications. Some solders are also available with the flux contained in its core, the amount of which may range from 0.5 to over 3.0, 2.2% being the most common. Soldering produces a weak joint, hence it should not be used where more strength is required or when the parts are subjected to vibration or heat.

14.1.1 Methods of Soldering

The following are the important soldering methods based on different modes of heating:

- Iron soldering
- Dip soldering
- Induction soldering
- Ultrasonic soldering
- Torch soldering

- Wave soldering
- Resistance soldering
- Furnace soldering

14.1.2 Iron Soldering

The iron soldering method uses a soldering iron and is the most commonly used method of manual welding. The copper tip of the soldering iron can be heated electrically or by oil, coke or gas burners. But external heating consumes more time, and hence it is heated by electrical resistance of the solder tip. The solder is melted by touching it to the tip of the soldering iron and is applied where the joint is required to be made. Generally the tip of the soldering iron is coated with iron to increase its life.

14.1.3 Torch Soldering

Torch soldering uses oxyacetylene or air-acetylene torch for heating and is used widely for soldering copper tubing to copper fittings.

14.1.4 Dip Soldering

Dip soldering uses a solder-bath maintained at a desired temperature and is used when soldering needs to be performed in one go since the parts to be joined are dipped in the solder-bath. It finds extensive application in joining wire ends in electronic industry, soldering automobile radiators and tinning.

14.1.5 Wave Soldering

Wave soldering uses the wave pattern of the molten solder (solder-bath) for joining the parts and is used mostly in electronic industry for the production of electronic circuit boards.

14.1.6 Induction Soldering

Induction soldering uses the heat produced by induced current and is used when large numbers of identical parts are required to be soldered.

14.1.7 Resistance Soldering

Resistance soldering uses the resistance of the workpiece to get the desired heat.

14.1.8 Ultrasonic Soldering

Ultrasonic soldering uses high-frequency vibrations as heating source and is used to solder the return bends to sockets of aluminium air-conditioner coils.

14.1.9 Furnace or Oven Soldering

Gas or electric furnaces are used in soldering only when a number of joints are required to be produced to increase the production rate and the assembly is complex making other methods of heating the joints impractical.

14.1.10 Advantages of Soldering

- Soldering can be used to join various metals and thicknesses.
- It can be easily automated.
- Automation increases rate of production.
- The process is economical.

14.1.11 Limitations

- It is not suitable for joining parts which require much strength or the parts that are subjected to vibration or heat, as the solder is comparatively weak and has a low melting point.
- Cleaning of the surfaces is a mandatory requirement.
- Butt joint cannot be made because of small faying surfaces.
- The operation requires skill and is time-consuming.
- Aluminium and stainless steels are difficult to solder because of the formation of strong, thin-oxide film (after cleaning). They require special soldering processes such as flux soldering (using special fluxes), friction soldering and ultrasonic soldering for removing the oxide film.

14.1.12 Suitability and Applications

Soldering is used extensively in electrical and electronics industries for joining wires and light articles made from steel, copper and brass.

14.2 Brazing

Brazing, like soldering, is a process of joining metals without melting the parent metal. In this process, the filler metal is placed in the gap between the parts to be joined and is heated to melt it. The molten filler metal completely fills the gap by capillary action and on cooling and solidification a strong joint results. The strength of the joint is attributed to three sources: atomic forces between the metals at the interface: alloying that occurs at elevated temperature due to the diffusion of metals (filler metal and base metal) and intergranular penetration.

The filler metal is usually a nonferrous alloy and has melting point more than 450 °C but lower than the melting temperature of the parent metal. Alloys of copper, silver and aluminium are the most common brazing filler metals. They are used in many forms such as ring, washer, rod or other shape to fit the joint brazed.

The basic types of joints found in brazing are the lap, butt and scarf (Fig. 14.1). Among all these, the lap joints are the strongest one because it offers greater contact area. Butt joint provides a smooth joint of minimum thickness. The scarf joint is a compromise between lap joint and butt joint but its alignment is more difficult.

The length of the lap required for a better joint is given by the following relations.

$$\text{For a flat joint,} \quad l = \frac{\sigma t n}{\tau} \tag{14.1}$$

$$\text{For a tubular joint,} \quad l = \frac{\sigma t(d - t)n}{\tau} \tag{14.2}$$

where

l = length of the lap (mm)
σ = tensile strength of the thinner member (Pa)

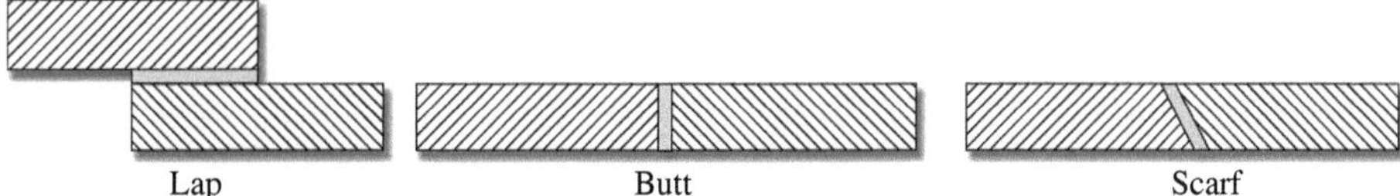

Fig. 14.1 Common brazing joints

t = wall thickness of the thinner member (mm)
τ = shear stress of the filler metal (Pa)
n = factor of safety
d = diameter of the lap (mm)

To achieve the maximum joint efficiency, the length of the lap should be at least three times the thickness of the thinner member.

The strength of the brazed joint depends on the joint design and the adhesion at the interface of the workpiece and filler metal. Proper adhesion is achieved if the faying surfaces are cleaned off all dirt, oil or oxides. Fluxes are usually used for cleaning the surfaces. They can be applied in paste form or liquid because of the ease of application to small parts and their adherence in any position. Controlled atmosphere (CO_2, CO, N_2, H_2 or inert gases like Ar and He) or vacuum are sometimes used to prevent oxidation during the process. They eliminate the need for flux. Vacuum is particularly suitable for reactive metals like titanium, zirconium, molybdenum, columbium and tantalum.

14.2.1 Methods of Brazing

The heat for brazing is obtained from a number of sources, and accordingly brazing methods have been classified.

14.2.2 Torch Brazing

Torch brazing is one of the oldest and most widely used methods of heating in brazing. In this process, the parts to be joined are first heated by an oxyacetylene flame (neutral flame) and then filler metal is placed in the gap between them. Thus it is the parent metal, not the flame, that transfers the heat to filler metal. The process is widely used in production applications and repair works. But difficulty arises in controlling the temperature and maintaining uniformity of heating.

14.2.3 Furnace Brazing

Furnace brazing is carried out in a furnace. The parts to be joined are precleaned and have the filler metal preplaced in the joint before being placed in the furnace. The furnace may be batch or continuous type. Batch type furnace is used for complex shapes, whereas continuous type is used for large-scale production of small parts with simple joint design. Vacuum furnaces can be used for reactive metals and are widely used in aerospace and nuclear applications.

14.2.4 Induction Brazing

Induction brazing is based on induction heating. In this process, the parts to be joined preloaded with filler metal are placed near a coil generating a very high-frequency (2×10^5 to 5×10^6 Hz) current. This current induces opposing current in the parts, which by electrical resistance, develop the heat. It is a rapid method of heating the parts. Desired temperature is achieved in 10–60 s. Good heat distribution, better heat control and uniformity of results are some of the major advantages of induction brazing. The process is suitable for continuous work.

14.2.5 Dip Brazing

In dip brazing, the assembled parts to be joined are dipped either in a molten filler metal bath or a molten salt bath containing flux. The bath is contained in a furnace and provides both the heat and braze metal for the joint as well as necessary protection from oxidation. The bath is maintained at a higher temperature than the filler metal being used. Due to uniform heating, the dip-brazed parts distort less than torch-brazed parts. It is preferred for joining small parts such as sheet, wire and fittings with multiple or hidden joints.

14.2.6 Infrared Brazing

This process utilizes the heat generated by infrared radiation energy obtained from infrared quartz lamps. Heat is concentrated at the desired spot by using reflectors. The process can be performed in air, inert atmosphere or vacuum. It is particularly suitable for joining very thin components, usually less than 1 mm thickness.

14.2.7 Resistance Brazing

In resistance brazing, the required heat for joining is obtained from the electrical resistance of the parts being joined. Unlike resistance welding, however most of the resistance and hence most of the heating is provided by the carbon or graphite electrodes. The process is used for connecting parts having electrical conductivity such as wires and cables.

14.2.8 Advantages of Brazing

- Due to higher temperature produced in brazing, the joint is more stronger than produced by soldering.

- Similar, dissimilar and difficult-to-weld metals can be easily joined.
- Any thinner sections or complex assemblies can be easily joined.
- Heat-affected zones and thermal distortion are minimum on account of low-temperature applications.
- The process is fast and economical and can be easily automated.

14.2.9 Limitations

- The surface cleaning is essential.
- The joint is adversely affected by heating and corrosion.
- The strength of the joint is lower than that of the base metal.

14.2.10 Suitability and Applications

Brazing is ideally suitable for dissimilar metals such as joining of ferrous to nonferrous, or metals with widely different melting points. It finds major applications in electrical industry, utensil manufacturing industry and maintenance industry. Typical applications include assembly of pipes to fittings, carbide tips to tools, radiators, heat exchangers, electrical parts and the repair of castings.

14.3 Braze Welding

In braze welding, the molten filler metal is not distributed at the joint by capillary action as it happens in brazing, but it is deposited at the point where the weld is required to be made as in the case of oxyfuel gas welding.

Figure 14.2 shows the difference between brazing and braze welding. As compared to brazing, more filler metal is used but the process is conducted at relatively low temperature, and hence warping and thermal distortion are minimum. It is mainly used for the repair of steel components and ferrous castings. Braze welding is always done with an oxyacetylene torch.

14.4 Adhesive Bonding

Adhesive bonding is a process of joining materials by means of an adhesive. Adhesion is generally due to molecular attraction between adhesive and workpiece. Since both metals and nonmetals can be bonded, adhesive bonding has grown significantly with the expanding role of plastics and composites.

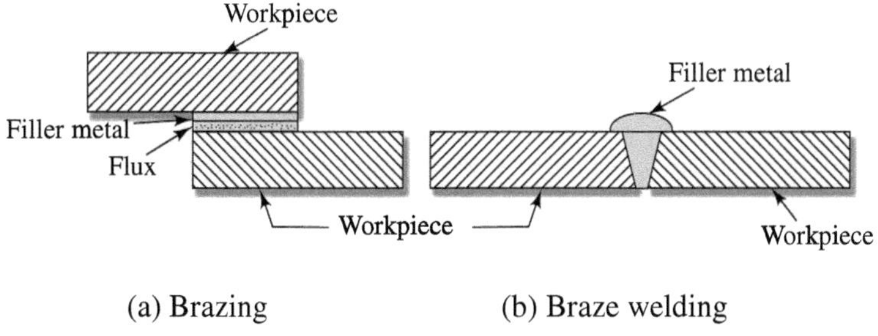

Fig. 14.2 Brazing versus braze welding

The adhesive is applied between the faying surfaces after cleaning them thoroughly by chemical or mechanical means. The strength of the joint greatly depends on the absence of dirt, dust, oil and various other contaminants. Contaminants affect the wetting ability of the adhesive adversely and prevent its spreading evenly over the interface.

Adhesives are available in different forms such as liquid, paste, solid, tape and film. They can be applied at the required place by brushing, spraying, roller coating or dipping. Pressure on the joint should be carefully applied after placing the adhesive between the faying surfaces. Pressure should be sufficient to achieve a uniform spread of adhesive but not enough to squeeze out the liquid and starve the joint.

Adhesives may be of structural or nonstructural type. Structural adhesives give higher bond strength, and the joint can be safely subjected to heavy loads for a long period of time. They are further classified into two groups: thermoplastics and thermosetting. Most commonly used thermoplastic adhesives are polyamides, vinyls and nonvulcanizing neoprene rubber. Thermosetting adhesives such as epoxies, anaerobics and silicones are generally preferred for elevated temperature service. They give strong, waterproof and heat-resistant joints. Epoxies are most widely used and can join most engineering materials, including metal, glass and ceramics. Nonstructural adhesives are household type of glues of vegetable or animal base, and some common examples are casein, rosin, shellac and asphalt.

The selection of a particular adhesive depends upon the materials being joined, the joint design and the operational requirements (strength, temperature variations, etc.). Design for adhesive bonds should be such that the joint is shear rather than tension. The ideal joint is a bevel lap joint shown in Fig. 14.3. Bevelling reduces stress concentration at bond edges. It is preferable to increase the width of the joint rather than the overlap for maximum strength per unit of bonded area.

When a thermosetting adhesive is being used, it will be necessary to cure the joint at an elevated temperature. Curing is usually done at a temperature of about 150 °C for about 30 min. Curing makes the joint more stronger.

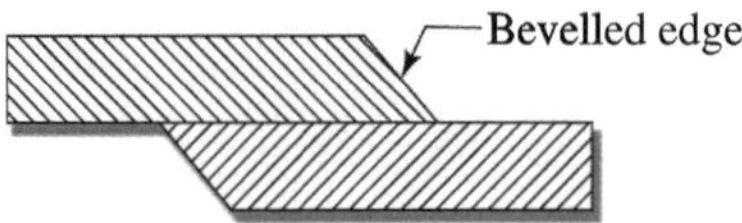

Fig. 14.3 A bevel lap joint

14.4.1 Advantages

- The process is versatile and offers joining of a number of similar and dissimilar materials in different thicknesses and shapes and sizes. For example, foils can be joined to each other or to thicker plates and very thin and fragile components can be bonded.
- Since the process is performed at not very high temperature, heat-sensitive materials can be joined without causing any thermal distortion.
- Heat-affected zones are usually missing in adhesive bonding.
- The elastomeric nature of some adhesives gives shock and vibration protection, and hence fatigue life is increased.
- Since the bond covers the entire joint area, the load is uniformly distributed, thereby eliminating stress concentrations. Also, sealing action can be achieved.
- The process is fast since large area can be bonded in a relatively shorter time.
- Adhesives are bad conductor of electricity, and hence there is no chance of electrochemical corrosion between dissimilar metals.
- Adhesive joints can withstand shear, compressive and tensile forces, and their strength is very much comparable to that produced by alternative methods of joining.
- It is an economical method of joining since adhesives are generally inexpensive.

14.4.2 Limitations

- The joint should not be subjected to peeling forces (Fig. 14.4). To increase the strength of the joint, additional fasteners at stress points are used.
- Adhesive joints are difficult to inspect once completed.
- The preparation of the surfaces is of paramount importance. The surfaces to be bonded must be chemically or mechanically scrubbed so that dirt, dust, oil and various other contaminants can be removed. The absence of these impurities helps to increase the strength of the joint. Cleanliness is usually more important than surface finish.
- Adhesive joint has limited reliability. Its quality deteriorates with time and hostile environmental conditions such as ultraviolet light, ozone, acid rain, moisture and salt adversely affect it.
- The joint cannot be subjected to elevated temperatures.
- Joint design and bonding methods require care and skill.

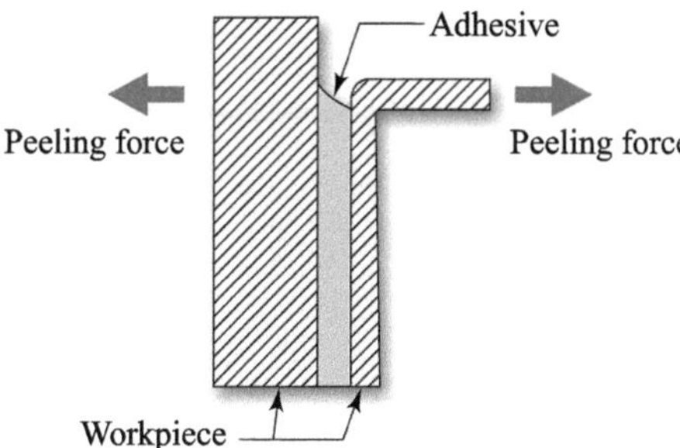

Fig. 14.4 Adhesive joint and peeling effect

14.4.3 Suitability and Applications

The process of adhesive bonding is very much useful for the formation of laminates and honeycomb structures. Plywood is the suitable example of lamination where several layers of wood are glued with adhesive. Packaging, bookbinding, house furnishings and footware are some common areas of application. Major users of this method are the aircraft and automobile industries. Typical applications include attaching of brake lining to brake shoes, fastening of stiffeners to aircraft skin and laminated windshield glass.

Short Answer Questions

1. What is soldering and where is it used?

Ans. Soldering is a process of joining metals and alloys without melting them using a tin-lead alloy called solder. It produces a weaker joint and is widely used in electrical and electronics industries to connect wire or lighter components.

2. How does brazing differ from braze welding?

Ans. The filler metal used in brazing is placed in the gap between the parts being joined and heated to melt and finally allowed to fill up the gap completely by capillary action, resulting in strong joint after solidification. On the other hand, braze welding is similar to gas welding where the work parts at the joint are heated to melt them along with filling the gap by filler metal.

3. What is the principle of adhesive bonding and where is it used?

Ans. Adhesive bonding is a joining process which uses an adhesive to join the parts. The joining takes place because of molecular attraction between adhesive and the parts being joined. It can be used to join metals, nonmetals, plastics and composites.

Multiple Choice Questions

1. The composition of a 60–40 solder is

   (a) 60% tin and 40% lead
   (b) 60% lead and 40% tin
   (c) 60% copper and 40% tin
   (d) 60% copper and 40% zinc.

2. Which of the following joining methods is based on capillary action?

   (a) Thermit welding
   (b) Gas welding
   (c) Adhesive bonding
   (d) Soldering.

3. Which of the following compounds is also known as killed spirit?

   (a) Zinc chloride
   (b) Ammonium chloride
   (c) Calcium carbide
   (d) Zinc carbide.

4. Which of the following soldering methods is most commonly used?

   (a) Dip soldering
   (b) Wave soldering
   (c) Iron soldering
   (d) Induction soldering.

5. Which of the following joints is the strongest one?

   (a) Lap joint
   (b) Scarf joint
   (c) Butt joint
   (d) All of the above.

Answers

1. (b) 2. (d) 3. (a) 4. (c) 5. (a)

Review Questions and Discussions

Q.1. Why is cleaning of parts essential in soldering?
Q.2. Why is soldering not suitable for a stronger joint?

Q.3. What are the different soldering methods?
Q.4. What is a solder?
Q.5. How does brazing differ from welding?
Q.6. What is the difference between soldering and brazing?
Q.7. What are the basic types of joints found in brazing?
Q.8. What are the advantages of paste brazing?
Q.9. Differentiate between brazing and braze welding.
Q.10. What is the principle of adhesive bonding?
Q.11. Where is adhesive bonding more effective?

15 Design Considerations in Joining Processes

Lap joint has increased strength because of higher contact area.

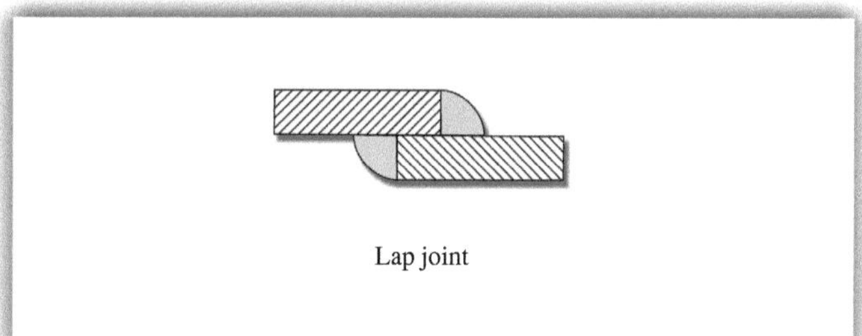

Lap joint

Learning Objectives
After reading this chapter, you will be able to answer some of the following questions:

- Why do edges need to be prepared in welding?
- Why is welding position an important consideration?
- How is strength of a joint affected by type of loading?

D. K. Singh, *Fundamentals of Manufacturing Engineering*,
https://doi.org/10.1007/978-981-99-8767-2_15

15.1 Introduction

Following points are considered in the design of a welding system:

- There should be minimum number of welds in the joint to affect the economy.
- Edges should be properly prepared before welding. Nonprepared or poorly prepared edges lack strength.
- Weld-bead size should be optimum to ensure adequate strength of the joint and also to reduce the wastage of filler material.
- Surfaces need to be cleaned before the start of welding to avoid possible oxidation and contamination.
- Study of heat-affected zone is important to know the structural changes during the welding.
- Welding current and welding speed are carefully selected.
- Welding position is very important. Overhead welding is difficult to perform, and also it produces more defects as compared to flat position welding.

15.2 Types of Welded Joints

Workpieces can be joined in several ways resulting in different types of joints as shown in Fig. 15.1. There can be many variations of each joint.

Selection of a joint depends upon the thickness and position of the workpieces being joined and type of welding process to be used. Butt joint is useful when workpieces need to be joined end-to-end. Circular or rectangular stocks are easily joined by butt joint. Lap joint is formed when some parts of the workpieces overlap each other. Lap joint has higher strength over butt joint because of more contact area. There can be single or double overlap. Tee and corner joints are used when the workpieces need to be positioned normal to each other.

15.3 Edge Preparation

Before actual start of a welding process, preparation of the edges to be welded is necessary. Edge preparation means straightening, cleaning and vee-ing (making in the V-shape) the edges.

The material being joined is given V-shape for the purpose of adding additional metal from the filler rod to increase the strength of the joint. The edges and the area adjoining them are cleaned off all the scale, rust, dust, paint, grease, moisture or any other foreign particles. Depending upon the degree of contamination, cleaning may be done with cotton

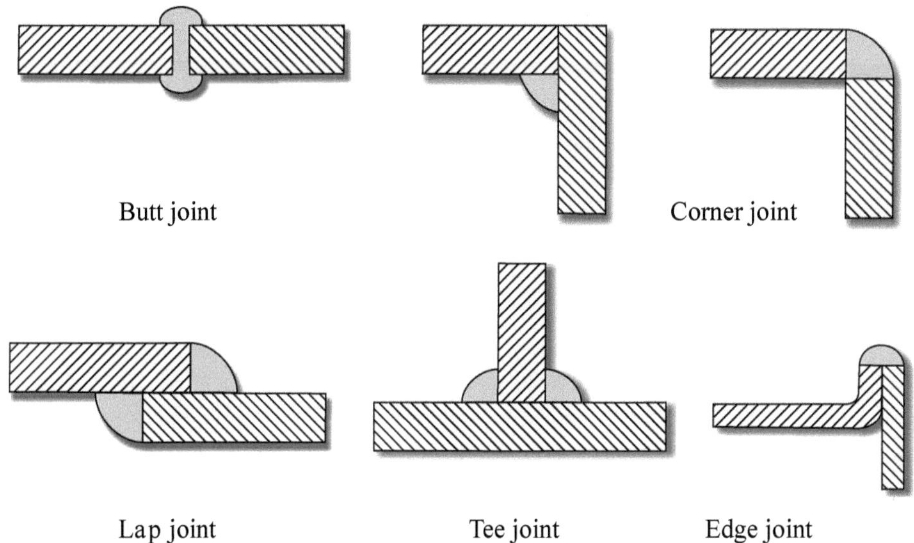

Fig. 15.1 Types of welded joints

waste, a stiff wire brush, a sand blaster, a degreasing solvent and pickling. Thorough cleaning helps producing homogeneous and strong welds.

The workpieces to be joined are positioned and spaced with respect to each other. Usually parts are placed in flat positions, because it is easier to carry out welding in such positions. Before welding an actual structure, it is always advisable to try on a scrap piece, in order to ascertain the optimum current.

The important factors affecting edge preparations include type of welding processes, type of metals to be joined, position of welding, access for arc and electrode, volume of deposited weld metal and shrinkage and distortion. Various types of edge preparations are shown in Fig. 15.10.

15.4 Types of Fusion Welds

There are four basic types of fusion welds:

- Groove weld
- Bead weld
- Fillet weld
- Plug weld.

Groove or butt welds are made between abutting plates in the same plane and are generally described by the way the edges are prepared. Edge preparation is required to form a groove between the abutting edges.

Bead weld, also called *surfacing weld*, is mainly used for joining thin sheets of metal (because of limited depth penetration) and building of surfaces.

Fillet welds are used to fill in a corner in tee, lap and corner joints. They are the most common welds in structural work. The size of the fillet is measured by the leg of the largest 45° right angle triangle that can be inscribed within the contour of the weld cross section.

Plug welds are used to attach one part on top of another, replacing rivets or bolts. All the four fusion welds are shown in Fig. 15.2. Best fillet-weld profiles are shown in Fig. 15.3. Acceptable and defective fillets are separately shown in Figs. 15.4 and 15.5, respectively. Figure 15.6 shows defective butt-weld profiles. Incorrect weld profiles may cause incomplete fusion or slag inclusions, thereby reducing the strength of the joint.

Excessive reinforcement creates notches which are the sources of stress concentration. It also increases the cost of operation without any gain in strength. Excessive reinforcement impairs the behaviour of the weldment under dynamic loads and is therefore undesirable.

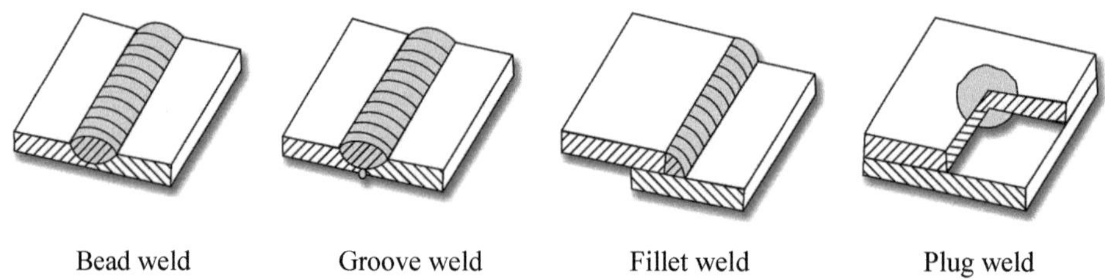

Fig. 15.2 Types of fusion welds

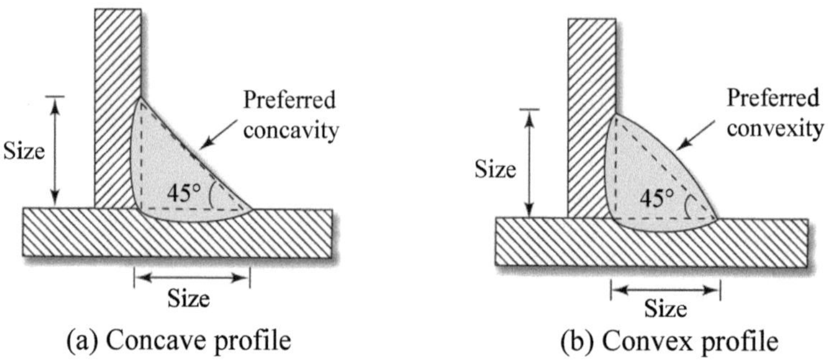

Fig. 15.3 Best design profiles of fillet welds

Fig. 15.4 Acceptable fillet-weld profiles

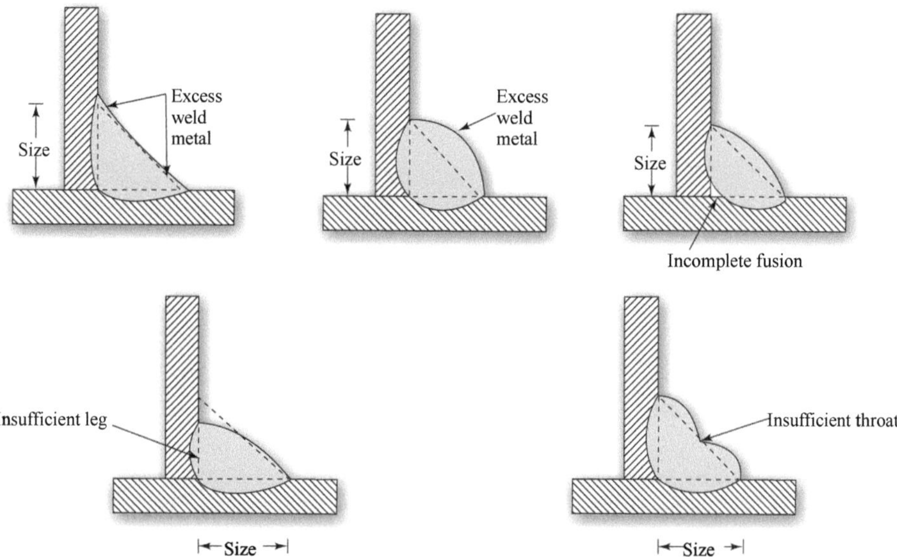

Fig. 15.5 Defective fillet-weld profiles

Fig. 15.6 Defective butt-weld profiles

15.5 Welding Positions

There are basically four welding positions, namely flat, horizontal, vertical and overhead. In flat or down hand position, face of the weld and weld bead both are horizontal. The plane of the workpiece is vertical, but the weld bead is horizontal in horizontal welding.

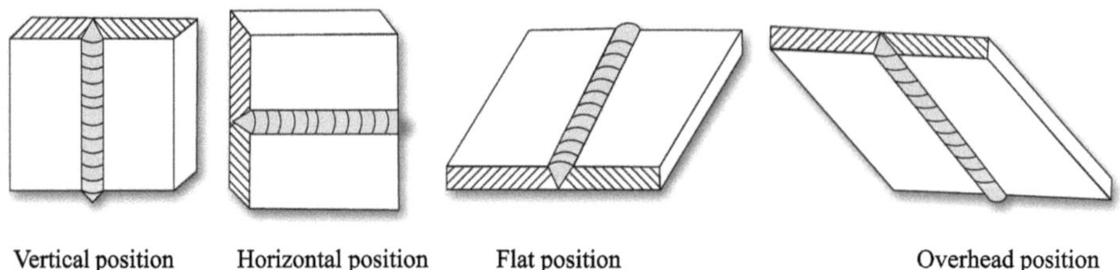

Fig. 15.7 Welding positions

In vertical welding, the plane of the workpiece and weld bead both are vertical. Overhead position is similar to flat position except with a difference that in flat position the welding is carried out from above, but is carried out from the underside in overhead position. The various welding positions are shown in Fig. 15.7.

15.6 Types of Load and Strength of the Joint

The load acting on the welded joint can be tensile, shear, bending or impact as shown in Fig. 15.8.

The weld joints should have sufficient strength so that the members do not fail by breaking or yielding, when subjected to usual operating loads or reasonable overloads. During the weld design, different formulae are used to find out the permissible strength (stress) under various loading conditions. The joint can be subjected to static as well as dynamic load. Static loads do not change during their application time. Dynamic loads

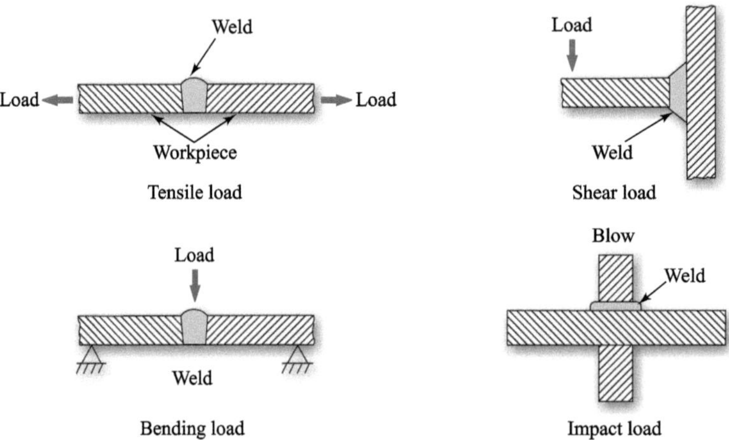

Fig. 15.8 Various types of loads acting on welded joint

change and may increase or decrease with time. An example of a dynamic load is an upright supporting steel beam in a highway bridge. The load on the beam increases or decreases as per the traffic requirements. The joint behaves differently under static and dynamic conditions. Accordingly, the strength of the joint is different in both cases.

A component or structure, which survives a single application of a static load, may fracture if the same load fluctuates. Such a failure is referred to as fatigue failure and is the most common type of failure (in 50 to 90% cases) of welded structures. Fatigue failure considerations are made in areas such as bridges, ships, pressure vessels, and automotive and aero components. Safety factors must be considered for any weld under stress. The working or allowable stress is obtained by dividing the failure stress by safety factor. But the fatigue failure cannot be simply offset by increasing the factor of safety or increasing the tensile strength of the base metal.

Although the stress distribution in welds is very rarely uniform and not as simple and accessible to accurate calculation as in the case of parent metal, experiments have shown that, for the sake of design calculations, uniform stress distribution is seriously influenced by conditions of design or loading. Different formulae for calculating the stresses in the weld are given in Fig. 15.9.

15.7 Design of Butt-Weld Joint

Edges are required to be prepared. The square-edge butt joint has straight edge and requires no preparation. The joint is often provided with a backing bar to take care of excessive root opening. It ensures full penetration of the weld metal, and the joint has the same strength as that of the base metal. Hence there is no need to calculate the strength of the weld. Design parameters for the butt joint are shown in Fig. 15.10.

15.8 Design of Fillet-Weld Joint

Lap joints, open corner joints and tee joints all use fillet-type welds. A fillet weld with T-joint is shown in Fig. 15.11. The fillet weld may have equal or unequal legs.

The effective throat is taken perpendicular to the weld face from the root of the joint (Fig. 15.12). Minimum fillet-weld size is given in Table 15.1. The weld size should not exceed the thickness of the thinner joined part. The strength of the fillet weld depends on the direction of load applied. The load may be applied in parallel or transverse direction

S.N.	Conditions of loading	Loading figure	Design remarks
1.	Tensile force, F	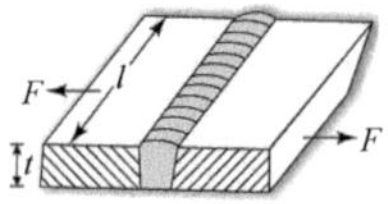	➢ Tensile stress (σ_t) is induced in the joint, given by $\sigma_t = \frac{F}{l \times t}$
2.	Bending Moment, M	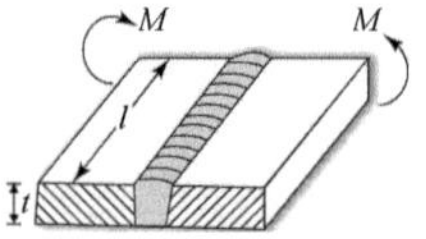	➢ Bending stress (σ_b) is induced in the joint, given by $\sigma_b = \frac{6M}{l \times t^2}$
3.	Tensile force, F	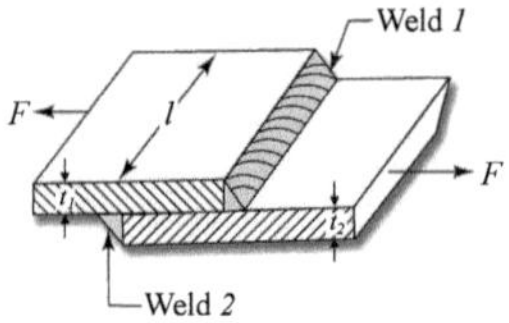	➢ Equal tensile stresses are induced in the two welds, given by $\sigma_{t_1} = \sigma_{t_2} = \frac{1.414F}{(t_1 + t_2)l}$
4.	Tensile force, F	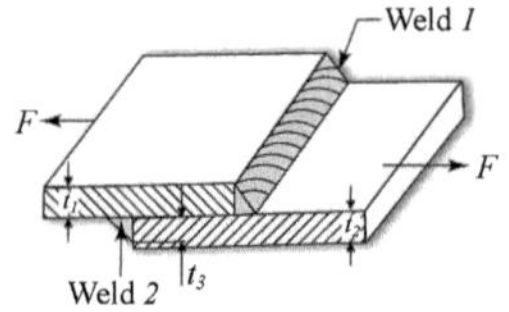	➢ Unequal tensile stresses are induced in the two welds, given by $\sigma_{t_1} = \frac{1.414Ft_2}{(t_1 + t_2)lt_3}$ $\sigma_{t_2} = \frac{1.414Ft_2}{(t_1 + t_2)lt_3}$
5.	Tensile force, F	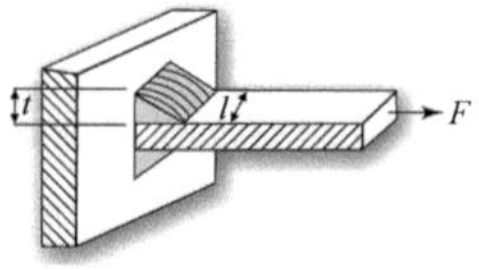	➢ Tensile stress is induced in the weld, given by $\sigma_t = \frac{0.707F}{lt}$
6.	Bending moment, M	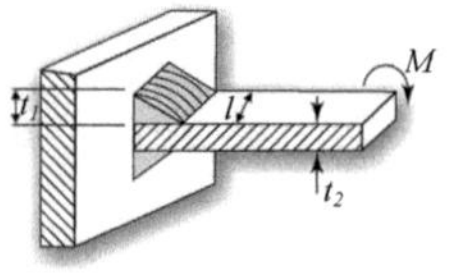	➢ Bending stress is induced in the weld, given by $\sigma_b = \frac{1.414M}{(t_1 + t_2)\, lt_1}$
7.	Bending moment, M	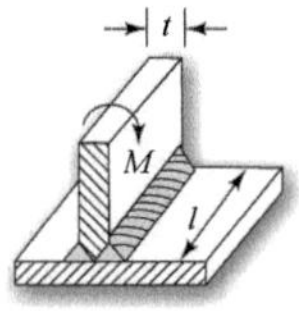	➢ Bending stress is induced in the weld, given by $\sigma_b = \frac{6M}{t^2 l}$
8.	Bending moment, M	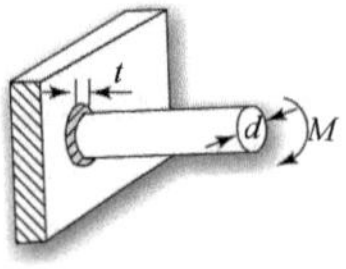	➢ Bending stress is induced in the weld, given by $\sigma_b = \frac{5.66M}{\pi d^2 t}$

Fig. 15.9 Stresses induced in weld joints

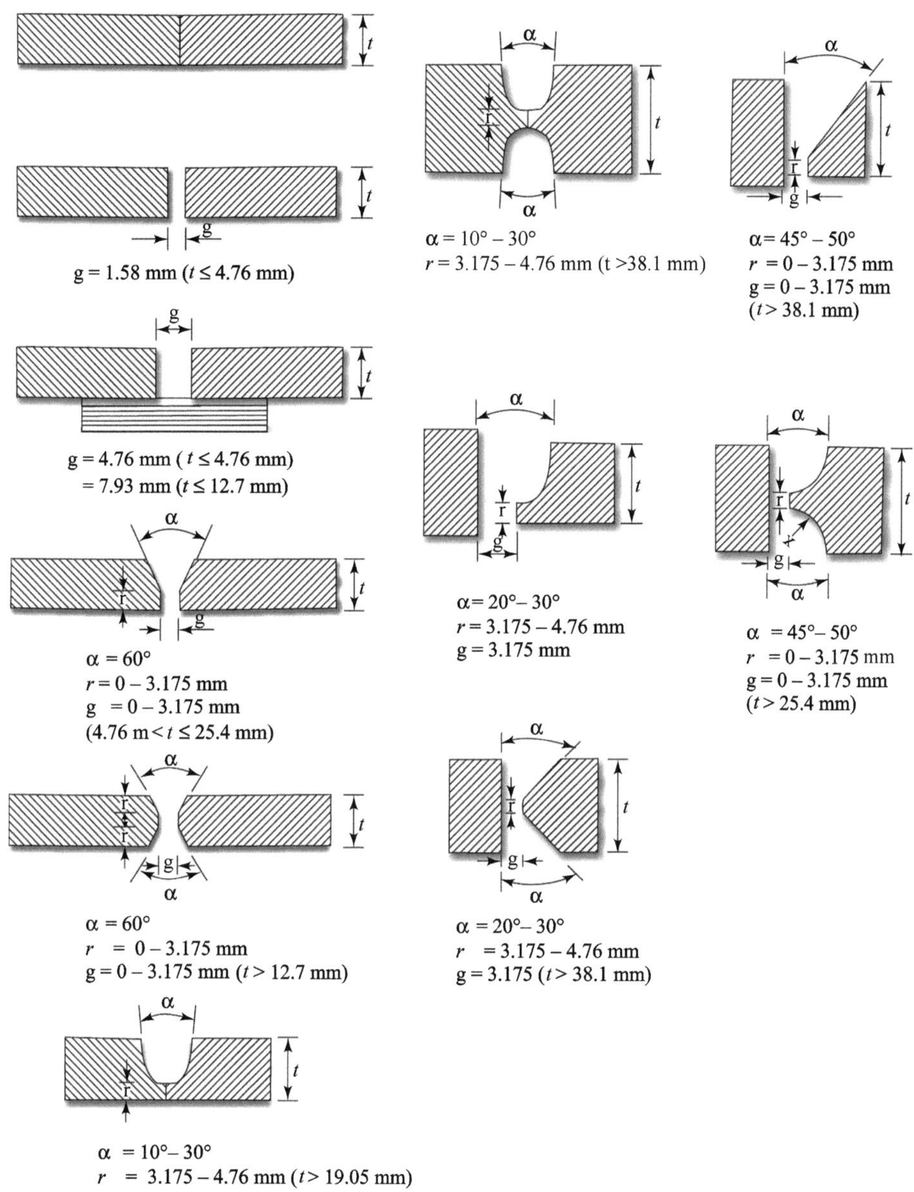

Fig. 15.10 Design aspects of butt joint

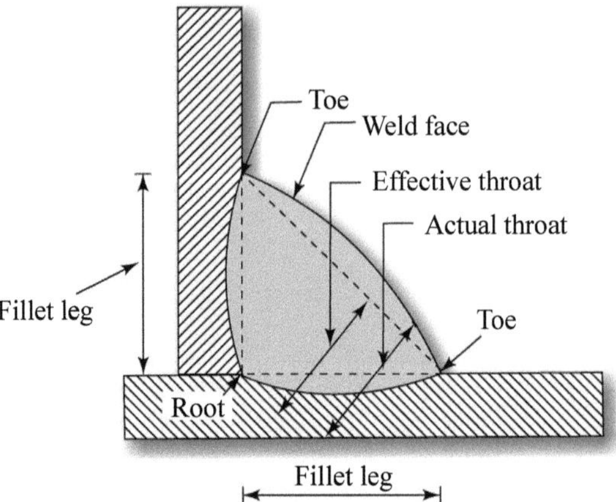

Fig. 15.11 Fillet-weld nomenclature

(Fig. 15.13). The joint fails due to shear in both cases along the plane which has maximum shear stress. For Fig. 15.13a, maximum shear occurs at an angle of 67.5° from the horizontal and for Fig. 15.13b, it is 45°. Hence, one-third more load can be applied in transverse direction than in parallel. Two 45° fillet welds with leg dimensions equal to 0.75 t (t = base metal thickness) will develop full strength of the plate for either type of loading, assuming weld metal thickness to be equal to that of base plate with average penetration condition.

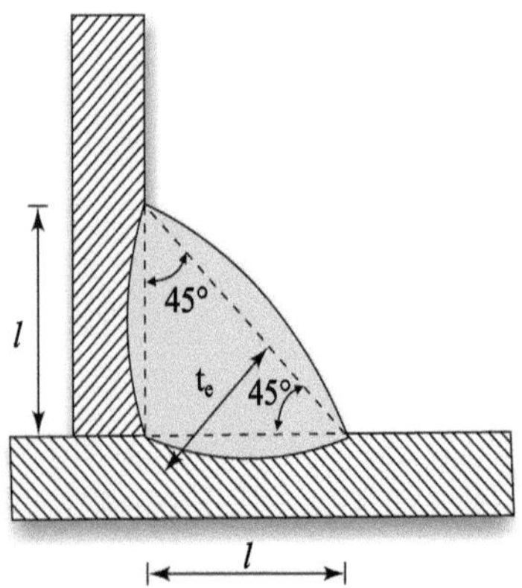

(a) Fillet weld with equal legs

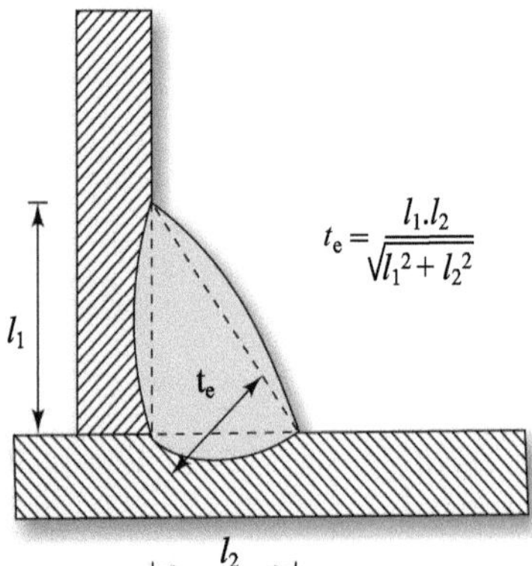

(b) Fillet weld with unequal legs

Fig. 15.12 Fillet weld with equal and unequal legs

Table 15.1 Minimum fillet-weld size

Material thickness (thicker part), mm	Fillet weld (l), mm
Up to 6.35	3.175
6.35–12.70	4.7625
12.70–19.05	6.35
19.05–38.10	7.9375
38.10–57.15	9.525
57.15–152.40	12.70
Over 152.40	15.875

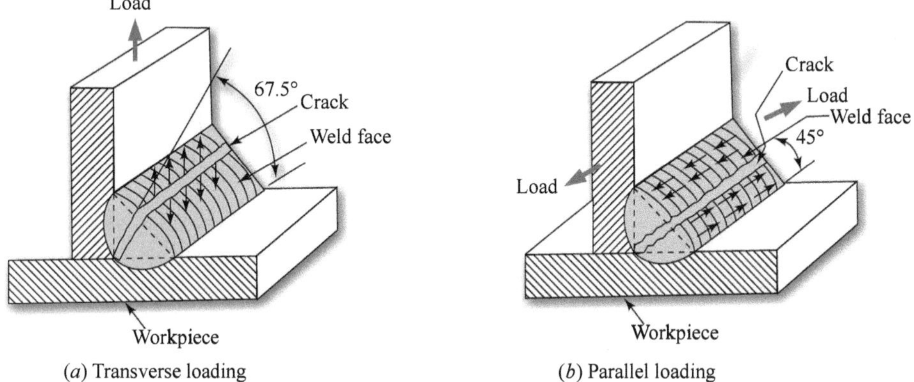

Fig. 15.13 Failure of fillet weld

15.9 Design for Brazing

Good brazing design is based on three important factors:

- Contact area
- Type of joint
- Joint clearance.

For a brazing joint to be stronger, there should be a higher contact area as compared to welding.

There are two important joints for brazing, namely lap joint and butt joint. Lap joint has more contact area and hence is more stronger. The strength of the lap joint can be equal to that of patent metal, if the joint is properly made. Proper amount of overlap is very important. Too little overlap will reduce the strength of the joint and excessive overlapping on account of insufficient penetration will produce poor joint. For maximum joint efficiency, amount of overlap should be equal to three times the thickness of the thinnest member. Brazed butt joint is not as strong as lap joint because it has limited brazing area.

Joint clearance is another important parameter in the design of brazed joint. Provision of optimal clearance between the parts is an essential requirement. Small clearance will prevent capillary action, and filler metal will not reach to the required place. Too much clearance allows overflow of filler metal again without reaching to the required place. In both cases, strength of the joint would be poor.

15.10 Design for Soldering

Design considerations w.r.t. contact area, type of joint and joint clearance in soldering are very much similar to that in brazing. For maximum strength of the soldered joint, the joint clearance should vary between 0.07 and 0.12 mm. For specific cases, e.g. for the joining of precoated metals, the joint clearance can be as low as 0.025 mm without affecting the strength of the joint. Table 15.2 gives the comparison of various joining processes.

Table 15.2 Comparison of various joining processes

Joining processes	Parameters				
	Strength	Design flexibility	Small-sized parts	Large-sized parts	Reliability
Adhesive bonding	*	***	***	**	**
Arc welding	***	**	*	***	***
Bolts and nuts	***	**	*	***	***
Brazing	***	***	***	***	***
Fasteners	**	*	*	***	**
Resistance welding	***	**	***	***	*
Riveting	***	**	*	***	***

Symbols: ***Very good
**Good
*Poor

Short Answer Questions

1. What is meant by edge preparation? Why do some parts being joined require edge preparation?

Ans. Edge preparation means making the edges of the parts straight and flat or vee-shaped, and cleaning them properly. It is required before the start of the joining process to make intimate contact between the parts for effective joining as in case of resistance butt welding.

2. Name the basic welding positions. Which welding position offers more difficulty and why?

Ans. The four basic welding positions used in practice include flat, horizontal, vertical and overhead. The overhead welding position offers more difficulty as there is a possibility of falling down of the molten weld pool and flux, causing injuries to the operator.

3. What is the difference between flat and horizontal welding positions?

Ans. In flat welding position, face of the weld and weld bead both are horizontal, whereas in horizontal welding position, plane of the workpiece is vertical, but the weld bead is horizontal.

4. Which type of joints require fillet welds? Where are fillet welds commonly used?

Ans. Fillet welds are used to fill in corners of tee, lap and corner joints. They are commonly used for structural work.

5. What are the important requirements for a good brazing joint?

Ans. A good brazing joint depends on three important factors, namely contact area, type of joint and joint clearance. A lap joint is more stronger than butt joint because of more contact area. Small clearance between parts prevents capillary action and does not allow filler metal to reach to the required place of the workpiece.

Multiple Choice Questions

1. Edges need to be prepared in a welding operation to

   (a) increase the strength of the joint
   (b) make the joint good-looking
   (c) reduce the cost of operation
   (d) make the operation faster.

2. Consider the following elements of edge preparation. It involves

   1. Straightening
   2. Cleaning
   3. Vee-ing

   Of these:

   (a) 1 alone is true (b) 1 and 2 are true
   (c) 1, 2 and 3 are true (d) 2 and 3 are true

3. Surfaces need to be cleaned before the start of a welding operation, because

   1. contaminants such as scale, rust, dust, paint, grease etc. reduce the penetration capacity of the electrode material to reach the required area.
   2. cleaning helps producing homogeneous and strong welds.
   3. it increases productivity.

   Of these:

   (a) 1 alone is true (b) 2 alone is true
   (c) 1 and 2 are true (d) 2 and 3 are true

4. Consider the following statements about a correct weld profile:

   1. It reduces the wastage of weld metal.
   2. It reduces the subsequent machining.
   3. It increases the strength of the joint.

   Of these statements:

   (a) 1 alone is true (b) 1, 2 and 3 are true
   (c) 1 and 3 are true (d) 3 alone is true

5. Which of the following welding positions is preferred?

   (a) Flat position (b) Vertical position
   (c) Horizontal position (d) Overhead position

6. Which of the following is the major cause of welding failure?

   (a) Creep (b) Fatigue
   (c) Bending (d) Shear

7. In transverse loading of a welded joint, failure occurs along the plane of maximum shear, which occurs at an angle of

(a) 30° from the horizontal
(b) 45° from the horizontal
(c) 60° from the horizontal
(d) 67.5° from the horizontal

8. Consider the following types of welded joints:

1. Butt joint
2. Corner joint
3. Lap joint
4. Tee joint

Of these:

(a) 1 and 2 are basic joints
(b) 1, 3 and 4 are basic joints
(c) 2, 3 and 4 are basic joints
(d) 1, 2, 3 and 4 are basic joints

Answers

1. (a) 2. (c) 3. (c) 4. (b) 5. (a) 6. (b)
7. (d) 8. (d)

Review Questions and Discussions

Q.1. What are the design parameters for welding?
Q.2. What is edge preparation?
Q.3. Why do some welding processes require edge preparation?
Q.4. How do groove welds differ from bead weld?
Q.5. Differentiate between butt joint and lap joint.
Q.6. Why is cleaning required before welding?
Q.7. Why is overhead welding difficult to perform?
Q.8. Why is arc welding difficult for small sized parts?
Q.9. Where are fillet welds used?
Q.10. How does the size of a fillet weld affect its strength?
Q.11. Why is the lap joint more stronger than the butt joint?
Q.12. How is effective throat different than actual throat?
Q.13. How is the joint efficiency related to the amount of overlap and thickness of the material being joined?

Theory of Metal Cutting

16

Taylor's tool life equation $VT^n = C$ gives relationship between cutting speed and tool life.

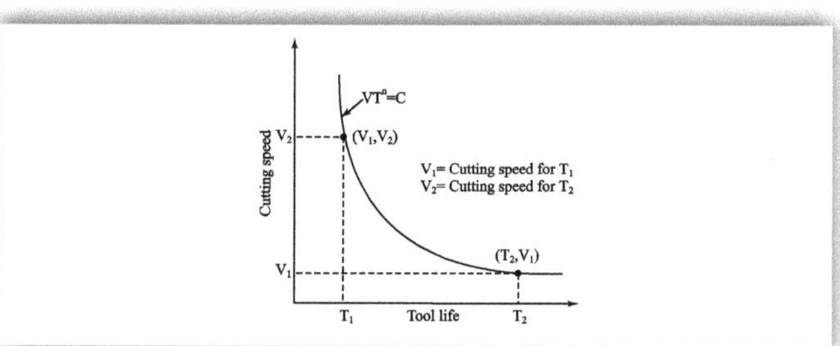

Learning Objectives

After reading this chapter, you will be able to answer some of the following questions:

- How does orthgonal cutting differ from oblique cutting?
- How is machinability affected by tool life?
- What is the function of cutting fluid?

D. K. Singh, *Fundamentals of Manufacturing Engineering*,
https://doi.org/10.1007/978-981-99-8767-2_16

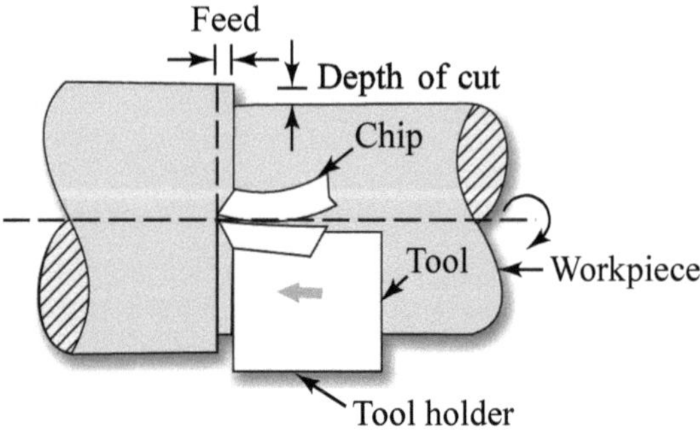

Fig. 16.1 A turning operation

16.1 Introduction

In machining, unwanted material is removed from a workpiece in the process of giving it the desired shape and size. The process is better known as metal cutting when the workpiece is made of metal. Single-point or multipoint cutting tools can be used for removing the material from the workpiece. A famous US mechanical engineer F. W. Taylor pioneered the metal cutting theory and gave the famous tool-life equation $VT^n = C$. A variety of shapes can be produced by machining. A model called orthogonal cutting is helpful in the understanding of machining processes.

Metal cutting principles are utilized in turning, milling, drilling and many other processes. Figure 16.1 shows the basic principle of turning operation.

16.2 Orthogonal Versus Oblique Cutting

In orthogonal cutting, the cutting tool edge is perpendicular to the direction of the cut, and there is no lateral flow of metal, nor is there chip curvature. All parts of the chip have the same velocity. It is a two-dimensional idealized model, and such a situation is realized when the cutting edge of the tool is straight and the relative velocity of the workpiece and the tool is perpendicular to the cutting edge. The cutting edge of the tool is inclined at an angle, with the normal to the cutting velocity vector. Hence to get an approximate result during the cutting process, orthogonal analysis is made. Examples include operations like sawing, parting off, etc.

In oblique cutting, the cutting edge and the cutting motion are usually not perpendicular to each other. Strictly speaking, almost all the machining operations are not orthogonal even when the cutting edge is perpendicular to the velocity direction. This is because of two cutting edges involved in a cutting operation: primary cutting edge (also called side cutting edge) and secondary cutting edge (also called end cutting edge). The cutting operation, where the primary cutting edge is normal to the cutting motion, is called restricted orthogonal machining. It is a three-dimensional cutting process. The oblique cutting is very near to practical approach and finds applications in shaping, drilling and milling.

16.3 Tool Geometry

The geometry of a single-point cutting tool with standard terminology is shown in Fig. 16.2, where the parameters appear in order. The various terms appearing in the figure are discussed below:

Shank is the main body of the cutting tool.

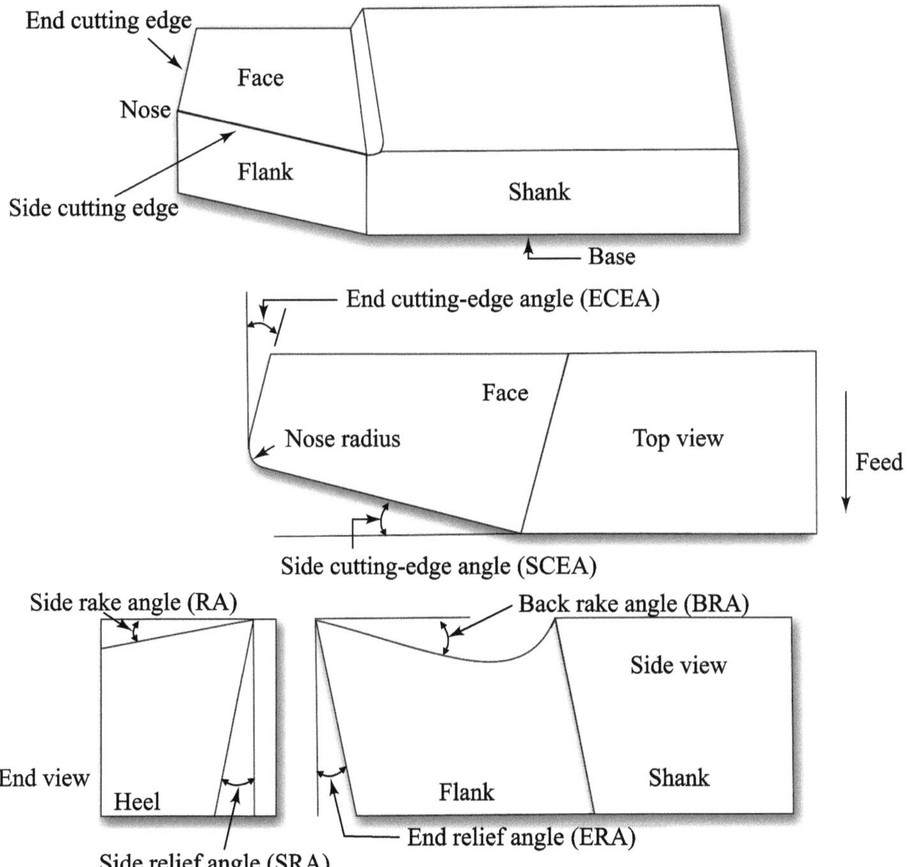

Fig. 16.2 Standard tool terminology for a single-point cutting tool

Flank forms the surface below and adjacent to the cutting edge.

Rake face is the surface on which chip slides during cutting or the face of the tool in contact with the chip.

Nose is the point of intersection of side cutting edge with end cutting edge.

Heel is the intersection of the flank and base of the tool.

Cutting edge is the sharp edge which removes material from the workpiece and lies on the face of the tool. There are two cutting edges as discussed in oblique cutting.

Side cutting-edge angle (SCEA) is the angle between the side cutting edge and the side of the tool shank. It prevents interference as the tool enters the workpiece and protects the tip of the tool from taking the initial shock of the cut. It can vary from 0° to 90°. On increasing SCEA, thickness of the chip decreases but its width increases.

End cutting-edge angle (ECEA) is the angle between the end cutting edge and a line normal to the tool shank. It provides a clearance to the trailing end of the cutting edge to prevent rubbing or drag between the machined surface and the trailing part of cutting edge. Its smaller value reduces the tool contact area with respect to the metal being cut. Larger value promotes vibration or chatter, thereby producing a poor surface finish and can initiate premature tool failure. For most operations, the ECEA is limited to 5°.

Side relief angle (SRA) and *End relief angle* (ERA) are the angles between the portion of the side or end flank immediately below the side or end cutting edge and a line perpendicular to the base of the tool, and measured at right angle to the side or end flank.

Relief angles are provided to make cutting possible and to ensure that there is no rubbing action between the work surface and flank surface of the tool. For general turning operation, relief angles may vary between 5° and 15°. Small relief angles give support to the cutting edge when machining hard and strong materials. To reduce the deflections of the tool and the workpiece and to provide good surface quality, larger relief angles are required.

Back rake angle (BRA) is the angle made by the rake face of the tool with the normal to the machined surface at the cutting edge. It controls the direction of the chip flow. It can be positive or negative. A positive rake angle reduces the cutting force and moves the chip away from the machined surface of the workpiece. Negative rake angle increases the cutting force required but provides greater strength at the cutting edge. This is particularly important in making intermittent cuts and in absorbing the impact during the initial engagement of the tool and work. In general, the rake angle is small for cutting hard materials and large for cutting soft ductile materials and can be negative for carbide and diamond tools. Generally speaking, the higher the hardness, the smaller the back rake angle.

Side rake angle (RA) is the angle between the tool face and a line parallel to the base of the tool. It is measured in a plane perpendicular to the base and the side cutting edge.

The side rake angle and the back rake angle combine to form the effective rake angle, also known as true rake angle or resultant rake angle of the tool. True rake inclination of a cutting tool has a major effect in determining the amount of chip compression and the shear angle. A small rake angle causes high compression, tool forces and friction resulting in a thick, highly deformed hot chip. Increased rake angle reduces the compression, the forces and the friction yielding a thinner, less deformed and cooler chip. In general, the power consumption is reduced by approximately 1% for each 1° in rake angle.

Nose radius is the rounded end that blends the side cutting-edge angle with the end cutting-edge angle. The nose radius has a major influence on surface finish. A sharp point at the end of a tool (i.e. no nose radius) will leave a groove in the path of cut. Increasing the nose radius usually decreases tool wear and improves surface finish. Generally for roughing, the largest possible nose radius should be selected so as to obtain the strongest point. A larger nose radius permits higher feeds, but must be checked against any vibration tendencies.

The introduction of coated tools has led to the development of improved tool geometries (low force groove geometries). Such geometry reduces the total energy consumption and breaks up the chips into small pieces. It increases the rake angle, which ultimately lowers the cutting force and power. This means that higher cutting speeds or lower cutting temperatures (and better tool lives) are possible.

16.4 Tool Signature

Seven parameters are required to completely define a cutting tool and are called tool signature for a given tool. These parameters in order are: back rake, side rake, end relief, side relief, end cutting edge, side cutting edge and nose radius. The first six parameters are angles to be expressed in degrees, and the seventh parameter, the nose radius is expressed either in inch or mm. A typical tool signature for a right-hand single-point cutting tool is shown in Fig. 16.3, where the parameters appear in order. The first six parameters are angles to be measured in degree and the last parameter in mm. They should appear in order starting with back rake angle and ending with nose radius.

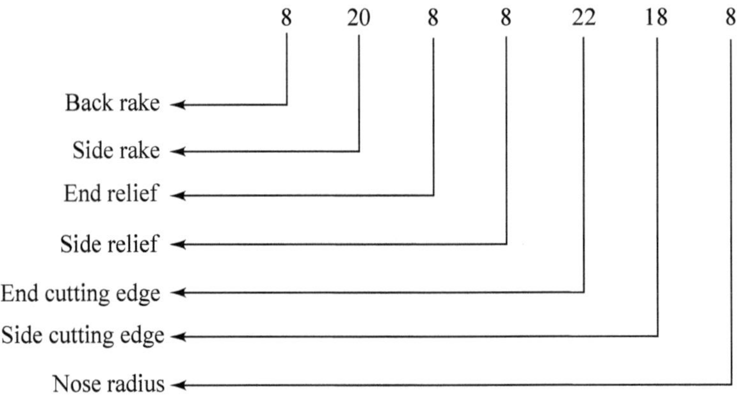

Fig. 16.3 A tool signature

16.5 Mechanics of Orthogonal Cutting

Cutting processes remove material from the surface of a workpiece by producing chips. The fundamental mechanics governing the removal of metal by a cutting tool is very complex.

During orthogonal cutting, the cutting tool moves to the left along the workpiece at a constant velocity and the workpiece is stationary. In the other situation, the tool is considered stationary and the workpiece moving. Both situations can be possible without changing the concept. A chip is formed ahead of the tool by deforming and shearing the material continuously along the shear plane. The details of orthogonal cutting process are shown in Fig. 16.4(a).

For the sake of analysis, let us suppose that the workpiece passes the tool with a velocity V, the cutting speed. This can also be the relative velocity of the tool with respect to the workpiece. The metal is sheared along the shear plane. The chip produced by the shearing action experiences two velocity components V_C and V_s along the tool face (rake face) and the shear plane respectively. V_s is the velocity of the chip relative to the workpiece and is known as *shear velocity*. It is the velocity at which shearing takes place. V_C is called the chip velocity. The velocity diagram associated with orthogonal machining is shown in Fig. 16.4(b).

Shear Plane (BC) is the plane along which cutting takes place.
Shear angle (ϕ) is the angle made by shear plane with the surface of the workpiece.
Uncut thickness (t) is actually the feed in certain machining operations, hence also known as depth of cut.

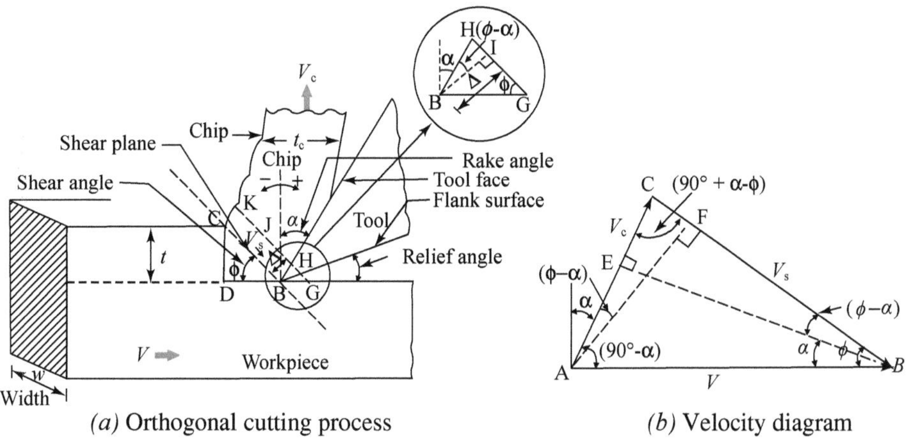

(a) Orthogonal cutting process (b) Velocity diagram

Fig. 16.4 Mechanics of orthogonal cutting

Chip thickness is t_c.

The *chip thickness ratio, r* is defined by the following expression:

$$r = \frac{t}{t_c} = \frac{\mathrm{BC}\sin\phi}{\mathrm{BC}\cos(\phi-\alpha)} = \frac{\sin\phi}{\cos(\phi-\alpha)} \tag{16.1}$$

where BC is the length of shear plane from the tool tip to the free surface of the workpiece. The ratio r is less than unity because $t_c > t$. The reciprocal of r is known as the chip compression ratio. Another way of expressing chip thickness ratio is based on the volume of chip. During orthogonal cutting process, density of the work material remains constant and the volume of the chip is equal to the volume of the metal removed.

$$tL = t_c L_c \tag{16.2}$$

where

L_c = length of chip

L = corresponding length of material removed from the workpiece.

Equation (16.2) can be rewritten as

$$\frac{t}{t_c} = \frac{L_c}{L} \tag{16.3}$$

Hence, the chip thickness ratio can be expressed as:

$$r = \frac{t}{t_c} = \frac{L_c}{L} \tag{16.4}$$

The vector sum of V_c and V_s is equal to V. The ratio r can also be expressed as a ratio of the two velocities.

$$r = \frac{t}{t_c} = \frac{V_c}{V} \tag{16.5}$$

The chip thickness ratio is a measure of deformation. Higher value of r is indicative of more deformation. It depends upon the rake angle of the cutting tool and the material being but. Use of cutting fluids increases r and hence reduces the energy required for cutting.

The shear angle, ϕ is expressed as

$$\tan\phi = \frac{r\cos\alpha}{1 - r\sin\alpha} \tag{16.6}$$

The ratio of V_s to V is

$$\frac{V_s}{V} = \frac{\cos\phi}{\cos(\phi - \alpha)} \tag{16.7}$$

The velocities V, V_c and V_s are important in power calculations, heat determinations and vibration analysis during chip formation.

The shear strain, γ can be expressed as

$$\gamma = \frac{GH}{\Delta} = \frac{GI}{\Delta} + \frac{IH}{\Delta}$$

or

$$\gamma = \cot\phi + \tan(\phi - \alpha) \tag{16.8}$$

where Δ is the thickness of an element *GJCB* of the undeformed material. For shear strains to be more, shear angle should be low and rake angle low or negative. Shear strains of 5 or large have been observed in actual cutting operations. Compared to forming and shaping processes, therefore, the material undergoes greater deformation during cutting. Also, deformation in cutting takes place generally within a very narrow deformation zone.

The shear angle has great significance in the mechanics of cutting operations. It influences chip thickness, force and power requirements and temperature produced as a result of work deformation.

The *Ernst* and *Merchant theory* was based on the assumption that the shear angle adjusts itself to minimize the cutting force, or that the shear plane is a plane of maximum shear stress. According to this theory, the following relationship was obtained for shear angle, ϕ.

$$2\phi + \beta - \alpha = 90^{\circ} \tag{16.9}$$

where β is the angle of friction and is related to the coefficient of friction, μ at the tool–chip interface by the following expression

$$\mu = \tan \beta \tag{16.10}$$

The *Merchant theory* was found suitable for certain nonmetallic materials but did not work well with steel. This theory is based on the relationship of shear stress and normal stress acting on the shear plane of the workpiece. The modified equation can be expressed as

$$2\phi + \beta - \alpha = C_m \tag{16.11}$$

where C_m is known as machining constant that depends on the workpiece material. For steel, the value of C_m was found to be 80°. Usually the value of C_m lies between 70° and 80°.

The *Lee and Shaffer theory* is based on the stress–strain distribution behaviour in perfectly plastic material. According to this theory, the expression for ϕ is

$$\phi = 45^{\circ} + \alpha - \beta \tag{16.12}$$

16.6 Cutting Forces and Power in Orthogonal Cutting

Cutting forces are determined for measuring power requirements during cutting operations, energy loss and to know the distortion of the workpiece and the cutting tool by these forces.

The various forces acting on the chip, cutting tool and workpiece are shown in Fig. 16.5. If the chip above the shear plane is considered as a rigid body moving with a constant velocity, the resultant of the forces must be zero, since the resultant force R acting on the back of the chip is equal and opposite to the resultant force R' acting on the shear plane. The resultant force R consists of friction force F and normal force N acting along and normal to the rake surface, respectively, on the tool–chip interface contact area. Hence

$$\frac{F}{N} = \mu \tag{16.13}$$

where μ is the coefficient of friction on the tool–chip interface area.

The Equation (16.13), using Equation (16.10), can be rewritten as:

$$\mu = \frac{F}{N} = \tan \beta \tag{16.14}$$

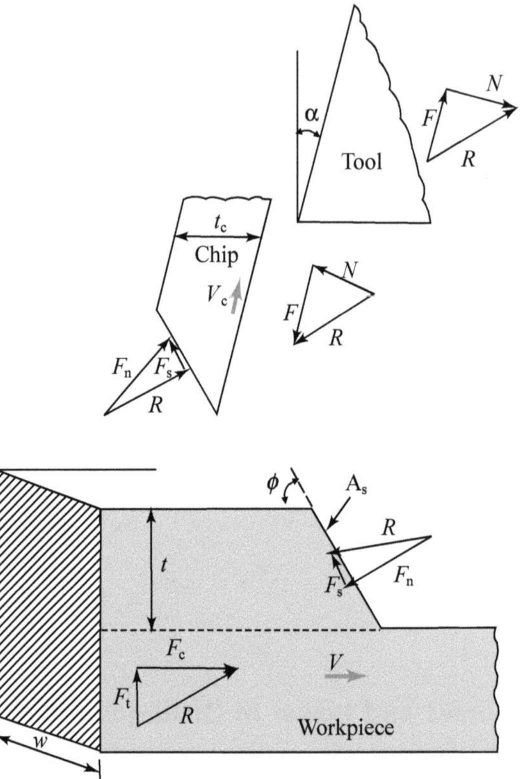

Fig. 16.5 Forces acting on workpiece, tool and chip

The resultant R' is composed of a shear force F_s and normal force F_n acting on the shear plane area, A_s. Determination of these two resultant forces R and R' is difficult; hence measurement of forces acting on the tool is made with a dynamometer. Electronic load dynamometers are frequently used to measure these forces. As it is impossible to measure cutting forces at the point of the tool and workpiece, the reactions are measured away from the cutting point. These reaction forces are combined in such a way that only two components of it, a cutting force F_c and a tangential (normal) force F_t represent all other above-mentioned forces. A Merchant's circle diagram shown in Fig. 16.6 is then used to express the forces F_s, F_n, F and N in terms of the measured dynamometer components, F_c and F_t.

Using Figure 16.6, the following relations can be obtained:

$$F = F_c \sin\alpha + F_t \cos\alpha \tag{16.15}$$

$$N = F_c \cos\alpha - F_t \sin\alpha \tag{16.16}$$

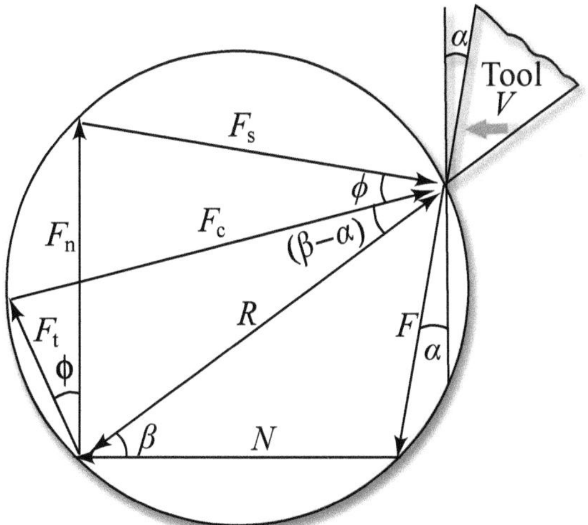

Fig. 16.6 A Merchant's circle diagram

and the resultant force, R is

$$R = \sqrt{F_c^2 + F_t^2} \tag{16.17}$$

When rake angle, α is zero, then $F = F_t$ and $N = F_c$ and hence the friction and normal forces can be measured directly by the dynamometer.

The shear force, F_s and normal force, F_n acting on shear plane are expressed as

$$F_s = F_c \cos\phi - F_t \sin\phi \tag{16.18}$$

$$F_n = F_c \sin\phi - F_t \cos\phi \tag{16.19}$$

The shear stress acting on the shear plane can be shown to be

$$\tau_s = \frac{F_s}{A_s} \tag{16.20}$$

But the shear area, A_s can be expressed as:

$$A_s = \frac{t \cdot W}{\sin\phi} \tag{16.21}$$

where

W = width of the workpiece
t = uncut chip thickness

Equation (16.20) can now be expressed as

$$\tau_s = \frac{F_s \cdot \sin\phi}{tW} \tag{16.22}$$

On substituting Equation (16.18) in Equation (16.22), we get

$$\tau_s = \frac{\sin\phi}{tW}[F_c\cos\phi - F_t\sin\phi] \tag{16.23}$$

For a given polycrystalline metal, the shear stress is constant and is not sensitive to variations in cutting parameters, tool materials or the cutting environment.

The power consumption during machining is given by

$$P = F_c \cdot V \tag{16.24}$$

where V is the cutting velocity.

Equation (16.24) can be further expressed as

$$P = F_c.V = VWt\,\tau_s\cos(\beta - \alpha)\left[\frac{1}{\sin\phi\cos(\phi + \beta - \alpha)}\right] \tag{16.25}$$

This power is dissipated mainly in the shear zone in the process of shearing the material and on the rake face due to tool–chip interface friction. About 95% of the power is consumed in the shear zone, and just 5% is used in friction and producing residual stresses in the workpiece.

The power dissipated in the shear plane and friction are given by

$$P_s = F_s V_s \tag{16.26}$$

$$P_f = F V_c \tag{16.27}$$

The specific energy is defined as the energy consumed per unit volume of material removal. It is a very convenient basis for judging the power required for a given process. The corresponding specific energies are expressed as:

$$u_s = \frac{F_s V_s}{tWV} \tag{16.28}$$

$$u_f = \frac{F V_c}{tWV} = \frac{Fr}{tW} \tag{16.29}$$

where $r = \frac{V_c}{V}$ = Chip thickness ratio

The total specific energy is given by

$$u_t = u_s + u_f$$

or

$$u_t = \frac{1}{tWV}[F_s V_s + F V_c] \tag{16.30}$$

The sharpness of the cutting tool greatly influences the power consumed during machining. The duller the tool, the higher are the forces and power required.

16.7 Cutting Speed and Feed

Cutting speed is the speed at which the cutting edge of the tool machines the surface of the workpiece. It is a measure of the peripheral speed as indicated by the following expression.

$$V = \frac{\pi D N}{100} \tag{16.31}$$

where

V = surface cutting speed of the workpiece (m/min)
D = diameter to be cut (mm)
N = revolutions per minute (rpm)
π = constant

This equation is valid for turning, drilling, boring, milling and other processes. In turning operation, the diameter D is the outermost dimension of the work being turned. In facing or cut off (parting) operations, the outer diameter is used, even though the diameter is continuously reducing during the cut. For boring, the diameter D is the initial dimension of the hole before being bored. In milling, the outermost diameter of the cutter serves as the diameter D, since it is rotating. In turning, the workpiece rotates; whereas in milling, the tool rotates and the workpiece is stationary.

Feed is the rate at which a cutting tool or grinding wheel advances along or into the surface of the workpiece. When the workpiece rotates as in the case of turning, feed is expressed in mm/rev; when the tool or workpiece reciprocates, feed is expressed in mm/stroke; and for stationary workpiece and rotating tools, it is expressed in mm/rev of the tool. For milling, feed is expressed in mm/tooth.

16.8 Types of Chips

Chips are formed largely by shearing action and compressive stresses on the metal in front of the tool. The compressive stresses are greatest at the points which are farthest from the cutting tool and are balanced by tensile stresses in the zone nearest the tool; hence the chip curls outwardly or away from the cut surface. Chips produced greatly influences the surface finish produced and overall cutting operations such as tool life, vibration and chatter. They can be broadly classified into three groups.

- Discontinuous chip
- Continuous chip
- Continuous chip with a built-up edge.

16.8.1 Discontinuous Chip

Discontinuous or segmented chips (Fig. 16.7) are produced in short pieces while machining brittle materials such as cast iron and bronze.

They are also produced when the cutting speed is very low or very high, the depth of cut is large, the cutting tool has low rake angle or the machining process is not using suitable cutting fluid.

The rupture of metal takes place when the metal directly above the cutting edge gets compressed to such an extent that the deformed metal starts sliding along the face and the magnitude of compressive stress reaches the fracture limit of the metal.

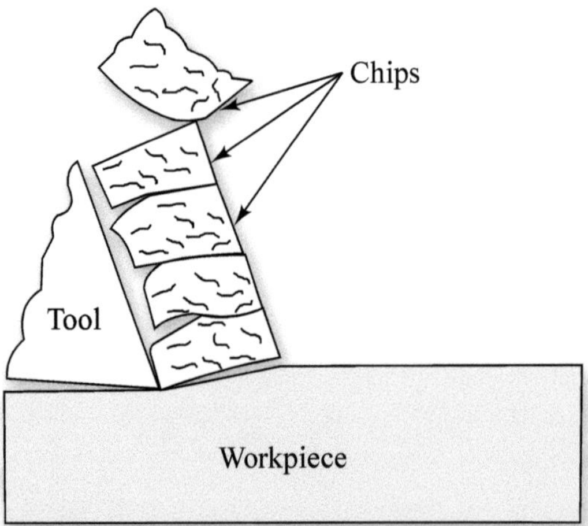

Fig. 16.7 A segmented chip

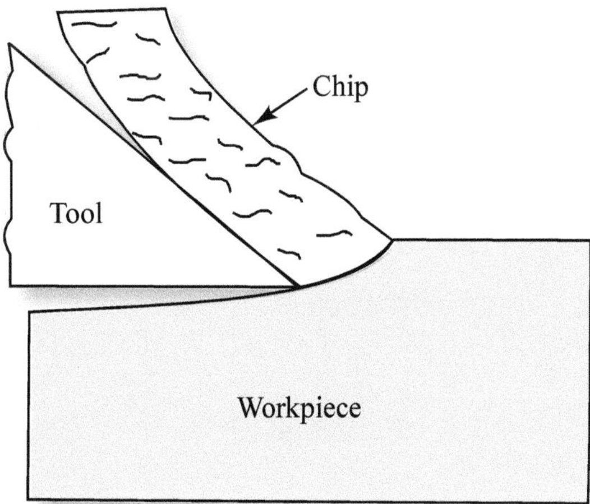

Fig. 16.8 A continuous chip

Because of their short lengths, discontinuous chips are easier to handle and dispose. They produce better surface finish because of their noninterference with the workpiece. However the tool may vibrate and chatter due to varying forces acting on it during chip formation.

16.8.2 Continuous Chip

Continuous chips (Fig. 16.8) are produced in the form of long coils due to continuous plastic deformation of the metal along the shear plane. They are produced when ductile materials are machined with carbide tools operating at higher cutting speeds. These chips are the most desirable ones. However because of their tendency to get curled around the tool, they need to be broken sometimes. Chip breakers are used to break them into short pieces for easy handling and disposal.

16.8.3 Continuous Chip with a Built-up Edge

During the machining process, part of the material removed gets attached at the tip of the tool and is known as built-up edge (BUE). Periodically, a small amount of this BUE is broken and carried away with the chip, while the rest adheres to the surface of the workpiece making it rough. Such types of chips (Fig. 16.9) are formed due to excessive pressure at the cutting edge while machining ductile materials with high coefficient of friction, which results in stronger adhesive bond between tool and workpiece materials.

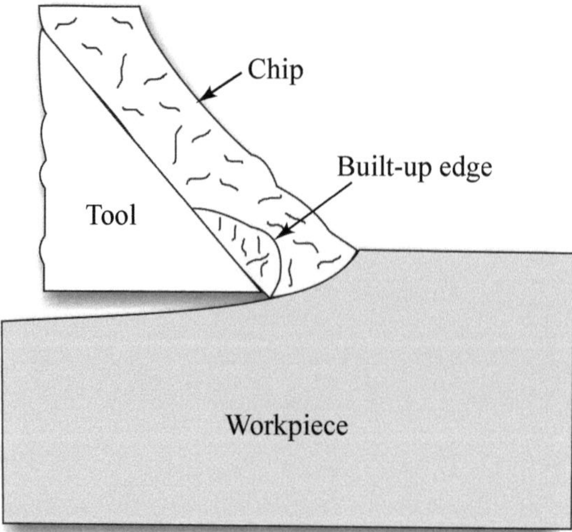

Fig. 16.9 A continuous chip with a built-up edge

The other factors which help in the formation of BUE include low rake angle, low cutting speed, large depth of cut and blunt tool. These chips are commonly observed in practice. Although BUE is generally undesirable because it often results in an inferior surface finish, but a thin and stable BUE is usually regarded as desirable, because it protects the tool's surface and reduces wear. The hardness of the stable BUE has been estimated to be 2 to 3 times higher than that of material being machined.

16.9 Heat Generation and Temperature in Metal Cutting

In metal cutting, the power consumed (F_c V) into the process is mostly (almost 99%) converted into heat energy. Once generated, the heat may remain in the workpiece, be carried away by the chips, or be transferred to the tool. The three elements which receive the heat are known as heat sinks. As cutting speed increases, a larger proportion of the heat generated is carried away by the chip and little heat goes into the workpiece. In general, the heat distribution in the chip, tool and work material as reported by *Schmidt* is in the ratio of 80: 10: 10 respectively (Fig. 16.10). Since most of the heat generated at higher cutting speeds is taken by the chip, hence it is usual practice to keep the machining speeds high. Although machining at higher cutting speed is desirable, for higher productivity, the faster tool wear due to excessive temperature puts a limit to the cutting speed. As speed increases, the time for heat dissipation decreases and thus temperature rises.

There are three sources of heat generation (Fig. 16.11). The major source of heat generation occurs in the shear zone where the material undergoes plastic deformation and is known as primary heat source. The second source of heat is the tool–chip interface in which the chip slides on the rake surface of the tool. This is known as secondary heat source. There is another source of heat called tertiary heat source where the work surface rubs against the surface of the tool. But contribution from this source is insignificant when sharp tools are used during machining.

The maximum temperature is reached in the tool due to heat generated in the cutting zone. Higher temperature at the cutting edge leads to the failure of the tool, both by softening and thermal stresses. It has been shown by *Trigger and Chao* that the highest temperature is never maximum at the cutting edge, but is slightly away from the edge and at about middle of the tool–chip contact length. A typical temperature distribution in the cutting zone is shown in Fig. 16.12.

Temperature increases with the strength of the workpiece material, cutting speed and depth of cut, and it decreases with increasing specific heat and thermal conductivity of the workpiece material. Cutting speed has maximum influence on the tool temperature. Cutting fluids can be used to reduce the temperature in the cutting zone. However the cutting fluid is more effective at low cutting speeds because of its maximum ability to reach the tool–chip interface.

Knowledge of heat distribution is useful in knowing the adverse effect of temperature rise on the strength and hardness of the cutting tool as well as workpiece. Due to heat intake, there may be dimensional changes in the workpiece. Also the machined surface may get adversely affected.

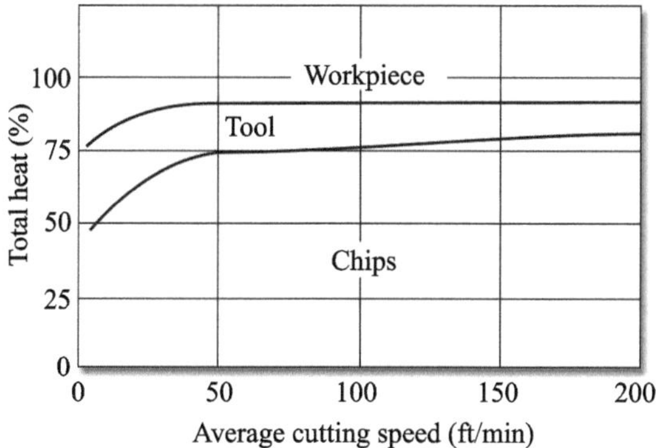

Fig. 16.10 Heat distribution in chips, tool and work

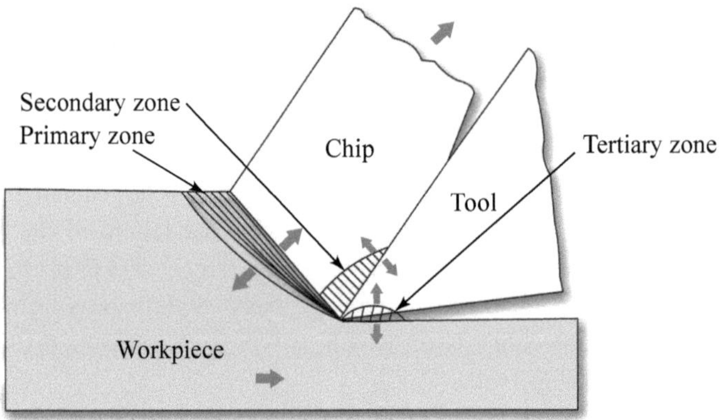

Fig. 16.11 Sources of heat generation

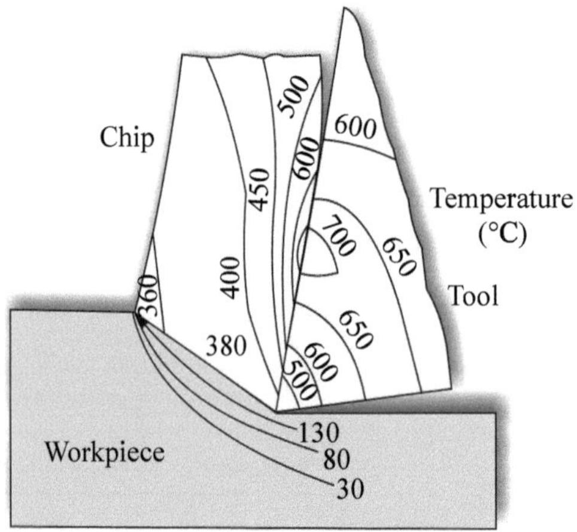

Fig. 16.12 Typical temperature distribution in the cutting zone

16.10 Temperature Measurement (Experimental Methods)

16.10.1 Tool–Work Thermocouple Method

The tool–chip interface acts as a hot junction of the tool–work thermocouple, and this method can be used to know the temperature in the cutting zone. A small voltage, called thermal emf, is induced in the circuit which is measured by a suitable milli-voltmeter and is then calibrated for the measurement of temperature. This method is suitable for carbide tools where induced thermal emf may be in the range of 10–15 mv. During the machining of carbon steels using HSS tools, the emf is lower (2–3 mv.).

16.10.2 Embedded Thermocouple Method

In this method, a thermocouple can be introduced in the tool and/or the workpiece. There is no need for calibration. If thermocouple does not reach close to the rake or flank surface, there will be error in the measurement of temperature. Alternatively, if the thermocouple is brought too close to the rake, the tool wear might result in the breaking of the thermocouple arrangement. An adjustment has to be made keeping in mind the two situations discussed above. For plotting the temperature distribution, a large number of thermocouples are required to be distributed over the rake face.

16.10.3 Infrared Technique

This technique uses radiation pyrometer to measure the infrared radiations coming from the cutting zone. However, this method indicates only surface temperatures, and the accuracy of the results depends on the emissivity of the surfaces, which is difficult to determine accurately.

16.10.4 Calorimetric Method

This method was introduced by *Sawwin* in 1912 for the evaluation of temperatures while performing machining on a lathe. It is based on the principle of energy conservation. The average chip temperature T_{cav} can be shown to be

$$T_{c_{\mathrm{av}}} = \frac{T + m_w(T - T_w)}{m_c \cdot C_c} \tag{16.32}$$

where

T = temperature of the calorimeter system with water and chip
T_w = initial temperature of water
m_w = weight of water
C_c = thermal capacity of the chip

Later on, the scope of this method was increased and was used by *Schmidt* for drilling and milling. It is an old method and is now no longer used.

16.10.5 Analytical Method

The temperature rise in the primary deformation zone is given by

$$\theta_p = \frac{(1-\Delta)W_p}{\rho C V t w} \tag{16.33}$$

where W_p = rate of heat generation in the primary zone
$= (F_c V - FrV)$

Since $W = W_p + W_s$
= total rate of heat generation

Also $W = F_c V$

and $W_s = FV_c = FrV$
= rate of heat generation in the secondary zone

Δ = fraction of primary heat which goes to the workpiece

ρ = density of the work material

C = specific heat of the work material

t = uncut chip thickness

w = width of cut

V = velocity of workpiece, this may be the relative velocity of workpiece and tool

r = chip thickness ratio

V_c = chip velocity

The maximum temperature rise, when the material passes through the secondary zone along the rake face of the tool is estimated as:

$$\theta_s = 1.13\sqrt{\frac{z}{l}}\left(\frac{W_S}{\rho C V w t}\right) \tag{16.34}$$

where

z = a nondimensional quantity
$= \frac{\rho C V t}{K}$

K = thermal conductivity of the material

l = contact length between chip and tool

$$= t_c[1 + \tan(\phi - \alpha)] \tag{16.35}$$

t_c = chip thickness

Substituting l in the expression for θ_s, we get

$$\theta_s = 1.13\sqrt{\frac{1}{\rho V C t K[1 + \tan(\phi - \alpha)]}} \cdot \frac{W_s}{W_t} \tag{16.36}$$

The average temperature rise in the secondary zone can be shown to be

$$\theta_{s_{av}} = \frac{W_s}{\rho C V\, wt} \tag{16.37}$$

The final temperature in the cutting zone is given by

$$\theta = \theta_0 + \theta_p + \theta_s \tag{16.38}$$

where θ_o is the initial temperature of the workpiece.

16.11 Tool Life

Tool life may be defined as the length of period for which a tool can be used satisfactorily for cutting operations. It is measured in several ways. Some of the different ways for expressing tool life are:

- Actual cutting time
- Number of pieces machined per tool
- Time interval between two consecutive tool resharpenings between which tool performs satisfactorily.

F. W. Taylor in 1907 established a relationship between tool life and cutting speed. The relationship is in the form of an empirical law and is given as:

$$VT^n = C \tag{16.39}$$

where

V = cutting speed (ft/min)

This is the velocity of the workpiece in relation to the tool.

T = tool life (min)
n = an exponent
C = constant (if $T = 1$ then $C = V$).

It must be understood, however, that the law is true only if other possible variables such as depth of cut, feed, rake angle, etc. are kept constant. Any change in these factors has a marked effect upon the value of C.

The Taylor's law is shown graphically in Fig. 16.13. It is usually plotted on log-log paper and is displayed as a log T versus log*V* graph. This graph is better known as tool-life curve. Such curves are usually linear over a certain range of cutting speeds but are rarely so over a wide range. The exponent *n* can be negative at low cutting speeds. For a given depth of cut and feed rate, the method to find the value of *n* for a HSS tool is shown in Fig. 16.14.

The slope of the curve (line) gives the value of *n,* and intercept of the line on *y*-axis gives the cutting speed for 1 min. of tool life equal to *C*. The position and slope of the line are affected by the workpiece material, tool material and feed rate. The line on the left has higher feed rate (Fig. 16.15). The recommended cutting speed for a HSS tool is generally the one that gives a tool life of 60–120 min (for carbide tools, 30–60 min). Cutting speeds may vary significantly according to the workpiece hardness and the particular operation.

The tool-life curves can be used for two purposes. Firstly, it can be used to determine the values of constants, *n* and *C* of tool-life equation. Secondly, it can be used to predict the values of tool life at other cutting speeds.

Values of *n* and *C* can be found only from cutting tests. The most comprehensive cutting test data available are those published in the American Society of Mechanical Engineers (ASME) handbook and the Production Engineering Research Association data sheets. Each combination of workpiece and tool materials has its own *n* and *C*. Some values of *n* and *C* for a particular work–tool combination are given in Table 16.1.

Constants *n* and *C* depend on the tool and work material, tool geometry and cutting conditions (except cutting speed).

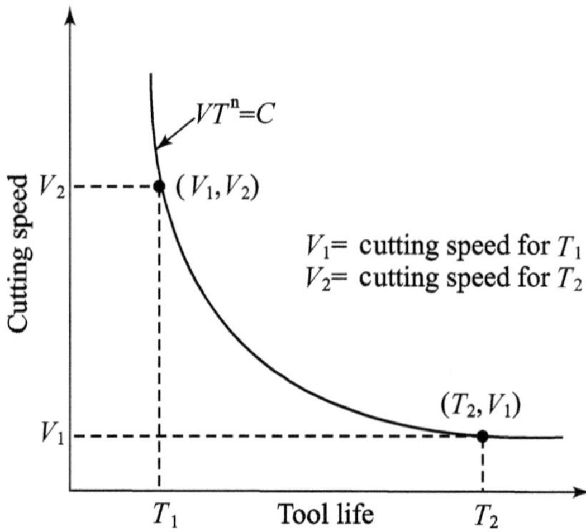

Fig. 16.13 Relationship between cutting speed and tool life

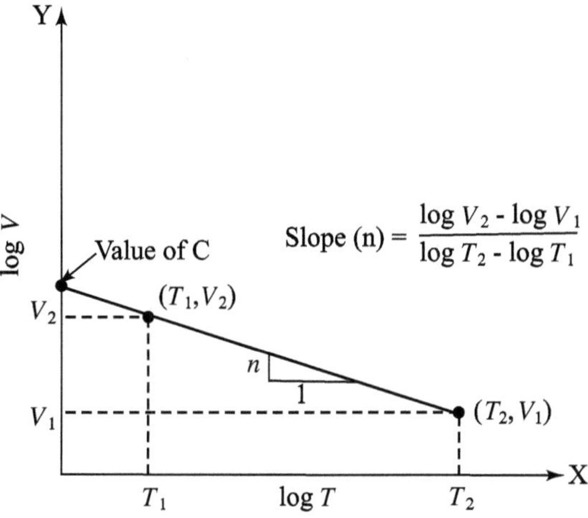

Fig. 16.14 Cutting speed and tool-life relationship on log scale

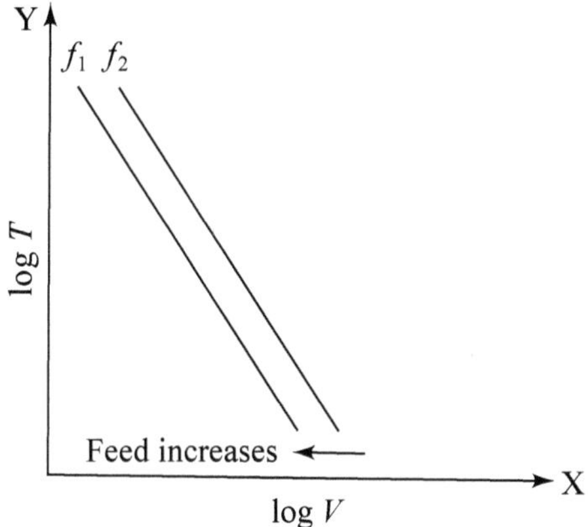

Fig. 16.15 Effect of feed on tool-life curve

Table 16.1 Values of n and C

Tool materials	HSS		Tungsten carbides	
Work materials	C	n	C	n
Stainless steel	170	0.08	400	0.16
Medium-carbon steel	190	0.11	150	0.20
Grey cast iron	75	0.14	130	0.25

The factors affecting tool life include microstructure of the material being cut, cutting speed, feed, depth of cut, tool material and its geometry and cutting fluids. However, there is rapid (exponential) decrease in tool life as cutting speed increases. It has been observed that the effect of changes in feed on tool life is relatively smaller than that of proportionate changes in cutting speed.

16.12 Tool Life and Machinability

Machinability is a broad term. Simply speaking, it is the ease of machining. The factors to be considered for the definition of machinability include forces and power requirement during cutting, tool wear and tool life, surface finish and integrity of the machined part and machining cost. Machinability may be defined in the following ways:

- The life of the tool before failure or tool resharpening,
- The quality of the machined surface, and
- The power consumption per unit volume of material removed.

But tool life is the most important criteria for defining machinability. Higher the tool life, the better is the machinability. Values of machinability given in the handbooks are based on tool life.

The measurement of machinability ratings for different materials is relative. The American Iron and Steel Institute (AISI) steel B 1112 with ultimate tensile strength of 566 MPa and Brinell's hardness of 179 is chosen as the standard material for defining machinability. It is assigned a machinability rating of 100 because it is easily machinable. The machinability rating of other materials are measured in terms of *B* 1112. The ratings are based on actual cutting tests conducted under standardized conditions. Tool-life curves are used to develop machinability ratings. The *B*1112 steel has a tool life of 60 minutes at a cutting speed of 100 feet per minute (fpm). For example, suppose any material, *X* has a 70% rating, it means that *X* has a cutting speed of 70% of *B*1112 for equal tool lives.

The tests conducted for knowing the machinability rating indicate the resistance of the material being cut and are influenced by several factors including material's characteristics such as its composition, hardness, grain size, microstructure, work-hardening nature and size; type and rigidity of equipment, coolant properties, feed and depth of cut and the kind of tool.

16.13 Tool Failure

The failure of a cutting tool may be due to one or a combination of the following modes:

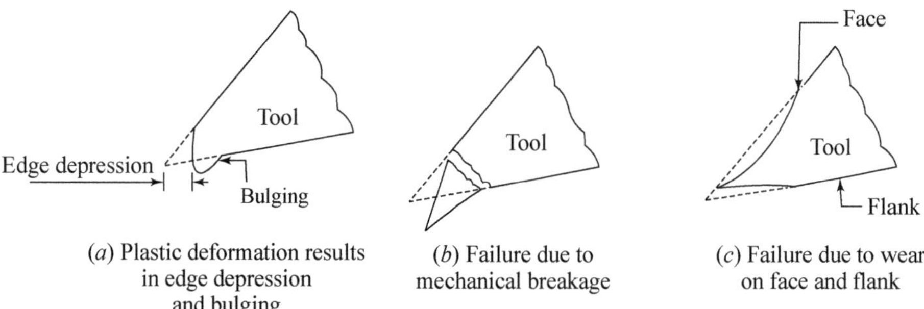

(*a*) Plastic deformation results in edge depression and bulging

(*b*) Failure due to mechanical breakage

(*c*) Failure due to wear on face and flank

Fig. 16.16 Modes of cutting tool failure

- Plastic deformation of the tool due to high temperature and large stress (Fig. 16.16(a)).
- Failure due to mechanical breakage because of large force applied and insufficient strength of the tool (Fig. 16.16(b)).
- Failure of the tool in the form of bursting (wear) of the cutting edge of the tool on its both surfaces, face and flank (Fig. 16.16(c)).

Dotted line shown on the left is the part of the original tool shape. Tool after failure is shown in solid lines.

The first two modes of failure may arise due to excessive temperature in the cutting zone, or due to large force acting on the cutting tool or due to lack of tool strength. The large area of intimate contact results in substantial friction between cutting tool and workpiece, causing the temperature to rise rapidly.

The cutting tool then loses its form stability and fails rapidly because of further softening of the tool material in the cutting zone due to increased temperature level. Sudden loads caused by dropping the tool or the rapid engagement into a large depth of cut may cause fracture. Inadequate application of coolant can result in localized cooling, increasing the level of thermal stresses and hence becomes one of reasons of tool failure. By a proper selection of the cutting tool material, tool geometry and cutting conditions, plastic deformation and mechanical failure can be prevented.

16.14 Tool Wear

Wear is defined as the progressive loss or removal of material from a tool's surface. Wear generally alters the surface topography and may result in severe surface damage. Tool wear is generally a gradual process and is always unavoidable and hence requires constant attention.

The following are the important types of wear:

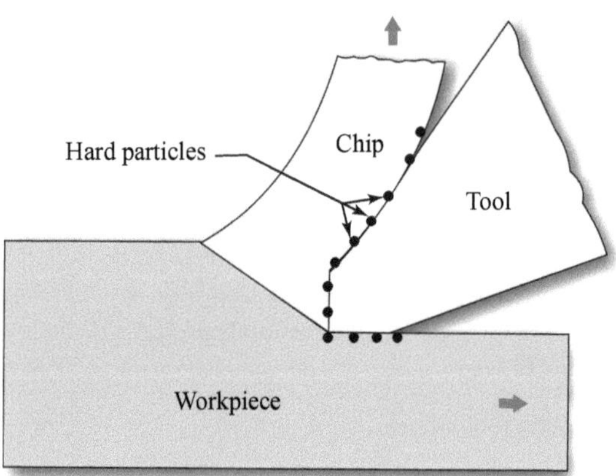

Fig. 16.17 Abrasive wear

- Abrasion wear
- Adhesion wear
- Diffusion wear
- Corrosive wear
- Fatigue wear

16.14.1 Abrasion Wear

Abrasion wear (Fig. 16.17) is very common and is caused by a hard and rough surface or a surface containing hard, protruding particles-sliding across a surface, i.e., it occurs in the tool when workpiece rubs against the tool surface.

This type of wear removes particles by forming microchips or slivers, thereby producing grooves or scratches on the softer surface. The cutting edge's ability to resist abrasion is largely dependent on the tool hardness. Many steels and cast irons, for example, may contain carbides, oxides and nitrides, whose hardness will exceed that of the martensitic matrix of a HSS tool.

16.14.2 Adhesion Wear

Adhesion wear (Fig. 16.18) is also known as attrition wear. It occurs mainly at low machining temperatures on the chip face (rake face) of the tool. When asperities of the tool and chip meet, they can deform and weld owing to localized heating and pressures resulting in the formation of a built-up edge (BUE). It is a dynamic structure, with successive layers from the chip being welded and hardened, becoming part of the edge. The

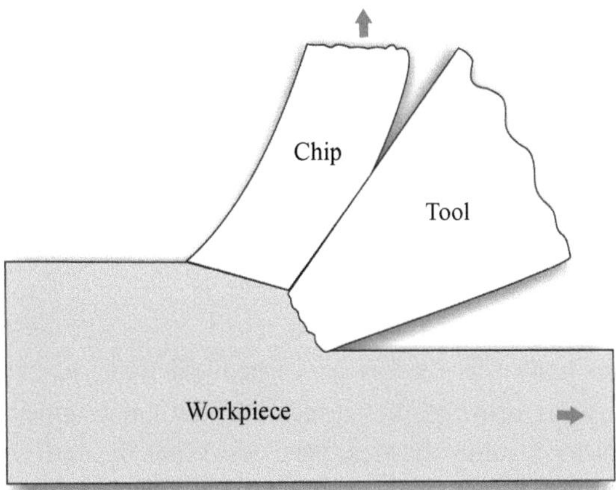

Fig. 16.18 Adhesion wear

BUE can be sheared off and commence build-up again or cause the cutting edge of the tool to break away in small pieces or fracture. The materials commonly subjected to adhesion wear include steel, aluminium and cast iron.

16.14.3 Diffusion Wear

Diffusion wear (Fig. 16.19) is more affected by the chemical bonding during the cutting process. In metal cutting, where intimate contact between the work and tool materials occurs and high temperature exists, diffusion wear can occur when atoms move from the tool material to the work material.

This results in a gradual deformation of the tool surface. The chemical properties of the tool material and its affinity towards workpiece material will decide the development

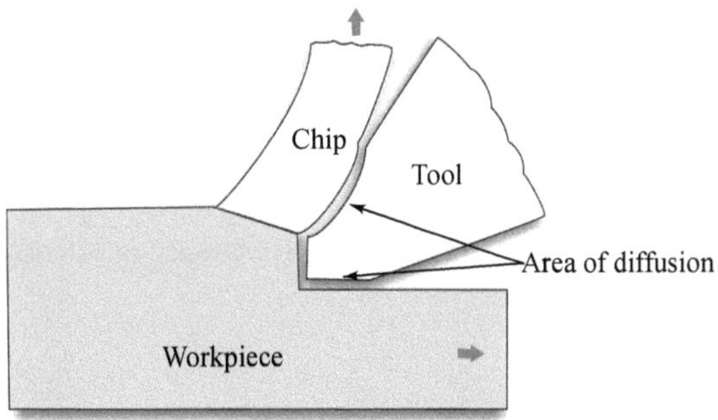

Fig. 16.19 Diffusion wear

of the diffusion wear mechanism. For example, tungsten carbide and steel have affinity towards each other and hence can produce diffusion wear. This results in the formation of a crater on the chip face of the tool. This mechanism strongly depends on temperature; the rate of diffusion increases exponentially with rise in temperature and is thus greatest at high cutting speeds.

16.14.4 Corrosive Wear

Corrosive wear, also known as oxidation or chemical wear, is caused by chemical or electrochemical reactions between the surface and the environment. The fine corrosive products on the surface constitute the wear particles. When the corrosive layer is destroyed or removed, as by sliding or abrasion, another layer begins to form, and the process of removal and corrosive-layer formation is repeated. It occurs at high temperatures, where due to oxidation of the carbide in the cutting tool, porous oxide films are formed which are more easily rubbed off by the chip. As a result, strength of the cutting tool is decreased and the tool wear results which appears in the form of typical notches. Also, the surface of the cutting tool gets eroded when a cutting fluid is introduced between the tool and workpiece. But this mechanism is quite uncommon.

16.14.5 Fatigue Wear

Fatigue wear is often a thermomechanical combination. Temperature fluctuations and the loading and unloading of cutting forces can lead to cutting edges cracking and breaking. It may result from hard or strong workpiece materials, very high feed rates or when the tool material is not hard enough.

16.15 Regions of Tool Wear

There are two basic regions of wear in a cutting tool: flank wear and crater wear. They are shown in Fig. 16.20.

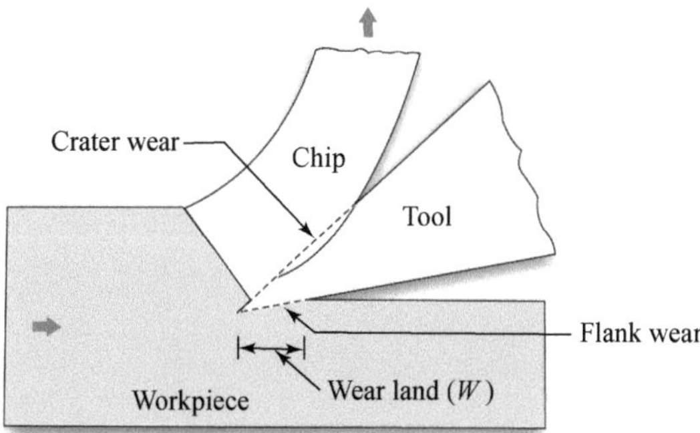

Fig. 16.20 Crater and flank wear

16.15.1 Flank Wear

Flank wear, as the name suggests, takes place on the flanks of the cutting edge. Abrasion and adhesion are primarily responsible for the flank wear.

The worn area, called wear land, is usually taken as a measure of the amount of wear and can be readily determined by means of a tool maker's microscope. It extends from the top to a distance on the flank surface, which is abraded away as cutting continues. The allowable wear land for various conditions is given in Table 16.2. For improved dimensional accuracy and surface finish, the allowable wear land may be made smaller than the values given in Table 16.2.

Flank wear is caused by friction between the newly machined surface of the workpiece and the contact area on the tool flank. This is usually the most common type of wear. Excessive flank wear produces poor surface texture and inaccuracy and increases friction as the edge changes shape. It should be noted that the wear on the flank face is seldom uniform along the active cutting edge; it is usually greater at the ends than in the centre.

Figure 16.21 shows the variation of tool flank wear with cutting time. This curve generally consists of three stages with different slopes.

Table 16.2 Allowable average wear land for cutting tools in various operations

Operations	Allowable wear land (mm)	
	High-speed steels	Carbides
Turning	1.5	0.4
Face milling	1.5	0.4
End milling	0.3	0.3
Drilling	0.4	0.4
Reaming	0.15	0.15

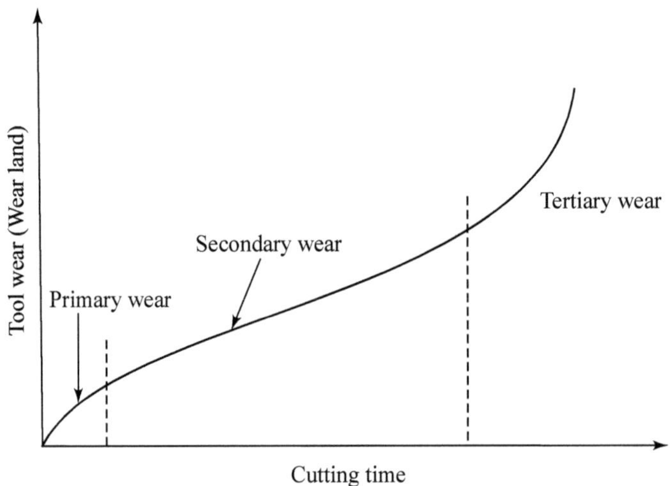

Fig. 16.21 Flank wear versus cutting time

- Initial wear stage
- Steady wear stage
- Final wear stage.

In the first stage, also called primary wear zone, there is rapid breakdown of the initially sharp cutting edge and a finite wear land is developed. This is due to plastic deformation and the elevated temperature in the cutting zone.

In the second stage, also called secondary wear zone, the wear rate stabilizes giving a nearly uniform wear rate which remains constant for a considerable period. The uniform wear zone constitutes the major portion of the period of usability of a cutting tool.

In the third stage, also called tertiary wear zone, tool flank wear accelerates rapidly and the tool fails soon after reaching this stage. In this stage, the presence of a large wear land drastically increases the tool temperatures, causing rapid deterioration of the cutting edge. It is advisable to replace or resharpen the cutting tools prior to the time the tool wear reaches third stage, where rapid breakdown occurs. This condition can be identified when the surface finish of the machined workpiece begins to deteriorate and temperature rises excessively. Figure 16.22 shows different wear curves for various cutting speeds. V_4 is the fastest cutting speed and therefore generates the highest interface temperatures and fastest wear rates result and hence tool life, T_4 is shortest at that speed. When the amount of wear reaches the value W_f, the permissible tool wear on the flank, the tool is said to be worn out. W_f typically varies between 0.63 and 0.76 mm.

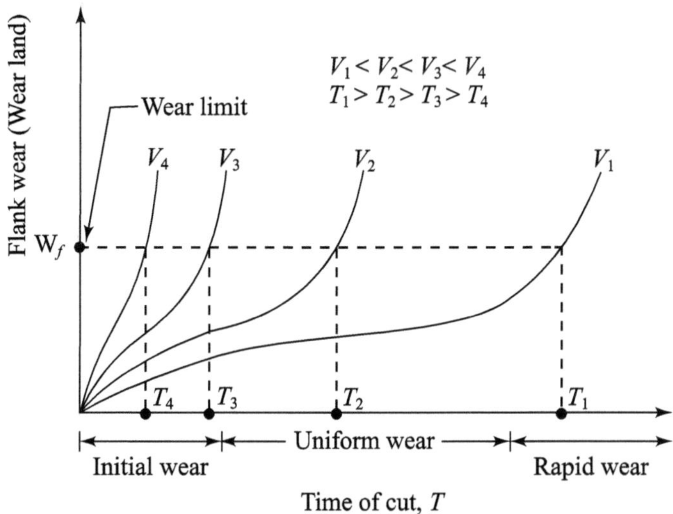

Fig. 16.22 Flank wear curves at different cutting speeds

16.15.2 Crater Wear

Crater wear occurs on the rake face of the tool in the form of a small crater or depression. It changes the chip–tool interface geometry, thus affecting the cutting process. The growth of a crater wear is a more complex phenomenon. Formation of crater is due to the abrading action (friction) of the chip as it passes over the tool face. Diffusion plays an important role in the development of crater wear in which movement of atoms across the tool–chip interface takes place.

Crater wear is largely a temperature-dependent phenomenon, and the temperature in the rake face is much higher than that in the flank face. The temperature at the chip–tool interface may be of the order of 1000 °C because of increased cutting speed. Since diffusion rate increases with increasing temperature, crater wear increases as temperature increases.

Flank wear can be easily measured experimentally, but crater wear is difficult to measure. The tool failure can be determined by measuring the crater depth (Fig. 16.23).

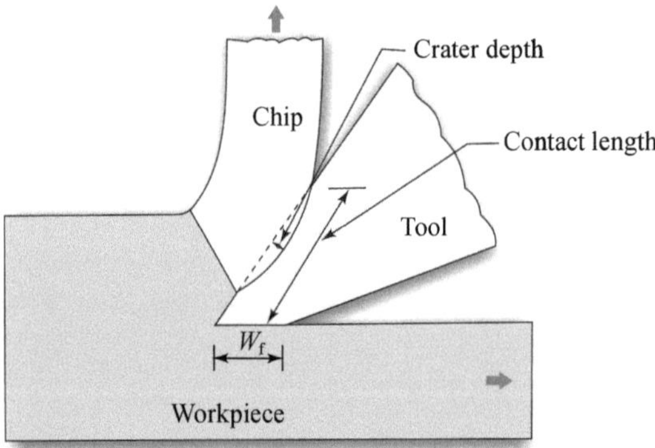

Fig. 16.23 Crater wear and crater depth

16.15.3 Cutting Fluids

Cutting fluids are used extensively in machining operations. They act primarily as a coolant and secondly as a lubricant. It should also not affect the nearby surrounding. Additionally, the cutting fluid must not cause rust or corrosion to the workpiece but should rather protect it. The machine must not be damaged, and the cutting fluid must not constitute a health risk for the operator. Cast iron should always be machined dry. Brass, bronze and aluminium may be cut dry or wet, and steel should always be machined with a lubricant. Following are the important functions of a cutting fluid:

- It reduces the friction at tool–chip interface.
- It washes away chips and dust.
- It reduces the temperature of tool and work.
- It promotes higher cutting speeds.
- It reduces cutting forces and power consumption.
- It increases tool life.
- It protects the newly machined surfaces from environmental corrosion.

Figure 16.24 shows the effect of lubrication between two contact surfaces. On using lubrication, the surfaces tend to slide rather than rubbing against each other. Also, the temperature level during the cutting process is reduced (Fig. 16.25).

Taylor used water as the cutting fluid during his work on tool life and found tremendous improvement in the process. This improvement increased the cutting speeds and allowed them to be doubled or tripled. Water is an excellent coolant; however the major demerit of it is that it causes rusting of workpieces and machine components and is a poor lubricant. During broaching or tapping (internal threading), lubrication of part is more important

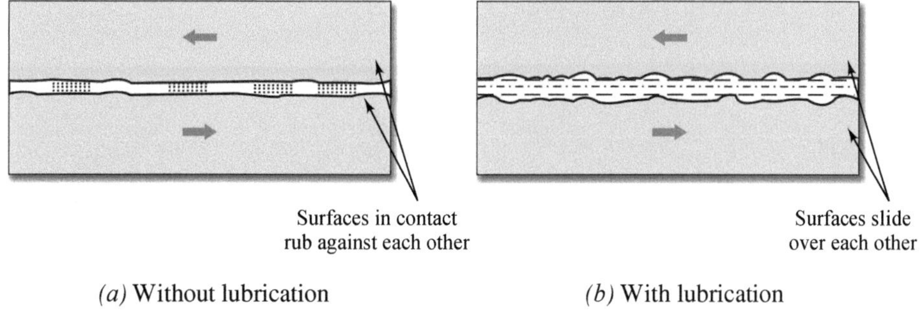

Fig. 16.24 Shows how the contact surfaces behave under lubrication and no-lubrication condition

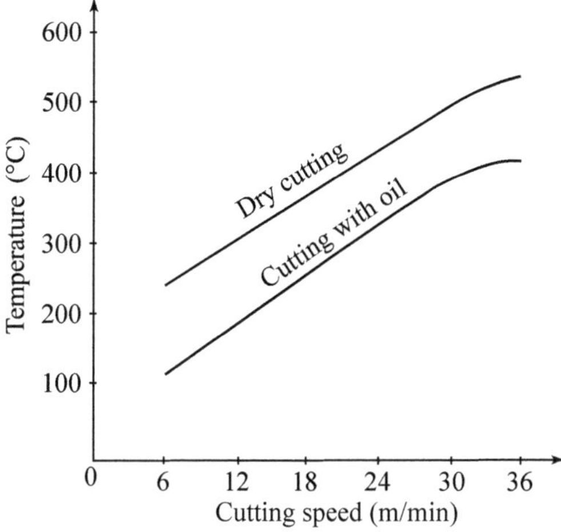

Fig. 16.25 Shows temperature and cutting speed relationship under lubrication and dry conditions

than its cooling. Oils offer less cooling effect but do not cause rust and have lubricating capacity.

An effective lubricant can modify the geometry of chip formation so as to yield a thinner, less deformed and cooler chip. Lubrication reduces the tendency for built-up edge (BUE) formation on the tool and thus improves surface finish. The lubricating properties of the cutting fluid reduce the wear between the workpiece and tool by separating the surfaces from one another as it is evident from Fig. 16.24. Reduction in wear is due to reduction in friction, if cutting fluid is introduced between the surfaces. Cutting fluid is drawn into the tool–chip interface by the capillary action of the interlocking network of surface asperities.

16.15.4 Types of Cutting Fluids

Cutting fluids can be divided into the following groups:

- Straight oils
- Cutting fluid mixed with water
- Synthetic cutting fluid
- Semi-synthetic cutting fluid.

Straight oils are undiluted neat oils. Wax or oil are sometimes mixed with water to form emulsion. They are frequently used. Straight oils are used for lubrication purpose and oil–water emulsions for cooling. The emulsion group is used mainly for metal cutting and grinding operations. It has the twin benefits: having the cooling properties of water and lubrication properties of oil. The PH value for oil emulsions should lie between 8.5 and 9.3. Lower PH values increase the tendency to corrode and the growth of bacteria but, on the other hand, high PH values involve the risk of skin irritation.

Synthetic cutting fluids contain no mineral oil and were used previously solely for grinding operations. However, the development of synthetics with improved lubrication properties and improved rust protection has increased their field of application considerably in recent years. The use of synthetic cutting fluids has economic advantages over oil-based fluids and offers quick heat dissipation, cleaning properties, simple preparation and provides good protection against rust.

Semi-synthetic cutting fluids contain a small amount of mineral oils plus additives to further enhance the lubrication properties. They are more suitable than oil emulsions for grinding operations because of small amount of oil present in them. Oil often tends to cause overloading of the grinding wheel. In semi-synthetic fluids, oil particles are smaller than in oil emulsions which make them more effective.

16.15.5 Methods of Application of Cutting Fluid

The method of application of a cutting fluid is very important. Following three methods are used:

- Flood cooling method
- Mist cooling method
- High-jet cooling method.

Flood cooling is the most common method. In this method, the cutting fluid is directed at the desired point through a nozzle. ***Mist cooling*** is used particularly with water-base cutting fluids. By using a specially designed nozzle, cutting fluids are converted into fine droplets (mist). This mist is directed at the cutting zone at high velocity. The droplet size may vary between 3 and 30 μm. This method is more effective as compared to flood cooling. In the ***high-jet method***, a powerful jet of cutting fluid (very high pressure) is directed to the cutting zone. It is the latest and most effective technique which also helps in breaking the chips during the cutting process.

16.16 Economics of Machining

The tool life directly influences the cost of tooling, and Taylor's equation on tool life suggests that if the machinability is carried out at higher cutting speeds, the tool life decreases sharply (exponentially). This necessitates frequent changing of tools at short intervals, thereby increasing the machine down time. As a result, the cost of tooling increases. On the other hand, if the operation is carried out at lower speeds, there is no need to change the tool very frequently and machine down time is also reduced but the production rate decreases. Hence, there is a need to come to a compromise that takes care of both the situations discussed above. At some intermediate cutting speed, the total cost of production inclusive of tooling cost is at a minimum. The tool life corresponding to this cutting speed gives economical tool life.

The total cost per operation is comprised of four individual costs apart from the cost of the material being used. These are machining costs, tool costs, tool changing costs and handling costs. The machining costs include the cost of operating the machine, cost of labour involved and overhead charges. Tool costs include the cost of the tool and cost of tool regrinding. The variation of these costs with cutting speed is shown in Fig. 16.26. All the costs are expressed on unit piece basis. It can be seen that machining cost decreases very sharply with increasing cutting speed. The tool cost goes up rapidly as the speed increases since the tool wear is high at higher cutting speeds. Rate of increase of tool changing cost is low with increasing cutting speeds. The handling costs are independent of cutting speeds. Adding up each of these individual costs results in a total unit cost curve, which is observed to go through a minimum point.

To find the cutting speed that will yield the minimum cost per piece, a total cost equation consisting of all the four costs is written and the individual costs are expressed as a function of cutting speed. This equation is then differentiated with respect to the cutting speed and set equal to zero. The optimum cutting speed for minimum cost is given by the equation

$$V_{\text{opt}} = \frac{C}{\left[\left(\frac{1}{n} - 1\right)\left(t_c + \frac{z_2}{z_1}\right)\right]^n} \tag{16.40}$$

where

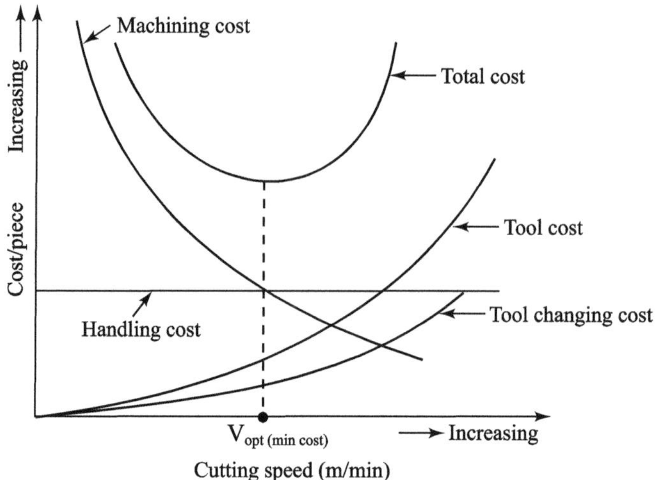

Fig. 16.26 Effect of cutting speed on various costs

n and C = constants of tool-life equation
t_c = tool changing time (min)
z_1 = direct labour and overhead cost per minute
z_2 = tool cost per grind including depreciation.

The optimum speed for maximum production criteria is found to be

$$V_{\text{opt}} = \frac{C}{\left[\left(\frac{1}{n} - 1\right)t_c\right]^n} \tag{16.41}$$

The expression $\left(\frac{1}{n} - 1\right)$, sometimes called *costs ratio,* shows the significance of n in relation to the economics of cutting. While deriving above equations, the depth of cut, d and feed, f are assumed to be constant. Since n is not significantly changed by normal variations of d and f, the costs ratio relationship is valid regardless of the particular values of d and f. For HSS tools, n is approximately 1/8 and the cost ratio for economic considerations is 7; for tungsten carbide tools, n is approximately 1/5 and the ratio is 4. These values can be taken as guidance for the start of machining.

Effect of cutting speed on different time components during the cutting process is shown in Fig. 16.27. Machining cost is proportional to the machining time as the cutting speed increases. Variation in tool changing time is similar to the variation in tool changing cost.

The difference in the cutting speeds for minimum total cost per piece and for maximum production rate is termed as high efficiency (Hi-E) range (Fig. 16.28). This concept was developed by *William. W. Gilbert* of General Electric in 1950s. Any speed selected within

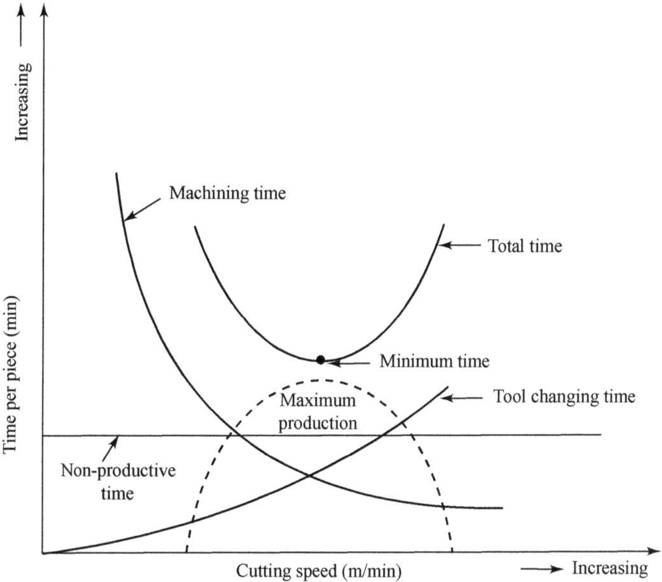

Fig. 16.27 Effect of cutting speed on machining time, tool changing time and total time

this range will give the optimum speed. The process is not economical beyond this range both cost-wise and productivity-wise.

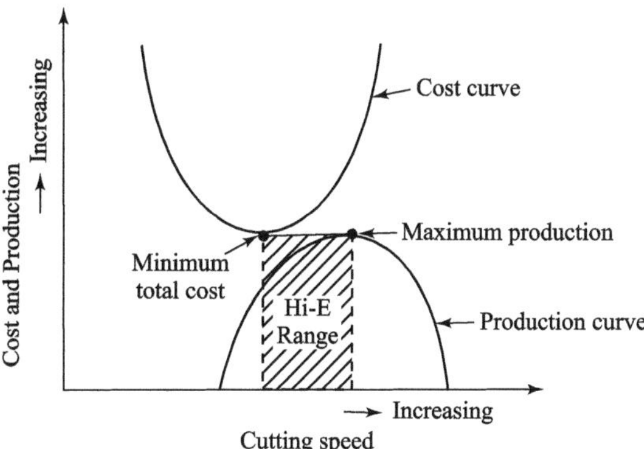

Fig. 16.28 High efficiency range

16.16.1 Solved Numerical Problems

Problem 16.1 For $n = 0.4$ and $C = 400$, find the percent increase in tool life, if the cutting speed is reduced by (a) 20% and (b) 40%?

Solution Taylor's equation is represented as:

$$VT^{n} = C$$

Let T_1, V_1 and T_2, V_2 are the initial and final conditions respectively.

$$V_1 T_1^{0.4} = 400 \quad \text{(i)}$$

$$V_2 T_2^{0.4} = 400 \quad \text{(ii)}$$

Dividing Eq. (ii) by Eq. (i), we get

$$\frac{V_2}{V_1} = \left(\frac{T_1}{T_2}\right)^{0.4} \quad \text{(iii)}$$

But $V_2 = 0.8\ V_1$.

Substituting V_2 in Equation (iii), we have

$$0.8 = \left(\frac{T_1}{T_2}\right)^{0.4}$$

$$\frac{T_2}{T_1} = 1.747$$

or $T_2 = 1.747 T_1$

Hence, the tool life is increased by 74.7%.

(b) When $V_2 = 0.6\ V_1$

Equation (iii), on substituting V_2, gives

$$\frac{T_2}{T_1} = 3.586$$

or $T_2 = 3.586\ T_1$

Hence, the tool life in increased by 258.6%. **Ans.**

Problem 16.2 For an orthogonal cutting operation, the rake angle is 10° and the coefficient of friction is 0.5. Using Merchant's equation, find the percentage increase in chip thickness when the friction is doubled.

Solution Given, $\alpha = 10°$

$$\mu_1 = 0.5$$

$$\mu_2 = 2 \times \mu_1$$

$$t_{c_2} = ?$$

The Merchant equation is

$$2\phi + \beta - \alpha = 90°$$

$$2\phi = 90° + \alpha - \beta$$

$$\phi = \frac{90° + \alpha - \beta}{2} \quad \text{(i)}$$

But $\mu = \tan \beta$.
$\beta_1 = \tan^{-1} \mu_1 = 26.56°$.
Using Equation (i)

$$\phi_1 = \frac{90° + 10° - 26.56°}{2} = 36.717°$$

The chip thickness ratio is given as:

$$r = \frac{t}{t_c} = \frac{\sin \phi_1}{\cos(\phi_1 - \alpha)}$$

or

$$\frac{t}{t_{c_1}} = \frac{\sin 36.717°}{\cos(36.717° - 10°)} = 0.6692 \quad \text{(ii)}$$

$$\beta_2 = \tan^{-1} \mu_2$$
$$= \tan^{-1}(1.0)$$
$$= 45°$$

Again use Equation (i) to get ϕ_2 as:

$$\phi_2 = \frac{90° + 10° - 45°}{2} = 27.5°$$

Now

$$\frac{t}{t_{c2}} = \frac{\sin 27.5°}{\cos(27.5° - 10°)} = 0.4841 \quad \text{(iii)}$$

Dividing Equation (ii) by Equation (iii), we get

$$\frac{t_{c2}}{t_{c1}} = \frac{0.6692}{0.4841} = 1.382$$

or $t_{c2} = 1.382 t_{c1}$

Hence, the chip thickness is increased by 38.2%. **Ans.**

Problem 16.3 In an orthogonal cutting operation on a workpiece of width 2.5 mm, the uncut chip thickness was 0.25 mm and the tool rake angle was zero degree. It was observed that the chip thickness was 1.25 mm. The cutting force was measured to be 900 N and the thrust force was found to be 810 N.

(a) Find the shear strength of the workpiece material.
(b) If the coefficient of friction between the chip and the tool was 0.5, what is the machining constant, C_m?

Solution

The chip thickness ratio, r is found to be

$$r = \frac{0.25}{1.25} = 0.2$$

The shear angle, ϕ is obtained as:

$$\begin{aligned}\tan\phi &= \frac{r\cos\phi}{1 - r\sin\phi}\\ &= \frac{0.2\cos 0°}{1 - 0.2\sin 0°} \quad \text{(Assuming } \alpha = 0s\text{)}\\ &= 0.2\end{aligned}$$

or $\phi = 11.31°$

The shear force, F_s is found as:

$$\begin{aligned}F_s &= F_c\cos\phi - F_t\sin\phi\\ &= 900\cos 11.31° - 810\sin 11.31°\\ &= 723.66\ \text{N}\end{aligned}$$

The shear area, A_s is given as:

$$A_s = \frac{t \cdot W}{\sin\phi}$$

$$= \frac{0.25 \times 2.5}{\sin 11.31^\circ}$$
$$= 3.187\ \text{mm}^2$$

Now the shear strength is obtained as:

$$\tau_s = \frac{723.66}{3.187} = 227 \frac{\text{N}}{\text{mm}^2} \qquad \textbf{Ans.}$$

The angle of friction is

$$\beta = \tan^{-1}(\mu) = 26.56^\circ$$

The machining constant, C_m is found by using the relation

$$C_m = 2\phi + \beta - \alpha$$
$$= 2 \times 11.31^\circ + 26.56^\circ - o^\circ$$
$$= 49.18^\circ. \qquad \textbf{Ans.}$$

Problem 16.4 In an orthogonal cutting process, the following observations were made:

Depth of cut = 0.25 mm
Chip thickness ratio = 0.45
Width of cut = 4 mm
Cutting velocity = 40 m/min

Cutting force component parallel to cutting

Velocity vector = 1150 N
Feed force = 140 N
Rake angle of the tool = 18°

Determine the resultant cutting force, power of cutting, shear angle, friction angle and force component parallel to shear plane.

Solution The resultant cutting force is found as:

$$\sqrt{F_c^2 + F_t^2} = \sqrt{1150^2 + 140^2}$$
$$= 1158.5\ \text{N} \qquad \textbf{Ans.}$$

The power consumed in cutting is

$$F_c \times V = 1150 \times \frac{40}{60}$$

$$= 766.67 \text{ W} \qquad \textbf{Ans.}$$

The shear angle, ϕ is obtained as:

$$\tan\phi = \frac{r\cos\alpha}{1 - r\sin\alpha}$$
$$= \frac{0.45\cos 18^\circ}{1 - 0.45\sin 18^\circ} = 0.4971$$

$\therefore \phi = 26.43^\circ$ **Ans.**

The friction force, F and normal force, N are found as:

$$F = F_c \sin\alpha + F_t \cos\alpha$$
$$= 1150 \sin 18^\circ + 140 \cos 18^\circ$$
$$= 488.52 \text{ N}$$

$$N = F_c \cos\alpha - F_t \text{Sin}\alpha$$
$$= 1150 \cos 18^\circ - 140 \sin 18^\circ$$
$$= 1050.45 \text{ N}$$

Now $\mu = \frac{F}{N} = \frac{488.52}{1050.45}.$

$= 0.465$

The friction angle, β is obtained as:

$$\beta = \tan^{-1}(\mu)$$
$$= \tan^{-1}(0.465)$$
$$= 24.94^\circ \qquad \textbf{Ans.}$$

Finally the force component parallel to shear plane is found as:

$$F_s = F_c \cos\phi - F_t \sin\phi$$
$$= 1150 \cos 26.43^\circ - 140 \sin 26.43^\circ$$
$$= 967.5 \text{ N} \qquad \textbf{Ans.}$$

Problem 16.5 Consider the following data for an orthogonal machining operation:

Cutting speed	= 90 m/min
Feed	= 0.15 mm/rev
Depth of cut	= 5 mm
Rake angle	= 10°
Chip thickness	= 0.35 mm

Clearance angle	$= 8°$
Tangential force	$= 220$ Kgf
Feed force	$= 120$ Kgf

Find the chip velocity and the specific energy consumption.

Solution The chip thickness ratio, r is given as:

$$r = \frac{0.15}{0.35} = 0.43$$

The shear angle, ϕ is found as:

$$\begin{aligned}\phi &= \tan^{-1}\left(\frac{r\cos\alpha}{1 - r\sin\alpha}\right)\\ &= \tan^{-1}\left(\frac{0.43\cos 10^{\circ}}{1 - 0.43\sin 10^{\circ}}\right)\\ &= 24.6^{\circ}\end{aligned}$$

Now the chip velocity, V_c can be obtained as:

$$\begin{aligned}V_c &= r \times \mathrm{V}\\ &= 0.43 \times 90\\ &= 38.7\ \mathrm{m/min} \quad \textbf{Ans.}\end{aligned}$$

The power consumption, P during cutting is

$$\begin{aligned}P &= F_c \times V\\ &= \frac{220 \times 90 \times 9.81}{60}\\ &= 3237.3\ \mathrm{W}\end{aligned}$$

The volume of material removed during cutting is expressed as:

$$\begin{aligned}&\text{Feed} \times \text{Depth of cut} \times \text{Cutting speed}\\ &= 0.15 \times 10^{-3} \times 5 \times 10^{-3} \times \frac{90}{60}\\ &= 1.125 \times 10^{-6}\ \frac{\mathrm{m}^3}{\mathrm{s}}\end{aligned}$$

Now the specific energy consumption is found as:

$$= \frac{3237.3}{1.125 \times 10^{6}}$$

$$= 2.877 \times 10^9 \frac{\text{J}}{\text{m}^3} \quad \textbf{Ans.}$$

Problem 16.6 During an orthogonal cutting of steel with a HSS tool having a rake angle of 20°, it was found that at a speed of 45 m/min, a feed of 0.3 mm/rev and a depth of cut of 4 mm, the chip thickness was 0.6 mm. Calculate the shear plane angle and the tool life, making suitable assumptions.

Solution The chip thickness ratio, r is found to be

$$r = \frac{0.3}{0.6} = 0.5$$

The shear plane angle, ϕ is obtained as:

$$\phi = \tan^1\left(\frac{r\cos\alpha}{1 - r\ \sin\alpha}\right)$$

$$= \tan^1\left(\frac{0.5\cos 20^\circ}{1 - 0.5\sin 20^\circ}\right)$$

$$= 29.54^\circ \quad \textbf{Ans.}$$

Assuming workpiece material of medium-carbon steel, we use
$C = 190$ and $n = 0.11$.
Substituting these values in the Taylor's equation, we have

$$VT^{\text{n}} = C$$

$$45 \times T^{0.11} = 190$$

Solving for T, we get
$T = 483791.96$ min **Ans.**

Problem 16.7 A cylinder of 155 mm diameter is to be reduced to 150 mm diameter in one turning cut with a feed of 0.15 mm/rev and a cutting speed of 150 m/min. Find the spindle speed, feed rate and metal removal rate.

Solution The average diameter, D_{av} of the cylinder is

$$D_{av} = \frac{155 + 150}{2}$$

$$= 152.5 \text{ mm}$$

The cutting speed, V is obtained as:

$$V = \frac{\pi D_{av} N}{1000} \text{ m/min}$$

- It gives spindle speed, N as:

$$N = \frac{150 \times 1000}{\pi \times 152.5} = 313 \text{ rpm} \quad \textbf{Ans.}$$

- The feed rate, f_m is obtained as:

$$f_m = f_r \times N = \frac{0.15 \times 313}{60} = 0.7825 \text{ mm/s} \quad \textbf{Ans.}$$

- The depth of cut, d is expressed as:

$$d = \frac{155 - 150}{2} = 2.5 \text{ mm}$$

- The metal removal rate can be obtained as:

$$= \frac{\pi \cdot f_r \cdot N \cdot D_{av} \cdot d}{60} \text{mm}^3/\text{s} = \frac{\pi \times 0.15 \times 313 \times 152.5 \times 2.5}{60} = 937.22 \text{ mm}^3/\text{s} \quad \textbf{Ans.}$$

Problem 16.8 During milling of a slot 20 mm wide, 10 cm long in a rectangular plate 10 cm × 20 cm, the following cutting conditions are observed :

Cutting speed	= 60 m/min
Diameter of end mill	= 20 mm
Number of flutes	= 8
Feed	= 0.01 mm/flute
Depth of cut	= 3 mm

Find the cutting time for the operation.

Solution The rpm (N) of the spindle can be obtained by using cutting speed equation as:

$$N = \frac{V \times 1000}{\pi \times D}$$
$$= \frac{60 \times 1000}{\pi \times 20} = 954.92.$$

The milling cutter feed rate is

$$f_m = f_t \times n \times N$$
$$= 0.01 \times 8 \times 954.92$$
$$= 76.4 \text{ mm/min}.$$

The cutter approach to the work is

$$\Delta L = \sqrt{d(D - d)}$$
$$= \sqrt{3(20 - 3)}$$
$$= 7.14 \text{ mm}.$$

The cutting time for the operation is obtained as:

$$t = \frac{L + \Delta L}{f_m}$$
$$= \frac{100 + 7.14}{76.4}$$
$$= 1.4 \text{ min} \quad \textbf{Ans.}$$

Short Answer Questions

1. Why is a cutting tool with negative or small rake angle used for machining harder materials?

Ans. Negative or small rake angle helps to keep the cutting forces in compression more favourably than in shear which is advantageous because of high compressive strength of harder materials.

2. Why is a positive rake angle generally desirable for a cutting tool?

Ans. A cutting tool with positive rake angle reduces the cutting forces as well as the temperature and power requirement during cutting operation.

3. What is the significance of nose radius in a cutting tool?

Ans. A cutting tool without nose radius has a very pointed edge which can damage the work surface or leave indentation marks on it during machining.

4. How does feed differ from feed rate?

Ans. Feed indicates the linear distance moved by the cutting tool in one revolution, whereas feed rate indicates the linear distance moved by the cutting tool in one minute. Feed rate (mm/min) is equal to the product of rotational speed (rpm) and feed (mm/rev).

5. Name the most widely used cutting tool in drilling and also state the typical value of its point angle.

Ans. The most commonly used cutting tool in drilling is twist drill for which the point angle is 118°.

6. What is tool life? How are tool life and machinability related to each other?

Ans. Tool life is the length of period of the cutting tool during which it can be used satisfactorily for cutting operations. Higher the tool life, the better is the machinability.

Multiple Choice Questions

1. Orthogonal cutting is a

   (a) one-dimensional process
   (b) two-dimensional process
   (c) three-dimensional process
   (d) six-dimensional process.

2. Oblique cutting is a

   (a) one-dimensional process
   (b) two-dimensional process
   (c) three-dimensional process
   (d) six-dimensional process.

3. Which of the following operations is an example of orthogonal cutting?

   (a) Shaping (b) Turning

(c) Drilling (d) Milling

4. Consider the following statements:

    1. In orthogonal cutting, the cutting edge of the tool is perpendicular to cutting direction.
    2. In oblique cutting, the cutting edge and the cutting motion are not normal to each other.
    3. Majority of the machining operations are based on orthogonal cutting.
    4. The chip slides along the rake face of the tool.

    Of these statements:

    (a) 1 and 2 are true (b) 1, 3 and 4 are true
    (c) 1, 2 and 3 true (d) 1, 2 and 4 are true

5. Nose radius is expressed in

    (a) degree (b) radian
    (c) mm (d) metre

6. A tool signature consists of

    (a) five elements (b) six elements
    (c) seven elements (d) eight elements

7. The heat distribution in the chip, tool and work bears the following ratio:

    (a) 60: 20: 20 (b) 80: 10: 10
    (c) 70: 20: 10 (d) 75: 15: 10

8. Continuous chips are produced

    1. at high cutting speeds
    2. when ductile materials are machined
    3. by a tool with low rake angle
    4. when brittle materials are machined

    Of these:

    (a) 1 and 4 are true (b) 3 and 4 are true
    (c) 1 and 2 are true (d) 2 and 3 are true

9. Discontinuous chips are produced

1. at higher cutting speeds
2. when ductile materials are machined
3. by a tool with low rake angle
4. when brittle materials are machined

Of these:

(a) 1 and 4 are true (b) 3 and 4 are true
(c) 1 and 2 are true (d) 2 and 3 are true

10. Consider the following chips:

1. Continuous chip
2. Discontinuous chip
3. Continuous chip with a built-up edge

Of these, which is more commonly observed in actual practices?

(a) 1 alone (b) 2 alone
(c) 1 and 2 (d) 3 alone

11. The favourable conditions for the formation of continuous chip with a built-up edge include

1. Higher coefficient of friction 2. Low rake angle
3. Higher cutting speed 4. Lower depth of cut

Of these:

(a) 1 and 2 are true (b) 2 and 3 are true
(c) 3 and 4 are true (d) 1 and 4 are true

12. Feed can be expressed in

(a) mm/stroke (b) mm/rev
(c) mm/tooth (d) All of the above

13. Which of the following stresses is important for chip formation?

1. Tensile stress 2. Shearing stress
3. Compressive stress 4. Bending stress

Of these:

(a) 1 and 2 are true (b) 2 and 3 are true
(c) 2 and 4 are true (d) 3 and 4 are true

14. The cutting speed of a tool is
    (a) directly proportional to the diameter being cut
    (b) directly proportional to the square of the diameter being cut
    (c) inversely proportional to the diameter being cut
    (d) independent of the diameter being cut.

15. Chip thickness ratio is a ratio of
    1. Chip thickness to uncut thickness
    2. Uncut thickness to chip thickness
    3. Cutting speed to chip velocity
    4. Chip velocity to cutting speed

    Of these:

    (a) 1 alone is true (b) 2 alone is true
    (c) 2 and 4 are true (d) 1 and 3 are true

16. The value of chip thickness ratio is

    (a) greater than one (b) greater than two
    (c) less than one (d) less than zero

17. The correct relationship between shear angle, ϕ; rake angle, α and chip thickness ratio r is

    (a) $\cot\phi = \frac{r\sin\alpha}{1-r\cos\alpha}$ (b) $\tan\phi = \frac{r\sin\alpha}{1-r\cos\alpha}$
    (c) $\tan\phi = \frac{r\cos\alpha}{1-r\sin\alpha}$ (d) $\tan\phi = \frac{r\cos\alpha}{1-r\cos\alpha}$.

18. The shear strain is expressed as:

    (a) $\tan\phi + \tan(\phi-\alpha)$ (b) $\cot\phi - \tan(\phi-\alpha)$
    (c) $\cot\phi + \tan(\phi-\alpha)$ (d) $\cot\phi + \tan(\phi-\alpha)$

19. According to Ernst and Merchant theory, the relationship between shear angle, (ϕ), friction angles, (β) and rake angle, (α) is

(a) $2\phi + \beta + \alpha = 90°$ (b) $2\phi + \beta - \alpha = 45°$

(c) $2\phi + \beta - \alpha = 90°$ (d) $\phi + 2\beta - \alpha = 90°$

20. The Taylor's tool-life equation is

(a) $V^n T = C$ (b) $VT^n = C$

(c) $V^2 T^n = C$ (d) $\sqrt{V} T^n = C$.

where V = cutting speed, T = tool life and C = constant

21. Consider the following statements:

1. Tool life increases with increase in cutting speed.
2. Tool life decreases with increase in cutting speed.
3. Carbide tools have higher value of n as compared to HSS tools in Taylor's tool-life equation.
4. Higher tool life is indicative of better machinability.

Of these statements:

(a) 1 and 3 are true (b) 2, 3 and 4 are true

(c) 3 and 4 are true (d) 1 and 4 are true

22. Consider the following statements:

1. Crater wear occurs on the rake face of the tool.
2. Wear land is related to flank wear.
3. The temperature on the flank face of the tool is much higher than on its rake face.
4. Cutting fluid is more effective at low cutting speeds.

Of these statements:

(a) 1 and 3 are true (b) 2 and 3 are true

(c) 1, 2 and 3 are true (d) 1, 2 and 4 are true

23. Consider the following statements:

1. Crater wear is caused due to friction when the chip passes over the tool face.
2. Crater wear decreases as temperature increases.
3. Crater wear increases as temperature increases.
4. Excessive flank wear produces poor machinability.

Of these statements:

(a) 1, 2 and 4 are true (b) 1, 3 and 4 are true
(c) 3 and 4 are true (d) 1 and 3 are true

24. Consider the following methods of application of cutting fluid:

1. Flood cooling 2. Mist cooling 3. High-jet cooling

Of these methods:

(a) 1 is the most effective one (b) 3 is the most effective one
(c) 1 and 3 are equally effective (d) 2 is the most effective one

25. Consider the following statements:

1. Machining cost increases with increase in cutting speed.
2. Machining cost decreases with increase in cutting speed.
3. Handling cost is independent of cutting speed.
4. Tool cost increases with increase in cutting speed.

Of these statements:

(a) 1 and 4 are true (b) 1, 2 and 3 are true
(c) 2, 3 and 4 are true (d) 3 and 4 are true

26. Match List I with List II and select the correct answer using the codes given below the lists:

List I	List II
(Tool wear)	**(Related to)**
A. Corrosive wear	1. Thermomechanical process
B. Adhesion wear	2. Chemical wear
C. Fatigue wear	3. Protruding particles
D. Abrasion wear	4. Attrition wear

Codes:

	A	B	C	D
(a)	2	4	3	1
(b)	2	4	1	3
(c)	4	2	1	3
(d)	2	3	4	1

27. Consider the following statements about lubrication:

    1. It reduces the friction at tool–chip interface.
    2. It does not affect cutting forces and power consumption.
    3. It increases tool life.
    4. It does not affect cutting speed.

    Of these statements:

    (a) 1 and 2 are true (b) 1 and 3 are true
    (c) 1, 3 and 4 are true (d) 2 and 4 are true

28. Consider the following statements:

    1. Neat oils are used mainly for lubrication.
    2. Neat oils are used mainly for cooling purpose.
    3. Synthetic cutting fluids offer good protection against rust.
    4. Mist cooling is related to water-based cutting fluids.

    Of these statements:

    (a) 1 and 4 are true (b) 2 and 3 are true
    (c) 1, 3 and 4 are true (d) 2, 3 and 4 are true

29. The cost ratio for HSS tools for economic consideration is

    (a) 4 (b) 5
    (c) 6 (d) 7

Answers

1. (b)	2. (c)	3. (b)	4. (d)	5. (c)	6. (c)
7. (b)	8. (c)	9. (b)	10. (d)	11. (a)	12. (d)
13. (b)	14. (a)	15. (c)	16. (c)	17. (c)	18. (d)
19. (c)	20. (b)	21. (b)	22. (d)	23. (b)	24. (b)
25. (c)	26. (b)	27. (b)	28. (c)	29. (d)	

Review Questions and Discussions

Q.1. Differentiate between orthogonal and oblique cutting.
Q.2. Why is orthogonal cutting difficult to achieve?

Q.3. Rake angle is sometimes negative. Why and in which case?
Q.4. What is meant by tool signature?
Q.5. Which tool angle has the maximum effect on the cutting force?
Q.6. What is the need of nose radius?
Q.7. What is Merchant's theory?
Q.8. What are different types of chips?
Q.9. What is ideal chip? How is it produced?
Q.10. Why does not the built-up edge form on the cutting tool, when operating at low or high speeds?
Q.11. How does the shear angle affect heat generation in machining?
Q.12. How does the temperature affect tool life?
Q.13. What are the effects of cutting speed, feed and depth of cut on the heat generation?
Q.14. Why are chemical stability and inertness important in cutting operations?
Q.15. What can happen if a dull tool is used for a cutting operation?
Q.16. What is Taylor's equation on tool life?
Q.17. Why is it not always advisable to increase the cutting speed in order to increase the production rate?
Q.18. What is the importance of n in Taylor's equation?
Q.19. How is machinability connected to tool life?
Q.20. What are the reasons for a tool failure?
Q.21. How does cutting speeds affect crater wear?
Q.22. What are the functions of a cutting fluid?
Q.23. What are the important characteristics of a cutting fluid?
Q.24. Why is water not preferred as a cutting fluid?
Q.25. What is the importance of tool-life curve?
Q.26. How does flank wear differ from crater wear?
Q.27. What are the different methods of application of a cutting fluid? Which is most effective and why?
Q.28. How is the efficiency range defined? How does the cutting speed affect it?

Machining Operations 17

Turning is an important machining operation and finds applications in reducing diameters of the workpiece.

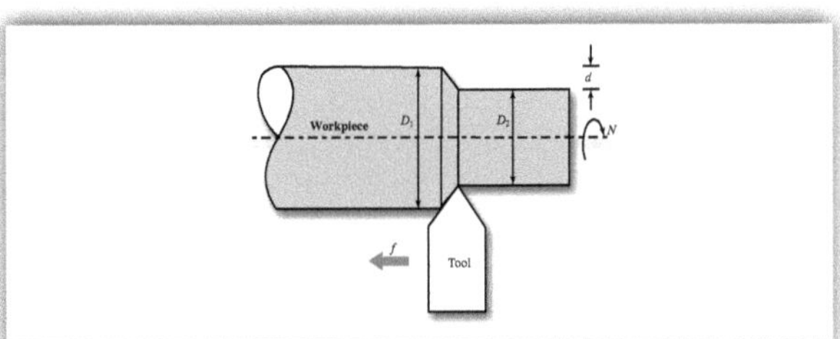

Learning Objectives

After reading this chapter, you will be able to answer some of the following questions.

- What are the factors affecting mental removal rate (MRR)?
- How dose up milling differ from down milling?
- How does dieing differ from tapping?

D. K. Singh, *Fundamentals of Manufacturing Engineering*,
https://doi.org/10.1007/978-981-99-8767-2_17

17.1 Introduction

Machining processes are used to remove unwanted material from a workpiece in order to give it proper shape and size and the required finish. This is accomplished by using various machine tools such as lathe, shaper, milling cutter, etc. Lathe is the oldest and most common machine tool, which removes material by rotating the workpiece against a single-point cutter. A number of operations can be performed on a lathe to produce the desired shape and size.

The operations performed on a lathe can be categorized into two groups:

(a) Standard operations.
(b) Special operations.

Standard operations include plain and step turning, eccentric turning, facing, taper turning, drilling, reaming, boring, knurling, threading and parting off.

Special operations include grinding, milling, copying or duplicating, spinning, tapping and dieing. All the above listed operations are discussed briefly giving important parameters such as cutting speed, cutting time, metal removal rate, etc.

17.2 Turning

In turning, the workpiece rotates while it is being machined. It is the most straight-forward metal cutting method and is the most widely used one. It can be used to produce straight, conical and curved surfaces on the workpiece. Turning is the combination of two movements; one is the rotation of the workpiece and the other is the feed movement of the tool. The feed movement of the tool can be along the axis of the workpiece, which means the diameter of the workpiece will be turned down to a smaller size. Alternatively, the tool can be fed towards the center, at the end of the part, which means the length of the workpiece will be faced down. Often feeds are combinations of these two directions, resulting in tapered or curved surfaces.

Plain turning, also called *straight turning*, is the simplest turning operation in which external diameter of the workpiece is reduced and in *step turning*, the reduction in diameter takes place in step lengths. These two processes result when the tool is fed longitudinally i.e. along the axis of the workpiece. If the tool is fed at an angle to the axis of rotation of the workpiece, an external conical surface results and the process is called ***taper turning***. If the tool is fed at 90° to the axis of rotation, using a tool that is wider than the width of the cut, the operation is called ***facing***, and a flat surface is produced at the end of the part, such as parts that are attached to other components. The three basic turning operations are shown in Fig. 17.1. The principle of turning is used to produce parts such as shafts, spindles, pins, handles and various machine components.

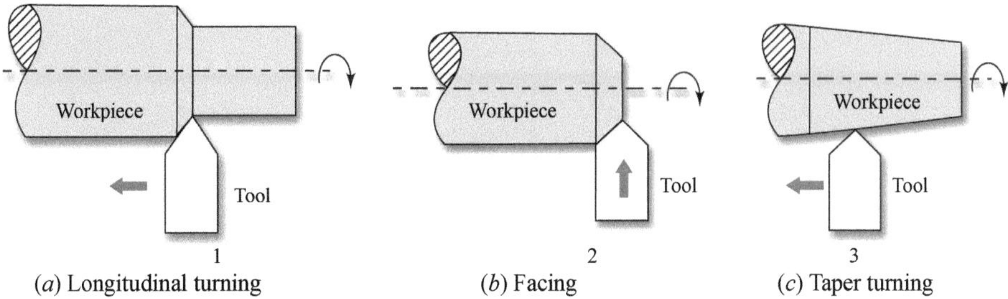

Fig. 17.1 Turning operations

If good finish and accurate size are desired, one or more roughing cuts are usually followed by one or more finishing cuts. ***Roughing*** means cutting the materials without much consideration to tolerances and surface finish. ***Finishing*** means cutting to obtain final dimensions with acceptable tolerances and surface finish. Roughing cuts have large depths of cut. Finishing cuts are light, usually being less than 0.38 mm in depth, with the feed as fine as necessary to give the desired finish. In most cases one finishing cut is sufficient but if more accuracy is required, two or more finishing cuts may be used.

17.2.1 Cutting Speed

In turning, the primary cutting motion is rotational with the tool feeding parallel to the axis of rotation. It should be noted that the cutting speed is only constant for as long as the spindle speed or part diameter remains the same. In a facing operation, where the tool is fed in towards the centre, the cutting speed will change progressively if the workpiece rotates at a fixed spindle speed. Mathematically, the cutting speed, V is given as:

$$V = \frac{\pi D N}{1000} \text{ (m/min)} \tag{17.1}$$

where

D = diameter to be cut (mm)
N = rotational speed of the workpiece (rpm)
π = a constant

As the diameter of the workpiece changes, the cutting speed also changes. The maximum cutting speed is at the outer diameter.

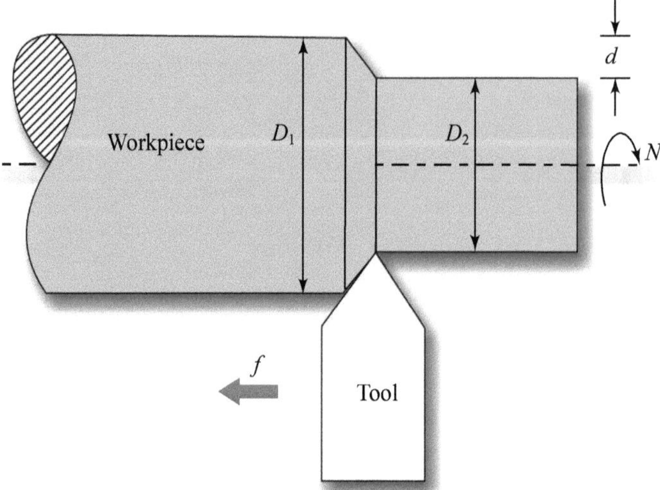

Fig. 17.2 Depth of cut and feed in turning operation

17.2.2 Depth of Cut

It is the difference between uncut and cut surface and is equal to one-half of the difference in uncut and cut diameter (Fig. 17.2).

$$d = \frac{D_1 - D_2}{2} \tag{17.2}$$

where

d = depth of cut
D_1 = uncut diameter
D_2 = cut diameter

The depth of cut is always measured at right angles to the feed direction of the cutting tool and not to the cutting edge.

17.2.3 Feed or Feed Rate

Feed is defined as the distance the tool travels per unit revolution of the workpiece. It is expressed in mm/rev.

17.2.4 Length of Cut

Length of cut is the distance travelled by the cutting tool parallel to the axis of rotation of workpiece plus some allowance in the form of overrun, to allow the tool to enter or exit the cut.

17.2.5 Cutting Time

The cutting time, t is given as:

$$t = \frac{L + A}{f_r N} \text{ (min)} \tag{17.3}$$

where

L = length of cut (mm)
A = allowance (mm)
f_r = feed (mm/rev)
N = rpm

The rpm, N is calculated from the cutting speed relationship as:

$$N = \frac{V \times 1000}{\pi D_1} \tag{17.4}$$

17.2.6 MRR

The metal removal rate (MRR) is given as:

$$\begin{aligned} \text{MRR} &= \frac{\text{volume of the material removed}}{\text{cutting time}} \\ &= \left(\frac{\frac{\pi}{4}(D_1^2 - D_2^2)L}{\left(\frac{L+A}{f_r N}\right)} \right) \\ &= \frac{\pi}{4} f_r N (D_1^2 - D_2^2) \left(\frac{\text{mm}^3}{\text{min}}\right) \text{ (using } A = 0) \end{aligned} \tag{17.5}$$

Substituting N in the above equation, the equation of MRR becomes

$$\text{MRR} = \frac{\pi}{4} f_r \cdot \frac{1000V}{\pi D_1} (D_1^2 - D_2^2)$$

$$= 1000V\ f_r \cdot \frac{(D_1 - D_2)}{2} \cdot \frac{(D_1 + D_2)}{2D_1}$$

But $\frac{D_1 - D_2}{2} = d =$ depth of cut

and $\frac{D_1 + D_2}{2D_1} \simeq 1$

$$\text{Hence,}\quad \text{MRR} \simeq 1000\ \text{V} f_r d \left(\frac{\text{mm}^3}{\text{min}}\right) \tag{17.6}$$

It should be noted here that Eq. (17.6) is an approximate equation for MRR in which depth of cut d is small compared to uncut diameter D_1.

The MRR can also be expressed in terms of average diameter of the workpiece and depth of cut.

$$\text{MRR} = \pi f_r N \cdot \frac{(D_1 + D_2)}{2} \cdot \frac{(D_1 - D_2)}{2}\ (\text{using } A = 0)$$

$$= \pi f_r N \cdot D_{\text{av}} \cdot d \left(\frac{\text{mm}^3}{\text{min}}\right) \tag{17.7}$$

where $D_{\text{av}} =$ average diameter of the workpiece

$$= \frac{D_1 + D_2}{2}$$

17.3 Facing

Facing (Fig. 17.1) is a turning operation in which a flat surface is produced by feeding the cutting tool normal to the axis of rotation of the workpiece. The cutting speed continuously decreases as the axis is approached, since N is constant.

$$\text{Length of cut,}\quad L = \frac{D}{2}\ (\text{for a circular solid workpiece of diameter } D)$$

$$= \frac{D_1 - D_2}{2}\ (\text{for a tube like workpiece})$$

where

D_1 = diameter of the workpiece before facing
D_2 = diameter of the workpiece after facing

The cutting time, t is given as:

$$t = \frac{\frac{D}{2} + A}{f_r N} \text{ (min)} \tag{17.8}$$

The metal removal rate, MRR is found to be

$$\text{MRR} = 500\, V f_r d \left(\text{mm}^3/\text{min}\right) \text{ (using } A = 0) \tag{17.9}$$

17.4 Drilling

Drilling is the process of producing a hole in an object by forcing a rotating twist drill against it (Fig. 17.3). Alternatively, holes can be produced by holding the drill stationary and rotating the work, such as drilling performed on a lathe with the workpiece held and rotated by a chuck.

Drills are usually made of high-speed steels, and many are now coated with titanium nitride (TiN) for increased wear resistance. Carbide-tipped or solid-carbide drills are used for hard and high-temperature materials such as concrete and bricks.

Drilling is among the most important operations in manufacturing. In drilling deep holes, the drill should be withdrawn occasionally to facilitate the removal of the chips and aid in getting coolant to the cutting edges.

The rpm of the rotating drill is given by the equation

$$N = \frac{1000V}{\pi D} \tag{17.10}$$

where

D = diameter of the drill (mm)

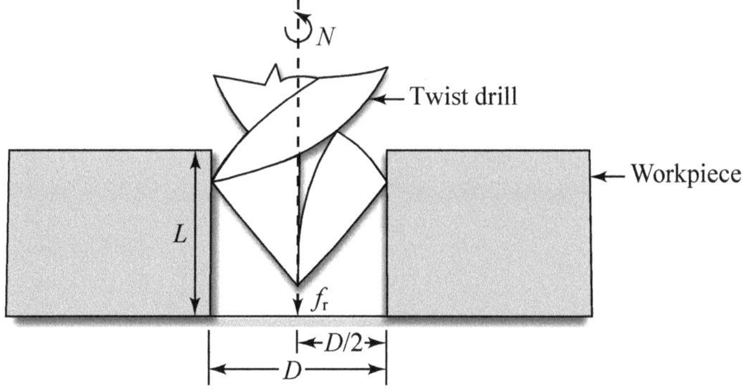

Fig. 17.3 A drilling operation

The depth of cut in drilling is equal to uncut thickness or one-half of the feed.

$$d = \frac{f_r}{2}$$

The length of cut in drilling is equal to the depth of hole L plus an allowance A for approach and for the tip of drill. The usual value of A is one-half of drill diameter.

Drilling time, t for a hole is given as

$$t = \frac{L + A}{f_m} \tag{17.11}$$

where f_m = feed (mm/min) = $f_r \cdot N$

The metal removal rate MRR is given as:

$$\text{MRR} = \frac{\pi D^2 f_r N}{4} \left(\frac{\text{mm}^3}{\text{min}}\right) \text{ (using } A = 0) \tag{17.12}$$

Substituting N, the MRR is found to be

$$\text{The metal removal rate (MRR)} = 250\, V f_r D \left(\frac{\text{mm}^3}{\text{min}}\right) \tag{17.13}$$

17.5 Boring

Boring is the process of enlarging a hole that has already been drilled. Hence boring does not create the hole rather, machines or opens the hole up to a specific size. Although conventional drills of 50 mm diameter and more are available, however above approximately 30-40 mm they become more difficult to handle, so holes bigger than this size are usually cut using a boring tool with a single cutting edge rotating at the radius of the required hole. While performing boring on a lathe, the work is usually held in the chuck or the faceplate which rotates during the process and the tool is forced against the workpiece. However, boring can also be done using a rotating tool with the workpiece remaining stationary. Boring sometimes, called internal turning, is done with cutting tools that are similar to those used in turning. The tool is fed parallel to the axis of rotation of the workpiece (Fig. 17.4).

The cutting time for boring can be obtained by using Eq. (17.3).

The metal removal rate (MRR) can be found by using Eqs. (17.5) and (17.6). In these equations,

D_1 = diameter to be bored

D_2 = initial hole diameter

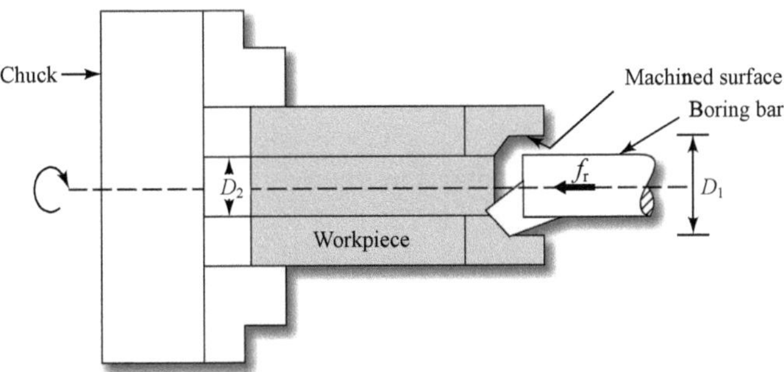

Fig. 17.4 A boring operation

17.6 Parting

Parting, also called cutting off, is the process of cutting a piece from the end of a part. The tool is fed normal to the axis of rotation as in the case of facing (Fig. 17.5). Parting uses thin tools with considerable overhang, hence accuracy of the process is difficult to achieve.

The length of cut, L is given as:

$$L = \frac{D}{2}, \text{ for a solid bar of a diameter } D$$
$$= \frac{D_1 - D_2}{2}, \text{ for a tube}$$

where

D_1 = outside diameter of the tube
D_2 = inside diameter of the tube

The equations for cutting time and metal removal rate are similar to those used for facing.

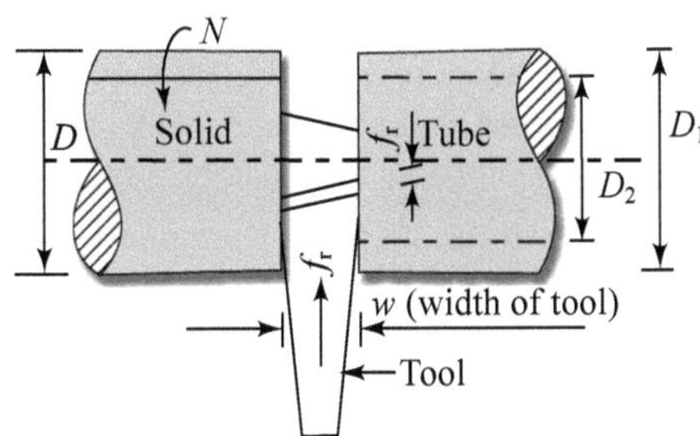

Fig. 17.5 A parting operation

17.7 Knurling

Knurling is the process of producing regularly shaped roughness (indentations, called knurls) on cylindrical surfaces. This is a chipless and cold-forming process and is performed by using a tool, called knurling tool on a lathe. The tool consists of a straight shank fitted with one or two knurling wheels in its front. These wheels are made of hardened tool steel and carry teeth on outer surfaces. The tool is fed across the axis of the work and pressed against its surface. Straight, diagonal and diamond knurlings are in common use.

17.8 Reaming

Reaming is a process used to produce accurate holes. It merely imparts to the previously drilled hole the necessary smoothness, parallelism, roundness and accuracy in size. The cutting tool used in the process is called reamer. It is a multiple-cutting-edge tool with straight or helically fluted edges. Reaming is performed at a slow speed and the speed is usually two-thirds of the speed for drilling the same materials. However, for close tolerances and fine finish, speeds should be slower. But the feeds are usually much higher than those for drilling and depend upon material to be reamed.

A number of reamers are available for performing the operation, important among them include hand, machine (chucking), shell, expansion and adjustable reamers. Hand and machine reamers are basic types of reamers. Each reamer is used for very specific application. Reamers are usually made of high-speed steels or solid carbides or have carbide cutting edges.

No special machines are built for reaming. The drilling machine can be used for reaming just by changing the cutting tool. A new development in reaming consists of a tool which combines drilling and reaming; the tip of the tool produces a hole by drilling and the rest of the same tool performs a reaming operation.

17.9 Milling

Milling is the process of cutting away material by feeding a workpiece past a rotating multiple-tooth cutter. Sometimes the workpiece remains stationary, and the cutter is fed to the work. The machined surface may be flat, angular, curved or combination of these. Since several teeth are engaged in the cutting operation, high metal removal rates are possible. Tool life is also extended. Milling is particularly well suited and widely used for mass-production work. The machine for holding the workpiece, rotating the cutter and feeding it is known as a milling machine.

There are many types of milling cutter and many ways in which they can be used. Milling cutters are available in many standard and special types, forms, diameters and widths. The teeth may be straight (parallel to the axis of rotation) or at a helix angle. The cutter may be right hand (to turn clockwise) or left hand (to turn counterclockwise). Important milling operations are discussed below.

17.9.1 Slab Milling

Slab milling, also called peripheral milling or plain milling, generates flat surfaces by using the teeth located on the periphery of the cutter body. The axis of cutter rotation is parallel to the workpiece surface to be machined. The diameter and the width of the cutter depend on whether a part is to be slab milled or requires a narrow width slot (Fig. 17.6). The cutter has straight or helical teeth, and each teeth acts as a single-point cutting tool.

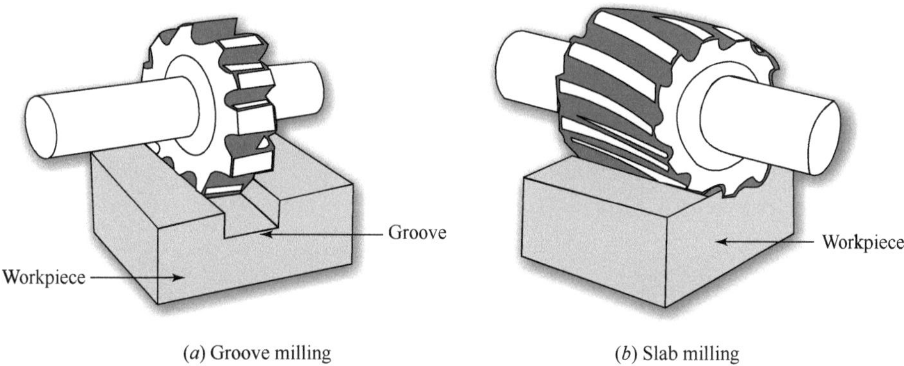

(*a*) Groove milling (*b*) Slab milling

Fig. 17.6 Types of plain milling

Less force is required, vibration and chatter are reduced, and a better quality of surface finish is produced with a helical-tooth cutter than with a straight-tooth cutter. This is possible because of the fact that the helical teeth are continuously engaged in comparison with the intermittent cutting action of a straight-tooth cutter. Milling cutters up to 18 mm wide generally have straight teeth and cutters over 18 mm wide usually have helical teeth. Angle of helix ranges from 45 to 60° or steeper. Straight cutters are used for light duty and helical for heavy duty.

17.9.2 Side Milling

Side milling uses side milling cutters similar to plain milling cutters. However, in addition to teeth around the periphery, other teeth are formed on one or both sides. The teeth may be either straight or helical. Most of the cutting is done by the teeth around the periphery. The side cutting teeth cut the side of a workpiece. Side mills are not recommended for milling slots because of their tendency to mill wider than the specified cutter width.

17.9.3 End Milling

End milling is the process of machining horizontal, vertical, angular and irregular-shaped surfaces. The cutting tool is called an end mill. The end milling process is shown in Fig. 17.7.

End mills are coarse-tooth cutters and are made of high speed steels or have carbide inserts. They are subjected to severe torsional and bending stresses in use. These limit the size of cut that can be taken. With a cut equal to the full width of the cutter, the maximum recommended depth of cut is 0.6D, D being the diameter of end mill. If, however, the cutting action is the cleaning up of the edge of a component, with the cut only 10% of the diameter, the depth can be increased to 1.5D.

The process can be used to mill grooves, slots, keyways and large surfaces. It is also widely used for profile milling in die making.

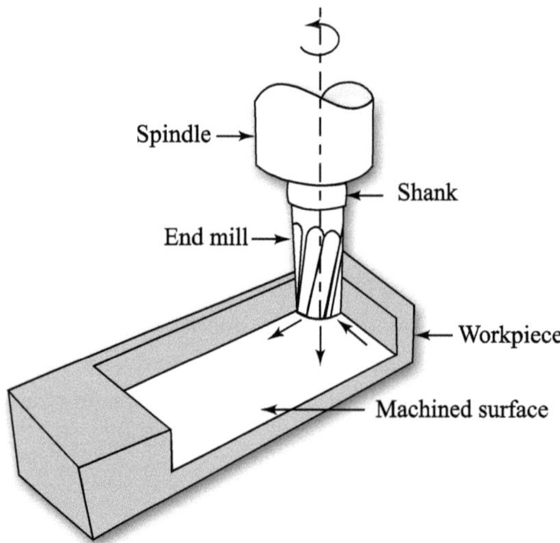

Fig. 17.7 An end milling operation

17.9.4 Face Milling

Face milling may be termed as the variation of end milling where the cutter has large diameter with several cutting teeth. The cutter diameter is usually 6 in or more. The teeth are bevelled or rounded at the periphery of the cutter.

Face milling cutter (Fig. 17.8) are made of high-speed steels, cast alloys or carbides, and are heavy-duty cutters. Heavy cuts, coarse feeds and high cutting speeds are essential. The cutter is mounted on a spindle having an axis of rotation speeds are essential. The cutter is mounted on a spindle having an axis of rotation perpendicular to the workpiece surface (Fig. 17.9). Face milling is used to produce flat surfaces in a wide variety of applications, e.g. the machining of cylinder heads.

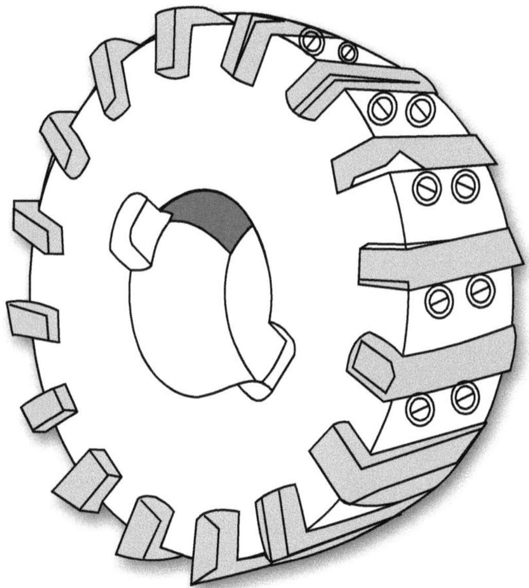

Fig. 17.8 A face milling cutter

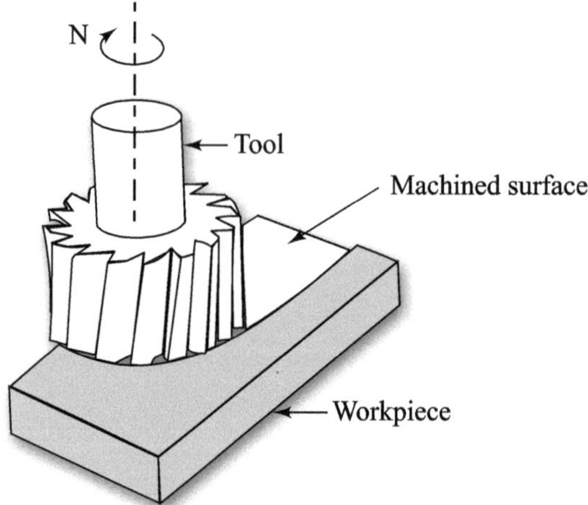

Fig. 17.9 A face milling operation

17.9.5 Method of Feeding the Work on Milling Machine

Surfaces in milling can be generated by the following methods:

- Up milling
- Down milling

17.9.6 Up Milling

Up milling is the oldest and conventional method of milling. Here, the workpiece is fed in opposite direction to that of the cutter rotation at the point of cutting. Cutting forces tend to lift the workpiece up from the table, hence the name up milling, and thus proper clamping is important.

17.9.7 Down Milling

In down milling, also called climb milling, the workpiece is fed in the direction of cutter rotation at the point of cutting. The cutting forces tend to hold the workpiece down onto the machine table, hence the name down milling and thus lower clamping fixture is required. It eliminates the rubbing action between cutter and workpiece as in the case of up milling and provides better surface finish with less tool chatter. Down milling is mostly preferred. Up and down milling processes are shown in Fig. 17.10.

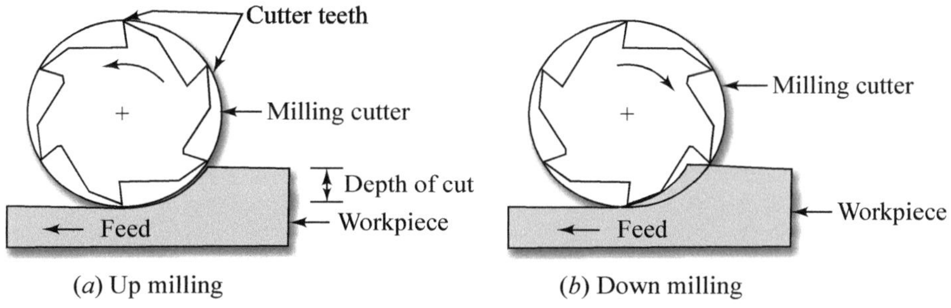

Fig. 17.10 Up and down milling operations

17.9.8 Milling Cutter Parameters

17.9.8.1 Cutting Speed

The cutting speed, V in milling, is the peripheral speed of the cutter, given by

$$V = \frac{\pi DN}{1000}\ \text{(m/min)} \tag{17.14}$$

where

D = cutter diameter (mm)
N = rotational speed of the cutter or rpm of the spindle

17.9.8.2 Depth of Cut

The depth of cut, d is the distance between the old and new machined surfaces.

17.9.8.3 Chip Thickness

The chip thickness is variable in milling due to relative longitudinal motion between cutter and workpiece, and often complicated to determine. In face milling, the maximum chip thickness is where up milling changes to down milling. For a straight-tooth cutter, the approximate undeformed chip thickness, t_c is given as:

$$t_c = 2f_t\sqrt{\frac{d}{D}} \tag{17.15}$$

where

d = depth of cut (mm).
f_t = feed of the cutter (mm/teeth).

17.9.8.4 Feed Rate

The milling cutter feed rate is given as:

$$f_m = f_t \times n \times N \tag{17.16}$$

where

n = number of teeth in the cutter.

17.9.8.5 Cutting Time

The cutting time, t is given by the expression

$$t = \frac{L + \Delta L}{f_m} \text{ (min)}$$

where

$$\begin{aligned} \Delta L &= \text{cutter approach to the work (mm)} \\ &= \sqrt{d(D-d)} \\ &\simeq \frac{D}{2} \end{aligned} \tag{17.17}$$

17.9.8.6 MRR

The metal removal rate is expressed as:

$$\begin{aligned} \text{The metal removal rate (MRR)} &= \frac{LWd}{t} \\ &= Wd\,f_m \left(\frac{\text{mm}^3}{\text{min}}\right) \end{aligned} \tag{17.18}$$

where

$\frac{L}{t}$ $= f_m$

W = width of the cut (mm)

= width of the cutter or the work (mm)

17.10 Tapping

Tapping is the process of producing internal .threads in a workpiece as in the case of a nut. The cutting tool, called a tap, is simply a hardened tool-steel screw with lengthwise grooves, called flutes, milled or ground across the threads. The flutes form a series of teeth and provide chip room along the entire length of the threaded portion of the tap. A tap nomenclature is shown in Fig. 17.11. The tap is used to cut threads in a plain hole generally drilled somewhat larger than the minor thread diameter. When the tap is turned into the hole, the teeth cut into the wall of the hole and remove materials to form threads of the same pitch as the threads of the tap.

There are many types of taps available. The solid tap, shell tap, expansion tap, inserted chaser tap, and collapsible taps are a few examples. A solid hand tap is used in majority of tapping operations. It was originally designed to be used by hand but now is almost exclusively used in machines. Spiral-point hand taps overcome the problems caused by chips accumulating in a hole while threading and also causing damage to the tap on reversal. It can also be used in blind holes provided there is sufficient depth beyond the threads to accommodate chips. It is highly recommended for through-hole machine tapping in most materials.

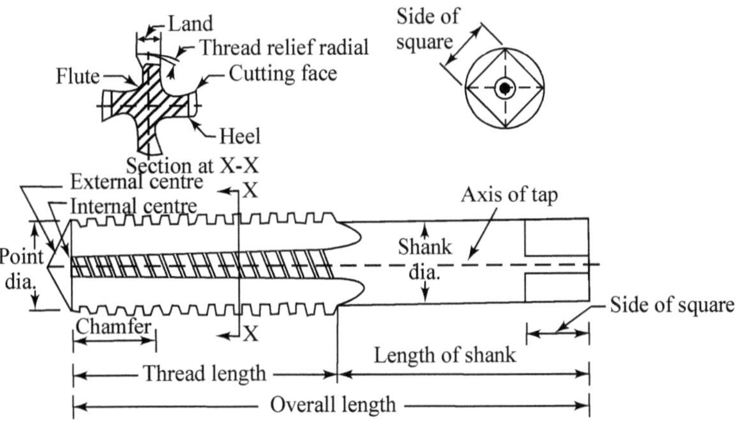

Fig. 17.11 Tap terminology

17.11 Dieing

Dieing is the process of producing external threads on a solid bar as in the case of a bolt. The cutting tools used for dieing are called dies. They are made of hardened tool or high speed steels in different shapes and sizes. Dies are of solid and adjustable type. The adjustable split die is the most popular one (Fig. 17.12). The adjustment is made by turning the adjusting screws in the die holder. The screws are adjusted until there is a slight drag between the die and the threaded part.

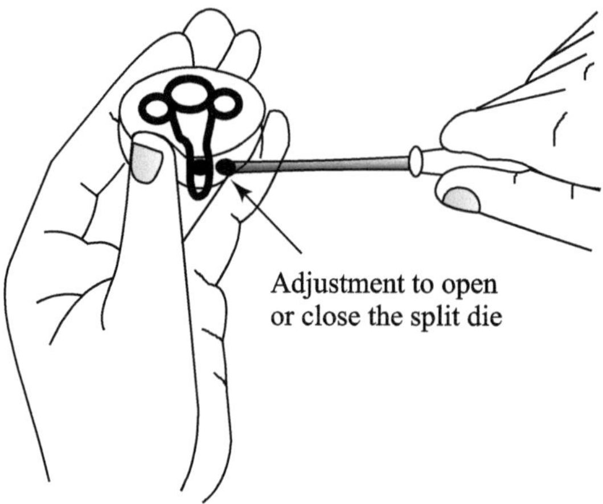

Fig. 17.12 An adjustable split die

17.12 Broaching

Broaching is one of the most important basic machining processes. It can produce extremely good surface finish with higher accuracy. The process is performed by means of cutting tool, called broach which has a series of teeth, each tooth standing slightly higher than the last. This rise per tooth also known as step or the feed per tooth determines the amount of material removed. A large broach can remove material as deep as 38 mm in one stroke.

The cutting tool has roughing, semifinishing and finishing teeth combined into one tool, hence the tool can machine any type of surface and that too in one pass. The last teeth on the cutting tool conform to the final dimension. Broaching operations are classified as internal and external, depending upon whether the surfaces are machined internally or externally. Internal broaching can cut various contours in holes already made by drilling

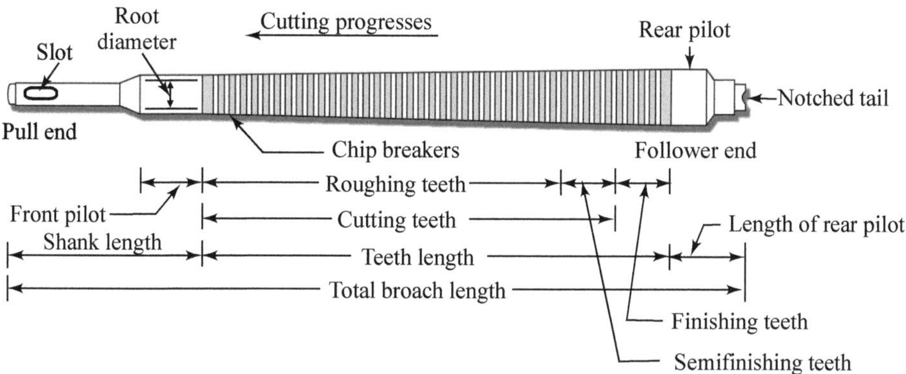

Fig. 17.13 A conventional pull broach

and other operations. External surface broaching competes with milling, shaping, planing and similar operations.

In most machines, the broach is moved past the work, but equally effective results are obtained if the tool is stationary and the work is moved. Broaches are available with various tooth profiles, including some with chip breakers.

Broaching machines either pull or push the broaches and are made horizontal or vertical. As the name suggests, in pull broaching, the broach is pulled and in push broaching it is pushed through the stationary work. Push broaches are usually shorter. On the other hand, pull broaches are longer and used for longer strokes. Horizontal machines can be used for longer strokes. Broaching machines are mostly hydraulically driven because of the large force requirements. A conventional pull broach is shown in Fig. 17.13.

17.12.1 Broaching Parameters

17.12.1.1 Pitch

The pitch is defined as the distance between two consecutive teeth, and is given by the expression

$$p = C\sqrt{l_w} \tag{17.19}$$

where

l_w = length of the workpiece being cut

C = constant

= 1.76 for l_w in mm

17.12.1.2 Length of Broach

The total length of broach for a pull broach is found as:

$$l_b = l_s + \left(n_r + n_s + n_f\right)p + L_{\mathrm{rp}} \text{ (mm)} \tag{17.20}$$

where

l_s = shank length (mm)
n_r = number of roughing teeth
n_s = number of semifinishing teeth
n_f = number of finishing teeth
l_{rp} = length of rear pilot (mm)

In a typical broach, there are usually three to five semifinishing and finishing teeth.

The number of roughing teeth, n_r is obtained as:

$$n_r = \frac{d - n_f t_f - n_s t_s}{t_r} \tag{17.21}$$

where

t_f = rise per tooth for finishing teeth (mm)
t_s = rise per tooth for semifinishing teeth (mm)
t_r = rise per tooth for roughing teeth (mm)
d = depth of cut (mm)

$= t_r + t_s + t_f$

The effective broach length is given as:

$$\begin{aligned} l_e &= l_b - l_w (\text{mm}), \text{ if the broach moves past the work} \\ &= l_b + l_w (\text{mm}), \text{ if the work moves past the broach} \end{aligned}$$

17.12.1.3 Cutting Time

The cutting time, t is obtained by using the formula

$$t = \frac{l_e}{1000V} \text{ (min)} \tag{17.22}$$

where V = Cutting speed (m/min). The cutting speed in broaching normally lies between 1.5 and 15 m/min.

17.12.1.4 Broaching Force

The minimum force, F required to operate a broach is given by

$$F = KndW(\text{Kgf}) \tag{17.23}$$

where

n = total number of teeth involved in cutting

$= n_r + n_s + n_f$

K = constant

= 400 for mild steels and cast iron

= 200 for aluminium, zinc and brass

17.12.1.5 MRR

The metal removal rate is found to be

$$MRR = 1000t_r \, \mathrm{W}_b \, Vn_r \, \left(\frac{\text{mm}^3}{\text{min}}\right)$$

where

W_b = width of the broach tooth (mm)

n_r = number of roughing teeth involved in cutting

$\simeq \frac{l_w}{p}$

Majority of the material is removed by roughing teeth and hence MRR is greatly dependent on its number.

17.12.1.6 Power

The power required for broaching is given as:

$$\text{Power} = \frac{FV}{60 \times 75} \, (HP) \tag{17.24}$$

where

F = cutting force (kgf)

17.13 Shaping and Planing

Shaping and planing are the oldest methods of machining and now replaced by milling and broaching. The major difference between the two is that, in shaping, the primary (cutting) motion is provided to the tool and the feed is given to the workpiece, whereas in planing, it is just the opposite (Fig. 17.14). A single-point cutting tool is used in both processes. The cutting motion is reciprocating and discontinuous; takes place only during forward stroke and the return stroke goes wasted. That is why these processes are limited to small productions. Shaper and planer can produce flat surfaces (horizontal, vertical, inclined or their combinations).

17.13.1 Machining Parameters

17.13.1.1 Cutting Speed

The cutting speed, V in shaping or planing is given by the following relation:

$$V = \frac{N(L + A)(1 + n)}{1000} \text{ (m/min)} \tag{17.25}$$

where

N = number of complete strokes per minute
L = length of stroke (mm)
A = approach length (mm)
$n = \dfrac{\text{time taken in return stroke}}{\text{time taken in forward stroke}} < 1$

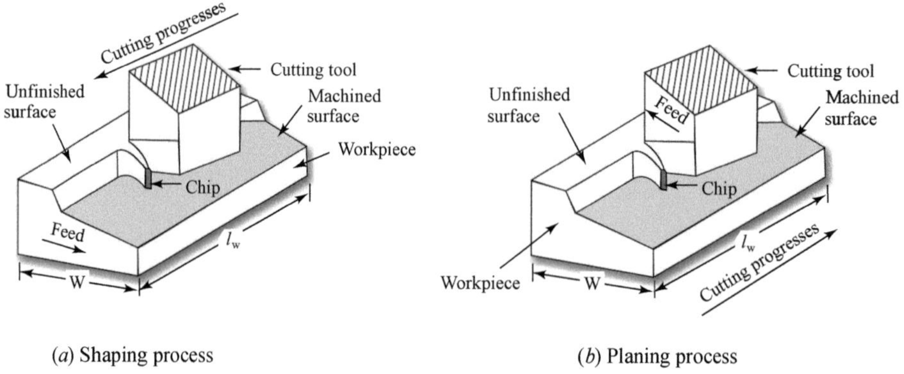

Fig. 17.14 Shaping and planing operations

When $A = 0$ and $n << 1$, the velocity equation reduces to

$$V = \frac{NL}{1000} \text{ (m/min)} \tag{17.26}$$

17.13.1.2 MRR

The metal removal rate is given as:

$$MRR = fd\,Nl_w \left(\text{mm}^3/\text{min}\right) \tag{17.27}$$

where f = Feed (mm)

17.13.1.3 Cutting Time

The cutting time, t is given by the relation

$$t = \frac{W}{Nf} \text{ (min)} \tag{17.28}$$

17.14 Sawing

Sawing is basically a cutting operation performed by means of a saw. Saw has a number of small teeth and each teeth takes part in the cutting operation. Sawing is a bulk-removal process and is used to produce various shapes using ripping, contour cutting, stack cutting and shaping. It can also produce internal and angular cuts.

The cutting tool, saw consists of a blade containing a number of cutting teeth. The saw blades are usually made of carbon and high-speed steels. Carbide blades are used for harder materials.

Following are the basic sawing equipment:

- Power hacksaws
- Band saws
- Circular saws
- Abrasive disk saw

Hacksaw is the simplest metal cutting tool in the form of a straight blade with a number of cutting teeth. It can be manually or power operated. If power operated, it is known as *power hacksaw*. It is generally used for cutting off bars, rods and structural shapes.

Band saws have a long, flexible and continuous blade that runs like a pulley belt on the rims of two wheels and gives a continuous cutting action. They may be classified as vertical and horizontal depending on the direction of blade travel whether it is vertical or horizontal. Vertical band saws are used for straight as well as contour cutting of flat

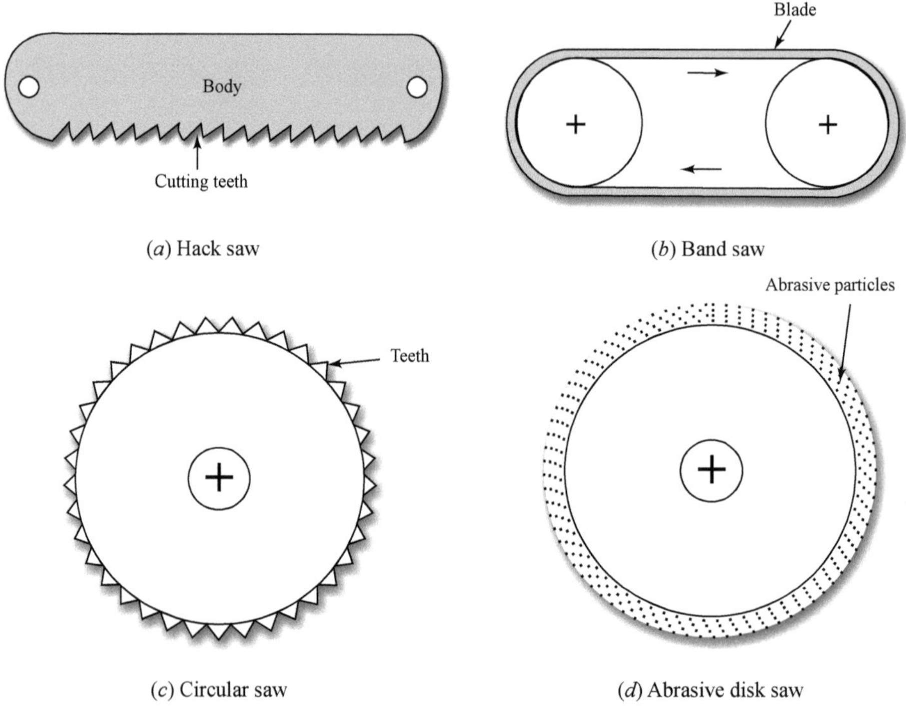

Fig. 17.15 Types of sawing equipment

sheets and other parts, which are supported on a horizontal platform. Horizontal band saw combines the flexibility of reciprocating power hacksaw and the continuous cutting action of vertical band saw, and has higher productivity than power hacksaw. They may be made CNC operated to increase the production run.

Diamond band cutting is a variation of band sawing. Diamond-impregnated band sawing cuts glass, carbide, ceramic, dies, and hard semiconductor materials.

Circular saws for cutting metal are often called cold saws. Cutting off with circular saws produces surface that are comparable in smoothness and accuracy with surface made by slitting saws in a milling machine or by a cut-off tool in a lathe. The saws are fairly large in diameter and rotate at low rotational speeds and are hydraulically operated.

Abrasive disc saws have no cutting teeth as in the case of power hacksaw or circular saws but they contain abrasive grains on their periphery. Cutting action entirely depends upon these abrasive grains and is unaffected by any metal softening. It can be performed in either dry or wet condition and can be used for economical cutting of plate, bar, castings, forging and tubing. It can cut harder materials. All the above stated four sawing operations are shown in Fig. 17.15.

17.15 Grinding

Grinding is the most common abrasive machining process in which material is removed by friction. The abrasive particles, known as grits, act as cutting points. The grits have the capacity to machine extremely hard materials. An abrasive is a small, nonmetallic hard particle having sharp edges with irregular shape. Abrasive particles can be made of silicon carbide (SiC), aluminium oxide (Al_2O_3), cubic boron nitride (CBN) and diamond. The diamond abrasive is the hardest. Next to diamond is CBN. Al_2O_3 is least hard among the given list but is a most widely used abrasive. Abrasive machining processes can be used to produce smoothest surface in which little material is removed. At the same time they can also be used as a stock removal process. All the abrasive machining processes are same as far as the metal removal processes is concerned, but they differ in the way that how many active grains are involved in the process. Since only a proportion of the grains actually cut, and the others absorb energy without cutting, the energy required is many times higher than cutting for a similar material removal, and there is considerable heat generation.

Unlike most cutting processes, in abrasive machining, the individual cutting edges have a random distribution and orientation. Since the particles providing the cutting edges are small and hence depth of engagement is small. As a result, very fine cuts are possible, and fine surface and close dimensional tolerances can be obtained. The scope of abrasive machining processes is tremendous and very hard materials such as hardened steel, glass, carbide and ceramics can be readily machined. These processes are finding extensive applications in automobile, space and aircraft industries.

Grinding is usually a finishing operation which can produce accurate works with tight tolerances. Hard and brittle materials can be machined without any damage.

Grinding wheel is an important machine tool in which many small abrasive particles bonded together are impregnated on its periphery. This rotating wheel if pressed against a workpiece will remove material from it in the form of minute chips, but the chip formation is variable on account of variable grain sizes. A typical vitrified grinding wheel consists of 50% abrasive particles, 10% bond and 40% cavities (porosities), all on volume basis. The higher the porosity, the greater the chip clearance space between the grits and the more freely will coolant penetrate the wheel. Grit and wheel geometry for a grinding wheel is shown in Fig. 17.16.

When cutting on the periphery of the wheel, the situation is very much similar to up or down milling, depending upon the direction of work speed with respect to wheel rotation (Fig. 17.17). When grinding on the face of the wheel, the situation is similar to face milling.

The following are the basic differences between grinding and milling:

- Grinding uses grits as cutting tools while in milling there are milling cutters.
- Grinding produces chips of very small sizes as compared to that produced by milling.

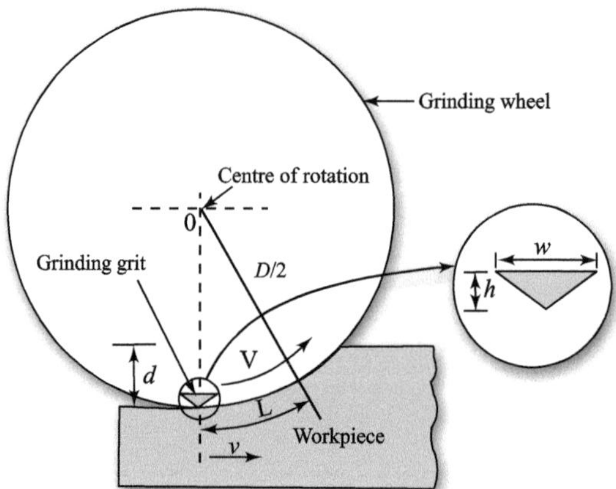

Fig. 17.16 A grinding operation

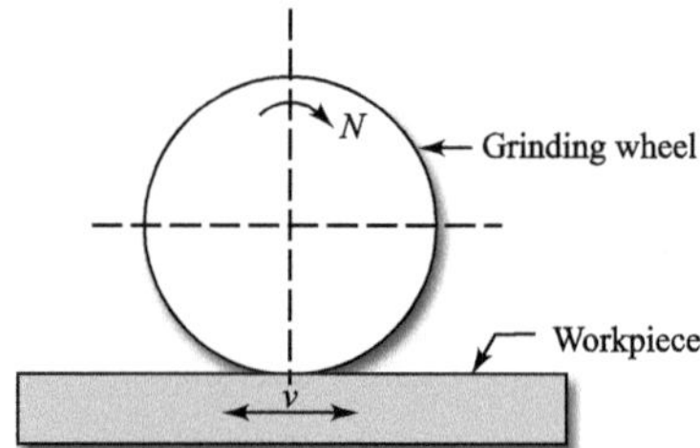

Fig. 17.17 Surface grinding on wheel periphery

- Grits have irregular geometry as compared to the regular geometry of milling cutter.

A sharp grit, i.e. the grit having a sharp edge, will cut a chip, but a dull grit is pulled away from the grinding wheel and washed away. The loss of grains from the wheel means that the wheel is changing size. The grinding ratio, in short *G* ratio, is the ratio of amount of material removed from the workpiece to the amount of material lost from the wheel both expressed in cubic inches. In conventional grinding, this ratio may vary between 20:1 and 80:1 and is a measure of grinding efficiency. It reflects the amount of work a wheel can do during its useful life.

17.15.1 Grinding Parameters

17.15.1.1 Grain Size

Grain size is measured in terms of number of grits per inch length on the periphery of the grinding wheel. Number of grits per inch varies between 8 and 240. 8–16 grit per inch is very coarse; 16–30 grit is coarse; 36–60 grit, medium; 80–120 grit, fine; and 150–240 grit, extra fine. In general, coarse grits are used for soft materials and fine grits for hard materials.

17.15.1.2 Chip Dimensions

(a) The length of the chip, L is given as:

$$L = \sqrt{Dd} \ \text{(mm)} \tag{17.29}$$

where

D = diameter of grinding wheel (mm)
d = depth of grind (cut) (mm)

(b) The average chip thickness for surface grinding is found to be

$$t = \sqrt{\frac{4v}{Vcr}\left(\frac{d}{D}\right)^{1/2}} \tag{17.30}$$

where

t = chip thickness (mm)
v = work speed (m/min)
V = peripheral cutting speed of the wheel (m/min)
= $\frac{\pi DN}{1000}$ N being the rpm of grinding wheel
r = $\frac{w}{h}$
w = width of the grit (mm)
h = height of the grit (mm)
c = number of active grits per m^2

The parameter c is measured by rolling the grinding wheel on a piece of smoked glass and counting, under a microscope, the marks left where grit points pierce the smoke film.

(c) The number of chips per unit of time, n is given by

$$n = \frac{VWc}{1000} \tag{17.31}$$

where
W = width of wheel (mm)

17.15.1.3 MRR

The metal removal rate is found to be

$$MRR = 1000v\,Wd\ \left(\text{mm}^3/\text{min}\right) \tag{17.32}$$

Cutting forces in grinding are difficult to measure because of random geometry of the grits. So the measured value will only give approximate value.

17.16 Finishing Operations

Various machining processes such as turning, boring, drilling, milling, etc. produce workpieces to their accurate size, but do not have a high degree of surface finish, hence they need to be subjected to one or more further operations to obtain the desired finish. Lapping, honing and superfinishing are examples under this category.

These processes are often referred to as *microfinishing* because the surfaces produced through these processes are specified in microunits such as microns.

17.16.1 Lapping

Lapping is an abrasive finishing operation used on flat, cylindrical or spherical surfaces by means of lap. The lap is a relatively softer material in which abrasive particles are embedded on its surface. Alternatively, loose abrasives (SiC, Al_2O_3 or diamond in fine screened sizes) may be carried between the lap and the work surface through carriers such as oil, grease or water. The type of abrasive and carrier are dependent on the materials to be lapped. Soft materials are lapped with aluminium oxide and hard materials with diamond or silicon carbide grit. As the lap is rubbed against the work, the abrasive particles in the lap remove materials from the surface of the workpiece. The lap is usually a soft porous metal such as cast iron or copper. It can be even leather or cloth.

Because of the small size of abrasive particles, the material removal rate is low. It is a slow process which is used only to produce geometrically true surface, or to correct minor surface imperfections (scratch marks left by grinding or honing) or to obtain very flat and

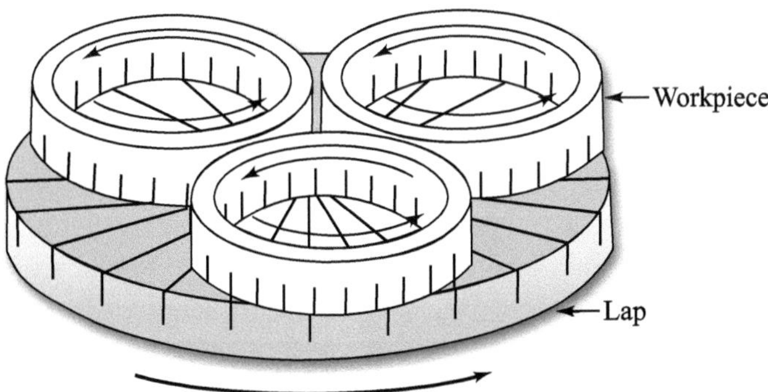

Fig. 17.18 A lapping operation

smooth surfaces. Extreme accuracy in both tolerance and geometry are the hallmarks of lapping. Materials (metals as well as nonmetals) of almost any hardness can be lapped.

Various types of lapping machines are available. In its most common industrial form, the lap is a rotating table and the parts being lapped have a planetary movement, resulting in a uniform surface of excellent flatness (Fig. 17.18). For lapping curved surfaces such as spherical objects and glass lenses, the lap has to be of three-dimensional shape. Since lapping is essentially a finishing operation, parts should not be far from the expected size and geometry before lapping.

17.16.2 Honing

Honing is an abrasive-machining process in which cutting speed is much lower than that of grinding. It is primarily used to size and finish bored holes or remove common errors (tool marks) left by cutting tools. The honing tool consists of a set of Al_2O_3- or SiC-bonded abrasives, called ***stones*** mounted on a metal mandrel (Fig. 17.19). The shoes and stones are uniformly spaced to produce even cutting forces. Material removal is very small typically about 0.127 mm or less. Because of low cutting speed used in the process, heat and pressure are less. As a result, there is no distortion of the workpiece and the process produces a nearly perfect diameter. The tool is inserted into the bore and adjusted to bear against the walls. The work is given a slow reciprocating motion as the mandrel rotates, thus generating a straight and round hole. A cutting fluid is used to flush away chips and to keep temperatures low. Sulpherized mineral base or lard oil mixed with kerosene is generally used as a cutting fluid.

Honing is also done on external cylindrical or flat surfaces and to remove sharp edges on cutting tools and inserts. Typical tolerances that can be maintained are 0.005 mm for the diameter and 0.013 mm for roundness and straightness. The soft materials such as silver and brass as well as very hard materials, including the hardest alloys, carbides and

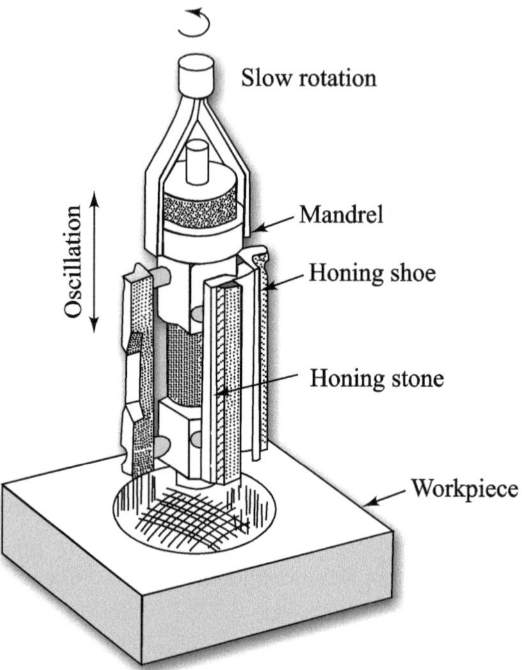

Fig. 17.19 A honing tool

ceramics can be honed. The use of microcrystal line CBN is especially advantageous for honing soft or medium-hard materials. Honing machines can be either horizontal or vertical. Horizontal honing machines are used for long holes such as canon or rifle barrels. Honing can also be done on many general purpose machines such as lathes and drilling machines. Typical honed products include I.C. engine cylinders, bearings, gun barrels, ring gages, piston pins and shafts.

17.16.3 Superfinishing

Superfinishing is a variation of honing and is primarily used on outside flat surfaces. It is effective in removing chatter marks and amorphous material from the work surfaces and is not a bulk material removal process. Because of its capacity to produce extremely high degree of surface finish, it is called a superfinishing operation.

The lubricant used in the process establishes a lubricating film between mating surfaces and minute peaks on the work surfaces are cut away by the honing stone. The lubricating film further protects the surface from being damaged, and hence uniform smoothness is obtained. There are two types of superfinishing operations: cylindrical and flat. Superfinishing finds applications in automobile parts.

Short Answer Questions

1. What is the difference among drilling, reaming and boring?

Ans. Drilling is a process of making a circular hole in a workpiece by means of a drill, whereas reaming is a process of finishing a drilled hole by means of a reamer. Boring is a process of enlargement of a drilled hole.

2. How does tapping differ from dieing?

Ans. Tapping is a process of producing internal threads as in case of a nut using a tap, whereas dieing is a process of producing external threads on a solid bar as in case of a bolt using a die.

3. What is the difference between turning and boring?

Ans. Turning is performed on the outside diameter of a solid cylinder to reduce its diameter, whereas boring is performed on the inside diameter of an existing hole to increase its size.

4. What is Knurling and why is it used?

Ans. Knurling is a process of producing irregularly shaped roughness (indentations, called knurls) on cylindrical surfaces for better grip of the hand when using them.

5. What is the difference between up milling and down milling?

Ans. In up milling, the workpiece is fed in opposite direction to the rotating cutter at the point of cutting and tend to be lifted from the machining table, whereas, in down milling, the workpiece is fed in the direction of the rotating cutter at the point of cutting and tend to be moved down onto the table.

Multiple Choice Questions

1. Which of the following operations is also known as internal turning?

(a) Milling (b) Tapping
(c) Boring (d) Facing

2. Match List I with List II and select the correct answer using the codes given below the lists:

List I	List II
(Machining processes)	**(Related to)**
A. Boring	1. Multiple-tooth cutter
B. Milling	2. Cutting a piece into two parts
C. Reaming	3. Enlargement of a hole
D. Parting	4. Finishing of a hole

Codes:	A	B	C	D
(a)	3	2	4	1
(b)	3	1	4	2
(c)	4	1	3	2
(d)	4	2	3	1.

3. Which of the following operations can be regarded as chipless operation?

(a) Boring (b) Reaming
(c) Milling (d) Knurling

4. Consider the following statements:

   1. Using helical-tooth cutter, tool vibration and chatter are minimum.
   2. A helical-tooth cutter produces a smooth surface.
   3. A straight-tooth cutter produces continuous cutting.
   4. The depth of cut in drilling is equal to one-half of the feed. Of these statements:

(a) 2, 3 and 4 are true (b) 1, 2 and 4 are true
(c) 1 and 2 are true (d) 3 and 4 are true

5. Consider the following statements:

   1. In up milling, the workpiece is fed in opposite to cutter rotation.
   2. Chip thickness is minimum at the start and maximum at the end in up milling.
   3. All the milling operations are based on up milling.
   4. Better surface is obtained in down milling.

      Of these statements:

(a) 1 and 3 are true (b) 3 and 4 are true

(c) 1, 2 and 4 are true (d) 2, 3 and 4 are true

6. Match List I with List II and select the correct answer using the codes given below the lists:

List I **(Machining processes)**	List II **(Related to)**
A. Tapping	1. External thread
B. Grinding	2. Internal turning
C. Boring	3. Internal thread
D. Dieing	4. Abrasive particles

Codes:	**A**	**B**	**C**	**D**
(a)	3	2	4	1
(b)	1	2	3	4
(c)	1	4	2	3
(d)	3	4	2	1.

7. Which of the following is least hard?

(a) Diamond (b) Cubic boron nitride

(c) Aluminium oxide (d) Silicon carbide

8. Which of the following is the most widely used abrasive?

(a) Diamond (b) Cubic boron nitride

(c) Aluminium oxide (d) Silicon carbide

9. Which of the following operations can produce chips of very small size?

(a) Boring (b) Grinding

(c) Milling (d) Turning

10. Consider the following operations:

1. Reaming 2. Lapping

3. Milling 4. Honing

Of these operations, which can be regarded as microfinishing operation?

(a) 1, 2 and 4 (b) 2, 3 and 4
(c) 1, 3 and 4 (d) 2 and 4

Answers

1. (c)	2. (b)	3. (d)	4. (b)	5. (c)	6. (d)
7. (c)	8. (c)	9. (b)	10. (d)		

Review Questions and Discussions

Q.1. Why is reaming operation performed?
Q.2. Milling is a versatile machining process. How?
Q.3. Why is milling more suitable for producing flat surfaces as compared to shaping for large production?
Q.4. Differentiate between up and down milling. Discuss their suitability.
Q.5. How is face milling different from peripheral milling?
Q.6. When is conventional milling advantageous?
Q.7. How does taper turning differ from plain turning?
Q.8. What are the materials used for making a drill? Why are drills sometimes coated with titanium nitride?
Q.9. What is the difference between boring and reaming?
Q.10. How does tapping differ from dieing?
Q.11. How is grinding different from other machining operations?
Q.12. What is honing?
Q.13. What is lapping?

Cutting Tool Materials 18

The increased hot hardness of ceramics and carbides makes them suitable for high speed operations.

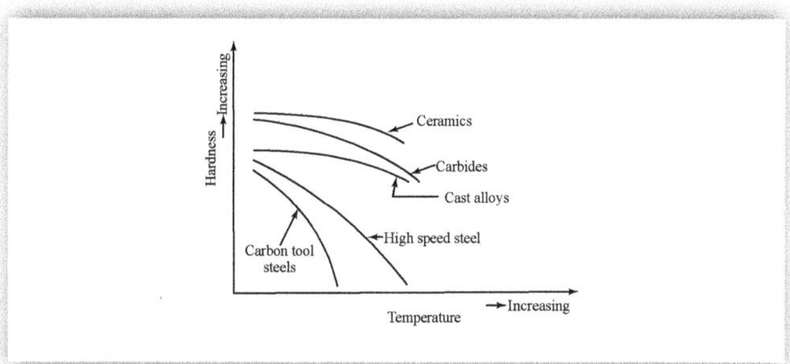

Learning Objectives

After reading this chapter, you will be able to answer some of the following questions.

- What are the important features of cutting tool materials?
- Why is high speed steel (HSS) most versatile cutting tool material?
- When is dimond used for cutting?

D. K. Singh, *Fundamentals of Manufacturing Engineering*,
https://doi.org/10.1007/978-981-99-8767-2_18

18.1 Introduction

The cutting tool material, cutting parameters and the selected tool geometry directly influence the productivity of the machining operation. Since machining is a deformation process, hence geometric stability of the cutting tool is very important. A wide range of cutting tool materials is available with a variety of properties, performance capabilities and cost. But the best material is one, which produces the machined part at the lowest cost.

18.2 Desirable Characteristics of a Cutting Tool Material

- The tool should maintain its hardness at increased temperatures. This property is known as red or hot hardness. During machining at high cutting speeds, very high temperatures are produced which can soften the tool, thereby affecting its hardness and strength badly. Elements such as cobalt, chromium, molybdenum, tungsten and vanadium form stable carbides at elevated temperatures and hence have better hot hardness as compared to HSS and high-carbon steel. The effect of temperature on the hardness of common tool materials is shown in Fig. 18.1.
- It should be tough enough to resist its fracture due to shock, vibration or impact loads. This property is useful during intermittent cutting operations, e.g. turning a splined shaft.
- It should have good abrasive resisting qualities so that it can resist wear before the tool is resharpened or replaced.
- The coefficient of friction should be low to reduce the frictional heat at tool–chip interface.

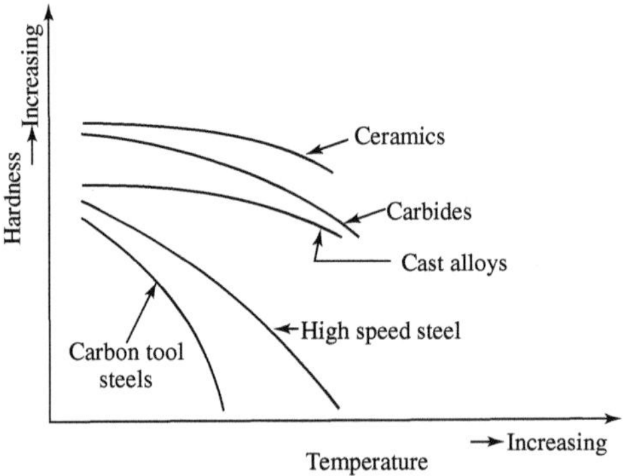

Fig. 18.1 Effect of temperature on the hardness of tool materials

- The thermal conductivity and the specific heat should be high to easily flow away the heat generated at the cutting edge.
- It should provide chemically inert atmosphere around the workpiece.

The following materials are used for making cutting tools:

- High-carbon steel
- High-speed steel (HSS)
- Cast nonferrous alloys
- Cemented carbides
- Coated cemented carbides
- Cermets
- Ceramics
- Coronite
- Cubic boron nitride (CBN)
- Diamond.

18.3 High Carbon Steel

Plain carbon steels containing carbon in the range of 0.6–1.5% were very much popular before the advent of HSS. They are principally used for drills, taps, broaches and reamers. High-carbon steel tools have poor hot hardness and hence lose their hardness even at low temperatures of around 200 °C. This limits their use for very low-speed cutting operations.

18.4 High-Speed Steel (HSS)

High-speed steels were once used for cutting at higher cutting speeds, and even today they are one of the important cutting tool materials. They contain mainly iron and carbon alloyed with tungsten, chromium, vanadium, molybdenum and cobalt. They are basically carbon steels with tungsten, chromium, vanadium, molybdenum and cobalt as alloying elements besides iron and carbon. It possesses excellent hardenability and has better hot hardness as compared to high carbon steels. Due to its good toughness (around twice) over cemented carbides, it is suitable for most of the cutting operations. The HSS tools are manufactured by three methods: cast, wrought and sintered (using powder metallurgy technique). HSS may be grouped in three important classes.

- 18-4-1 type contains 18% tungsten, 4% chromium and 1% vanadium. It is the most popular and standard type and is known for its good wear and heat resistance.

- 6-6-4-2 type contains 6% tungsten, 6% molybdenum, 4% chromium and 2% vanadium. It has excellent toughness and cutting ability.
- Super HSS is so named because of its higher cutting efficiency. Presence of cobalt (5–8%) increases its hot hardness. Tungsten, chromium and vanadium are additional elements present in it. It is primarily used for heavier cuts.

18.5 Coated High-Speed Steels

Coated high-speed steels were developed in the late 1970s to take care of difficult-to-machine materials. A thin layer (1–2 μm) of titanium nitride (TiN) with an equivalent hardness of RC 80–85 is coated on the HSS tools to increase their wear resistance. Physical vapour deposition (PVD) method has proved to be the best process for coating HSS. This method applies TiN at a much lower temperature, about 400 °C as compared to 1000 °C used in chemical vapour deposition (CVD) method. Also, it allows sharper corners and gives a lower coefficient of friction, both desirable properties.

18.6 Cast Nonferrous Alloys or Cast Cobalt Alloys

They were discovered by *EllwoodHynes* in 1914 and are also known as *stellites*. These are generally cobalt based, and other alloying elements are chromium, tungsten and carbon. The range of elements in these alloys is 40–50% cobalt, 15–35% chromium, 12–25% tungsten and 2–4% carbon. A typical analysis can be 45% cobalt, 35% chromium, 18% tungsten and 2% carbon. They do not contain iron except as an impurities. Cast cobalt alloys are produced only by casting and then dressed by grinding. They can't be forged or readily cut to shape when cold. The casting provides a tough core. They can maintain their hardness to much higher temperature beyond the point where HSS burns up. They offer 25% increase in cutting speed over HSS tools. These tools find application in the machining of cast iron, plain carbon steels, alloy steels and nonferrous alloys.

18.7 Cemented Carbides

Cemented carbides are also called sintered carbides because they are produced by powder metallurgy techniques. They are made of micron size hard carbide particles of tungsten, titanium and tantalum cemented together by a binder. The binder is mostly cobalt. A typical analysis of a carbide suitable for steel machining is 82% tungsten carbide, 10% titanium carbide and 8% cobalt. Straight tungsten carbides are the strongest and most wear resistant but are subject to rapid cratering, when machining steels. To improve resistance to cratering, tantalum carbide and titanium carbide are added to the basic composition, resulting in tungsten-tantalum carbide or tungsten-titanium carbide.

Cemented carbides are much harder and chemically more stable, have better hot hardness (the red hardness of carbide tool materials is superior to all others, with the exception of ceramic; it can maintain its cutting edge at temperatures more than 1200 °C), higher stiffness and lower friction, and operate at higher cutting speeds than do HSS. But carbide tools do not have as much resistance to shock as HSS and can't be shaped after sintering. Hence cemented carbide tools are available in insert form in many different shapes; squares, triangles, diamonds and rounds. Carbide inserts are mechanically attached to the tool shank so that if the cutting edge gets damaged or blunted it can be suitably replaced. The development of the disposable insert and electrolytic grinding techniques have made cemented carbides extremely useful for machining.

18.8 Coated Cemented Carbides

With the advent of coated cemented carbides, the scope of machining operations widened dramatically. Carbide tools may be coated with a thin layer (2–10 μm) of titanium nitride, titanium carbide, titanium carbonitride or aluminum oxide to a tungsten carbide base. Titanium carbide and aluminum oxide are very hard materials, provide wear resistance and are chemically inert, providing a chemical and heat barrier between tool and chip. Titanium nitride is not such a hard material but gives a lower coefficient of friction to the faces of the insert and better cratering resistance. Important changes that were noticed with coated cemented carbide over uncoated carbide included increased temperature bearing capacity, *i.e.*, increased red hardness, increased cutting speeds, increased tool life and increased wear resistance. Multilayer coatings ensure prolonged tool life.

Commercial application of uncoated carbide tool materials is declining. Coated cemented carbides are first choice grades for a majority of turning, milling and drilling operations in most materials. More than 75% of turning operations and more than 40% of milling are today performed with coated carbides. Coated carbides provided a cutting tool material much closer to the ideal, with new combinations being formed.

There are two methods used to provide coating on carbide tools: chemical vapour deposition (CVD) and physical vapour deposition (PVD). The CVD method is mostly used for carbide tools with multilayer and ceramic coatings.

18.9 Inserts

In order to prevent the total replacement of a cutting tool more frequently due to wear as in the case of carbon steel and HSS tools and to make the process faster and less time-consuming, individual cutting tools with a number of cutting points are mechanically attached to the shank of the tool. These individual cutting tools are called inserts. When one edge or corner becomes dull, the insert is rotated or turned over to expose to a new cutting edge. Inserts are usually made of carbide materials in order to get increased tool life. Carbide inserts are available in a variety of shapes, such as square, triangle, diamond and round.

18.10 Ceramics

Ceramics are produced by powder metallurgy technique. They contain aluminium oxide (Al_2O_3) as the major component along with certain additives of titanium, magnesium or chromium oxide mixed with a binder. Ceramic tools are in the form of disposable inserts.

Ceramic cutting tools are hard, with high hot hardness, and do not react with the workpiece materials. They have long tool lives and can machine at very high cutting speeds (two to three times the cutting speeds of tungsten carbide). The main application areas for ceramics are: grey cast iron, heat-resistant alloys, hardened steels, nodular cast iron and to some extent steel. Very high metal removal rate is possible with difficult-to-machine materials such as superalloys, chilled cast iron and high-strength steels. They are usually used without coolant. It has been claimed that ceramic will remove 20% more metal at up to 42 times the optimum volumetric metal removal rate of carbide. Ceramics have hardness comparable to carbides but are more brittle and hence are effective only at very high cutting speed for uninterrupted cutting operations, such as finishing or semifinishing by turning. They are not recommended for aluminium and titanium alloys because of oxidation problems.

There are two basic types of ceramics:

(a) Alumina (Al_2O_3)-based ceramics.
(b) Silicon nitride (Si_3N_4)-based ceramics.

18.10.1 Alumina (Al_2O_3)-Based Ceramics

Alumina-based ceramics have very high abrasive resistance and hot hardness and are chemically more stable than high-speed steels and carbides. They can be in the pure, mixed or reinforced form.

Pure alumina-based ceramics, also called white ceramics, contain Al_2O_3 up to 99% and are cold pressed. They lack strength and toughness, and their thermal conductivity is low.

Mixed alumina-based ceramics, also called black ceramics, contain titanium carbide and titanium nitride in the range of 20–40% and are hot pressed. They have increased hot hardness because of the presence of above two components and are suitable for high-temperature machining.

Reinforced alumina-based ceramics, also known as whisker-reinforced ceramics, are made from silicon carbide and are hot pressed. Whisker is a silicon carbide crystal of about one micron diameter and length more than 20 microns and functions in the same way as do fibres in fibre glass. The whiskers make up some 30% of the contents. The whisker reinforcement has dramatic effect. The toughness, strength and thermal shock resistance are considerably increased along with hot hardness and wear resistance. Recently, whisker-reinforced ceramic materials that have greater transverse rupture strength have been developed.

18.10.2 Silicon Nitride (Si_3N_4)-Based Ceramics

Silicon nitride is the newest and most promising of the cutting tool materials. An example of a silicon nitride-based ceramic is *Sialon*, so called because of the presence of silicon, aluminium, oxygen and nitrogen in its composition. Materials formerly being machined only by grinding can now be cut with silicon nitride. It has better thermal and mechanical shock resistance, and toughness than alumina-based ceramic. The increased toughness and shock resistance greatly reduce the problems of edge chipping and catastrophic failure. It is excellent at maintaining hot hardness at temperatures higher than those suitable for cemented carbides. Grey cast iron machining using silicon nitride is excellent in dry and wet conditions, and the cutting speed can be as high as 1500 m/min. It stands up well to intermittent cuts and when depths of cut vary. Because of chemical affinity, it is not suitable for machining steels.

18.11 Cermets

Cermets consist of CERamic and METals bonded together by powder metallurgy techniques. They are aluminium oxide-based materials containing titanium carbide, titanium nitride and titanium carbonitride. Cermets have higher hot hardness and higher oxidation resistance over cemented carbides. They are quite useful for higher cutting speeds involving lower feed and lower depth of cut and produce surfaces of higher accuracy and finish.

18.12 Coronite

The properties of coronite lie between cemented carbide and high speed steel. Its properties such as hot hardness and wear resistance are superior over high-speed steel. The toughness is comparable to high-speed steel but better than cemented carbide.

It contains higher percentage (almost 50%) of titanium nitride in the form of tiny particles, which is responsible for its increased hot hardness. The particle size may be 0.1 micron, compared to 1–10 microns in cemented carbide or HSS.

18.13 Cubic Boron Nitride (CBN)

Cubic boron nitride, developed in 1962 by General Electric Company of USA, is a material that is next to diamond in hardness. Because of its extreme hardness, it is also called artificial diamond. The CBN tool is manufactured by bonding a 0.5–1 mm layer of polycrystalline cubic boron nitride to a tungsten carbide insert (Fig. 18.2) under high temperature and pressure conditions.

CBN is relatively a brittle cutting tool material but bonding to the substrate gives stiffness and resistance to fracturing. It has excellent hot hardness up to very high temperatures (2000 °C), high abrasive wear resistance and has low chemical reactivity at the tool–chip interface. The CBN tools can be used to machine difficult-to-machine materials such as hardened steels, hard-chill cast iron and nickel- and cobalt-based superalloys. These materials can be machined at very high cutting speeds (about five times) and with a higher metal removal rate (about five times) than cemented carbide, and with superior accuracy, finish and surface integrity.

CBN inserts are excellent for finishing to close tolerances in operations with hard steel. Surface texture with Ra value of 0.3 and tolerances of ± 0.01 mm are turned with CBN. Much longer tool life is achieved as compared to cemented carbide and ceramics tools (about five to seven times that of a ceramic tool). The tool, however, needs to be properly supported and clamped during its use in order to prevent premature failure.

CBN tools have been used successfully for heavy interrupted cutting on hard materials using negative rakes. CBN should be applied to harder work materials. If the components are too soft, excessive too wear will result. The harder the material, the less tool wears. The grains in CBN are very small, and to avoid microchipping (one of the predominating

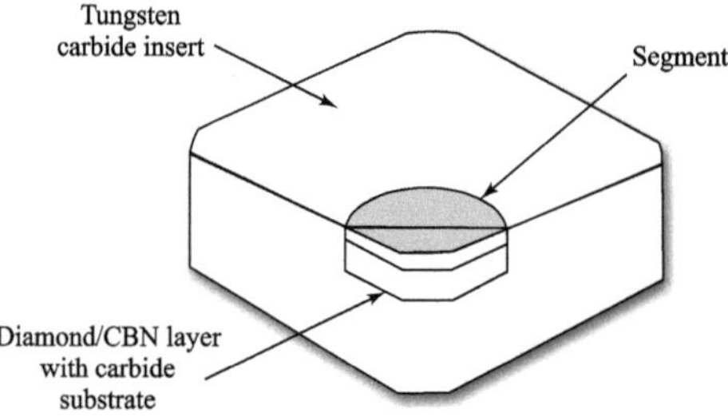

Fig. 18.2 Diamond or CBN layer on a tungsten carbide insert

wear modes of CBN tools), cutting edges are suitably chamfered. However, the CBN tools are costly.

18.14 Diamond

Diamond is the hardest known material and can be used for cutting operations. Diamonds may be natural or artificial. Natural diamond is monocrystalline and is very brittle and has the tendency to fail suddenly by edge chipping or cleavage. Artificial diamond is polycrystalline and has superior properties over natural diamond and hence finds wide applications as a cutting tool material.

Artificial diamond tool consists of a very thin layer of fine grain-sized diamond particles sintered together and metallurgically bonded to a cemented carbide substrate, similar to CBN tools, at very high temperature and ultra-high pressure. The cemented carbide provides the necessary elastic support for the hard and brittle diamond layer above it. The random orientation of the diamond crystals eliminates any direction for crack propagation. This results in hardness, toughness and wear resistance uniformly high in all directions.

The artificial diamond tools offer dramatic performance improvements over carbides. Tool life can be many times longer than cemented carbide tools. These tools are used on all nonferrous materials and a wide range of nonmetallic materials including composites, resins, rubber plastics, carbon, presintered ceramics, carbide and sintered carbide. They are especially recommended where abrasive wear is a strong factor.

Because of the very brittle nature of artificial diamond, very stable conditions such as rigid tools and machines and high speeds are necessary for machining with it. It cannot be used for ferrous applications due to strong chemical affinity between carbon in tool and work.

Diamond is also used as an abrasive in grinding and polishing operations and as coatings (coated tools). Typical applications include the precision boring of holes and the finishing of plastics or other abrasive materials.

Short Answer Questions

1. What is the chief constituent of ceramic tools and how are they manufactured?

Ans. Aluminium oxide (Al_2O_3) is the chief constituent of ceramic tools, which are manufactured by powder metallurgy techniques.

2. Differentiate between cemented carbides and coated carbides?

Ans. Cemented carbides consist of tungsten carbide (WC), titanium carbide (TiC), tantalum carbide (TaC) and cobalt (Co), which acts as a binder. Coated carbides, on the other hand, are basically cemented carbides having a thin layer coating of titanium carbide, titanium nitride or aluminium oxide to increase their wear resistance.

3. What does 18–4-1 high-speed steel (HSS) indicate?

Ans. It indicates the composition of the high-speed steel, which contains 18% tungsten, 4% chromium and 1% vanadium.

4. What is hot hardness? What is its other name?

Ans. Hot hardness is the ability of a material to maintain its hardness at increased temperatures, which is a basic requirement for high-speed cutting tools. Its other name is red hardness.

5. Arrange the following cutting tool materials in increasing order of their hot hardness: Plain carbon steels, cemented carbides, ceramics, high-speed steels and cast cobalt alloys.

Ans. Plain carbon steels < High-speed steels < Cast cobalt alloys < Cemented carbides < Ceramics.

6. Why are coated carbide and ceramic cutting tools not suitable for intermittent and low-speed cutting operations? For what speed are they used?

Ans. Coated carbide and ceramic cutting tools are brittle and have low toughness, and hence cannot withstand vibration or impact that occurs during intermittent or low cutting speeds, and can fail prematurely. They are used for machining at higher cutting speeds.

7. What are inserts? What are they made of?

Ans. Inserts are individual cutting tools mechanically attached to the shank of the tool, and are easily replaceable in case one tool gets damaged or blunted by simply rotating or turning over to expose to a new cutting edge. They are usually made of carbide materials to ensure increased tool life.

8. Why is diamond cutting tool not suitable for the machining of ferrous materials?

Ans. Carbon in diamond cutting tool has strong chemical affinity with ferrous materials making work materials brittle and unsuitable for machining.

9. What type of cutting conditions are required for diamond cutting tool?

Ans. Diamond cutting tools require very stable conditions such as rigid tools and machines and are operated at high cutting speeds because of their very brittle nature.

10. What is the composition of carbide tools? How are they manufactured?

Ans. Carbides tools are made of micron size hard carbide particles of tungsten, titanium and tantalum cemented together by a binder which is mostly cobalt. They are manufactured by powder metallurgy techniques.

11. What are the important properties of carbide tools?

Ans. Carbide tools are much harder and chemically more stable, have better hot hardness which helps it to maintain its cutting edge at temperatures more than 1200 °C, have higher stiffness and offer cutting speeds higher than high-speed steel (HSS).

Multiple Choice Questions

1. The cutting tool material should be harder than work material. Approximately it is
   (a) 5% more harder than work material
   (b) 10% more harder than work material
   (c) 20% more harder than work material
   (d) 50% more harder than work material

2. Consider the following statements about cutting tool materials:
   1. Their red hardness is more.
   2. They have high coefficient of friction.
   3. Their thermal conductivity and specific heat are high.
   4. They have lower toughness.

   Of these statements:

   (a) 1 and 2 are true (b) 1 and 3 are true
   (c) 3 and 4 are true (d) 1, 2 and 3 are true

3. Consider the following elements:

1. Cobalt
2. Tungsten
3. Vanadium
4. Magnesium
5. Nickel.

Of these elements, which one is responsible for hot hardness of the cutting tool?

(a) 1 and 2
(b) 2, 4 and 5
(c) 1, 2 and 3
(d) 1, 3 and 4

4. Which of the following cutting tool materials loses its hardness at increased temperature?

(a) Ceramics
(b) HSS
(c) Carbides
(d) Stellites

5. 18-4-1 HSS contains

(a) 18% tungsten, 4% chromium and 1% carbon
(b) 18% molybdenum, 4% tungsten and 1% vanadium
(c) 18% tungsten, 4% chromium and 1% vanadium
(d) 18% chromium, 4% tungsten and 1% carbon

6. Which of the following elements is the chief constituent of super HSS?

(a) Nickel
(b) Cobalt
(c) Copper
(d) Titanium

7. Coated HSS contains a thin layer of

(a) Cobalt carbide
(b) Copper carbide
(c) Titanium nitride
(d) Vanadium nitride

8. Which of the following cutting tool materials is also known as stellites?

(a) Cermets
(b) Cast alloys
(c) HSS
(d) Cubic boron nitride (CBN)

9. Which of the following elements is the chief constituent of stellites?

(a) Tungsten (b) Vanadium
(c) Titanium (d) Cobalt

10. Inserts are usually made of

(a) HSS (b) Carbon steels
(c) Ceramics (d) Cemented carbides

11. Consider the following cutting tool materials:

1. Carbon steels 2. Cemented carbides
3. HSS 4. Ceramics

Which of these materials is processed by powder metallurgy?

(a) 1 and 2 (b) 2 alone
(c) 3 and 4 (d) 2 and 4

12. Consider the following properties of a cutting tool material:

1. Increased hardness 2. Increased toughness
3. Better hot hardness 4. Increased wear resistance

Cemented carbides have the following properties:

(a) 1 and 2 (b) 1 and 3
(c) 1, 3 and 4 (d) 1, 2, 3 and 4

13. Sialon consists of

(a) Sulphur, iron, aluminium, oxygen and nitrogen
(b) Silicon, aluminium, oxygen and nitrogen
(c) Silicon, aluminium, oxygen and nickel
(d) Sulphur, iron, argon, oxygen and nitrogen

14. Which of the following materials is the major constituent of coronite?

(a) Tungsten carbide (b) Titanium carbide
(c) Titanium nitride (d) Vanadium carbide

15. CBN stands for

(a) Carbon boron nitrogen
(b) Cubic boron nitride
(c) Carbon bromine nitrogen
(d) Carbon beryllium nitrogen

16. The chief constituents of cermets are:

(a) Aluminium carbide and titanium nitride
(b) Aluminium oxide and titanium carbide
(c) Tungsten carbide and tantalum carbide
(d) Aluminium bromide and titanium carbide

Answers

1. (d)	2. (b)	3. (c)	4. (b)	5. (c)	6. (b)
7. (c)	8. (b)	9. (d)	10. (d)	11. (d)	12. (c)
13. (b)	14. (c)	15. (b)	16. (b)		

Review Questions and Discussions

Q.1. State the desirable properties of cutting tool materials.
Q.2. Why are HSS tools often preferred over carbide tools?
Q.3. What is the purpose of coating a cutting tool?
Q.4. Which tool materials would be suitable for interrupted cutting operations? Why?
Q.5. Why are ceramic and coated carbide tools not so popular?
Q.6. What is the composition of Sialon?
Q.7. What are the benefits of coated high speed steels?
Q.8. Why do ceramic tools maintain their hardness at increased temperatures?
Q.9. Where do cemented carbide tools find applications?
Q.10. What are inserts?
Q.11. Compare cubic boron nitride (CBN) and diamond tools?
Q.12. Arrange the following tool materials in order of their increasing hot hardness. Ceramics, cermets, high-speed steels and cemented carbides.
Q.13. Why are cemented carbide tools not suitable for low cutting speeds?

19 Forming Operations

Forging is a very useful forming operation with numerous applications.

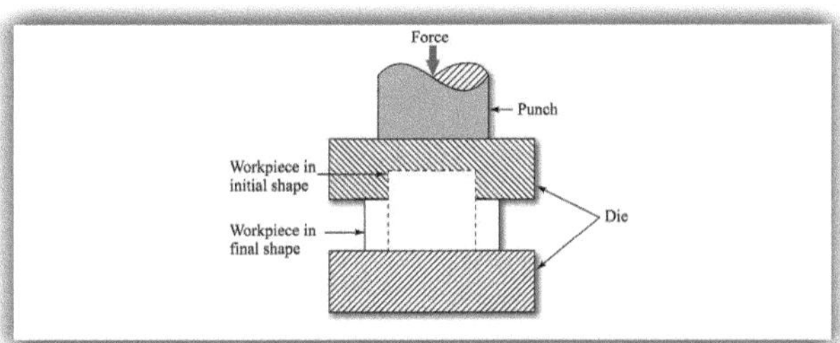

Learning Objectives
After reading this chapter, you will be able to answer some of the following questions.

- How dose extrusion differ from drawing?
- What is the function of mandrel?
- How does blanking differ from piercing?

D. K. Singh, *Fundamentals of Manufacturing Engineering*,
https://doi.org/10.1007/978-981-99-8767-2_19

19.1 Introduction

Forming processes are used to produce the desired shape and size through the plastic deformation of the materials. Plastic deformation takes place by means of a slip mechanism. The overall usefulness of metals is largely due to the ease by which they can be formed into useful shapes. The deformation takes place due to changes in thickness or cross sections of the workpiece. In sheet metal forming operations, on the other hand, thickness and surface area remain relatively constant. Metal deformation involves simple shearing, simple or combined bending or their combinations.

Forming process can be classified into two groups: hot working and cold working process. Recrystallization temperature differentiates between these two processes. If the process is performed above the recrystallization temperature of the metal, it is referred to as hot working and the process performed at room temperature is called cold working. When metal is hot-worked, the deformation forces are less, greater deformation is possible without fracture and mechanical properties are relatively unchanged. Higher deformation forces are required in cold working, and the strength and hardness of the metal are increased due to strain hardening.

19.2 Formability

Formability is defined as the ability of a material to be plastically deformed without fracture. The face-centred cubic (fcc) crystals provide exceptionally high formability because they have the greatest tendency to slip. Formability depends upon the ductility which in turn depends upon the grain size of a material. Large grain size materials have good ductility, and hence they have high formability. Besides grain size, it is the strain rate (defined as the speed at which deformation takes place), which affects the formability of a material.

19.3 Rolling

Rolling is used to reduce the thickness or change the cross section of a given workpiece by means of rolling mills. It is the most widely used operation in modern manufacturing and is performed in hot or cold states.

Hot rolling is carried out above recrystallization temperature of the material. The metal should be heated uniformly in order to get uniform deformation during the process. The process refines the grain structure of the material and is used to produce flat plate, sheet and foil.

During the process, heated workpiece is passed between two rolls that rotate in opposite directions (Fig. 19.1). Rolls are mode of cast iron, cast steel and forged steel. Forged

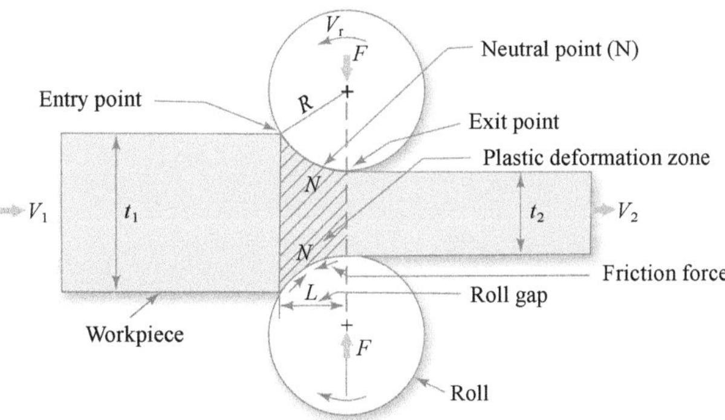

Fig. 19.1 A rolling operation

rolls have more strength, stiffness and toughness than cast rolls. If more reduction in thickness is desired, more than one pass can be used.

The workpiece enters the rolls with a velocity V_1 and comes out with a velocity V_2. The rolls rotate with a surface velocity $V_{r.}$ In practice, $V_2 > V_r > V_{1.}$ The frictional force generated due to the motion of the rolls and acting along the contact interface pulls the workpiece forward. There is a point in the roll gap, where the workpiece velocity and the roll velocity are equal. This point is known as neutral point N. The neutral point has significance in rolling. The material to the left of the neutral point has a velocity less than the roll velocity and the frictional force acts in the direction of roll velocity. On the other hand, the material to the right of the neutral point is moving at a velocity greater than the roll velocity and the frictional force acts in opposite direction to that of roll velocity. The frictional force changes its direction after the workpiece crosses the neutral point as shown in Fig. 19.9.1.

In rolling, the volume of metal before the start of the process and at the end of the process remains the same.

$$Q_1 = Q_2$$
$$\text{or } t_1 W V_1 = t_2 W V_2$$
$$t_1 V_1 = t_2 V_2$$
$$V_2 = V_1 \left(\frac{t_1}{t_2} \right) \tag{19.1}$$

where

Q_1 = initial volume of metal before it enters the rolls, per unit time
Q_2 = final volume of metal when it comes out of rolls, per unit time
A_1, A_2 = initial and final cross sections of the workpiece respectively
t_1, t_2 = initial and final thicknesses of the workpiece respectively

W = width of the workpiece (= constant)

The surface or peripheral velocity of the rolls is given by

$$V_r = 2\pi NR \tag{19.2}$$

where

N = rotational speed of the rolls, rpm
R = roll radius

The draft is defined as the difference between the initial and final thicknesses of the workpiece and depends on the coefficient of friction between contact surfaces and roll radius.

$$t_1 - t_2 = \mu^2 R \tag{19.3}$$

where μ = coefficient of friction

It means that more friction and higher roll radius will give more reduction in thickness.

The roll force F is found to be

$$F = LWY_{av} \tag{19.4}$$

where L = roll gap or contact arc length

$$\simeq \sqrt{R(t_1 - t_2)}$$

Y_{av} = average yield point stress

The above equation is just an approximation where the coefficient of friction is assumed to be zero. If the contact friction increases, the situation significantly differs and more force is required than given by equation (19.4).

The power required per roll is given as:

$$P = \frac{2\pi FLN}{60}\ (\mathrm{W}) \tag{19.5}$$

In Eq. (19.5), L is in metre and F in newton. Smaller diameter rolls have smaller length of contact as compared to larger diameter rolls for a given reduction (Fig. 19.2). Consequently, lower forces and power are required to bring changes in the shape. But they have the risk of being separated apart by the metal coming through the rolls. Hence number of rolls can be increased. Exact determination of forces and power requirement is difficult because of difficulty in the analysis of contact zone. To make the analysis simpler, the following assumptions are made.

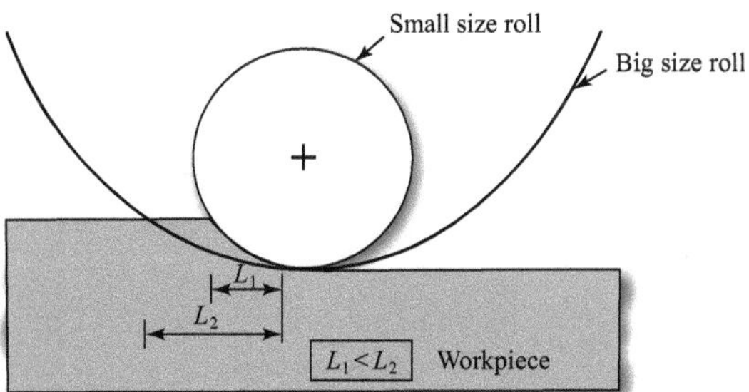

Fig. 19.2 Effect of roll size on contact arc length for a given reduction

- The rolls are rigid.
- There is no lateral spread of the material; i.e., width of the workpiece is assumed to be constant.
- The coefficient of friction is low and is constant in the contact zone.
- The yield stress of the material remains constant in the contact zone.

Because of high-temperature application in hot rolling, oxidation and scaling take place on metal surfaces and it becomes essential to descale them. Also, dimensional changes occur in the tools due to heating. Lubricants may be used during hot rolling to cool the rolls and wash the scales from the rolled surfaces. Ferrous alloys usually do not require lubrication. Oils, emulsions and fatty acids can be used during hot rolling of nonferrous alloys.

Cold rolling is performed at room temperature, and it produces close dimensional tolerances and good surface finishes. Also, descaling is not necessary because of the absence of temperature.

Sometimes the rolled sheet lacks the proper flatness because of a number of variable factors. To make the sheet up to the required level of flatness, the sheet is passed through a series of levelling rolls. Mild steel, for example, is a material which does not produce flat sheet during rolling due to the yield point elongation effect which results in Lueder's bands (surface irregularities). To overcome this problem, temper rolling is carried out which has a 0.5 to 1.5% reduction capacity.

19.4 Drawing

Drawing is a plastic deformation process and is used mainly for reducing the diameter of bars, wires or tubes by pulling them through a die (Fig. 19.3). It can be used in both hot and cold forms. To control the inside diameter during tube drawing, usually a mandrel is used (Fig. 19.4). Tube drawing can also be performed without mandrel using a round die,

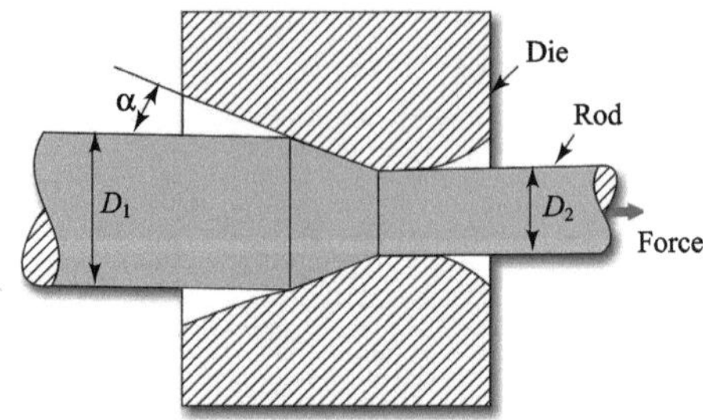

Fig. 19.3 A drawing operation

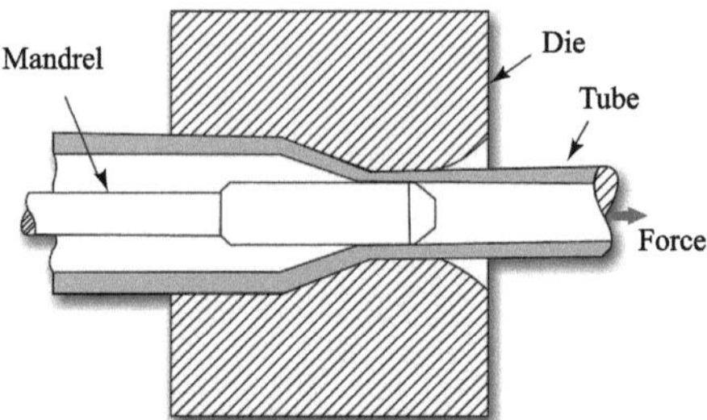

Fig. 19.4 Tube drawing using a mandrel

and the process is called sinking. Drawing is done on drawbench. One end of the material to be drawn is reduced by swaging to allow it to pass through the die, and this reduced portion is attached to the drawing mechanism, which can be chain driven.

In large reductions, the operation may be performed in a number of passes with annealing in between to restore the ductility so that further deformation can take place. The reduction in area is usually restricted to between 20 and 50% since higher reduction will require higher tensile forces that end to exceed the tensile strength of the reduced product, which is not desirable.

The drawing force required to pull a rod or wire is found to be

$$\frac{\sigma_{t_2}}{\sigma_Y} = \left(\frac{1+c}{c}\right)\left[1 - \left(\frac{D_2}{D_1}\right)^{2C}\right] \tag{19.6}$$

where

σ_{t_2} = tensile drawing stress corresponding to tensile force at the outlet
σ_Y = yield stress of the work material
c = $\mu \cot \alpha$
μ = coefficient of friction between contact surfaces

α = semi-die angle
D_1 = diameter at the inlet
D_2 = diameter at the outlet

The drawing force F may be expressed as:

$$F = \sigma_{t_2} \times \frac{\pi}{4} D_2^2 \tag{19.7}$$

Die materials for drawing are usually tool steels and carbides. Diamond dies are used for drawing fine wire with diameters ranging from 2μm to 1.5 mm.

Proper design of die and use of lubricants are very much important for obtaining a product with good quality and surface finish, dimensional accuracy and strength.

19.5 Deep Drawing

Tube or rod drawing is essentially different from deep drawing, in which sheet metal is given hollow shapes. Deep drawing is also known as cup or radial drawing because of its ability to produce cup-shaped objects. The depth produced is usually more than the diameter, although shallow parts can also be produced by using this method. Typical products made out of deep drawing include domestic pots, pans, food containers and automobile fuel tanks.

In this process, a flat sheet metal blank kept under a blankholder is forced into a die cavity by means of a punch (Fig. 19.5). The force on blankholder should be such that it allows material to slide into the die cavity, but must be great enough to prevent wrinkling of the sheet as it is drawn in. The force on the blank is given through the punch. The punch transmits the force through the walls of the cup as the flange being drawn into the die cavity. As the punch forms the cup, the amount of material in the flange decrease.

The deformation produced is measured by limiting drawing ratio (LDR) defined as:

$$LDR = \frac{\text{maximum blank diameter}}{\text{punch diameter}} = \frac{D_b}{D_p}$$

It is sometimes also expressed as percentage reduction, given by

$$\%\ \text{reduction} = \frac{D_b - D_p}{D_b} \times 100$$

A percentage reduction of 48% is considered excellent on the first draw. Succeeding draws are smaller. There should be no appreciable change in the thickness of the material between the blank and the finished part.

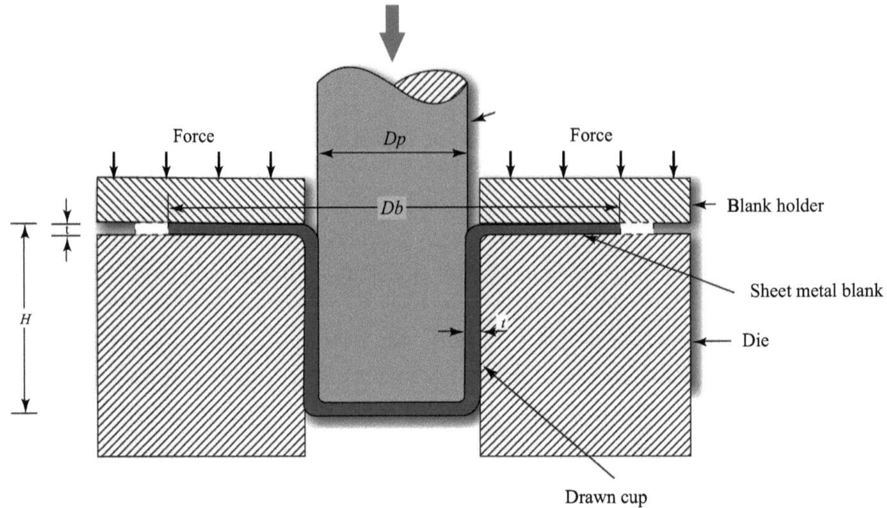

Fig. 19.5 A deep drawing operation

Results of deep drawing are mostly empirical in nature, and research has been confined almost exclusively to the drawing of cylindrical cups. For other shapes theoretical analysis is too much complicated and has no practical significance.

The diameter of the blank required to draw a given cup may be obtained approximately by equating surface areas.

$$\frac{\pi}{4}D_b^2 = \frac{\pi}{4}D_p^2 + \pi D_p H$$

$$D_b = \sqrt{\left(D_p^2 + 4D_p H\right)} \tag{19.8}$$

where H = height of the cup

The calculation is based on inside diameter of the sup drawn which equals the punch diameter.

The maximum punch load can be empirically approximated to

$$P_{\max} = \sigma_t \pi D_p t \tag{19.9}$$

where

σ_t = ultimate tensile strength of the sheet metal
t = blank thickness

The above equation is valid for a drawing ratio of 2 and assuming that cup tends to fail by tearing of the wall near the bottom when drawing ratio exceeds 2; and drawing load increases approximately linearly with respect to the drawing ratio.

For other drawing ratios, the punch load is given as:

$$P = \sigma_t \pi t (D_b - D_p) \tag{19.10}$$

Because of the presence of friction between the cup and die wall, punch load needs to be increased usually by 30%. Considering this factor, the equation (19.10) modifies to

$$P = 1.3\sigma_t \pi t (D_b - D_p) \tag{19.11}$$

The equation for blank diameter and punch load is based on the assumption that during the process metal thickness does not change and clearance between punch and die is about 5% larger than the blank thickness.

The failure in deep drawing generally results from thinning of the cup walls under high longitudinal tensile stresses.

19.5.1 Ironing

If the gap between punch and die is large, the drawn cup will have nonuniform wall thickness. In ironing, the gap is made less than the original wall thickness of the incoming material so that the cup wall is thinned and elongated simultaneously but uniformly. The thickness of the base remains unchanged (Fig. 19.6).

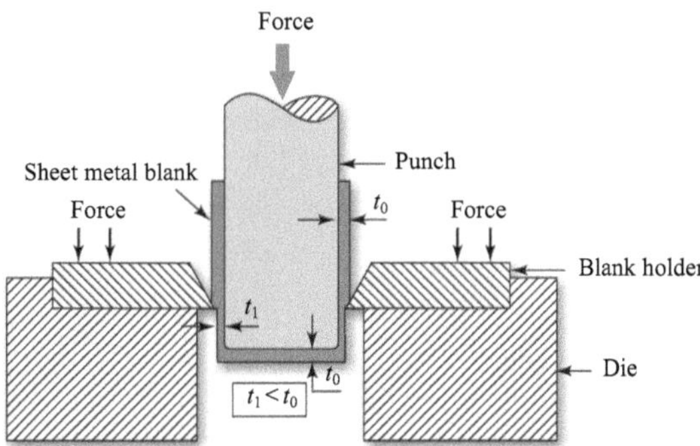

Fig. 19.6 An ironing operation

19.5.2 Redrawing

When it is difficult to obtain the desired depth of cup in one draw, then redrawing is performed (Fig. 19.7). Direct or reverse redrawing methods can be used.

Conventional or direct redrawing is shown in Fig. 19.8 (*a*). The pressure pad prevents wrinkling of the metal. In reverse redrawing, the cup is subjected to bending in the direction opposite to its original bending configuration as shown in Fig. 19.8 (*b*). It has the advantage of working the material more uniformly than direct redrawing.

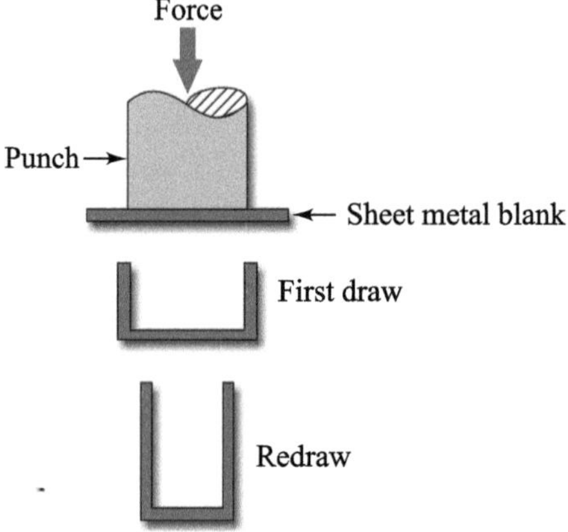

Fig. 19.7 A redrawing operation

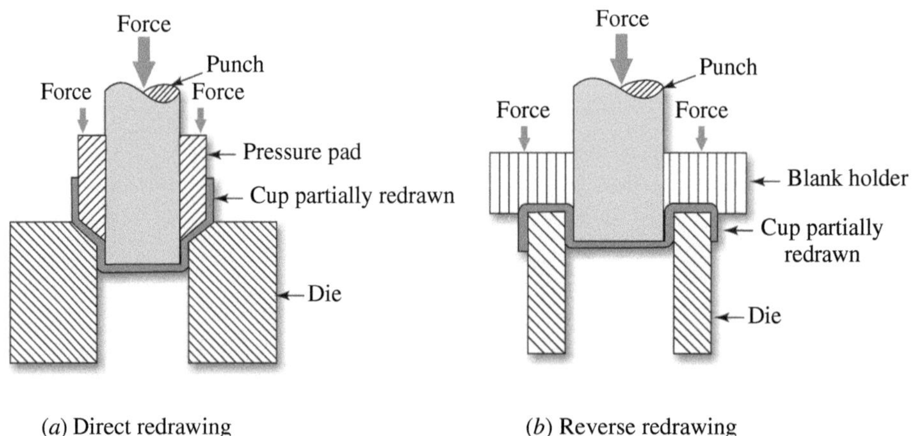

Fig. 19.8 Direct and reverse redrawing

19.6 Embossing

Embossing produces impressions on metal sheets. The thickness remains unchanged during the process and one side is the reverse of the other. The mating die conforms to the same configuration as the punch (Fig. 19.9). Embossing finds application in making nameplates, medals or aesthetic designs on thin sheet metal or foil.

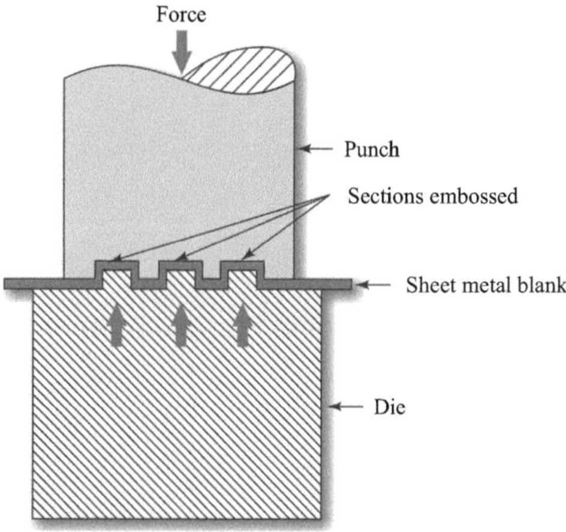

Fig. 19.9 An embossing operation

19.7 Bending

Bending produces plastic deformation of metal about its neutral axis where compression occurs on one side and tension on another side (Fig. 19.10). Both tensile and compressive stresses are below the ultimate strength of the material. During the process, position of the neutral axis changes; it shifts more towards compression side. Bars, rods, wires, tubing and structural shapes as well as sheet metal are bent to many shapes in cold conditions through dies.

Folder can be used for simple bending of thin sheet where angle of bend can be very large. Bends in heavier sheet or more complex bends in thin material are generally made on press brakes and are the commercial method of production.

The bending force F required to bend a sheet metal is empirically approximated to

$$F = \frac{K \cdot \sigma_u L \cdot t^2}{W} \tag{19.12}$$

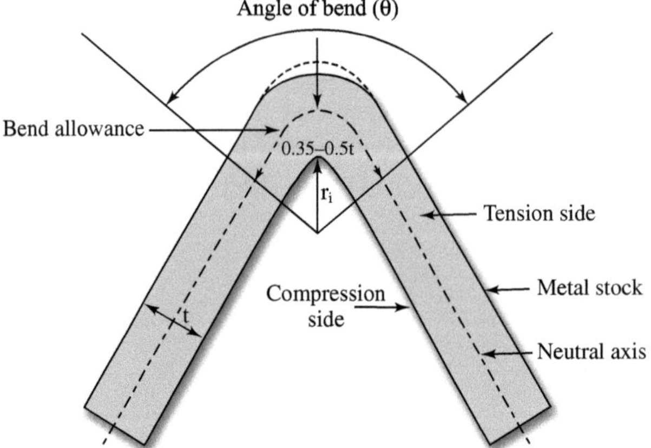

Fig. 19.10 Principle of bending operation

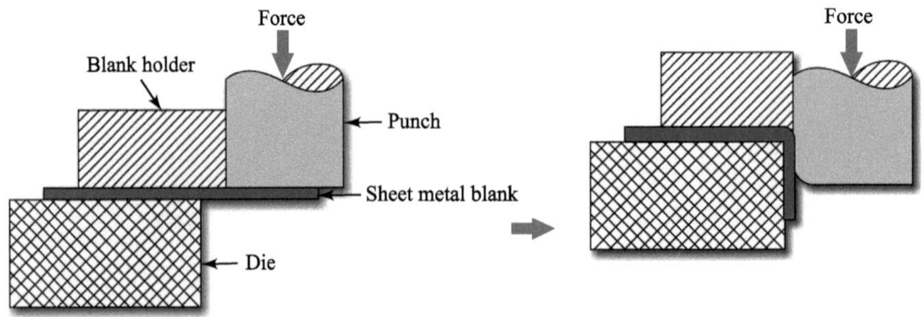

Fig. 19.11 Bending produced by a wiping die

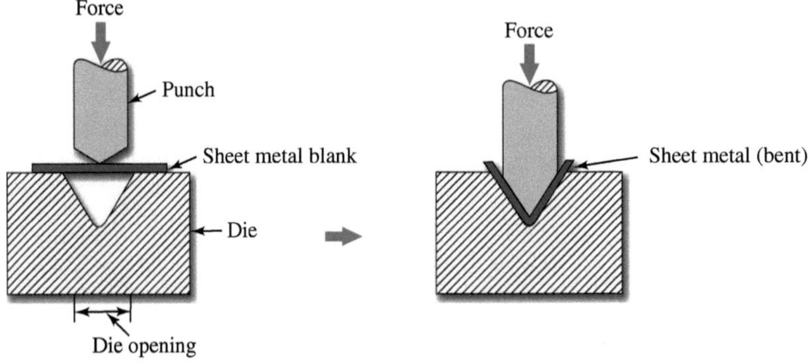

Fig. 19.12 Bending produced by using a V-die

where

σ_u = ultimate tensile strength of the work material
L = length of the bend
t = metal thickness
W = width of V-die or U-die, known as die opening
= length between punch and die radii inclusive of sheet thickness in case of a wiping die
K = constant
= 0.3 for wiping die (Fig. 19.11)
= 0.7 for a U-die
= 1.33 for a V-die (Fig. 19.12)

Thickness of the metal is slightly decreased at the point of bending, i.e., where the bending load acts (Fig. 19.10).

The bend allowance is defined as the arc length of the bend and is given by

$$L = \frac{\theta}{360} 2\pi (r_i + Kt) \tag{19.13}$$

where

L = bend allowance (inch)
θ = angle of bend (degree)
r_i = inside radius of bend (inch)
t = metal thickness (inch)
K = constant
= 0.33for $r_i \leq$ 2t
= 0.50for $r_i >$ 2t

The bent metal possesses some elasticity, which it tries to recover when the punch load is removed. This phenomenon is known as spring back. The final bend angle after spring back is smaller, and the final bend radius is larger than before (Fig. 19.13).

The bend radius due to spring back can be approximately equated to

$$r_2 = \frac{r_1}{\left[4\left(\frac{r_1 \sigma_y}{Et}\right)^3 - 3\left(\frac{r_1 \sigma_y}{Et}\right) + 1\right]} \tag{19.14}$$

where

r_1 = bend radius before spring back
r_2 = bend radius after spring back
E = modulus of elasticity of the work material
σ_y = yield stress of the work material

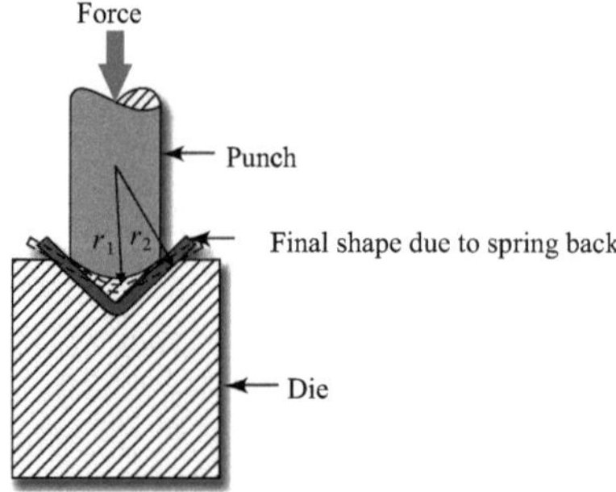

Fig. 19.13 Spring back effect

19.7.1 Roll Bending

Bending of a metal sheet or plate into particular curvature is commonly performed by roll bending. It is also known as roll forming. The process consists of passing the metal sheet through a series of three rolls of equal diameter, two of them are fixed in a particular position and the third one is adjustable, which controls the degree of curvature (Fig. 19.14).

Continuous roll forming is used for bending continuous lengths of sheet metal. It is a fast process and is used for mass production work. The process uses a series of forming rolls to gradually change the shape of the metal. The number of rolls to be used in the process is decided by the intricacy of the shape, the size of the section, the thickness and the type of material used. The rolls are generally made of carbon steel or grey iron and may be chromium plated to increase their wear resistance. Typical products include channels, gutters, panels and frames.

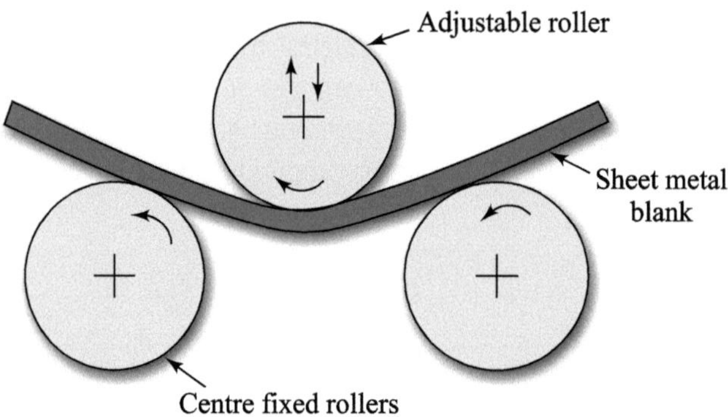

Fig. 19.14 A roll forming operation

19.8 Stretch Forming

Stretch forming consists of stretching the sheet metal gripped by two or more sets of jaws that stretch it and wrap it around a form block (Fig. 19.15). The process requires the metal to be stretched beyond its yield strength but less than its tensile strength. It is very useful in making prototype models of aircraft and automobile parts by using aluminium or stainless steel sheet because of its ability to produce compound curves in sheet stock. It is also used for producing large panels for truck and trailor bodies.

The stretching pressure P required during the process is empirically found to be

$$P = 1.25\sigma_y A(lb) \tag{19.15}$$

where

σ_y = yield strength of the work metal (psi)
A = cross-sectional area of the blank (inch^2)

In most operations, the blank is a rectangular sheet.

Since the metal is deformed by tensile stretching, there is less tendency to spring back. Thinning and strain hardening are inherent in the process but the thickness reduction should not exceed 5% of the original thickness. The form blocks are generally made of zinc alloys, steel, plastics or wood.

A combination of stretch and draw forming is shown in Fig. 19.16. It seems to be similar to deep drawing but differs in the sense that the ends of the stock are gripped firmly between the blank holder and the die.

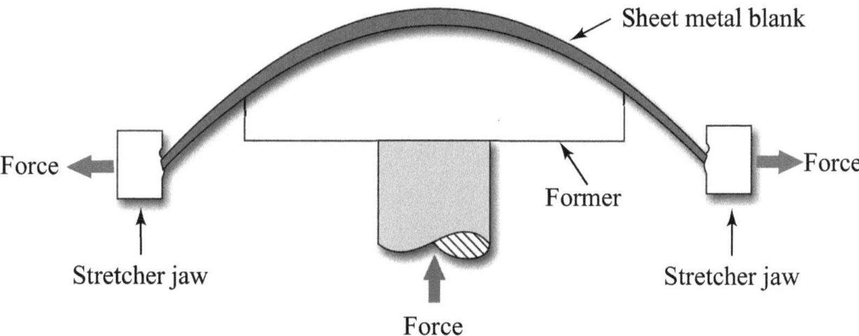

Fig. 19.15 A stretch forming operation

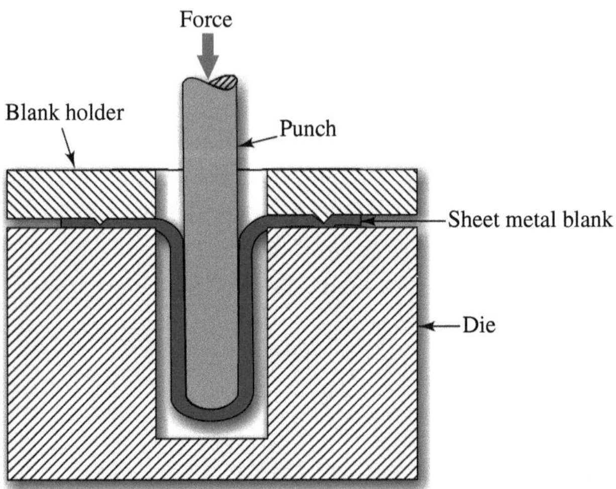

Fig. 19.16 Combination of stretch and draw forming operation

19.9 Forging

Forging is used to produce the desired shape in a material by using compressive forces, obtained from impact blows or steady press or a combination of the two. It is the oldest known metal-working process. Typical forged products include bolts and rivets, connecting rods, shafts for turbines, etc. The process may be performed at room temperature or at increased temperature. Accordingly, forging processes can be categorized as cold forging or hot forging. Warm forging is an intermediate stage and is carried out at a moderate temperature. Cold forging requires greater forces while hot forging smaller forces because the metals in hot state are relatively soft. But surface finish and dimensional accuracy of cold-forged products are superior to hot-forged products.

In general, forging offers the following advantages:

- Improves mechanical properties of the work material because of grain alignment, and as such the forged parts have increased strength, ductility, toughness and resistance to fatigue impact, shock and vibration.
- Imparts directional properties in the forged parts because of grain alignment in a particular direction.
- Provides close dimensional tolerances to the forged parts.
- Eliminates the wastage of materials as the forged parts are very close to the final shape.
- Improves weldability, and as such forged parts can be easily welded.

But forging also suffers from some of the following limitations:

- Irregular or intricate shapes which are very easy to produce in casting, cannot be produced by forging.
- Forging operations involve higher costs because of expensive forging dies.

There are many types of forging operations. Some of them are discussed below.

19.9.1 Hammer or Smith Forging

Hammer forging is the oldest of all the forging processes and was originally performed by a village blacksmith, hence the name smith forging. It still exists in practice. Here the heated workpiece is striked repeatedly with a hammer in order to give it a desired shape. The process is not accurate, and complex shapes cannot be produced. To improve and make the process faster, dies and semiautomatic hammer can be used. Accuracy of the process largely depends on the skill of the operator. The process can be performed in a closed or open die.

19.9.2 Open-die Forging

In open-die forging, the heated workpiece is placed between two flat dies (open) and then hammered to produce the desired shape (Fig. 19.17). It does not confine the flow of metal. It is a simple but a slow process. The process is also performed by using presses in addition to hammer (drop) mechanism. As the workpiece is hammered or pressed, it is repeatedly manipulated between the dies until the final shape is achieved. Open-die forged parts are usually rough and require machining. The process is mainly used to preshape the metal in preparation for further operations.

The force required during the open-die forging of a solid cylindrical piece is found to be

$$F = \sigma_y \pi r^2 \left(1 + \frac{2\mu r}{3l}\right) \tag{19.16}$$

where

F = forging force
σ_y = yield stress of work material
r = radius of the workpiece
l = length of the workpiece
μ = coefficient of friction between contact surfaces

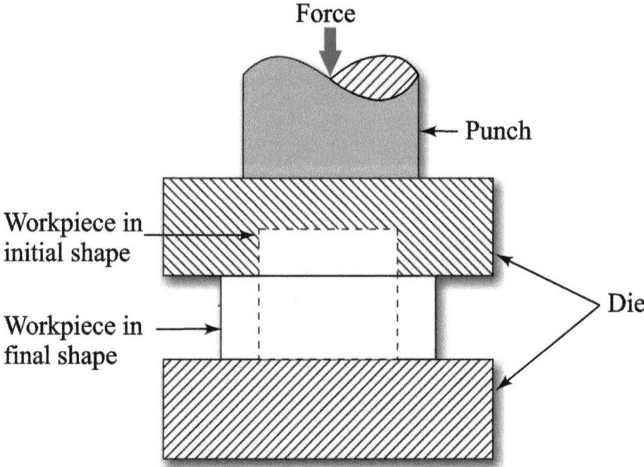

Fig. 19.17 An open-die forging process

19.9.3 Impression-Die and Closed-Die Forging

In impression-die forging, the heated workpiece is placed between two shaped dies and hammered or pressed. During hammering or pressing, the workpiece takes the die shape (Fig. 19.18). A small amount of material is forced outside the die impression, forming flash which is finally machined. The flash has an important role. It helps to build up pressure on the material between the dies which ensures the filling of die cavity.

Closed-die forging is the variation of impression-die forging and uses no flash. It has better utilization of material than open flat dies, better physical properties, closer tolerances, higher production rates and less operator skill. Right quantity as well as proper flow control of material between the dies are important for closed-die forging.

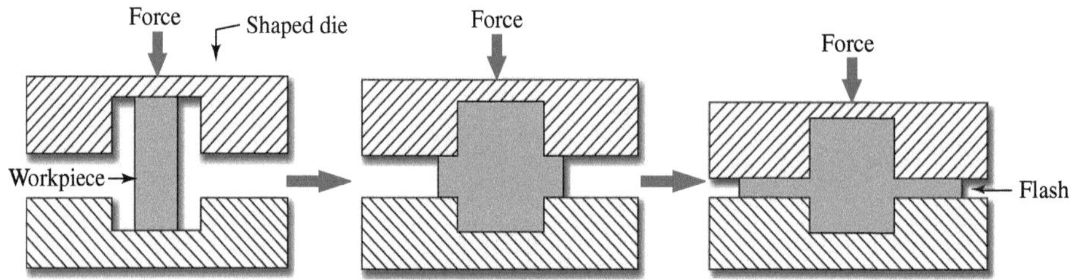

Fig. 19.18 Steps in closed-die forging process

19.9.4 Drop Forging

Drop forging uses matched dies with closed impression. The forging is produced by impact which forces the heated metal to conform to the die shape. To ensure proper flow of the metal during the intermittent blows, the operation is divided into a number of steps. Each step changes the shape of the workpiece progressively. For complex forgings more than one set of dies may be required.

19.9.5 Press Forging

Press forging involves a slow squeezing action produced by mechanically or hydraulically operated presses as compared to rapid impact blows of a hammer in drop forging. The slow squeezing action penetrates completely through the metal, producing a more uniform deformation and flow of the metal during the process. Open or closed dies may be used. Heated dies are usually used to reduce heat loss, promote surface flow and enable the production of finer details and closer tolerances. The work may be carried out hot or cold. Depending on the complexity of the workpiece, a set of dies may be required to obtain the final product.

Press forging has many advantages over drop forging. In press forging, a greater proportion of the total energy input is transmitted to the workpiece than in drop forging where much of the impact of the drop hammer is absorbed by the machine and foundation. Press forging gives higher dimensional accuracy and can be performed in a single operation.

19.9.6 Upset Forging

In upset forging, only a portion of the work material is deformed in a die, while the remaining part remains unaffected. The starting stock is usually a wire, rod or bar. The process is used to form heads on fasteners such as bolts, screws, rivets, nails and to shape valves, couplings and many other small components. Upset forging generally employs split dies that contain multiple positions or cavities and may be open and closed type. (Fig. 19.19). The heated bar or rod is positioned in the die and clamped. A hydraulic ram moves longitudinally against the bar, upsetting it into the die cavity.

In open upset forging, the unsupported length l of the rod does not exceed $3d$ to prevent its buckling, d being rod diameter. If $l > 3d$, then closed upset forging is preferred with die diameter $D \leq 1.5\ d$.

The maximum length of the stock that can be upset is empirically expressed as:

$$L = \frac{KP}{\pi} \tag{19.17}$$

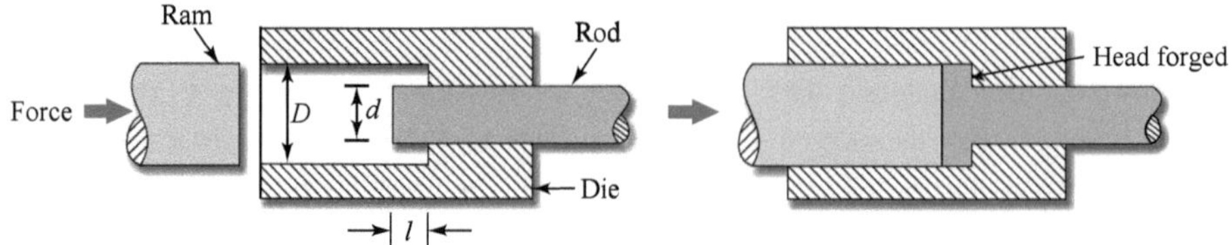

(*a*) Closed upset forging

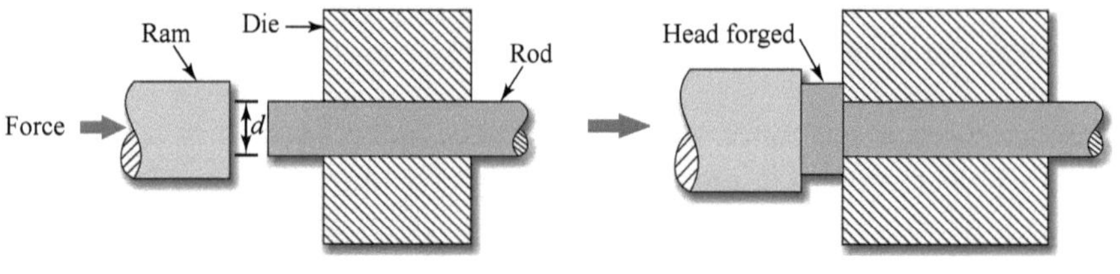

(*b*) Open upset forging

Fig. 19.19 Closed upset and open upset forging

where

L = maximum length of the stock to be upset (inch)
P = perimeter of cross section (inch)
K = constant (usual values 2 or 3)
= 2.6 for steel

19.9.7 Roll Forging

In roll forging, the cross section of round or flat bar stock is reduced or shaped at the cost of its length. A pair of cylindrical or semicylindrical rolls, each containing one or more shaped grooves according to the shaping required is used. The heated stock is placed between the rolls in an open position (Fig. 19.20). As the rolls rotate, the stock is gripped by the roll grooves and is progressively shaped. The stock is introduced repeatedly with 90° rotation between passes. Roll forging is used to produce tapered shafts, leaf springs, knives and hand tools.

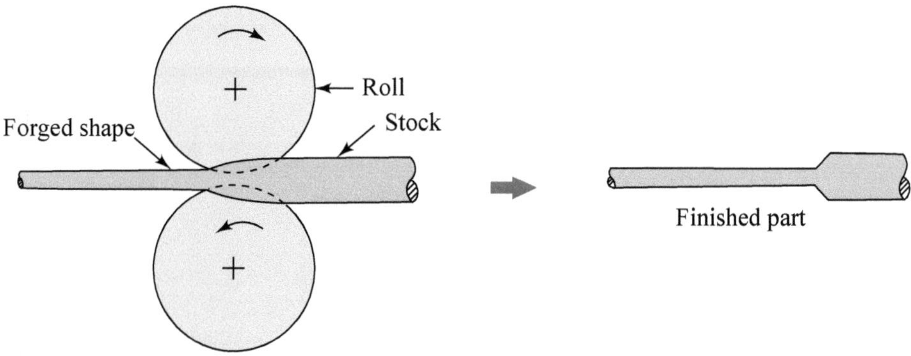

Fig. 19.20 A roll forging operation

19.9.8 Swaging

Swaging is also called rotary swaging or radial forging. It is generally used for reducing the diameters or tapering the end of bars and tubes through rotating dies which are set within a ring of rollers similar to a roller bearing (Fig. 19.21). The dies rotate about the workpiece and open and close rapidly. They exert radial impact forces on the workpiece. The die movements are obtained through the movement of rollers.

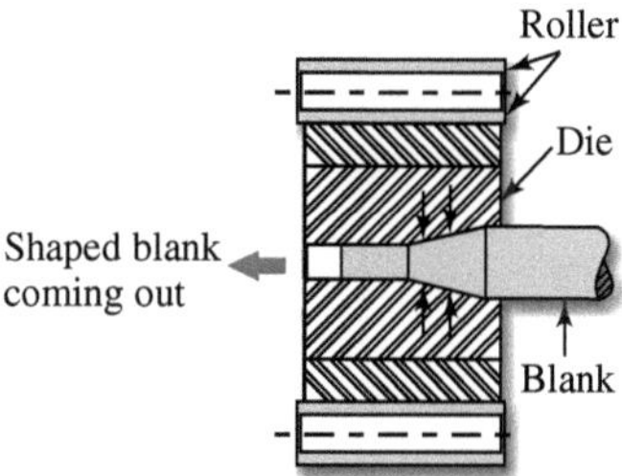

Fig. 19.21 A swaging operation

The swaging process can also be used to assemble fittings over cables and wires where fitting is swaged directly on the cable or reduce the ends of tubing to allow one tube to socket into the end of another.

19.9.9 Tube Swaging

Tube swaging is a variation of swaging in which thickness of the tube is controlled by means of a mandrel (Fig. 19.22). The change in the tube thickness is dependent on the

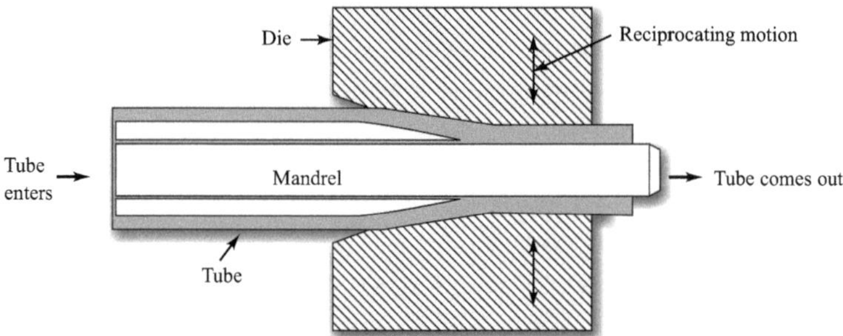

Fig. 19.22 Tube swaging using a mandrel

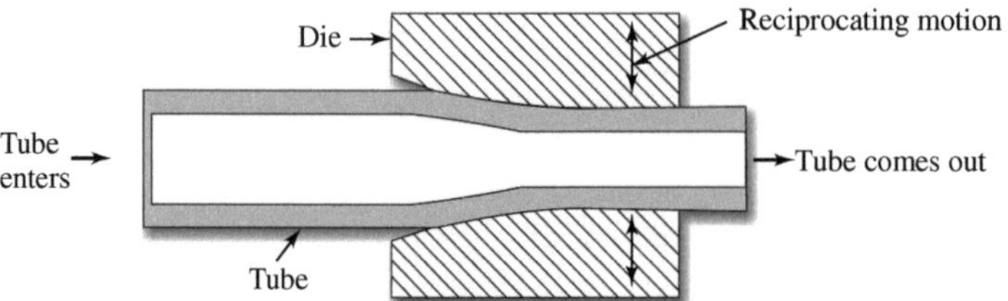

Fig. 19.23 Tube swaging without using a mandrel

mandrel diameter. Mandrel helps to achieve uniform thickness of the tube, although the process can also be performed without using mandrel (Fig. 19.23).

19.9.10 Precision Forging

Precision forging is employed to obtain a part which is very near to its final dimensions and that requires little or no additional finishing operations. Hence this process is also called net-shape forging or near-net-shape forging. Most conventional forgings require extensive machining because they cannot be pressed to finished dimensions. Finishing of the part is always a costly affair and there is a wastage of material. To minimize both the expense and waste, precision forging is employed.

Precision forged products have very less draft (between 0 and 1°) and smooth surfaces (3.175 μm or less). In this process, special dies are used. Aluminium and magnesium alloys are particularly suitable for precision forging because of their low forging load and temperature requirements.

19.9.11 Coining

Coining is a closed-die forging operation performed by die and punch at room temperature. The metal is confined in the die, and its lateral flow is prevented. The process is typically used in minting coins, medals, jewellery and other products where exact size and fine details are required. A single impact on the punch produces the desired design on each side of the piece (Fig. 19.24).

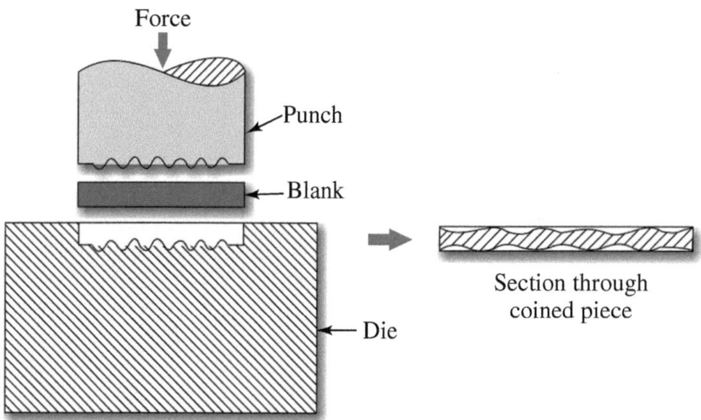

Fig. 19.24 A coining operation

19.9.12 Hubbing

Hubbing is used to produce multiple identical cavities economically which can be used during plastic moulding and die casting. In this process, a hardened steel form known as hob, which is exactly equal to the size of the piece to be moulded, is hydraulically pressed into the blank by a press ram (Fig. 19.25). Several pressings may be required for a satisfactory cavity. During the process, flow of metal in the blank is restrained from lateral movement by a heavy retaining ring around the block. The advantage of hubbing is that a number of identical cavities of good surface finish can be produced from a single hub.

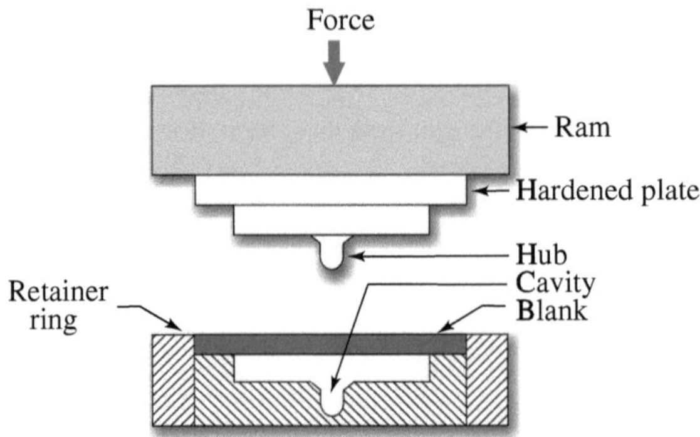

Fig. 19.25 A hubbing operation

19.9.13 Hot-Die Forging

Hot-die forging is also known as isothermal forging because both the workpiece and die are heated to the same temperature. Using hot dies, thinner sections with fewer die sequences can be achieved. Dimensional accuracy of isothermally forged parts is excellent, but the die life is short.

19.9.13.1 Piercing

Piercing is also called punching and it is a shearing operation. It is used to produce seamless tubing or to make holes in a metal sheet. To make a seamless tube, a heated cylindrical billet is passed between two conical-shaped piercing rolls which impart axial as well as rolling movement to the billet and force it over the mandrel. The alternate squeezing and bulging of the billet open up a seam in its centre, the size and shape of which are controlled by the piercing mandrel. The first pass makes a rather thick-walled tube, which is again passed over a tapered plug and through grooved rolls in a two-high mill where the thickness is decreased and the length increased. It is then passed through a reeler and sizing rolls to straighten it and reduce the diameter and/or wall thickness. Seamless tubes can also be expanded in diameter with thin walls by passing them over a larger mandrel.

Piercing differs from blanking (another shearing operation) in the end result part. In blanking, the piece being punched out becomes the workpiece (the useful product) and the remaining part is treated as waste. On the other hand, in piercing, the punched-out part is considered waste and the remaining part is the workpiece (Fig. 19.26).

The blanking and piercing operations are extensively used in the electrical and mechanical industries to produce parts for electric motors, electrical equipment, household appliances and so on.

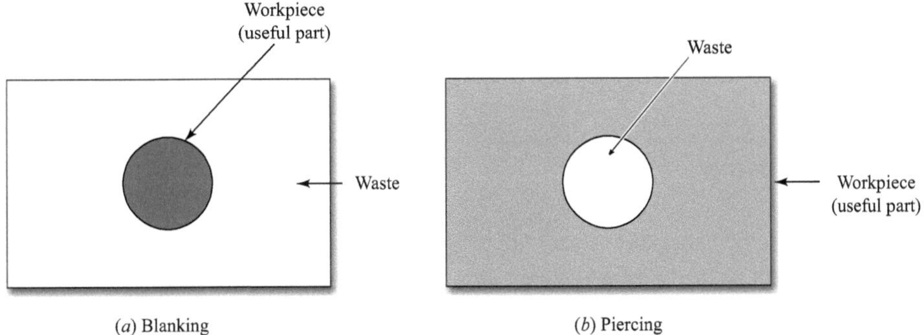

Fig. 19.26 Blanking and piercing operations

19.10 Forging Defects

Forging defects imply defects in the forged parts which can cause fatigue failure of the components. These defects may occur due to initial defects in the work material, incorrect design of forging dies that is responsible for nonuniform distribution of pressure to deform the work material in the forging die cavity, poor forging design, improper forging method, nonuniform heating and cooling of the work material that produces temperature gradients or microstructural changes in the work material because of phase transformation. Some defects appear on the surface of the forged parts and others occur internally. Following are the important forging defects.

- **Laps** develop on the surface of the forged parts in the form of small sharp cavities. They occur when thinner sections (webs) of the forged parts buckle due to insufficient metal flow in the die cavity during forging.
- **Scale pits** occur on the surface of the forged parts in the form of shallow surface depressions, which are produced when scale present in the die cavity gets attached to the surface of the part and creates pit on removal.
- **Cracks** may develop on the surface of the forged parts due to improper forging methods that do not apply uniform pressure. Internal cracks may develop in thicker sections (ribs) of the forged parts due to excessive metal coming out of the die cavity as flash during forging or due to nonuniform pressure distribution in the die cavity.
- **Unfilled sections** are produced when the work material does not spread uniformly to every section of the closed-die cavity because of low heating of the work material or poor die design.
- **Mismatched forgings** are produced because of nonalignment of the upper and lower forging die before the application of pressure.
- **Fins** and **rags** are projections or loose metal which are attached to the surface of the forged part.

- Burnt metal, overheated metal or decarburized steel occur because of excessive heating.
- Defects such as **abnormal structure**, **inclusions** or **seams** initially present in the original work material may remain included in the forged part.

19.11 Extrusion

Extrusion can be performed in either cold or hot state. Here the metal or alloy to be shaped is forced through shaped dies. Cold extrusion is carried out at room temperature. Because of the large forces required in extrusion, most metals are extruded in hot state. At the same time, strain hardening is also eliminated in hot extrusion. Hot extrusion is preferred when the metals lack ductility at room temperatures. Metals most easily extruded include lead, copper, aluminium, magnesium and their alloys. They have relatively low yield strengths and low hot working temperatures. Extrusion temperature for various materials is given in Table 19.1.

Extrusion is performed by means of high-capacity hydraulic presses which can shape the metals easily to the required shape, not possible by other methods. Longitudinal holes can be more easily produced by extrusion than by rolling. Billet-to-product cross-sectional area ratios can be as high as 100:1 for soft metals. Extrusion dies are relatively less expensive, and drastic reductions can be achieved in a single pass.

The continuous extrusion produces product of infinite length and complements and competes with drawing and shape rolling with respect to nonferrous products with small and uniform cross sections. Conventional extrusion is a discontinuous process of converting finite-length billets into finite-length products.

The pressure exerted by the ram on a solid cylindrical billet (Fig. 19.27) is found to be

$$P_x = \sigma_0\left[1.7\ln\frac{D_1}{D_2} + \frac{2X}{D_1}\right] \tag{19.18}$$

where

Table 19.1 Extrusion temperature for various materials

Materials	Temperature (°C)
Aluminium and its alloys	375–475
Copper and its alloys	650–975
Lead	200–250
Refractory alloys	975–2200
Steels	875–1300

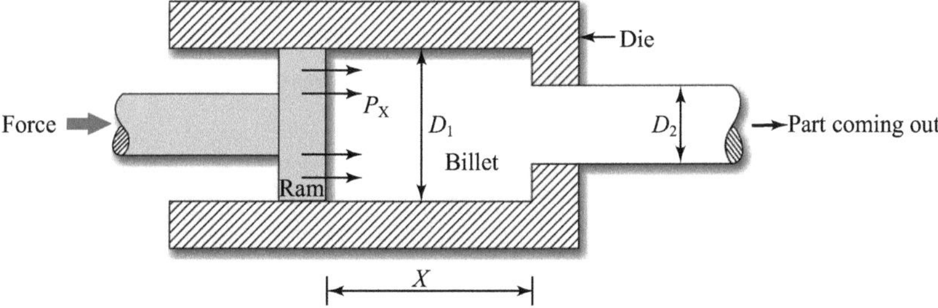

Fig. 19.27 An extrusion operation

P_x = pressure applied by the ram in x-direction (psi)
σ_0 = yield strength of the work material (psi)
D_1 = diameter of the billet before extrusion (inch)
D_2 = diameter of the extruded part (inch)
X = length of the billet to be extruded (inch)

The extrusion force is given by the following equation:

$$F = P_x \cdot \frac{\pi}{4} D_1^2 \tag{19.19}$$

19.11.1 Direct Extrusion

In direct or forward extrusion, metal confined in a chamber is forced through the die opening by applying pressure at the back of the billet by hydraulic ram (Fig. 19.28). The solid ram drives the entire billet out of the stationary die. The die opening is usually circular but can be of other shapes also. More force is required in this process since part of it goes wasted in overcoming friction between moving billet and the confining chamber.

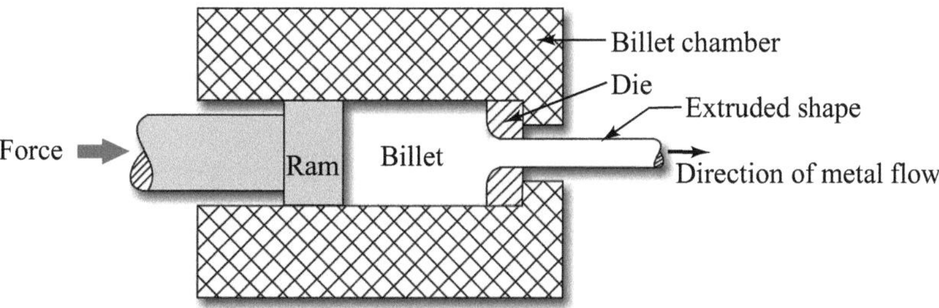

Fig. 19.28 A direct extrusion process

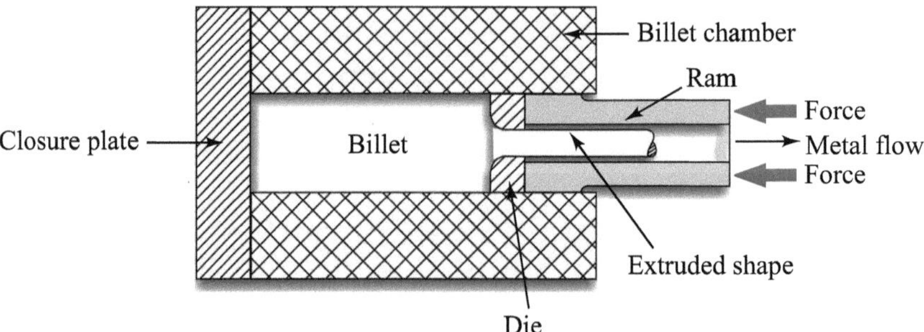

Fig. 19.29 An indirect extrusion process

19.11.2 Indirect Extrusion

In indirect or backward extrusion, the die is mounted on the face of a hollow ram and the material when forced comes out through the opening in the ram (Fig. 19.29). It requires less deformation force, when compared to direct extrusion. It is because of the absence of friction between billet and billet chamber.

19.11.3 Impact Extrusion

Impact extrusion is essentially a cold process which is mostly used for making collapsible medicine tubes, toothpaste tubes, shaving cream tubes and food cans from more ductile metals such as zinc, lead, tin and aluminium. During the process, the billet is placed in a die cavity and is given a strong single blow through the punch which causes the metal to flow plastically around the punch. The tube thickness is controlled by the clearance between the die and punch, and the outside diameter of the tube is equal to the die diameter. Impact extrusions are low in cost and have excellent surface finish. It is much simpler to produce a small deep canister in aluminium by impact extrusion than by deep drawing (Fig. 19.30).

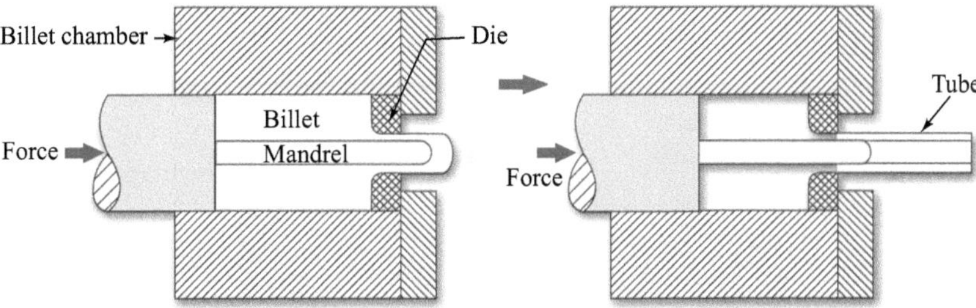

Fig. 19.30 An impact extrusion process

19.11.4 Hydrostatic Extrusion

It is based on the use of fluid pressure. The billet is completely submerged in the fluid which exerts pressure on it through a ram (Fig. 19.31). There is no physical contact between billet and its chamber. It is usually performed at room temperature.

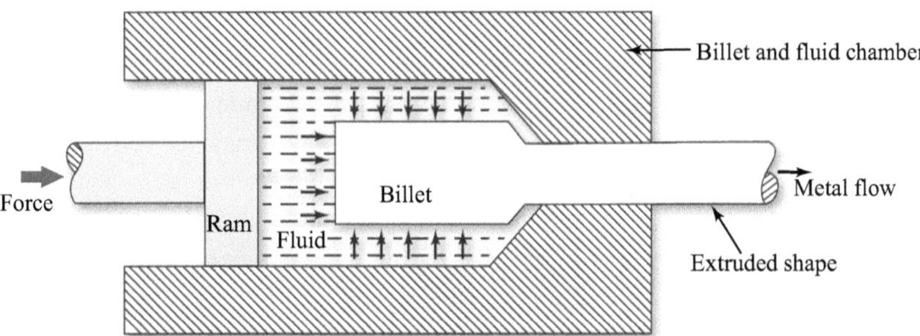

Fig. 19.31 A hydrostatic extrusion process

19.11.5 Tube Extrusion

Tube extrusion is used to produce tubular products with the help of a mandrel. During the process, the heated billet is placed in billet chamber and is pushed through a ram (Fig. 19.32). The pressure exerted by ram helps to flow the metal around the mandrel and come through the die opening. Because of additional tooling requirement, this process is more expensive than solid extrusion. It is an economical method of producing hollow sections.

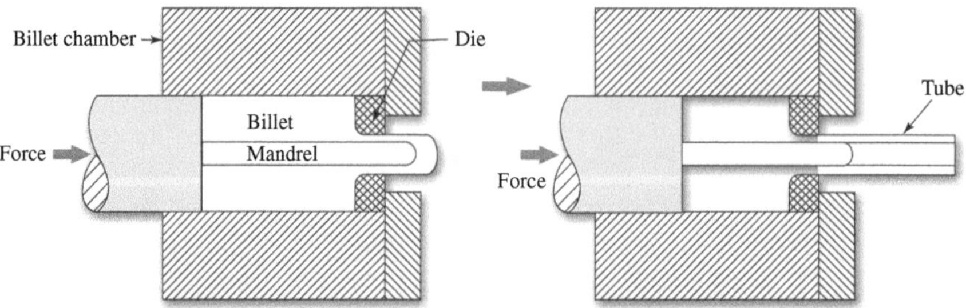

Fig. 19.32 A tube extrusion process

19.12 Spinning

Spinning is a cold-forming process of shaping sheet metal by forcing it against a former while it is rotating (Fig. 19.33). It is used to produce axisymmetric parts such as lamp reflectors, cooking utensils, bowls and bells of musical instruments from a single piece of metal which is usually of flat circular form. It resembles the process of forming clay on a potter's wheel.

The sheet metal blank is attached to the tailstock spindle while the former is attached to the main spindle of the lathe. As the spinning tool is gradually pressed against the rotating blank, it starts taking the desired form. The tools may be activated manually or by a computer-controlled hydraulic mechanism. Hand spinning uses hand tools. The process is usually performed at room temperature, although hot spinning also exists. The thickness of the finished part and that of the starting blank are essentially the same. Since

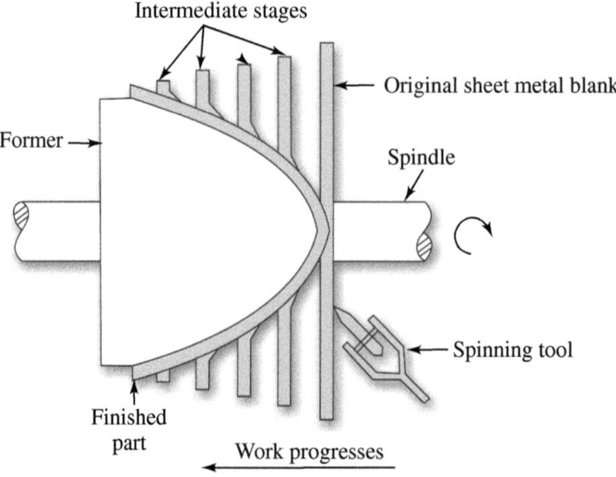

Fig. 19.33 A metal spinning operation

the former is not subjected to high pressure, it can be made of hardwood or even plastic. To reduce the friction of the spinning tool, lubricants such as soap, beeswax, white lead and linseed oil can be used.

19.12.1 Shear Spinning

Shear spinning, also called power spinning, uses power-driven roll formers in place of hand-operated spinning tools to force the sheet metal against the mandrel. Conical as well as cylindrical shapes can be formed by shear spinning. Fig. 19.34 shows the production of a conical shape from a flat metal sheet blank. The roll former forces the blank to conform to the shape of the mandrel. The metal flows entirely under shear action. Parts up to 3 m diameter can be formed. The blank thickness does not remain uniform but changes during the process and depends upon the angle of the particular region.

The final blank thickness t_f is given by the relation

$$t_f = t_o \sin \alpha \tag{19.20}$$

where

t_o = original blank thickness
α = semi-cone angle

Parts having semi-cone angle less than 30° require the operation to be performed in two stages with an intermediate anneal in between. Drastic reduction in wall thickness is

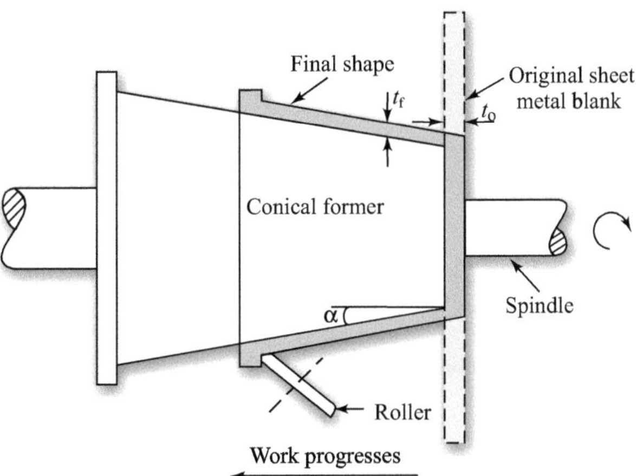

Fig. 19.34 A shear spinning operation

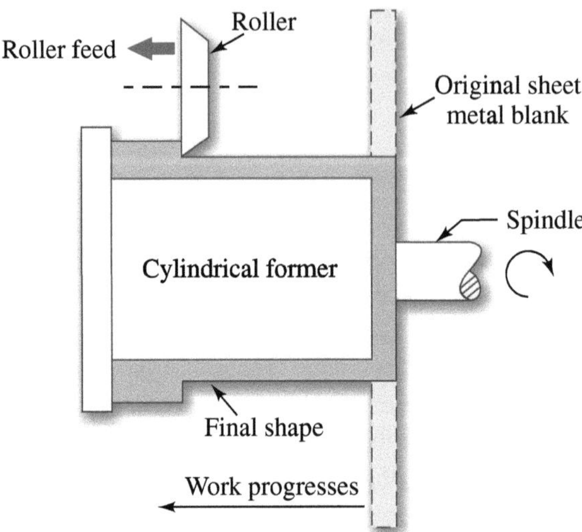

Fig. 19.35 A tube spinning operation

possible which can be one-fifth of the thickness of the original blank. During the process considerable heat is generated, hence usually water-based, are required.

19.12.2 Tube Spinning

Tube spinning is especially used to reduce the thickness of cylindrical parts by spinning them on a cylindrical mandrel (former) using rollers (Fig. 19.35). It uses preformed blanks. The process can reduce the tube thickness externally or internally. Typical products include pressure vessels, high-pressure gas cylinders and kitchenwares such as pots and pans. The parts have heavy bases but thinner walls.

19.13 Shearing

The shearing process involves a shear state of stress and is extensively used in industry to cut metal sheets and plates by placing them between two shearing blades or rolls with a clearance of 5–10 percent of the thickness of the work material.

Short Answer Questions

1. What is the difference between cold working and hot working operations? Give two examples of each operation.

Ans. Cold working operations are carried out below recrystallization temperatures (mostly room temperatures) and do not involve heating of the work materials. As such they require large deformation forces to deform the work material while shaping them, and the grain structures are distorted. Examples of cold working operations include drawing and rolling. Hot working operations, on the other hand, are performed above the recrystallization temperature of the work material. As a result they require small deformation forces, and the grain structures are refined. Examples of hot working operations include forging and extrusion.

2. What is strain hardening?

Ans. Strain hardening, also called work hardening, is a process in which a workpiece is worked upon in its cold state, and its strength increases because of strain produced in it.

3. What are recrystallization and recrystallization temperature? Why is it important?

Ans. Recrystallization is a process of formation of strain-free new grains, when a metal is subjected to increased temperatures. The temperature at which the process of recrystallization occurs is called recrystallization temperature, which is usually one-half of the melting temperature of the metal expressed in Kelvin. Heating of a material near its recrystallization temperature reduces force and power requirements needed for its shaping.

4. How does drawing differ from deep drawing?

Ans. Drawing is a process of reducing the diameter of bars, wires or tubes by pulling them through a die. Deep drawing, on the other hand, is a process to produce cup-shaped objects like domestic pots, pans, food containers and automobile fuel tanks from a sheet metal blank.

5. What is the difference between open-die forging and closed-die forging?

Ans. The work material is compressed between two flat open dies in open-die forging, whereas the work material is compressed between two shaped dies enclosing it completely in closed-die forging.

6. What is drop forging? What are its demerits?

Ans. Drop forging uses the impact energy to shape the work material. During drop forging most of the energy is taken by the anvil and other supporting devices, which goes as a waste and very few parts of the energy is utilized in shaping the work material.

7. What is the difference between direct extrusion and indirect extrusion?

Ans. In direct extrusion, the work material is forced from behind through a solid ram and it comes out through a die opening in the direction of motion of the ram. In indirect extrusion, the work material when forced comes out through a die opening in a direction opposite to the motion of the solid or hollow ram. Also there is more physical contact between the work material and the walls of the container in direct extrusion as compared to indirect extrusion, which results in increased friction and thus requiring more force.

8. What is the difference between a metal sheet and a plate?

Ans. Thickness of the metal sheet normally lies between 0.4 and 6 mm and that of plate exceeds 6 mm.

9. How does blanking differ from punching?

Ans. In blanking, the cut-out part, called blank, is a useful product and the remaining part is considered as a waste. In punching, the cut-out part, called slug, is a waste product and the remaining part is useful.

10. What is the difference between slotting and perforating?

Ans. Slotting is a punching operation to produce an elongated or rectangular hole. Perforating is also a punching operation to produce a number of similar size holes in a sheet metal blank.

11. What are coining and embossing? How do punch and die differ in both?

Ans. Both coining and embossing are sheet metal operations to make impressions or indentations on a sheet metal blank. Coining is used to produce raised sections in a part such as coin, whereas embossing is used to produce raised lettering as in case of name plates and medals. Punch and die used in embossing have essentially fitting cavity contours and the two sides of the embossed part are reverse of each other, whereas punch and die can have different cavity contours in coining.

12. What is roll bending? Name a few products made by roll bending.

Ans. Roll bending is a sheet metal operation to bend a sheet metal blank into the desired curvature by means of rolls. A few roll bended products include channels, gutters and structural sections such as frames.

13. What are the important characteristics of forged parts?

Ans. The forged parts have the following characteristics:

- They have directional properties because of grain alignments in a particular direction.
- They are close to final shape and do not require further machining, thereby preventing wastage of materials.
- They have improved strength and ductility, and are resistant to impact fatigue, shock and vibration.
- They have confined structures as their cavities are closed up by hammering.
- They have reduced cracks and blowholes.

Multiple Choice Questions

1. Hot working operations offer the following advantages:

    1. Deformation forces are reduced.
    2. There is no strain hardening.
    3. Mechanical properties are relatively unaffected.
    4. Problems such as oxidation and scaling are missing.

    Of these statements:

    (a) 1 and 2 are true (b) 2 and 3 are true
    (c) 1, 2 and 3 are true (d) 1, 2 and 4 are true

2. Recrystallization results in

    1. Grain refinement 2. Decrease in strength and hardness
    3. Increase in ductility 4. Relieve of internal stresses

    Of these:

    (a) 1 and 4 are true (b) 1 and 3 are true
    (c) 1, 2, 3 and 4 are true (d) 1, 3 and 4 are true

3. Recrystallization usually occurs at which of the following temperature?

    1. One-third of the melting point, expressed in absolute temperature, for pure metals
    2. One-half of the melting point, expressed in absolute temperature, for alloys
    3. One-third of the melting point, for pure metals
    4. One-half of the melting point, for alloys

    Of these:

    (a) 1 alone is true (b) 1 and 2 are true
    (c) 4 alone is true (d) 3 and 4 are true

4. The typical values of coefficient of friction for cold and hot rolling operations are respectively

(a) 0.2 and 0.3
(b) 0.1 and 0.3
(c) 0.1 and 0.4
(d) 0.2 and 0.4

5. Consider the following statements about the flat rolling:

   1. The friction force is more on the entrance side than on the exit side.
   2. The force is equal on both entrance and exit sides.
   3. Without friction, rolling is not possible and the work will slip rather than passing between the rolls.

   Of these statements:

(a) 1 alone is true
(b) 2 alone is true
(c) 1 and 3 are true
(d) 3 alone is true

6. In flat rolling, which of the following relationship is correct?

(a) $V_1 < V_r < V_2$
(b) $V_2 < V_r < V_1$
(c) $V_r < V_2 < V_1$
(d) $V_1 < V_2 < V_r$

V_1 = entrance velocity of the work
V_2 = exit velocity of the work
V_r = surface velocity of the rolls

7. The forward slip in rolling is expressed as:

(a) $\frac{V_2 - V_r}{V_2}$
(b) $\frac{V_2 - V_r}{V_r}$
(c) $\frac{V_r}{V_2 - V_r}$
(d) $\frac{V_1 - V_2}{V_r}$.

8. In a rolling process, roll separating force can be decreased by

   (a) reducing the roll diameter
   (b) increasing the roll diameter
   (c) providing back-up rolls
   (d) increasing the friction between the rolls and the metal.

9. In order to get uniform thickness of the plate by rolling, one provides

(a) camber on the rolls
(b) offset on the rolls
(c) hardening of the rolls
(d) antifriction bearings

10. Bar drawing depends on the following parameters:

(a) Die angle
(b) Ratio of initial and final cross-sectional areas of the stock
(c) Friction at stock-die interface
(d) All of the above.

11. For bar drawing, semi-die angle may vary between

(a) 0 and 5°
(b) 20 and 25°
(c) 6 and 20°
(d) 25 and 30°

12. Consider the following state of stress:

1. Compressive stress in the flange
2. Tensile stress in the wall
3. Tensile stress in the bottom part

13. Which of the following factors promotes the tendency for wrinkling in drawing operation?

(a) Increase in the ratio of thickness to blank diameter of work material
(b) Decrease in the ratio of thickness to blank diameter of work material
(c) Decrease in the holding force on the blank
(d) Use of solid lubricants.

14. Consider the following statements about tearing in a drawn part. It occurs

1. When the tensile stress in the flange is high.
2. When the tensile stress in the wall is high.
3. When the drawn part is pulled over a sharp die corner.
4. When the compressive stress in the base is high.

Of these:

(a) 1 and 2 are true
(b) 2 and 3 are true
(c) 4 alone is true
(d) 2 alone is true

15. The maximum reduction for first redraw is

(a) 15% (b) 20%
(c) 25% (d) 30%

16. The maximum reduction for second redraw is

(a) 10% (b) 16%
(c) 20% (d) 25%

17. Which of the following forging operations is also known as precision forging?

(a) Open-die forging (b) Impression-die forging
(c) Flashless forging (d) Roll forging

18. Which of the following manufacturing processes requires the provision of gutters?

(a) Closed-die forging (b) Centrifugal forging
(c) Investment casting (d) Impact extrusion

19. Which of the following processes is most commonly used for the forging of bolt heads of hexagonal shape?

(a) Closed-die drop forging (b) Open-die upset forging
(c) Closed-die press forging (d) Open-die progressive forging

20. The forging defect due to hindrance to smooth flow of metal in the component called 'Lap' occurs, because

(a) the corner radius provided is too large
(b) the corner radius provided is too small
(c) draft is not provided
(d) the shrinkage allowance is inadequate.

21. Which of the following metals is best suitable for extrusion either hot or cold?

(a) Zinc (b) Magnesium
(c) Copper (d) Aluminium

22. Which of the following materials requires lubricant when hot extruded?

(a) Magnesium (b) Zinc
(c) Tin (d) Steel

23. The following operations are performed while preparing the billets for extrusion process:
    1. Alkaline cleaning
    2. Phosphate coating
    3. Pickling
    4. Lubricating with reactive soap

    The correct sequence of these operations is

    (a) 3, 1, 4, 2 (b) 1, 3, 2, 4
    (c) 1, 3, 4, 2 (d) 3, 1, 2, 4

24. The mode of deformation of the metal during spinning is

    (a) Bending (b) Stretching
    (c) Rolling and stretching (d) Bending and stretching

25. Tube spinning can
    (a) reduce external wall thickness of a cylindrical part
    (b) reduce internal wall thickness of a cylindrical part
    (c) reduce both external and internal wall thicknesses of a cylindrical part
    (d) increase the internal wall thickness of a cylindrical part

Answers

1. (c)	2. (c)	3. (b)	4. (c)	5. (c)	6. (a)
7. (b)	8. (a)	9. (a)	10. (d)	11. (c)	12. (a)
13. (b)	14. (b)	15. (d)	16. (b)	17. (c)	18. (a)
19. (b)	20. (b)	21. (d)	22. (d)	23. (d)	24. (d)
25. (c)					

Review Questions and Discussions

Q.1. What is recrystallization temperature?
Q.2. What is the difference between cold working and hot working processes? Which one is more easier to employ and why?
Q.3. How is formability defined? What are the factors affecting formability?
Q.4. Which structure has high formability and why?
Q.5. Why are multiple passes usually required in wire-drawing operations?
Q.6. What is springback effect? How is it taken care in bending?
Q.7. Why does a metal become thinner in the region of a bend?

Q.8. Why are open-die forged parts made larger than closed-die forged parts?
Q.9. Why are heated dies generally used in hot press forging operations?
Q.10. What is flash and why is it undesirable?
Q.11. What is the difference between press forging and drop forging?
Q.12. What is swaging? For what type of products is it useful?
Q.13. Aluminium and tin are extruded in cold state. Why?
Q.14. Why is hydrostatic extrusion better than the conventional extrusion process?
Q.15. What is strain hardening? How does the strength of a material increase due to strain hardening?
Q.16. Why are pure metals more easily cold-worked than alloys?
Q.17. How is a hydraulic press different from a mechanical press?

Q.18. How is a seamless tube produced?
Q.19. What is the difference between punching and blanking?

Nonconventional Machining Operations 20

Some profiles produced by photochemical blanking.

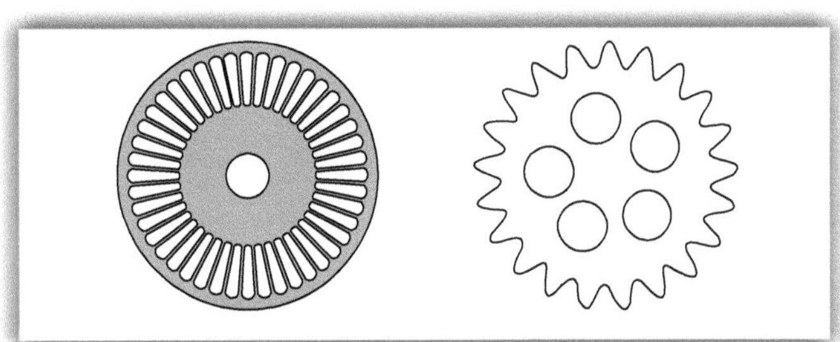

Learning Objectives

After reading this chapter, you will be able to answer some of the following questions.

- How dose chemical blanking differ from chemical milling?
- How dose water jet machining differ from abrasive water-jet machining?
- Why is vacuum required in electron beam machining?

D. K. Singh, *Fundamentals of Manufacturing Engineering*,
https://doi.org/10.1007/978-981-99-8767-2_20

20.1 Introduction

Nonconventional machining methods remove materials without chip formation and are generally nonmechnical. They do not use sharp cutting tools. Such processes offer considerable technical and economic advantages over traditional machining methods. They can be used to machine hard materials as well as difficult-to-machine metals. Complex internal and external profiles are easily produced by using such methods.

Nonconventional machining processes have the following features:

- Mechanical contact with the workpiece is missing.
- It is chipless and stress-free machining operation.
- Hardness of the tool is immaterial.
- It gives better surface finish and close tolerances.
- Very hard and fragile materials can be easily machined.
- Complex shapes are easily produced.

On the negative sides, these methods are costly. Their specific power consumption is very high. Aerospace and nuclear industries are the major users of nonconventional machining methods.

20.2 Chemical Machining

Chemical machining is the oldest of the nontraditional machining methods. Earlier it was used for engraving metals and hard stones. The scope of the process has widened to include the production of printed circuit boards and microprocessor chips.

In this process, the part to be machined is dissolved in aqueous solutions of salt, alkalies or acids such as NaOH, KOH, NH_4OH, HNO_3, H_2SO_4 and NaCl. These solutions are termed as etchants. The etchant corrodes the part and removes a layer of material either from its entire surface or selected portions. If the material is removed selectively, then the remaining portion needs to be protected by a thick layer of maskant such as wax, paint or polymer. No masking is required for uniform weight reduction. Complex shapes can be produced by repeating masking and machining alternatively several times until the desired shape is achieved.

Metal in chemical machining may be removed either by electro-etching or electroless-etching.

In electro-etching method, the metal removed from the part gets deposited on the electrode during the passage of current through the electrolytic solution. On the other hand, in electroless-etching, the metal removed from the part is chemically converted into metallic salt to be carried away during the replacement of the etchant. The metal removal strongly depends on the length of time during which the part is immersed in electrolytic solution. The process is shown in Fig. 20.1.

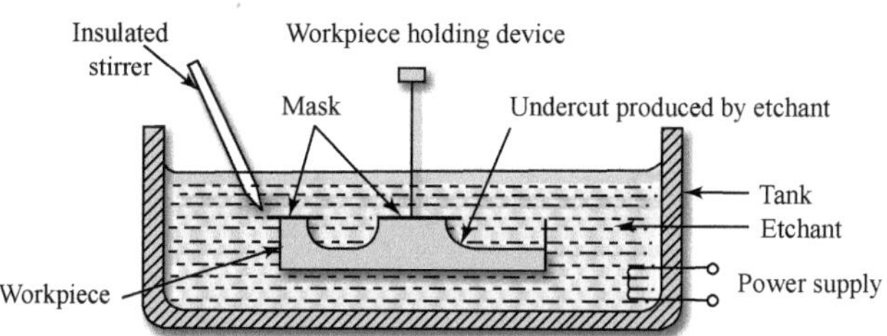

Fig. 20.1 A chemical machining process

20.2.1 Advantages

- Chemical machining is very useful and economical for weight reduction.
- The structure of the metal remains intact and is free of internal stresses in contrast to conventional machining.
- Very thin parts with close dimensional tolerances can be produced. Also, the parts have goods surface finish.
- Equipment cost as well as operational cost are lower compared to conventional machining methods.
- Very intricate shapes be it external or internal, which are otherwise difficult to produce, can be easily produced.

20.2.2 Limitations

- Because of limited depth of cut, holes and sharp corners are difficult to produce.
- The process is slow on account of slow action of the etchant.
- Local environmental pollution problem may arise due to gases generated in the process.
- Material removal is not uniform in case of cast or welded parts because of porosity and differential grain structures in their different sections.
- Undercutting is an inherent problem where sides are eroded by the etchant. The ratio of side penetration to depth of cut is called etch factor (Fig. 20.2).

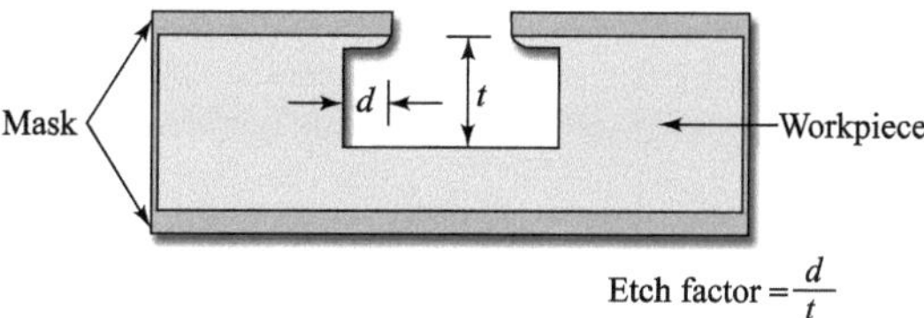

Fig. 20.2 Etch factor in chemical machining

20.2.3 Chemical Milling

Chemical milling is mainly used in the removal of excess metal to reduce the weight of aircraft components in aircraft industry. It replaces the conventional milling by chemical action, hence is so named. It is used to remove the metal selectively and not thoroughly.

20.2.4 Chemical Blanking

Chemical blanking is fundamentally different from conventional blanking. In chemical blanking, the material is removed by chemical dissolution whereas in conventional blanking, material is removed by shearing. It is an electroless-etching process where work material is removed in the form of metallic salts. Parts manufactured by chemical blanking include printed circuit boards, decorative panels and thin sheet parts.

20.2.5 Photo Chemical Blanking

Photochemical blanking is the extension of chemical blanking, where work material is removed by using a photographic technique.

The process starts with the making of a photographic negative which contains the accurate image of the work part. The negative is used to contact-print the component images onto a metal sheet that is covered with a photographic resist coating. When exposed to ultraviolet light, the coating on the metal panel becomes a polymerized layer and acts as a barrier to the etching solution. Now the panel is developed in a spray to remove the coating, except in the areas of the workpiece that have been converted into etch-resistant images. After spraying, both sides of metal are washed and dried. The photoresist may or may not be removed from the parts. The sequence of operations is shown in Fig. 20.3. Some of the parts produced by photochemical blanking are shown in Fig. 20.4.

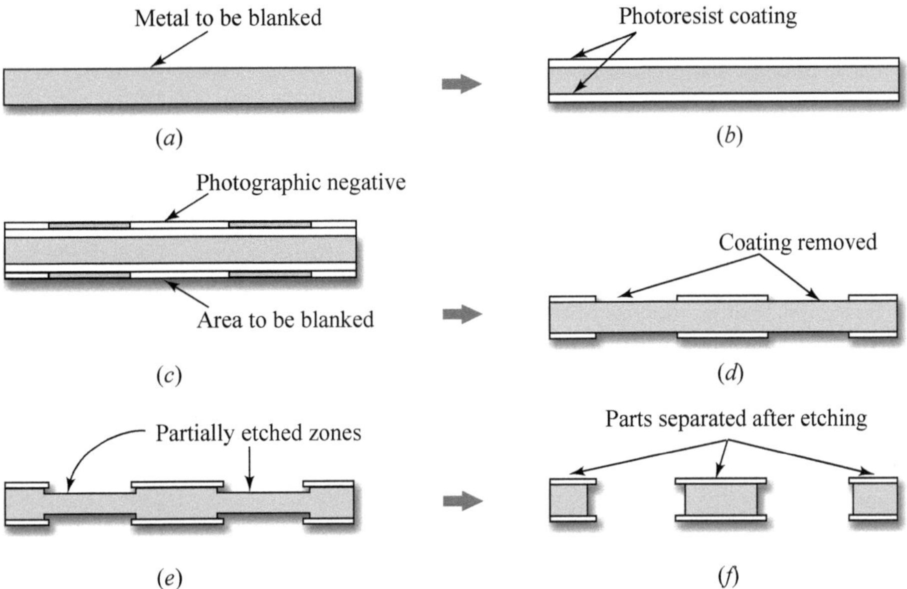

Fig. 20.3 Steps in photochemical blanking

Fig. 20.4 Some profiles produced by photochemical blanking

20.2.6 Advantages

- Extremely thin metal can be machined without distortion.
- Fragile parts and materials (e.g., hard and brittle), which are difficult to machine by conventional methods, are easily processed by photochemical blanking.
- It can produce very small parts.
- It produces burr-free and stress-free parts.

- It involves low equipment cost.
- Design flexibility offers any change in the part dimension.
- Close dimensional tolerances can be maintained.
- It can be easily automated.

20.2.7 Limitations

- The process is suitable only for very low metal thickness.
- It requires skill on the part of the operator.
- Since etchants are harmful chemicals and are also corrosive in nature, they require careful handling.
- Inherent undercutting problem puts limitations on the minimum size of the holes produced.

20.2.8 Suitability and Applications

The process is suitable for extremely thin metal (most blanking is under 1.5875 mm thick). Typical applications of photochemical blanking include fine screens, printed circuit cards, electric motor limitations and masks for colour television.

20.3 Electrical Discharge Machining (EDM)

Electrical discharge machining removes the metal from the workpiece by the spark erosion principle. The spark is produced between a contour-shaped electrode which acts as a tool and workpiece. The tool electrode is made cathode and the workpiece anode.

In this process, the workpiece and the electrode end is submerged in a dielectric fluid as shown as Fig. 20.5 (*a*). They are separated from each other by a well maintained gap, known as spark gap. This gap is also filled with the dielectric fluid. For maintaining the gap, servo control is used. The capacitor used in the circuit delivers the energy for the spark to be produced in the gap. As the condenser is energized, its potential rises rapidly to a value sufficient to overcome the dielectric fluid in the gap and thus producing the spark. The spark generated causes the melting and vaporization of the electrode materials and thus producing craters on both the electrodes as shown in Fig. 20.5 (*b*). Filters are used to remove metallic particles from the circulating dielectric fluid.

Equal amount of materials are not removed from both electrodes. It has been observed that if both the electrodes are made of the same material, the electrode connected to the positive terminal generally erodes at a faster rate. For this reason, the workpiece is normally made the anode. EDM usually uses graphite as tool electrode because it has low wear rate, high electrical conductivity and is relatively easy to machine. Copper alloyed

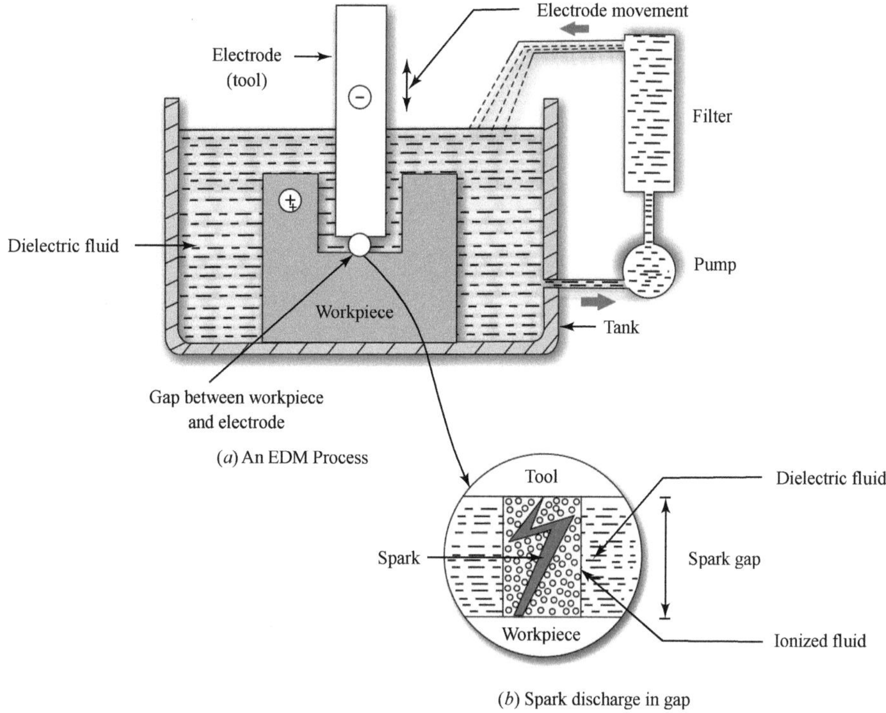

Fig. 20.5 An electrical discharge machining (EDM) process

electrodes give large metal removal rate for the same wear. Copper–tungsten electrodes are used for machining hard materials.

Kerosene oil, paraffin and mineral oil are important dielectric fluids of which kerosene oil is most widely used. The dielectric fluid has the following functions:

- It acts as an insulator between tool and work.
- It acts as a flushing medium and washes away particles of eroded metal from the workpiece or tool.
- It acts as a coolant and takes away the intense heat generated by the spark.
- It acts as a spark conductor.
- It prevents particles of workpiece adhering to the tool electrode and increases metal removal rate as compared with operating in air.

Metal erosion is more with liquid dielectric than air or gas.

Though the surfaces appear smooth, asperities and irregularities are always present. With increase in spark frequency, surface finish improves.

20.3.1 Calculations of Important Parameters

The voltage across the gap, known as discharge voltage, V_d is given as:

$$V_d = V_s\left[1 - e^{\frac{t}{RC}}\right] \quad \text{(volt)} \tag{20.1}$$

V_s = source voltage (volt)
R = resistance (ohm)
C = capacitance (coulomb/volt)
t = charging time (sec)

For maximum power delivery, the discharge voltage is usually three-fourth of the supply voltage.

Spark frequency is the number of sparks produced per sec, given by

$$f = \frac{1}{t} = \frac{1}{RC \cdot \log_e\left(\frac{V_s}{V_s - V_d}\right)} \tag{20.2}$$

The energy released per spark is given by

$$E = \frac{1}{2}C{V_d}^2 \tag{20.3}$$

The metal removal rate (MRR) can be expressed as:

$$MRR \simeq \frac{K.{V_d}^2}{2R\log_e\left[\frac{V_s}{V_s - V_d}\right]} \tag{20.4}$$

where K = constant of proportionality. It denotes the fraction of power effectively used in the material removal.

The *MRR* depends on discharge current, pulse duration and pulse frequency. High metal removal rate produces poor surface finish but low rate gives better finish. With increased gap, metal removal rate is more.

20.3.2 Advantages

- Fragile parts can be machined without distortion because no mechnical work is involved.
- Holes of desired size can be easily made in hard and brittle materials.

- The hardness of the cutting tool is immaterial and the tool need not be harder than the workpiece, because of no mechanical contact between tool and workpiece.
- Carbide tools and dies as well as complex shapes can be made economically.

20.3.3 Limitations

- Only electrically conducting materials can be machined. Materials like glass, ceramics or other nonconducting materials cannot be machined.
- It is a slow process compared to conventional methods.
- The surface finish is poor because of presence of small craters.
- Power consumption is more.

20.3.4 Suitability and Applications

- The process is most suited to the sinking of irregularly shaped holes, slots and cavities.
- EDM is extensively used in the manufacture of tools and dies.
- It is also used in the manufacture of hobs for gear cutting.

20.4 Travelling-Wire EDM

It is a variation of electrical discharge machining (EDM). The tool electrode used in this process is in the form of a continuously moving conductive wire which travels through the workpiece from a feed reel to a takeup reel. The spark is produced in the gap between the moving wire and the workpiece (Fig. 20.6).

The wire is usually made of brass, copper or tungsten and is fed at a constant velocity. Deionized water is used as a dielectric which gives faster cutting than oil and is also fire proof. But oil gives better surface finish and increased die life.

Computer numerical control (CNC) can be used to control the cutting path of the wire thereby increasing the accuracy of the work.

Advantages

- The tool electrode need not be shaped to the required contour of the workpiece.
- The travelling wire can produce very fine and intricate die openings.
- The surface finish is good.
- The process is economical over conventional EDM.
- Thicker parts have higher cutting rates.
- Geometric and dimensional tolerances are tight.

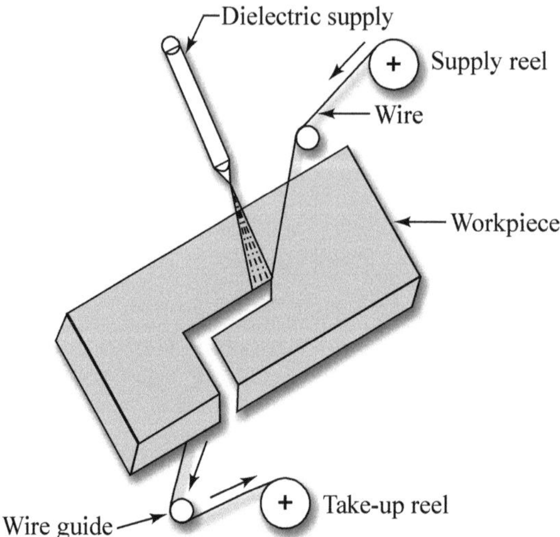

Fig. 20.6 A wire EDM process

20.4.1 Suitability and Applications

The process can be used for machining and cutting both. It is widely used for the manufacture of punches, dies and tools from hard materials. Dies to be used for several applications such as press stamping, extrusion, powder metallurgy etc. can be machined easily by using this method. It also finds application in the machining of precision parts to be used in electronic industry and for repair works.

20.5 Electrochemical Machining (ECM)

Electrochemical machining is the reverse of electroplating. The process uses electrolytic action to dissolve the workpiece metal and is based on the Faraday's law of electrolysis. It is basically a deplating process in which workpiece acts as an anode and the tool as a cathode. The process is shown in Fig. 20.7. The tool electrode is contour-shaped according to the workpiece. The electrode accuracy is important since the surface finish of the tool electrode is reproduced in the surface of the workpiece. The tool is fed gradually towards the workpiece. Tools are usually made of copper or brass and sometimes stainless steel.

The important electrolytes used in the process include NaCl, KCl and $NaNO_3$ of which NaCl is most commonly used. A low voltage and high current is passed through the electrolyte. The workpiece is surrounded by the electrolyte. The electrolyte flows through the gap between the tool and the workpiece under pressure and is then pumped back to the working zone. The penetration rate of the tool is proportional to the current density.

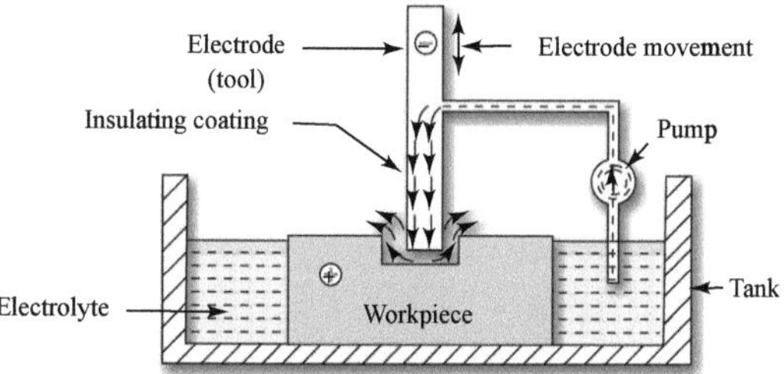

Fig. 20.7 An electrochemical machining process

More current density produces higher penetration. But current density should be constant if a uniform gap is needed. The electrical resistance is lowest (and hence the current is highest) in the region where the tool and workpiece are closest and it is the zone of highest metal removal from the workpiece. Increase in temperature of electrolyte although improves the surface finish but the size of the gap is increased on account of accelerated rate of metal removal. This changes gap resistance. As a result, less current flows and ultimately the metal removal rate is lowered to normal. When operating parameters are properly chosen, small variations are self-adjusting.

The amount of metal removed can be estimated by using Faraday's first law of electrolysis. It is given by the expression

$$m = \frac{1}{96500} \cdot I \cdot t\left(\frac{A}{n}\right) \quad (20.5)$$

and the volume of metal removed is given by

$$V = \frac{1}{96500} \cdot I \cdot t\left(\frac{A}{n}\right)\frac{1}{\rho}\eta \quad (20.6)$$

where

I	= current flowing through the gap (A)
t	= time of current flow (sec)
A, n and ρ	= atomic weight, valency and density of the work material respectively
η	= current efficiency

$$= \frac{\text{actual metal removal per unit time}}{\text{theoretical value of metal removal per unit time}} \times 100$$

Current efficiency increases with the increase in current density and also depends on the type of electrolyte. With NaCl, current efficiency is close to 100%, but for nitrate sulphate solutions, it is somewhat lower.

The specific metal removal rate can be expressed as:

$$S = \frac{1}{96500} \cdot \left(\frac{A}{n}\right) \cdot \frac{1}{\rho} \cdot \eta \quad \left(m^3/\text{Asec}\right) \tag{20.7}$$

and the feed rate of electrode is given as:

$$f = \rho_c \times S \tag{20.8}$$

where ρ_c = crrent density (A/m^2)

$$= \frac{E}{\rho_s h} \tag{20.9}$$

E = machining voltage (volt)
ρ_s = mpecific resistance (resistivity) of electrolyte (ohm-m)
h = work-tool gap, also known as machining gap (m)

Equation (20.8), after substituting ρ_c from Eq. (20.9) and S from Eq. (20.7), becomes

$$f = \left[\frac{E}{\rho_s \cdot h}\right] \cdot \left[\frac{1}{96500} \cdot \left(\frac{A}{n}\right) \cdot \frac{1}{\rho} \cdot \eta\right] \quad (\text{m/sec}) \tag{20.10}$$

20.5.1 Advantages

- Metal removal rate only depends on ion-exchange rate and is independent of work parameters such as its strength or hardness. It helps to machine fragile and difficult-to-machine materials.
- It produces no stress in the workpiece. Secondly, workpiece suffers no metallurgical changes because the process involves low temperatures.
- High surface finish can be achieved.
- There is no physical contact between the cutting tool and the workpiece, which implies that hardness of the tool does not matter and also it rules out the possibility of any tool wear.
- It also gives constant rate of penetration even for harder materials but it decreases for conventional methods with workpiece hardness (Fig. 20.8).

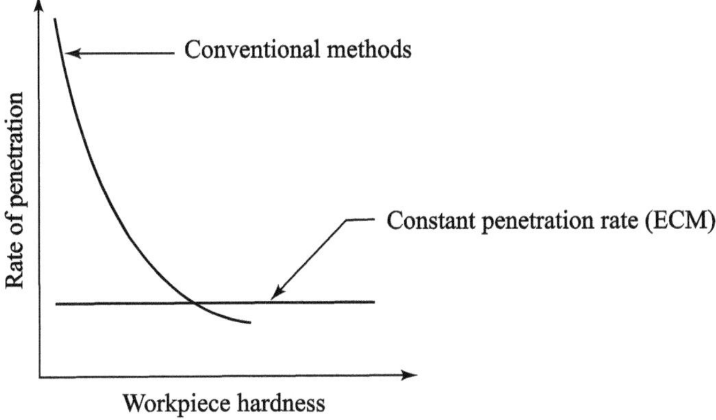

Fig. 20.8 Workpiece hardness versus penetration rate for ECM

- Complex shapes can be produced in one go eliminating a sequence of operations such as milling, grinding, deburring etc.
- It produces burr-free surfaces.
- Metal removal rate is higher for hard materials.
- Close tolerances are possible.

20.5.2 Limitations

- Only electrically conducting materials can be machined.
- Sharp square corners or flat bottoms are difficult to achieve because of erosion by electrolyte.
- The process consumes large amount of power.
- The equipment cost as well as operation cost are very high.
- Corrosion and rusting of the workpiece, tooling and equipment can take place because of the corrosive nature of most common electrolyte, sodium chloride.

20.5.3 Suitability and Applications

Electrochemical machining is well suited for mass production of complex shapes in high strength and difficult-to-machine materials, particularly in the aerospace industry for the production of the turbine blades, jet-engine parts and nozzles. The process can also be used for producing small holes. It is a very promising method of producing high-strength components in the shape of solids of revolution. The production method employed is similar to the electrochemical turning set-up; i.e., the workpiece is rotated between two moving cathodes which generated the required shape.

20.6 Electrochemical Grinding (ECG)

Electrochemical grinding, also called electrolytic grinding, is a variant of electrochemical machining. It combines electrochemical machining (ECM) and conventional grinding both to remove the metal from the workpiece. About 90% of the metal is removed by electrolytic action, while the rest 10% by the abrasive action.

In this process, the workpiece is placed near a rotating grinding wheel embedded with fine abrasive particles. The abrasive particles are nonconducting materials such as aluminium oxide, diamond or borazon (CBN). The grinding wheel acts as tool electrode (cathode). The workpiece forms another electrode (anode). A low-voltage high current is used in this process. A thin film of electrolyte is directed to flood the space between the grinding wheel and the workpiece. Electrolyte is fed in the form of a jet on the face of the wheel in the direction of rotation of the wheel. The ECG process is shown in Fig. 20.9. The electrolyte also acts as a coolant for the wheel and workpiece, hence it keeps the process cool. The abrasive action of the grinding wheel combined with the flushing action of electrolyte serves to remove the products of decomposition (metal oxides and hydroxides) from the surface of the workpiece. An average metal removal rate is 2.7 mm^3/sec per 100A.

20.6.1 Advantages

- Since the material removed by the abrasive action of the wheel is small, hence tool (wheel) wear is very small. The saving in wheel wear also constitutes a major economy.
- The process is fast because of grinding wheel used in the process.
- It produces less tool wear even in case of hard materials.
- It gives a burr-free operation and there are no residual stresses.
- Coupled with numerical controls, it can improve the accuracy and repeatability of the process and offer increased productivity.
- Surface finish is very good.

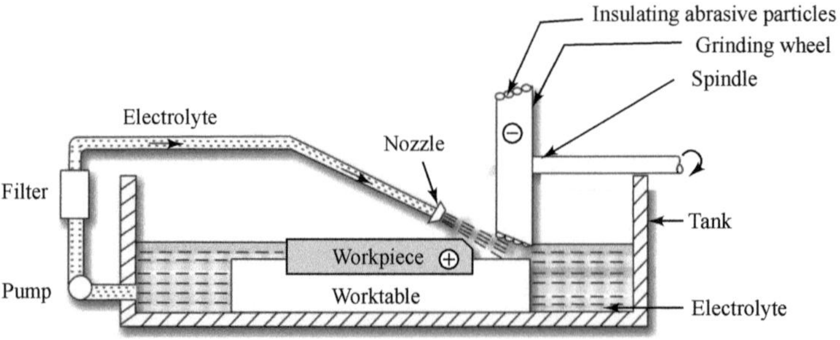

Fig. 20.9 An ECG process

20.6.2 Limitations

- Only electrically conducting hard material can be machined.
- The equipment cost is more as compared to conventional grinding process.
- The operating cost is high because of the expensive electrolyte.
- Corrosive nature of electrolyte can damage the set-up especially its tools and piping systems.

20.6.3 Suitability and Applications

ECG has been used very successfully in the grinding of hard materials like cemented carbides, stellites and refractory materials. The process is also used for shaping and sharpening carbide cutting tools, which cause high wear rates on expensive diamond wheels in normal grinding. ECG greatly reduces this wheel wear. Fragile parts (honeycomb structures), surgical needles and tips of assembled turbine blades have been successfully processed by ECG.

20.7 Laser Beam Machining (LBM)

LASER stands for Light Amplification by Stimulated Emission of Radiation. It is a device used for converting electrical energy stimulated into a highly directional, nearly parallel and coherent beam of monochromatic light. The energy of the laser is utilized to remove the material from the workpiece. Most of the material removal takes place by evaporation, although partially in a liquid state.

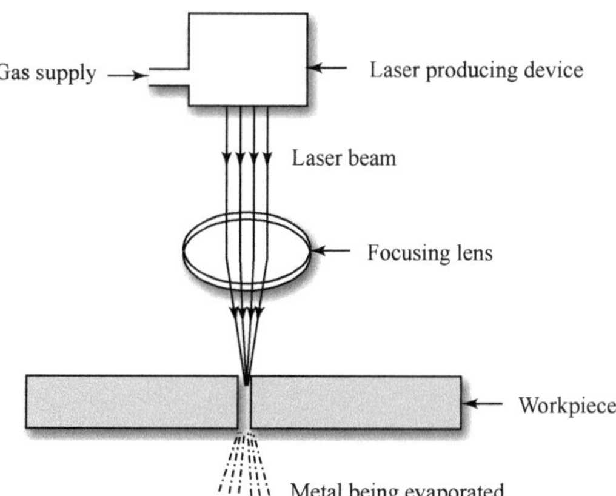

Fig. 20.10 A gas-assisted laser cutting process

The important lasers used for material removal include neodymium glass, neodymium-yttrium-aluminium-garnet (Nd-YAG), ruby and carbon dioxide. The first threes are solid-state lasers and the last one is a gas laser. Because of their higher power outputs, solid-state lasers are widely used in machining.

The Nd-YAG laser is the most commonly used solid laser. Laser beams may be used in combination with a gas stream, such as oxygen, nitrogen or argon. These gases blow away the molten and vaporized material from the surface of the workpiece. A focusing lens is used to concentrate the laser beam to a point on the workpiece. Workpiece is kept at the focal length of the lens. Laser cutting process is shown in Fig. 20.10. The material comes out in the form of dust.

20.7.1 Advantages

- Almost all the metals and many nonmetallic hard materials can be cut with the laser beam with varying cutting speeds.
- The heat-affected zone is minimum.
- Very small diameter holes can be easily produced in thinner materials.
- It can easily machine difficult-to-reach areas.
- Because of physical contact missing between workpiece and tool, work distortion does not occur.
- It can be easily automated.
- There is no need for vacuum as in the case of electron beam machining (EBM).
- The process is very fast.

20.7.2 Limitations

- It can not be used for cutting thicker work materials because of low cutting efficiency.
- The equipment cost is high.
- Laser beams are harmful and they need to be handled carefully.
- It is difficult to produce uniform circular hole. Tapered holes may be produced instead of circular ones.

20.7.3 Suitability and Applications

LBM is chiefly suitable for removing materials in smaller quantities from thinner sections. It is widely used in electronics and automotive industries in drilling microsize holes and cutting metals, nonmetals and composite materials. 0.05 mm diameter holes with depth-to-diameter ratio 20:1 are commonly drilled. Materials such as mild steel, stainless steels, nickel, titanium and some refractory metals are cut readily.

Welding is another important area of application for laser. CO_2 lasers are mainly used for thinner metals and ruby lasers for thicker materials. Laser (especially CO_2) can also be used for metal cutting.

20.8 Electron Beam Machining (EBM)

Like LBM, electron beam machining is a thermoelectric process. The process produces very high temperatures as a result of high-speed electrons. When these electrons are allowed to impinge on the workpiece, tremendous heat is generated which is sufficient to melt the metal and get it vaporized. This micromachining process is generally performed in a vacuum chamber (10^{-5} mm of Hg) to prevent the electrons from scattering. The electrons travel at about 50 to 80% of the speed of light. As they hit the surface of the metal, the kinetic energy is converted into thermal energy. Magnetic lenses are used to focus the electron beam on the workpiece. The process is shown in Fig. 20.11.

20.8.1 Advantage

- Extremely close tolerances can be maintained.
- Heat-affected zones are minimum.
- The beam can be concentrated on a very small area.
- It produces better surface finish and narrow kerf.
- Thermal distortion is least.
- The process is fast because it is entirely nonmechanical.

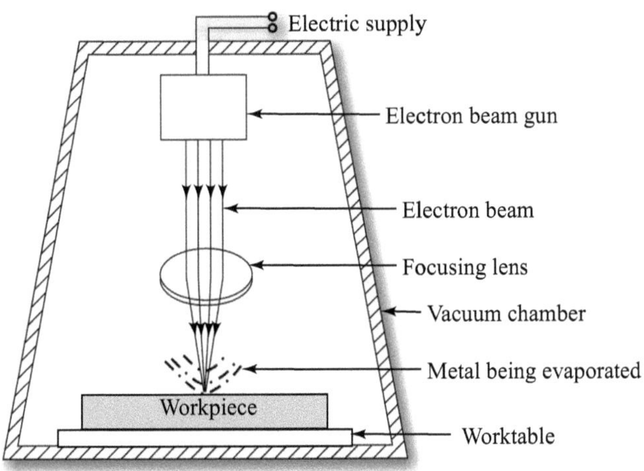

Fig. 20.11 An electron beam machining process

20.8.2 Limitations

- The equipment cost is very high.
- The interaction of the electron beam with workpiece surface produces hazardous X-rays, hence shielding is necessary.
- Vacuum is essentially required.
- Because of very low material removal rate, the process is economical only for small volume cuts.
- Skilled labour is required.
- Very high voltage is required to accelerate the electrons.
- The process can machine only thinner parts.

20.8.3 Suitability and Applications

EBM is particularly suitable for producing very small diameter holes. It is especially adapted for micromachining. Major applications of EBM include machining in thin materials, cutting of slots and drilling of holes with very high depth-to-diameter ratios, usually more than 100:1.

20.9 Plasma Arc Cutting

Plasma is a source of high energy density and thus is ideal for intense local heating. The plasma arc can produce temperatures as high as 33,000 °C. Plasma is considered as the fourth state of matter. The term *plasma* refers to a hot ionized gas which conducts electricity.

The plasma arc torch consists of a constricted arc that operates between the tungsten electrode and the workpiece. The electrode forms negative terminal and the workpiece positive terminal. The torch is directed onto the work material where the highly concentrated heat of the plasma melts the metal, and a high-velocity gas stream removes the metal from the kerf. Nitrogen, hydrogen and argon are important gases used for cleaning the cut. The process is shown in Fig. 20.12.

More recently, water-injection plasma cutting has been developed. The water flow is directed radially at the plasma through the nozzle to constrict the arc and cool the workpiece. Advantages of such process include narrower kerf, cleaner cut face, longer-lasting torch parts and a narrow heat-affected zone. Water injection plasma system is shown in Fig. 20.13.

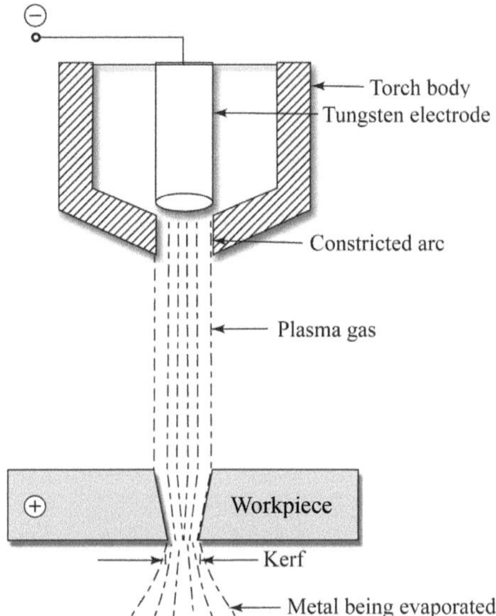

Fig. 20.12 A plasma arc cutting process

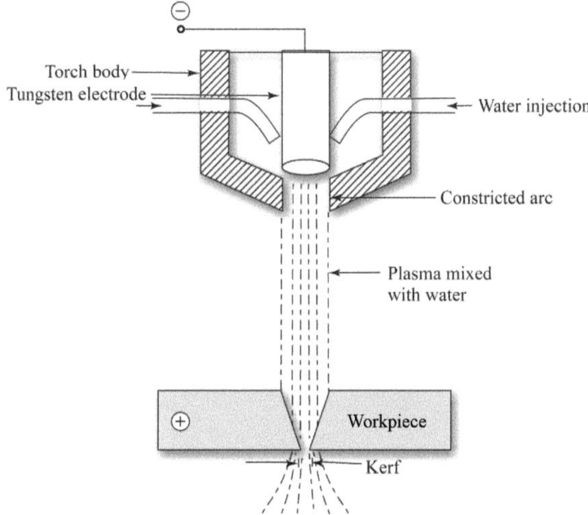

Fig. 20.13 A water injection plasma cutting process

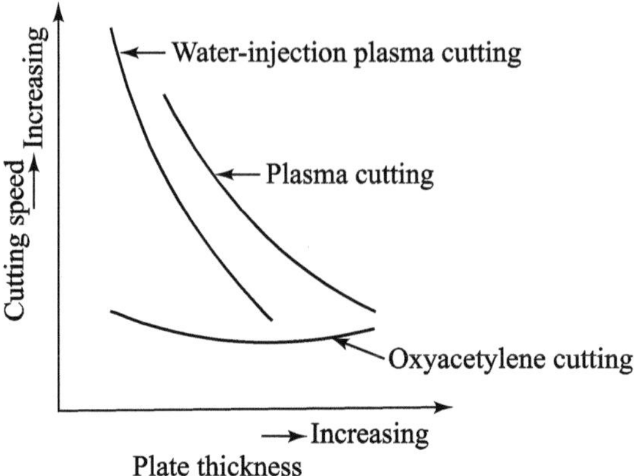

Fig. 20.14 Plate thickness versus cutting speed for plasma cutting, water-injection plasma cutting and oxyacetylene cutting

20.9.1 Advantages

- The material removal rates are much higher than in EDM and LBM.
- It can cut any material irrespective of its hardness.
- It produces narrower kerf.
- The heat-affected zones are minimum.
- The process is fast but the speed decreases with the thickness of the workpiece. Fig. 20.14 shows the variation of cutting speed with plate thickness for three processes namely, plasma, water-injection plasma and oxyacetylene cutting.
- Small size holes can be produced.
- The process can be easily automated with CNC applications.
- Due to its good reproducibility, the process can machine exotic metals at high rates.

20.9.2 Limitations

- Only electrically conductive metals can be cut. Oxidation and overheating can damage the surfaces.
- It requires high initial investment.

20.9.3 Suitability and Applications

Since the cutting process is independent of chemical reaction, it can be used on both ferrous and nonferrous metals. Profile cutting of metals, particularly of stainless steel and aluminium, are important commercial application. However, mild steel, alloy steel, titanium, bronze, and most metals can be cut cleanly and rapidly.

20.10 Water-Jet Machining (WJM)

Water-jet machining, also called hydrodynamic machining, uses a high-velocity stream of water as the cutting agent. Although water is used in most of the applications, other fluids such as alcohol, glycerin and cooking oil can also be used.

The water jet is directed to strike the workpiece at the point where cutting is desired. It acts like a saw and cuts a narrow groove in the material. The kerf may be only 0.07 mm wide. The diameter of the jet varies between 0.076 to 0.50 mm and it comes out of the orifice at a velocity in between 600 to 900 m/sec. This high velocity of water is enough to cut the material. The process is shown in Fig. 20.15.

20.10.1 Advantages

- The process provides a clean surrounding involving no atmospheric pollution.
- No heat is produced.
- No starting hole is required.
- Burr is minimally produced.
- The cut edge is very smooth and requires no further machining.
- Any contour be it internal cuts or inside corner radii can be cut.
- A relatively small volume of fluid is required.
- Good surface finish is obtained.

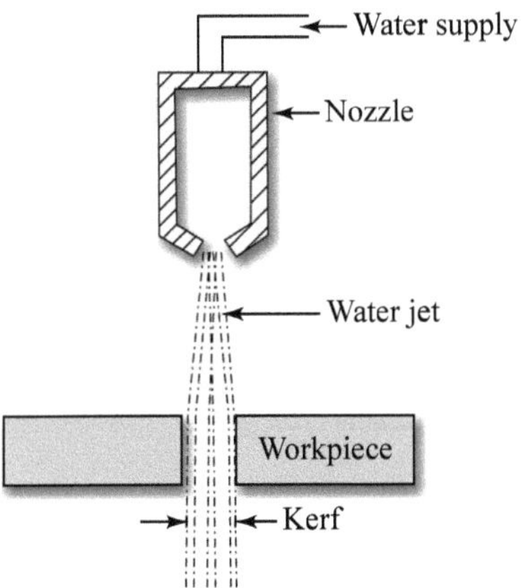

Fig. 20.15 A water-jet machining process

20.10.2 Limitations

- It requires a very high pressure pump.
- It produces a lot of noise.
- There is a possibility of evaporation if the gap between the nozzle and the workpiece is more.

20.10.3 Suitability and Applications

The materials cut by water-jet machining include plastics, fabrics, rubber, leather, wood products and paper. Since it is a very clean process, hence it also finds applications in food processing industry for cutting and slicing products.

20.11 Abrasive Water-Jet Machining

This process uses abrasive particles in the water jet to increase the metal removal rate. Abrasive particles may be silicon carbide and aluminium oxide. The abrasive is thoroughly mixed with water to increase the penetration power. The concentrated abrasive jet strikes at the desired point of workpiece, where the cut is required to be made (Fig. 20.16).

Abrasive water-jet machining can be used to cut metallic, nonmetallic and composite materials. The process is also useful for heat-sensitive materials.

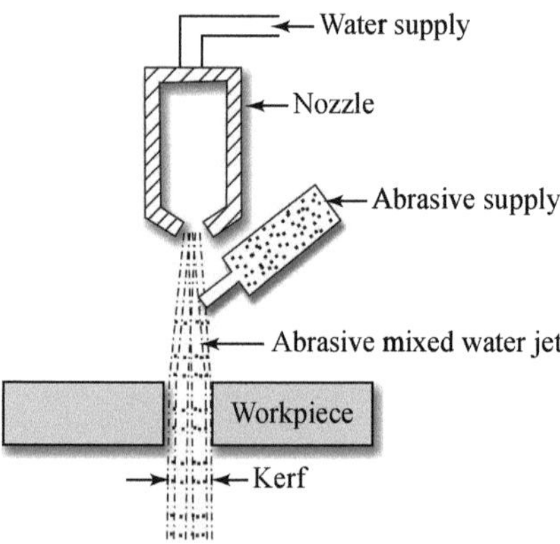

Fig. 20.16 An abrasive water-jet machining process

20.12 Abrasive Jet Machining (AJM)

Abrasive jet machining differs from abrasive water-jet machining in the sense that water used in the former is replaced by some gaseous medium to increase the metal removal rate. AJM is a finishing process and is not used for bulk metal removal.

In the AJM process, a high-velocity jet of dry air, nitrogen or carbon dioxide containing abrasive particles strikes the workpiece at the point where cutting or machining is desired. These gases act as the carrier medium for the abrasive particles. The abrasive jet-velocity may vary between 150 and 300 m/sec. The process uses much finer abrasives as compared to abrasive water-jet machining. The particle size of the abrasives used generally varies between 10 and 50 μm. Aluminium oxide, silicon carbide, boron carbide and diamonds are commonly used abrasives. The gap between the workpiece and nozzles is about 1 mm. The process is shown in Fig. 20.17.

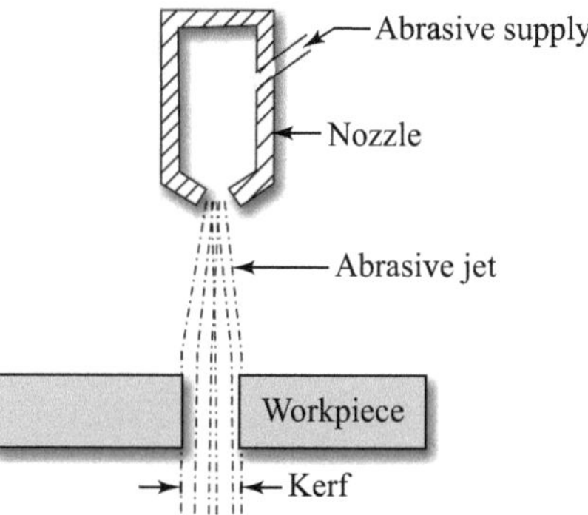

Fig. 20.17 An abrasive jet machining process

20.12.1 Advantages

- The process can produce small holes, slots or intricate patterns in very hard, thin, heat-sensitive or brittle metallic and nonmetallic materials. Even diamonds can be cut using diamond abrasives.
- Fragile materials can be easily machined without damage.
- Heat generation is almost nil.

20.12.2 Limitations

- The material removal rate is very low which makes the process extremely slow.
- Uniform holes are difficult to obtain.
- Sharp corners can not be obtained because abrasives can round off them.
- Abrasives can not be reused because of possible contamination. They can clog the system.
- The process involves environmental pollution because a lot of dust is generated by the chips and abrasive.
- It is not suitable for soft materials because they may get damaged by the abrasive particles.

20.12.3 Suitability and Applications

The process is very much suitable for electronics industry to make fine cuts on small nonmetallic components. The process can be used for removing different layers of resistors, capacitors, insulators and semiconductors from printed circuit boards in the process of preparation of integrated circuits. It can also be used for deburring or removing flash from the parts, cleaning the components with irregular shapes or removing oxides and other surface films from metal surfaces.

Short Answer Questions

1. How does nonconventional machining differ from conventional machining?

Ans. Conventional machining uses a sharp cutting tool to remove the material from the workpiece during machining process and there is a physical contact between cutting tool and workpiece. On the other hand, nonconventional machining does not use a sharp cutting tool, rather uses electrical, chemical, mechanical and thermal means to remove the material from the workpiece, and also there is no physical contact between tool and workpiece.

2. Why are abrasive particles not reused in abrasive jet machining? Name a few abrasive particles used in abrasive jet machining.

Ans. Abrasive particles usually get contaminated with different gases used in the process, which adversely affects their cutting efficiency. Aluminium oxide, silicon carbide and diamond are the most commonly used abrasives in abrasive jet machining.

3. How is water-jet machining different from abrasive water-jet machining? Which is more efficient?

Ans. Water-jet machining uses a high velocity water jet to remove the material from the workpiece, whereas abrasive water-jet machining uses an abrasive mixed water jet for cutting purpose. Abrasive water jet has higher cutting efficiency and hence removes more material.

4. What factors affect the cutting efficiency in ultrasonic machining? Where is ultrasonic machining widely used?

Ans. The cutting efficiency in ultrasonic machining increases with increase in frequency and amplitude of the vibrating tool, and is also affected by the concentration of the slurry used that consists of water and abrasive particles. Ultrasonic machining is used to machine hard and brittle materials like carbides, ceramics and glass.

5. How does electrical discharge machining (EDM) differ from wire EDM? Where is EDM widely used?

Ans. Electrical discharge machining (EDM) uses a spark (discharge) to melt the work material at a defined location, whereas wire EDM uses a moving thin wire to continuously produce spark for machining, and hence is faster than conventional EDM as well as more economical. EDM is particularly used to machine hard, brittle and difficult-to-machine materials and to produce irregularly or complex shaped holes and cavities.

6. How is chemical machining different from electrochemical machining? Where is chemical milling widely used?

Ans. Chemical machining uses a chemical agent called etchant to remove material from the workpiece without involving electricity. On the other hand, electrochemical machining removes material by the chemical action of an electrolyte using Faraday's law of electrolysis involving anode (workpiece) and cathode (cutting tool) of an electrical circuit. Chemical milling is extensively used in aircraft industry to remove excess material from the aircraft wings and similar other components to reduce the weight.

7. What are the typical applications and serious limitations of electron beam machining and laser beam machining?

Ans. Both electron beam machining and laser beam machining are used to produce microsize holes and for cutting grooves and slots in thinner parts. Their serious limitations include high equipment cost, high energy requirement, unsuitability for thicker parts, vacuum requirement in case of electron beam machining and smaller material removal from the workpiece.

8. What are micromachining operations? Which processes are categorized as micromachining operations?

Ans. Micromachining operations can produce very small size (micro) holes in the workpiece without affecting the surrounding areas. Electron beam machining, laser beam machining and photochemical machining are included in micromachining operations.

Multiple Choice Questions

1. Consider the following statements about the nontraditional machining methods:

   1. Hard materials can be easily machined without being damaged.
   2. Complex shapes are easily produced.
   3. They have low specific power consumption.
   4. Tools need not be harder than workpiece.

   Of these statements:

   (a) 1 and 3 are true
   (b) 1, 2 and 3 are true
   (c) 1, 2 and 4 are true
   (d) 2, 3 and 4 are true

2. Consider the following statements about the chemical machining:

   1. It is based on the use of etchants to remove certain layers from a material.
   2. It suffers from undercutting.
   3. The etch factor is the ratio of side penetration to depth of cut.
   4. It affects metal structure.

   Of these statements:

   (a) 1, 3 and 4 are true
   (b) 1, 2 and 3 are true
   (c) 2, 3 and 4 are true
   (b) 1, 2 and 3 are true

3. Chemical milling finds extensive applications in

   (a) Automobile industry
   (b) Aircraft industry
   (c) Food processing industry
   (d) All of above

4. Consider the following statements:

   1. Chemical milling removes material by dissolution.
   2. In EDM, the workpiece acts as cathode and tool as anode.
   3. The tool remains unaffected in EDM and no disintegration takes place.
   4. Kerosene oil, paraffin and mineral oil are important dielectric fluids used in EDM.

   Of these statements:

   (a) 1 and 2 are true
   (b) 1 and 4 are true
   (c) 2 and 3 are true
   (d) 3 and 4 are true

5. Match List I with List II and select the correct answer using the codes given below the lists:

List I	**List II**
(Nontraditional methods)	**(Related to)**
A. Chemical machining	1. Dielectric fluid
B. Electrical discharge machining	2. Faraday's law of electrolysis
C. Electrochemical machining	3. Maskant
D. Electron beam machining	4. Vacuum chamber

Codes:	A	B	C	D
(a)	1	3	2	4
(b)	3	1	2	4
(c)	2	4	1	3
(d)	2	3	1	4

6. In water-jet machining, typical value of stand-off distance is

(a) 5 mm (b) 10.5 mm
(c) 15.5 mm (d) 3.2 mm

7. In electric discharge machining, better surface finish is obtained at
   (a) low frequency and low discharge current
   (b) low frequency and high discharge current
   (c) high frequency and low discharge current
   (d) high frequency and high discharge current
8. The metal removal in electrochemical machining varies
   (a) inversely proportional to the gap between work and tool electrode
   (b) inversely proportional to the square of the gap
   (c) directly proportional to the square of the gap
   (d) directly proportional to the gap
9. What is the usual velocity of electrons in electron beam machining?
   (a) 25% of the velocity of light
   (b) 50% of the velocity of light
   (c) 75% of the velocity of light
   (d) equal to the velocity of light

10. Which of the following processes is termed as a micromachining operation?

    (a) Electric discharge machining
    (b) Electrochemical machining
    (c) Electron beam machining
    (d) Water-jet machining

11. The typical value of etch factor in chemical machining varies between

(a) 1 and 2	(b) 2 and 3
(c) 3 and 4	(d) 4 and 5

12. Match List I with List II and select the correct answer using the codes given below the lists:

List I	**List II**
(Machining processes)	**(Description)**
A. Abrasive jet machining	1. Uses abrasive particles such as aluminium oxide and diamond
B. Laser beam machining	2. Uses a water-jet containing abrasive particles
C. Electrochemical grinding	3. Uses a gaseous medium containing abrasive particles
D. Abrasive water-jet machining	4. Micromachining operation

Codes:	A	B	C	D
(a)	1	4	3	2
(b)	1	3	4	2
(c)	3	4	1	2
(d)	3	1	4	2.

13. Match List I with List II and select the correct answer using the codes given below the lists:

List I	**List II**
(Machining processes)	**(Description)**
A. Electrodischarge machining	1. Drilling microholes in very hard materials
B. Electrochemical machining	2. Drilling holes in glass
C. Ultrasonic machining	3. Die sinking
D. Laser beam machining	4. Machining contours

Codes:	A	B	C	D
(a)	4	2	3	1
(b)	3	4	1	2
(c)	4	3	2	1
(d)	3	4	2	1

14. During ultrasonic machining, the metal removal is affected by
    (a) hammering action of abrasive particles
    (b) rubbing action between tool and workpiece
    (c) high frequency sound waves
    (d) high frequency eddy currents

15. Consider the following machining disadvantages:

    1. Low metal removal rate
    2. Poor surface finish
    3. High tooling cost

    Which of these disadvantages is associated with ultrasonic machining (USM)?

    (a) 1 and 3
    (b) 1 and 2
    (c) 2 and 3
    (d) 1, 2 and 3

16. Which of the following methods requires vacuum?

    (a) Laser beam machining
    (b) Electron beam machining
    (c) Electric discharge machining
    (d) Electrochemical machining

17. Consider the following statements about electrochemical grinding (ECG):

    1. A rubber bonded alumina grinding wheel acts as cathode and workpiece as anode.
    2. A copper bonded alumina grinding wheel acts as cathode and workpiece as anode.
    3. The metal removal takes place due to the pressure applied by the grinding wheel.
    4. The metal removal takes place due to electrolysis.

    Which of the above statements is true?

    (a) 1 and 3
    (b) 2 and 4
    (c) 2 and 3
    (d) 1 and 4

18. Which of the following processes does not cause tool wear?
    (a) Ultrasonic machining
    (b) Electrochemical machining
    (c) Electric discharge machining
    (d) Anode mechanical machining.

Answers

1. (c)	2. (b)	3. (b)	4. (b)	5. (b)	6. (d)
7. (c)	8. (a)	9. (c)	10. (c)	11. (a)	12. (c)
13. (c)	14. (a)	15. (a)	16. (b)	17. (b)	18. (b)

Review Questions and Discussions

Q.1. How does chemical blanking differ from chemical milling?
Q.2. What is the principal advantage of using a moving-wire electrode in electrical discharge machining?
Q.3. What is the principle of electrical discharge machining?
Q.4. What is the difference between chemical machining and electrochemical machining?
Q.5. How does a tool wear in electrochemical machining?
Q.6. What is the importance of gap between the electrode and the workpiece in ECM?
Q.7. Differentiate between electrochemical machining and electrochemical grinding?
Q.8. Which laser is the best for cutting metals?
Q.9. What is the safety problem connected with electron beam machining?
Q.10. What is plasma? How does plasma arc cutting work?
Q.11. What is the difference between abrasive water-jet machining and abrasive jet machining?

21 Powder Metallurgy

Powder rolling is used for making electronic components.

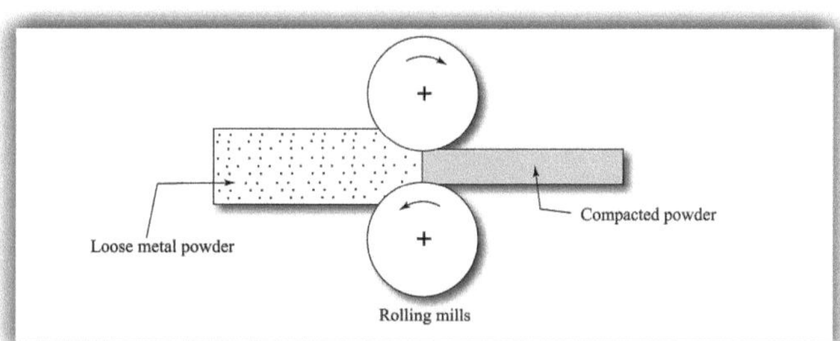

Learning Objectives
After reading this chapter, you will be able to answer some of the following questions:

- When is powder metallurgy technique useful?
- What is the purpose of sintering?
- What are the important characteristics of powder metallurgy products?

D. K. Singh, *Fundamentals of Manufacturing Engineering*,
https://doi.org/10.1007/978-981-99-8767-2_21

21.1 Introduction

In powder metallurgy (PM), parts are produced from metallic powders by applying pressure and heat. The powder particles physically interlock to produce sufficient strength during the application of pressure. The compact powders are then heated in a neutral or reducing atmosphere at a temperature below melting point of the powder. As a result, fine particles are bonded together, thereby improving the strength and hardness. The method of powder metallurgy is useful to produce desired properties and characteristics which are otherwise difficult to obtain. Typical PM products include gears, bearings, cams, cutting tools, piston rings and connecting rods.

21.2 Basic Processes in Powder Metallurgy

Following are the basic powder metallurgy operations:

- Powder production,
- Mixing or blending,
- Compaction and
- Sintering

21.3 Powder Production

There are several methods of producing metal powder. All metal powders, because of their individual physical and chemical characteristics, cannot be manufactured in the same way. Almost any metal, alloy or nonmetals such as ceramic and polymer can be converted into powder form.

Atomization is an excellent means of producing powders from many of the low-temperature metals, such as lead, aluminium, zinc and tin. Presently iron and steel are also processed by atomization. It is the commercial method for producing metal powders. In this process, a molten metal or alloy is injected through a nozzle, where it is atomized (fragmented into small size molten droplets) by jets of inert gas, air or water. Upon solidification, a wide range of particle sizes and shapes is formed. The size of the particles formed depends upon the temperature of the metal, the rate of flow, nozzle size and jet characteristics. With slow cooling and high surface tension, spherical shapes are formed before solidification. And with more rapid cooling (as in the case of water atomization), irregular shapes are produced.

Nickel-based superalloys are converted into powdered form by methods such as inert gas atomization, rotating electrode and soluble-gas atomization. The inert gas process is usually done with argon gas jets. The rotating electrode method consists of throwing the molten metal from the rotating electrode followed by crushing the droplets to the powdered form. In the soluble-gas process, molten metal is forced into a vacuum and the liquid erupts into particles.

The ***electrolytic deposition method*** is used to produce metal powders of very high quality. The process is similar to electroplating and is generally used for processing silver, tantalum, tungsten, molybdenum, nickel and cobalt. The method uses two electrodes (anode and cathode) and an electrolyte, usually an aqueous solution or fused salt. During the passage of high-ampere current through the electrolyte, powdery deposits are produced on the cathode, which is removed, washed and dried to produce the metal powders of very high purity.

The ***reduction method*** uses reducing agents such as carbon monoxide and hydrogen to reduce metal oxides into metallic form. The resulting metal is then crushed to produce powder.

21.4 Powder Mixing or Blending

Blending is carried out for the purpose of homogenizing the particle sizes so that a uniform mixture can be obtained.

Lubricants are added to the powder to reduce friction between the particles, thereby improving flow characteristics during compaction. Lubricants also ensure longer die life. Stearic acid, lithium stearate or powdered graphite are the principal lubricants used. Blending is done either dry or wet, where water or other solvent is used to improve mixing, reduce dusting and lessen explosion hazards. Dry blending is more common, and however in case of finely divided aluminium powders, wet blending is used.

21.5 Compaction of Metal Powders

In compaction, blended powders are pressed using mechanical or hydraulic presses. Lubricants are used to improve metal flow characteristics by reducing friction. Hydraulic presses have higher tonnage capacity and slower cycle time as compared to lower tonnage and faster cycle of mechanical presses. Compaction pressure varies from metal to metal and depends on the characteristics and shape of the particles, method of blending and lubrication. The pressed powder is known as green compact, and the process is generally carried out at room temperature.

There are two main methods of compacting metal powders: (1) with a punch and die and (2) isostatically. Selection of the method depends upon the complexity of the part to be produced. Mechanical pressing is mostly used for simple parts. High-density intricate parts can be made by isostatic compaction.

Isostatic compaction produces uniform pressure and is performed either at room temperature or elevated temperatures. ***Cold isostatic pressing*** **(*CIP*)** is performed at room temperature in a flexible mould made of rubber or other elastomers in a liquid (oil or water) pressurizing medium. ***Hot isostatic pressing*** **(*HIP*)** is performed at elevated temperatures in a rigid metal sheet mould. The pressurizing medium is inert gas such as helium or argon. Higher-density products with uniform and isotropic properties are obtained. Near-net-shaped parts are possible, thus eliminating the scope of costly machining operations.

21.6 Metal Injection Moulding

As in the injection moulding of plastics, metal powders can be compacted to near theoretical density in complex external or internal configurations. In injection moulding ultrafine metal, ceramic or carbide powders (< 10 μm) are blended with a polymer or a wax-based binder. The purpose of using binder is to increase the fluidity of powdered material so that complex shapes can be produced easily. The binder can be added up to 50% by volume. The mixture is then heated to make it pastelike and is injected into the mould cavity. The binder is now removed by controlled heating. The parts are then subjected to sintering, where they shrink by 15–25%. To compensate for the shrinkage, injection moulding products are made larger than the desired size.

Injection moulding is a useful method for producing very thin-walled small intricate parts as well as delicate cross sections which cannot be produced by conventional methods. It produces higher-density products. The major limitation of the process is its high cost which can only be justified for large-scale productions. Typical parts include components for watches, guns and automobiles.

21.7 Powder Rolling

Also known as roll compaction, it is one of the methods of producing high-density PM products. In this process, metal powders are fed between two rolling mills rotating in opposite directions. The process is used for making sheet metal for electrical and electronic components and for coins. Figure 21.1 gives the schematic representation of roll compaction.

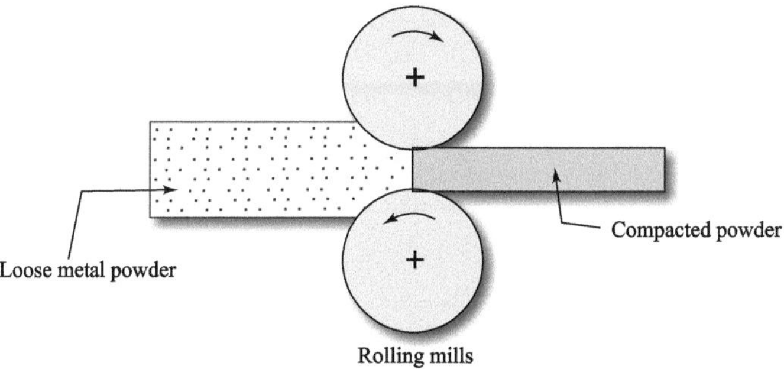

Fig. 21.1 A powder rolling process

21.8 Extrusion

It is another technique for producing high-density products. Rods, wires and small billets can be produced by the hot extrusion of encapsulated powder. Superalloy powders are hot extruded for improved properties.

21.9 In Situ Compaction

It is another means of producing a high-density shape from fine particles. In this process, molten metal is quickly atomized into powder and sprayed onto a collector plate. Depending upon the shape of the collector plate, the product obtained can be a finished part, a strip or plate, a deposited coating or a preform for subsequent operations, such as forging. It is used to produce medium-alloy steel, tool steel and superalloy parts of smaller size.

21.10 Pressureless Compaction

In this process, no pressure is involved and the die is filled with metal powder by the action of gravity force. The powder is heated (sintered) directly in the die. The density of the part is low and the process is mainly used for producing porous parts, such as filters.

21.11 Sintering

It is the process of heating the compact metal powders below their melting points. However, when the product is composed of more than one material, the sintering temperature may even be above the melting temperature of one or more components. The lower melting point materials then melt and can flow into the voids between the remaining particles. Sintering temperatures usually range between 0.7 and 0.9 times the melting point of the metal or alloy. For example, sintering temperatures for metals such as iron and copper are 1095 and 870 °C, respectively.

The sintering time varies from metal to metal. Higher melting point metals require more sintering time.

Protective atmosphere is important to prevent oxidation of metal powders. Vacuum sintering is frequently employed with stainless steel, titanium and the refractory metals. Reducing atmospheres of hydrogen and nitrogen is used to reduce the formation of oxides.

Sintering increases density of the compact powders because of their intimate contact during heating. But the increased density results in a contraction in product dimensions which is taken care of by oversizing the compaction dies.

In spark sintering, metal powders are pressed further (after green compact) and then subjected to heating by a high-energy electrical spark which also removes the surface contaminants from the powder particles. This process ensures good bonding during compaction at elevated temperature and gives more dense metal parts.

21.12 Secondary and Finishing Operations

Many PM parts are used in the as-sintered condition. However, many products utilize one or more secondary operations. The purposes of these operations are to impart dimensional accuracy to the sintered part and to improve its strength and surface finish to get special characteristics by additional densification. Some operations are discussed below.

Sizing is a pressing operation in which a sintered part is pressed in a die in order to make it more dimensionally accurate.

Coining improves dimensional accuracy of the part. It uses increased pressures (oversizing) resulting in increased density.

Impregnation is accomplished by immersing the sintered parts into heated oil or other liquids. The oil is absorbed in the part due to capillary action and is released under pressure or heat during the use of the part. This is termed as self-lubricating properties of the part. The most common examples are oil-impregnated bearings which contain upto 30% oil by volume.

Infiltration is the process of filling the pores of a sintered part with molten metal to decrease the porosity. The filler metal has melting point lower than powdered metals and is absorbed by capillary action. Infiltration increases density, strength, hardness, machinability and corrosion resistance of the PM product. The most common application is the infiltration of iron-based compacts with copper.

PM parts generally do not require *machining* since they are dimensionally accurate and of finished shapes. But making of grooves, threads or undercuts requires machining. Drilling, tapping and grinding are common machining operations performed on sintered metal parts by means of tungsten carbide tools.

21.13 Properties of Powder Metallurgy Products

High tensile strength and good machinability are two important properties of PM parts. Low-density parts have low strength. Irregularly shaped particles interlock on compacting and produce good green strength. PM parts require little or no machining.

HIP parts have strength comparable to that of wrought materials. Even the strength can increase if the density approaches its theoretical value.

Ductility and corrosion resistance of PM parts are lower due to the presence of pores but can be improved considerably by compaction followed by additional sintering. But on the positive side, porosity can provide sound and vibration damping characteristics.

21.14 Powder Metallurgy Products

Following are some of the prominent PM products, most of which are completed without machining.

- Porous or permeable products, such as oil-impregnated bearings and filters are the major PM products. These bearings are self-lubricating in nature and do not require any lubrication during their service life. Metal filters are used for filtering hot or cold fluids and air. They are used in gasoline tanks for separating moisture and dirt from the fuel system.
- Complex-shaped parts such as gears and cams require considerable machining if produced by methods other than powder metallurgy. These parts require no machining if made by powder metallurgy. The extremely hard and nonmachinable sintered or cemented carbide cutting tools can only be made by powder metallurgy.
- PM parts have superior properties because of their increased density. Such properties are useful for aerospace applications. Alnico magnets, produced from aluminium, nickel, cobalt and iron powders, are superior to cast magnets.

Numerous other parts manufactured by powder metallurgy include clutch faces, brake drums, welding rods, etc.

21.15 Advantages of Powder Metallurgy

Following are some of the major advantages of powder metallurgy:

- Since PM parts are produced with high dimensional accuracy and better surface finish requiring little or no machining, therefore machining cost is greatly reduced or eliminated.
- Some products cannot be made by any other method. For example, the sintered or cemented carbide tools and porous bearings are made only by PM technique and no other method can be used.
- Complex-shaped parts such as irregularly shaped holes, eccentrics splines, counterbores, gears and cams can be conveniently and economically produced by powder metallurgy.
- Various combinations are possible in powder metallurgy such as metal–metal, metal-alloy, or metal-ceramic. Immiscible materials can be combined.
- Compositional variation leads to changes in mechanical properties. It is possible to produce a part with different properties in its different sections. The excellent damping characteristics of PM parts find applications in several equipments such as dictating machines and air-conditioning blowers.
- Powder metallurgy technique reduces wastage of material because of near-net-shaped capability of the process. This is important for expensive materials.

21.16 Limitations of Powder Metallurgy

- The initial cost of the equipments is high. Even conversion of metal into powdered form is a costly affair. The higher cost can only be justified by higher rate of production.
- Mechanical properties of a powdered product are inferior because of residual porosity resulting due to nonuniform distribution of density in the product.
- Intricate parts are difficult to manufacture because of little flowability of metal powders during compaction.
- Low melting point materials such as tin, zinc and cadmium offer difficulty during sintering because their oxides cannot be reduced below their melting temperatures.
- Inert atmosphere is required in case of some metals such as aluminium, magnesium and iron because of explosion problems.

Short Answer Questions

1. What are the important properties of powder metallurgy? Name a few powder metallurgy products.

Ans. Powder metallurgy offers better surface finish, high-dimensional accuracy, excellent damping characteristics, no requirement of further machining and no wastage of material. A few important powder metallurgy products include self-lubricating bearings, gears, cams, connecting rods and cutting tools.

2. Differentiate between true density and bulk density?

Ans. True density is the density of the compact metal powders, whereas bulk density is the density of the metal powders in the loose state. Bulk density is smaller than true density.

3. What is sintering and why is it carried out? Why is temperature increased gradually in sintering?

Ans. Sintering is a process of heating the pressed (compact) powders of the metal below its melting temperature to increase its strength and hardness. Gradual increase of temperature reduces the possibility of cracks in the compact powder, as fast heating can produce high pressure of air entrapped in the pores of compact powder, producing cracks in them.

4. How do self-lubricating bearings work?

Ans. Voids in the self-lubricating bearings are filled with oil which comes out on their surfaces when load or pressure is applied, thus reducing the friction.

Multiple Choice Questions

1. Consider the following methods of metal powder production:

   (a) Atomization (b) Electrolytic deposition method
   (c) Reduction method

   Very high-quality metal powders can be produced by

   (a) 1 alone (b) 2 alone
   (c) 1 and 2 (d) 2 and 3

2. Which of the following metals requires wet blending?

   (a) Silver (b) Cobalt
   (c) Aluminium (d) Nickel.

3. Consider the following statements about injection moulding:

   1. It produces high-density products.
   2. It is useful for small-sized intricate parts.
   3. It is an economical process.
   4. It requires shrinkage compensation.

   Of these statements:

   (a) 1, 2 and 3 are true (b) 1, 2 and 4 are true
   (c) 2, 3 and 4 are true (d) 1 and 4 are true

4. Consider the following statements:

   1. Powder rolling is used to produce sheet metals for electrical and electronic components.
   2. Sintering is usually carried out at a temperature below the melting point of the metal powder.
   3. Atomization is the most widely used method of powder production for low melting metals.
   4. Extrusion produces low-density products.

   Of these statements:

   (a) 1 and 2 are true (b) 2 and 4 are true
   (c) 1, 2 and 3 are true (d) 2, 3 and 4 are true

5. Consider the following advantages of powder metallurgy:

   1. Net-shaped objects can be made.
   2. Wastage of material is minimum.
   3. Mechanical properties of the products are superior.
   4. Uniform density products are easily produced.

   Of these statements:

   (a) 2 and 3 are true (b) 1 and 2 are true
   (c) 1, 2 and 3 are true (d) 2 and 4 are true

6. Which of the following metals produces explosive powders?

(a) Aluminium (b) Magnesium
(c) Titanium (d) Zirconium

Of these:

(a) 1 and 2 are true (b) 2 and 3 are true
(c) 1, 2 and 3 are true (d) 1, 2, 3 and 4 are true

7. Most of the metals have the following size of their powder:

(a) above 500 μm (b) 75 mm–90 μm
(c) 500 mm–1000 μm (d) 1000 mm–1500 μm.

8. Aspect ratio is a ratio of

(a) largest dimension to the smallest dimension of the metal powder
(b) smallest dimension to the largest dimension of the metal powder
(c) surface area to volume of metal powder
(d) volume to surface area of metal powder.

9. For a spherical metal powder, the value of aspect ratio is

(a) zero (c) 1.0
(b) 0.5 (d) 10.

10. For flakelike or needlelike metal powders, the value of aspect ratio is

(a) zero (c) 1.0
(b) 0.5 (d) 10.

11. Which of the following materials requires minimum compaction pressure?

(a) Brass (c) Bronze
(b) Iron (d) Aluminium.

Answers

1. (b) 2. (c) 3. (b) 4. (c) 5. (b) 6. (d)
7. (b) 8. (a) 9. (c) 10. (d) 11. (d)

Review Questions and Discussions

Q.1. What are the different methods of producing metal powders?
Q.2. Why are metal powders blended?
Q.3. What is meant by green compact?
Q.4. What are the major objectives of compacting operation?
Q.5. What is isostatic pressing?
Q.6. Differentiate between cold and hot isostatic pressing.
Q.7. Why is a protective atmosphere required during sintering?
Q.8. What are the merits and demerits of hot isostatic pressing?
Q.9. How is the sintering temperature usually related to the melting point temperature of the material being sintered?
Q.10. What are the different shapes of metal powders?
Q.11. What is the advantage of irregular-shaped metal powder?
Q.12. Name the two most important characteristics of a metal powdered part.
Q.13. What is infiltration?
Q.14. Differentiate between impregnation and infiltration.
Q.15. Why are the lubricants used in powder metallurgy?
Q.16. How can you justify the higher cost of metal powdered products?

22 Metrology

A surface plate is used to measure flatness of an object.

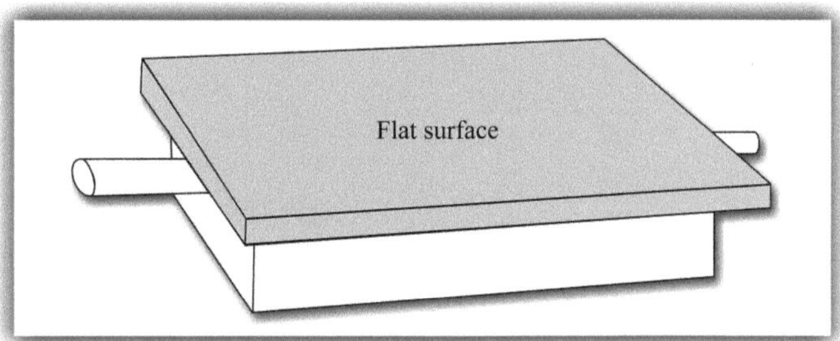

Learning Objectives

After reading this chapter, you will be able to answer some of the following questions:

- What is the difference between line and end measurements?
- What is *Ra* value?
- How does extrinsic measurements differ from intrinsic measurements?
- When is sine bar useful?

D. K. Singh, *Fundamentals of Manufacturing Engineering*,
https://doi.org/10.1007/978-981-99-8767-2_22

22.1 Introduction

Metrology deals with measuring linear and angular magnitudes of technical parts in standard units of length and angle. The basic purpose of dimensional measurements in production is to assure and to verify the agreement of the product with the specifications of the design for its economical production. Manufacturing engineering uses extensive applications of metrology. Quality control is the traditional sector of dimensional measurements. Industry's need for higher accuracy has been felt long back. The accuracy has risen from ten thousandth to the millionth part of an inch. Quality assurance systems, such as ISO 9000 developed by the International Standards Organisation, are being developed and used globally to guarantee that manufactured goods meet certain design, performance and quality specifications.

22.2 Precision Measurement Principles

The reliability of dimensional measurements may be regarded as a function of the occurring errors and of their assessments. Errors in measurements can be of two types: random and systematic.

Random errors may result from inaccurate scale reading, wrong specimen setting and mistakes in recording. These errors are difficult to predict, and statistical methods are used to determine them. Standard deviation is used for finding the range of scatter in the measured values from a mean value.

Assuming that the variations of the individual measurements are based on normal distribution (also called Gaussian distribution) curve, the standard deviation σ is defined as:

$$\sigma = \sqrt{\frac{\sum_{i=1}^{n} \left(X_i - \overline{X}\right)^2}{n}} \tag{22.1}$$

where

X_i = individual readings

$\overline{X}$ = mean of all readings.

$$= \frac{\sum_{i=1}^{n} X_i}{n} \tag{22.2}$$

n = total number of readings.

In order to assure with a greater probability that the reported measurement values are true, the precision of the process is often appraised by considering an uncertainty

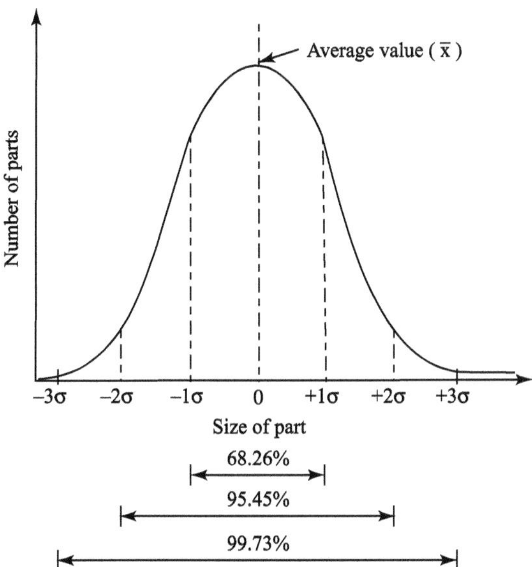

Fig. 22.1 A normal distribution curve

zone that extends over $\pm 1\sigma$, comprising theoretically 68.26% or 2, comprising theoretically 95.45% or $\pm 3\sigma$, comprising theoretically 99.73% of all readings (Fig. 22.1). Since production patterns do not follow the normal distribution exactly, the above approximations are often made. Most manufacturing processes result in products whose measurements of the geometrical features and size are distributed normally. Majority of the measurements are clustered around the average value $\overline{X}$, and $\overline{X}$ will be equal to the nominal value only if the process is accurate, i.e., perfectly centred.

Systematic errors are generally measurable and controllable and are associated with the measuring instrument's capabilities, environment and other extraneous conditions.

Variations in the environment always affect the results of dimensional measurements. Temperature in particular can seriously affect the reliability of the obtained results since linear coefficients of expansion of various materials are strongly temperature-dependent. The internationally agreed reference temperature for metrology laboratories is 68° F (20 °C) with ± 0.5° F variations. Air purity and optimum humidity are another factors to be considered. The introduced air is filtered (to remove foreign particles) and dehumidified (to remove the possibility of rust); the relative humidity varies between 35 and 45%.

22.3 Metrology Terminology

22.3.1 Calibration

Calibration is the process of comparing an unknown measuring device with a standard unit which is considered as the reference. It gives the amount of deviation from the standard value.

22.3.2 Accuracy

Accuracy means closeness to the exact size.

22.3.3 Error

Error is defined as the difference between the measured value and the true value of a measurement system.

22.3.4 Precision

Precision refers to the repeatability of the measurement process and is based on a number of measurements rather than a single measurement. An instrument is said to be precision if it is able to measure a given object correctly all the time and the difference between the readings is negligibly small. A common wooden ruler, for example, will expand or contract, depending on the environment, thus giving different and hence unreliable measurements and is therefore not a precision instrument.

22.3.5 Sensitivity

Sensitivity is defined as the ability to detect very minor differences in the measurement values.

22.3.6 Tolerance

Tolerance in the measurement system is defined as the difference between the maximum and minimum limit of the measured value. It gives the variation in size tolerated to cover reasonable imperfections in workmanship and varies with different grades of work.

Tolerances are important not only for proper functioning of products, but they also influence the manufacturing costs. Smaller tolerance means higher production costs. Tolerances may be unilateral or bilateral. Unilateral tolerances are represented as $35^{+0.00}_{-0.10}$ and bilateral tolerances as 35 ± 0.06 or $35^{+0.06}_{-0.05}$, i.e., with equal or unequal variations.

22.4 Line and End Measurement

Line measurement is connected with the measurement of distance between two lines or edges with the help of an instrument such as a steel rule.

End measurement is connected with the measurement of distance between surfaces by using end faces of gauge blocks, slip gauges, etc. This is the most common method of measurement in industrial practice because it can produce accuracy of higher order.

22.5 Measuring Instruments

All measuring devices can be broadly classified into two groups: direct and indirect. Direct measuring instruments can measure the dimension without the aid of other equipment, as in the case of a micrometer. Indirect-measuring tools require a standard of reference, as in the case of measurement through a dial indicator.

22.5.1 Line-Graduated Measuring Instruments

Line graduations commonly represent linear distances, but can display angular spacing as well. The measuring instrument has a line-graduated scale so that the size of the object being measured can be read directly on this scale.

22.5.2 Linear Measurements

The metre in SI system is defined in terms of wavelength of light. Eleventh General Conference of weights and measures held in Paris in 1960 defined metre as the length equal to 1,650,763.73 wavelengths of red-orange radiation of Krypton isotope 86, measured in a vacuum. But with the purpose of obtaining more accurate metre, it was redefined using speed of light as basis in the 17th General conference of weights and measures held in 1983. Metre was defined as the length of the path travelled by light in vacuum in 1/299,792,458 s. This new definition of metre is used today. The helium–neon laser is the best light source for measurement. It is capable for producing light that is far more monochromatic and more intense than other light sources, and therefore, extremely sharp fringe lines are produced. While using Krypton-86 as light source, spectral lines are so close that they are superimposed on each other and accurate measurement becomes difficult.

Light standard of length has the following advantages:

- Length is invariably constant.
- It is easily replaceable in case of damage.
- It gives the highest accurate value of metre.
- It is most easy to produce and can be produced at any time in any location.

Apart from temperature effects, errors in linear measurement can arise from the following sources:

- Flexure of contacting surfaces
- Errors of alignment
- Errors of reading, i.e., parallax effects and vernier acuity.

22.5.3 Precision Gauge Blocks

Gauge blocks are made of hardened steel and have square, rectangular or round shape (Fig. 22.2). They were first developed by a Swedish engineer C. E. Johansson in 1895 and consisted of 102 blocks. Nowadays tungsten carbide and chromium carbide gauge blocks are used, which have superior wear resistance characteristics.

Rectangular and square blocks are commonly used. Square blocks have central holes to keep them together by means of a tie rod. During object measurement, these blocks are wrung together by sliding one past another. The size of the gauge is defined as the distance between two gauging surfaces on opposite sides of the blocks.

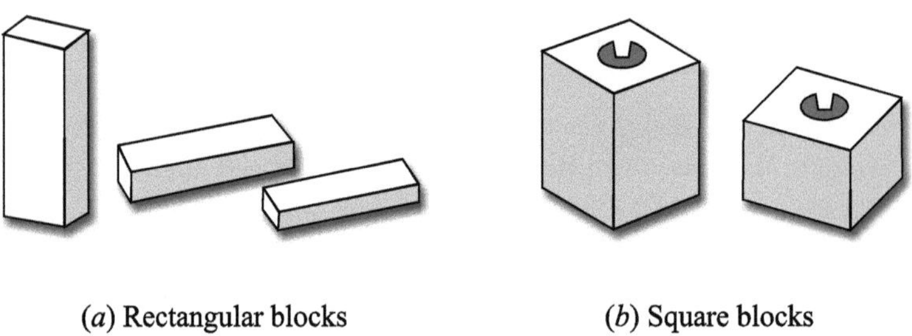

(*a*) Rectangular blocks (*b*) Square blocks

Fig. 22.2 Gauge blocks

Fig. 22.3 A rule

Gauge blocks are precision products made to rigid specifications. They are used for precise measurement of parts and for verifying measuring tools such as micrometres, comparators and various limit gauges. They are highly temperature-sensitive, and their accuracy is valid only at 68° F (20 °C). They can measure upto 0.00254 mm or even more accuracy.

22.5.4 Rule

Rule is the simplest and most widely used linear measuring instrument. It is made of steel and has line graduations with different degrees of accuracy to suit diverse requirements (Fig. 22.3). An accuracy upto 0.3968 mm can be measured by using such scales.

22.5.5 Vernier Caliper

Vernier caliper, also known as caliper gauge (Fig. 22.4) is an end-measuring instrument and can measure inside and outside dimensions. It can measure with accuracy up to 0.0254 mm. It consists of the main scale having line graduations and two jaws. One of the jaws is integral with the graduated scale, the other has marking to indicate the corresponding scale position.

During the measurement of outside dimensions, the part is held between the jaws and during inside measurement, the part is placed over the ends of the two jaws.

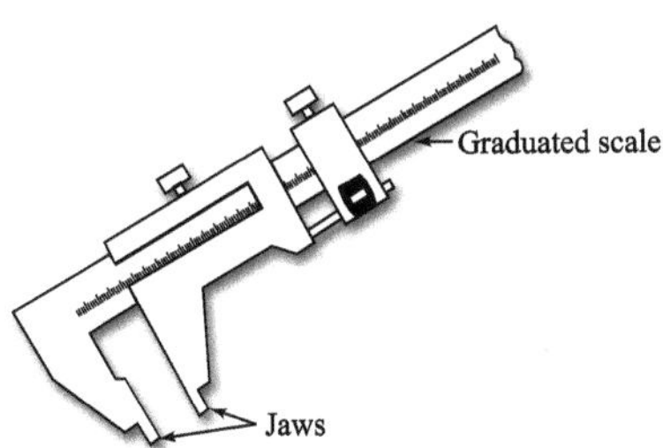

Fig. 22.4 A vernier caliper

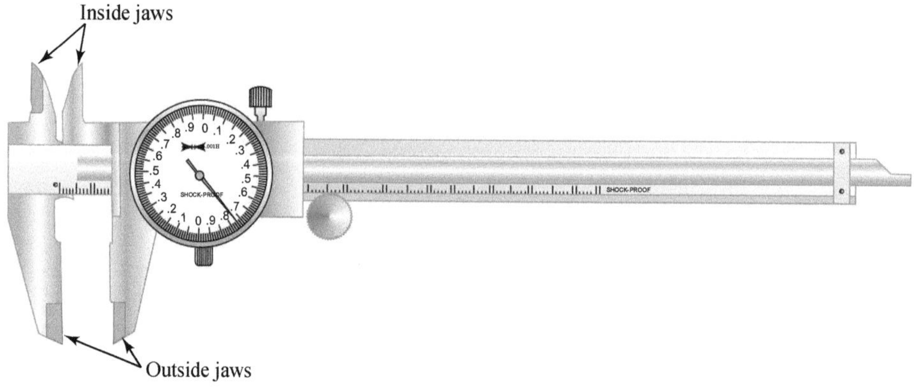

Fig. 22.5 A dial caliper

Dial caliper, another type of vernier caliper, is shown in Fig. 22.5. The extreme right part is used for depth and step measurements. It can give accuracy up to 0.0254 mm.

Digital caliper has LED screen which displays the measured dimensions. It is an improved version of vernier caliper. Digital caliper can be connected to a miniprocessor and printer for real-time data acquisition to be used in the analysis of statistical process control (SPC) and statistical quality control (SQC).

22.5.6 Height Gauge

Height gauge differs from caliper gauge in the sense that it has a single jaw, because the surface plate on which the instrument base rests functions as the reference plane. Vernier gauges usually have offset jaws whose contact surfaces can be brought to coincide with the reference plane when the slide position indicates zero height. Height gauge can be used to measure external and internal depth as well as the distance. It can also be used for checking perpendicularity, flatness, straightness, slot positions, centre locations and diameters. A digital height gauge with LED screen is shown in Fig. 22.6.

22.5.7 Micrometer

The micrometer, also known as micrometer calliper, is one of the most widely used precision measuring devices. It consists of a fixed anvil, a movable spindle, a graduated thimble which rotates during the measurement process and a ratchet which controls the pressure to be applied on the part being measured (Fig. 22.7).

Fig. 22.6 A height gauge

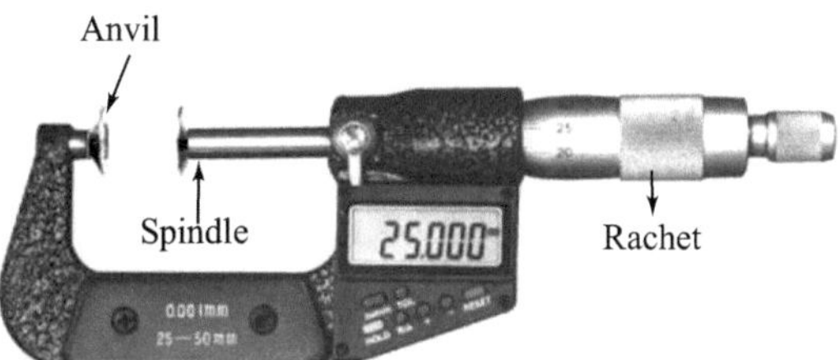

Fig. 22.7 A micrometer

The micrometer consists of two scales, a liner scale to measure directly the axial advance of the spindle and a circumferential scale on the thimble to measure the partial rotation of the screw in case when one turn is not complete. Micrometers designed for finer measurement have a vernier scale to permit the evaluation of fractions of circumferential graduations.

Micrometers can be used for inside measurements, depth reading and the measurements of screw threads. Micrometer readings can be accurate upto 0.0254 mm, and the accuracy can be further improved using new digital design.

Digital micrometer can measure upto 0.00127 mm accuracy. It has LED screen which displays the reading. A minicomputer can be attached to the digital micrometer to enhance the accuracy of the measurement. Digital micrometer can be used in SPC analysis.

22.6 Surface Texture

The surface texture in metal cutting is defined as the resulting irregularities arising from the plastic flow of metal during a machining operation. It varies mainly with the method of machining, type and condition of tool, workpiece material and overall stability. The surface texture has the following three components:

- Roughness
- Waviness
- Lay.

22.6.1 Roughness and Waviness

Roughness, also known as primary texture, is the smaller irregularities in the surface texture and is inherent in the production process. It shows the deviation of the surface of a component from a perfectly true and smooth plane. For example, tooling marks left on the surface during turning operation performed with a coarse feed on a lathe constitutes roughness. Waviness, also known as secondary texture, is a slow undulation in the component surface resulting from the machine or work deflection, sideway wear, vibrations and chatter. It is the larger irregularities upon which roughness irregularities are superimposed.

Waviness and roughness have no distinguishable boundary. Waviness is the characteristic form of topographical variations that are measurable as the profile of the part in an actual or imaginary cross section. It implies a repetitive and essentially regular occurrence of topographical features, and assumption that is based on the typical surface topography of machined surfaces. Waviness and roughness are shown in Fig. 22.8. Waviness width expresses the distance between adjacent crests. Waviness height is the distance, in a direction normal to the general surface, between the crests and the valleys.

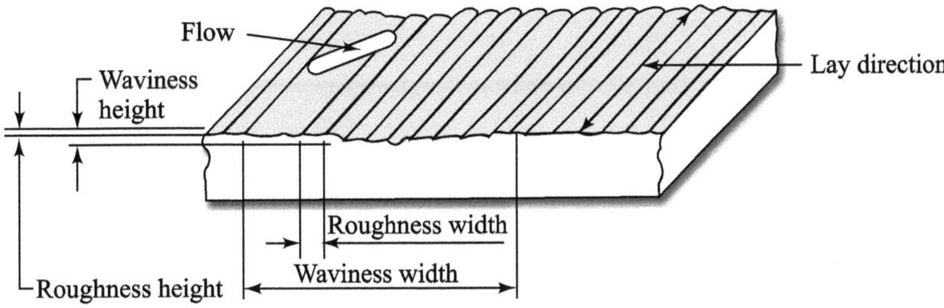

Fig. 22.8 Surface texture terminology

22.6.2 Lay

Lay is the orientation of surface pattern generated by the machining method.

22.7 Roughness Measurement

A number of instruments and techniques are in use for the purpose of roughness measurements. Most important among them includes methods such as stylus, tactile, light interference and replica. Stylus method is most commonly used. Numerical methods for roughness measurement include CLA, RMS, *Rz* and *Ra* values. The *Ra* value method is frequently used and has been recommended by ISO for expressing surface roughness.

22.7.1 *Ra* Value

The *Ra* value represents the arithmetic mean of the variation in the roughness profile from the mean line (M) and is measured in micrometers (μm). It does not give a clear indication of the shape of the physical surface. Its use is limited as very rough, fine, and short surfaces are not suitably assessed through the *Ra* value. The surface profile for Ra can be obtained through measuring with a stylus instrument. The mean line is drawn in such a manner that areas above and below it are equal in value, i.e., areas $A + C + E + G + I =$ areas $B + D + F + H + J$ (Fig. 22.9).

The *Ra* value is expressed as:

$$Ra = \frac{A + B + C + D + E + F + G + H + I + J}{L} \times \frac{1000}{V} (\mu\text{m}) \tag{22.3}$$

where

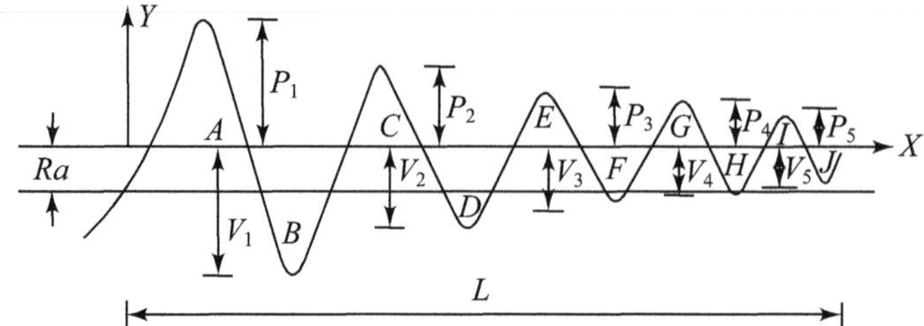

Fig. 22.9 *Ra* and *Rz* values

L = sampling length
V = vertical magnification of the graph.

The different areas are expressed in square millimetres and the length L in millimetres.

22.7.2 RMS Value

The RMS value is defined as the square root of the mean of the squares of the ordinates (heights) of the surface measured from a mean line within the evaluation length.

Mathematically,

$$h_{rms} = \sqrt{\frac{h_1^2 + h_2^2 + h_3^2 + \cdots + h_n^2}{n}}$$

$$= \sqrt{\frac{\sum_{i=1}^{n} h_i^2}{n}} \quad (22.4)$$

$$= \sqrt{\frac{1}{L}\int_0^L h^2 \mathrm{d}L} \quad (22.5)$$

where h_1, h_2 h_3, ..., h_n = ordinates from the mean line.

The RMS value is slightly greater than the Ra value for the same profile. The ratio of the two depends on the shape of the profile but in most of the cases it is slightly different from that for a sine wave, for which rms value is equal to 1.11 times the *Ra* value.

22.7.3 *Rz* Value

The *Rz* value represents mean peak-to-valley height. It is calculated as the arithmetic average of the five highest peaks and the five deepest valleys over the profile within the sample length. Due to ten points (5 peaks and 5 valleys) involved, it is also known as 'ten-point height of irregularities' method. Depending upon the profile form, extra assessment may be needed if there are insufficient peaks and valleys. The ISO *Rz* value is shown in Fig. 1.9 along with *Ra* value. This parameter is much easier to obtain from a profile graph *Ra* value.

Mathematically, *Rz* value is expressed as:

$$Rz = \frac{1}{5}[(P_1 + P_2 + P_3 + P_4 + P_5) - (V_1 + V_2 + V_3 + V_4 + V_5)] \times \frac{1000}{V} \quad (22.6)$$

where

P_1, P_2, P_3, P_4 and P_5 = peaks (mm)
V_1, V_2, V_3, V_4 and V_5 = valleys (mm).

22.7.4 *CLA* Value

The centre-line average (CLA) is defined as the arithmetic average of the ordinates drawn about a centre line of the profile within the sampling length.

Mathematically,

$$CLA = \frac{h_1 + h_2 + h_3 + \cdots + h_n}{n} \quad (22.7)$$

22.8 Flatness

Flatness represents the geometric concept of a plane. A theoretically perfect flat surface is one in which pairs of points selected randomly are connected by straight lines which are entirely contained in the surface.

The following instruments are used for flatness measurements:

- Optical flats
- Toolmaker flats
- Surface plates
- Straightedge.

22.8.1 Optical Flats

Optical flats are thick discs of either glass or quartz with parallel faces ground and polished flat to very high order of accuracy. Usually only one face is of specified flatness; it may be coated to increase the light reflected from the surface. Their surfaces are very near to a true plane.

During flatness measurement, optical flats are laid on the surface to be inspected is such a manner as to create interference bands (fringes) observable under monochromatic light; the resulting band pattern permits the object's flatness conditions to be evaluated (Fig. 22.10). This method of flatness measurement using light waves is called interferometry. The figure shows two fringes obtained with an incident light wavelength λ equal to 0.6 μm. The separation is found to be $2 \times \frac{1}{2}\lambda = 2 \times \frac{0.6}{2} = 0.6$ μm. The number of fringes that appear is dependent on the separation between the surface and the bottom face of the optical flat. The fringe is straight because the surface is flat as fringe always follows a line of constant distance between the surface and the lower face of the optical flat. They are, in a sense, contour lines. Straight fringes always lie parallel to the line of contact or line of minimum displacement of the optical flat from the surface. When surfaces are not flat, fringes are curved.

Optical flats are made in a wide range of sizes commonly from 25 to 300 mm in diameter, or larger, and in thicknesses adapted to the diameters, varying from 6 to 50 mm. They can measure with 0.00003 mm accuracy.

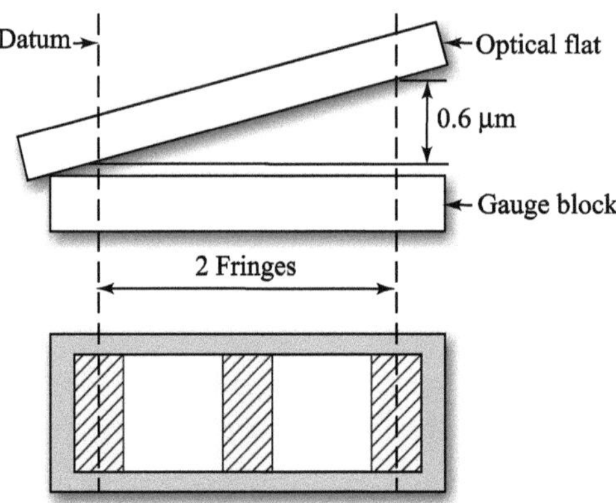

Fig. 22.10 Flatness measurement using optical flat

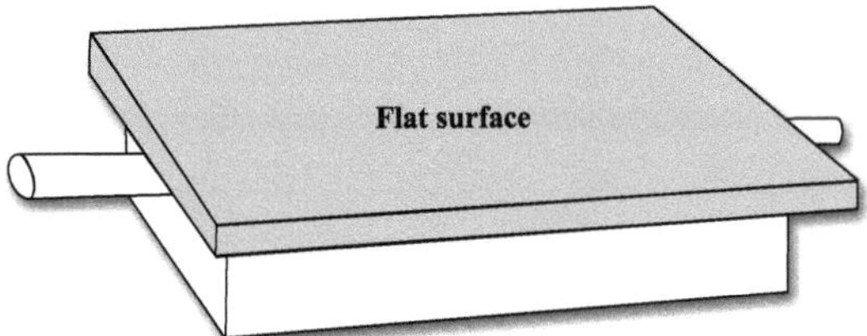

Fig. 22.11 A surface plate

22.8.2 Toolmaker Flats

Toolmaker flats are steel discs, which are lapped on both faces to a high degree of flatness, comparable to the measuring surface of a gauge block. The primary purpose of toolmaker flats is to serve as a mounting surface for gauge block assemblies and similar high precision elements of dimensional measurements. Frequently used sizes of toolmaker flats are 50 mm and 100 mm diameters, in thicknesses of 13 mm and 19 mm, respectively.

22.8.3 Surface Plate

A surface plate is an accurately machined casting or lapped granite block (Fig. 22.11). It has a surface of proved flatness. During flatness testing, the top of the plate should first be rubbed with a thin smear of a paste made up of red lead and oil. The surface to be tested should be wiped clean and then placed in contact with the surface plate and moved about. If it is reasonably flat, upon examination after this treatment, spots of the red lead will be visible all over it.

Surface plates are only infrequently used for measuring flatness directly. They are mostly used as a solid datum plane for the purpose of a very wide range of dimensional measurements.

22.8.4 Straightedge

A straightedge may be in the form of a steel strip or as a stiff casting with the edge straight (Fig. 22.12). For lengths greater than 300 mm, the ribbed cast iron pattern is to be recommended. The surface to be tested is compared with the straightedge in several directions. Any small discrepancy between the surface and the edge can be visible when the set-up is given light from one side. It can measure up to 1/200 of a millimeter.

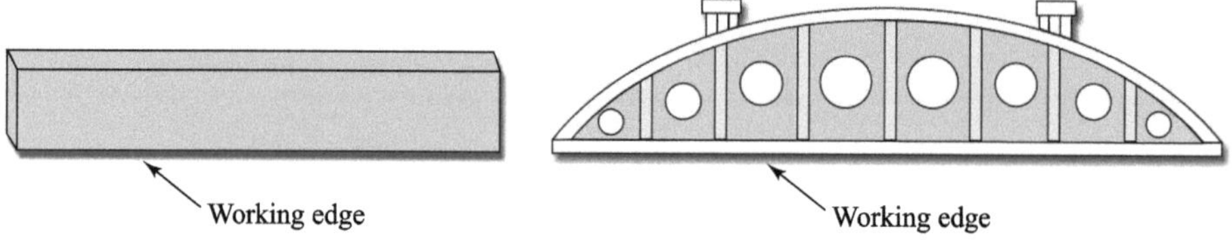

Fig. 22.12 Straightedges

22.9 Roundness

Roundness is the most important geometric form of manufactured parts and is important in the proper functioning of rotating shafts, pistons and cylinders and steel balls in bearings. Quantitatively, the degree of roundness is expressed by the dimension of out-of-roundness which is specified as the difference between the largest and the smallest measured radii, measured from a defined centre point. It is important to note that perfectly round part cannot be produced by any known means of manufacturing. Perfect roundness can be approached to varying degrees, and metrology plays here an important role giving the extent to which the object's condition actually approaches perfect roundness. The deviation of a hole with respect to roundness is normally caused by deflection, vibration, insufficient lubrication, wear, etc.

There are two basic systems of roundness measurements.

- Roundness measurement with intrinsic datum
- Roundness measurement with extrinsic datum.

22.9.1 Intrinsic Datum Roundness Measuring System

It is the conventional method of roundness checking and is more widely used. During testing, the object is positively supported on selected points of its surface, and another surface point is contacted by the probe of an indicator gauge of appropriate sensitivity. When the part is rotated while resting on the same support elements, variations in the location of the momentarily contacted points of the surface are shown by the indicator. The amplitude of these variations indicates the amount of roundness irregularities.

There are three methods under intrinsic datum system. They are discussed below.

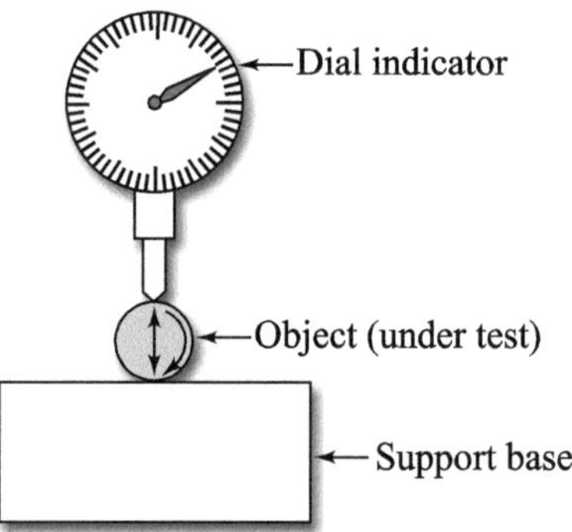

Fig. 22.13 Roundness measurement using diameter

22.9.2 Diametrical Measurement of Roundness

The diametrical measurement is the easiest method of inspecting roundness. For this purpose, the diameter of the part is tested in many different directions but confined to the same plane and normal to the axis of the part by using micrometer or callipers (Fig. 22.13). The greater the number of these measurements and the better they encompass the entire perimeter of the object, the more reliable are the inspection results.

If the test shows same diameter at every position, the part can be assumed to be round and if the diameters vary, the part is not round.

22.9.3 Roundness Measurement Using Vee-Block

In this method, the object to be tested is placed in a Vee-block and is slowly rotated and shifted to new positions. A dial indicator is used to record any variations in dimensions (Fig. 22.14). This process is widely used in general shop practice.

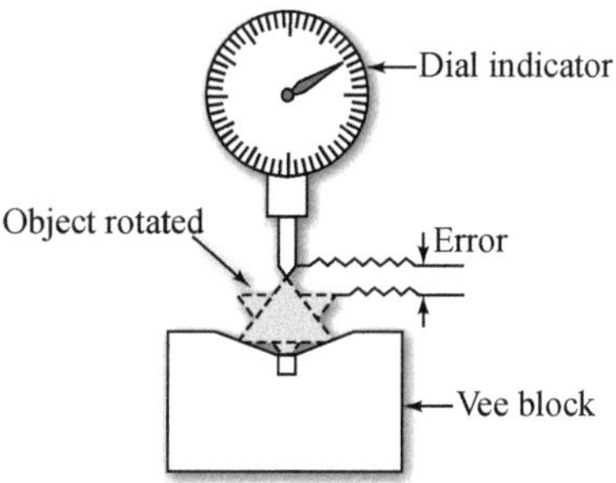

Fig. 22.14 Vee-block-assisted roundness measurement

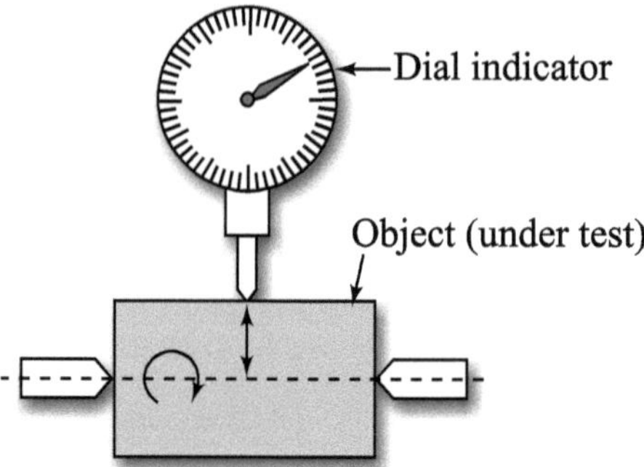

Fig. 22.15 Centre method of roundness measurement

22.9.4 Roundness Measurement Between Centres

The object to be tested is placed in the centres of the bench and clamped. The probe of an indicating instrument is now brought into contact with the selected element of the surface to be inspected and the object rotated, usually by hand. The range of indicator reading is indicative of radial out-of-roundness measurement. Figure 22.15 shows the roundness measurement of a cylindrical object.

22.9.5 Extrinsic Datum Roundness Measuring System

This system does not use any part of object as a datum. The referencing is accomplished from an external member, an ultraprecise spindle with almost perfectly runout free rotation.

This system has the following advantages:

- It requires minimum tooling.
- It is easier to operate; the positively supported object needs no centring or squaring adjustments.
- It is adaptable for continuous in process-measurements and for monitoring automated production.

22.10 Straightness

Straightness is related to a straight-line measurement. The shortest distance between two points is along a straight line. The method for checking the straightness of a surface element is by means of direct contact comparison with a tool of known and adequate straightness. The sensitivity of such direct comparison is, of course, limited. The straight edge, which can also measure flatness, represents the most commonly used tool for that type of limited sensitivity straightness measurement. It is discussed in the section of flatness measurement. Other types of tools for direct contact checking of straightness are the knife edge and the triangular straight edge. These tools have narrow measuring edges and are used for checking the straightness of a surface, or the coincidence of tangent lines to two parts of nominally equal lengths, by observing the light gap that may appear between the part surface and the knife edge at certain sections of the contact length indicating that the surface is not perfectly straight. With proper skill and adequate lighting, gaps of 0.00254 mm width can be distinctly detected.

Level method is used for testing the straightness of a horizontal surface by means of a sensitive level. Collimators are target instruments commonly used in conjunction with telescopes for alignment measurements.

22.11 Angular Measurements

Angular measurements are derived from the subdivision of a circle for which basic unit is the right angle obtained from the intersection of two mutually perpendicular straight lines. Right angle is divided into 90 units, each unit represents 1°, each degree into 60 min and minute into 60 s. This system is called sexagesimal. But mostly angle is defined in terms of relationship between the radius and arc of a circle and its unit is radian. When the arc PQ which subtends a certain angle θ at the centre becomes equal to radius of the circle, then the angle is equal to 1 radian (Fig. 22.16). Angular measurements do not need referencing.

There are two main sources of error in angular measurement.

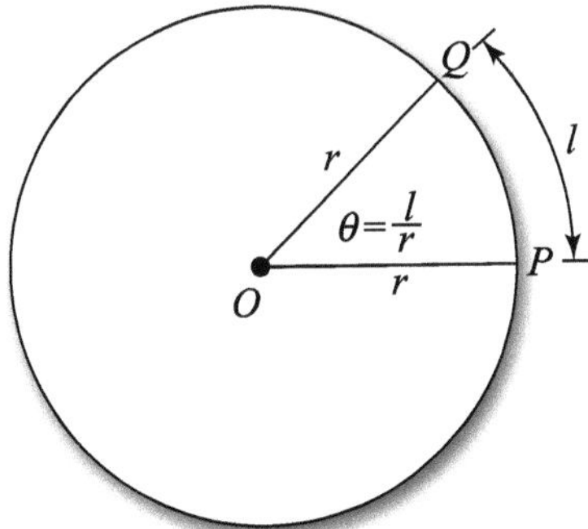

Fig. 22.16 Definition of radian

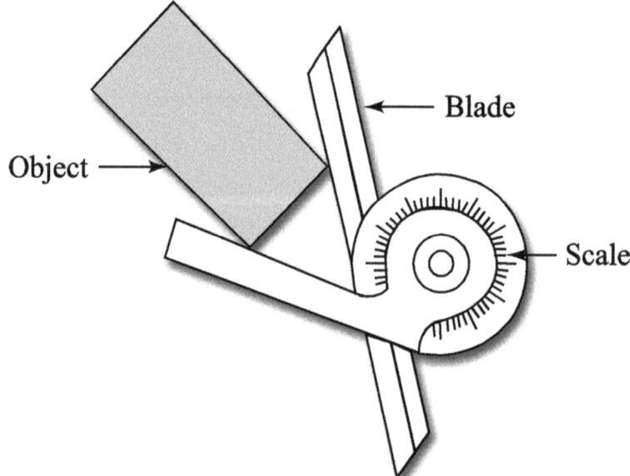

Fig. 22.17 A bevel protractor

- Error of centring, e.g., of a divided scale
- Error between the plane in which an angle is defined and the plane in which it is measured.

22.11.1 Bevel Protractor

Bevel protractor is an important angle–measuring instrument. The accuracy of the bevel protractor can be increased by incorporating a vernier, and the resulting instrument is called bevel vernier protractor. The two movable blades of protractor are brought into contact with the object being measured, and the angle can be measured directly on the vernier scale in degrees and minutes (Fig. 22.17).

22.11.2 Sine Bar

A sine bar is a hardened steel beam mounted on two rolls or hardened cylinders of equal diameter and mutually parallel at a known distance. This distance can be 5 inch or 10 inch, and accordingly it is known as 5 or 10 inch sine bar. The sine bar is used for measuring accurate angles and is generally used in conjunction with gauge blocks. The principle of operation of sine bar is based on the application of trigonometry according to which sine of an angle is the ratio of opposite side to the hypotenuse of the right-angled triangle.

During the measurement process, the object to be measured in placed on the sine bar and one end of it is raised through a stock of gauge blocks until the top surface of the object becomes parallel to the surface plate (Fig. 22.18). The entire set-up rests on

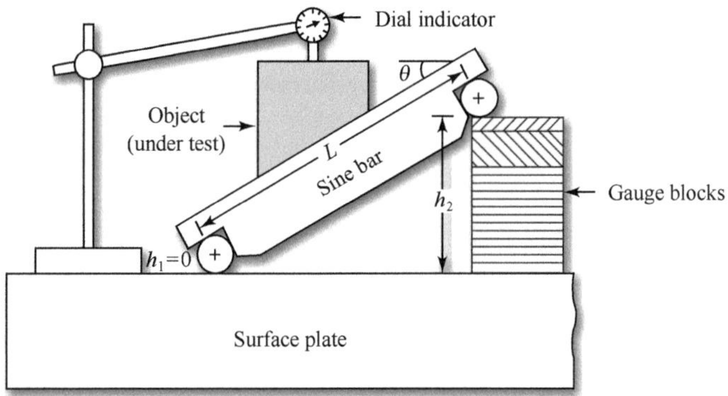

Fig. 22.18 Use of a sine bar

a surface plate made of cast iron or granite. The reading on the dial indicator gives the desired angle. The height of the stack directly determines the difference in height between two ends of the sine bar.

Mathematically,

$$\sin\theta = \frac{h_2 - h_1}{L} \tag{22.8}$$

In Fig. 22.18, $h_1 = 0$, because left end of the sine bar lies on the surface plate itself.

22.12 Coordinate Measuring Machines (CMM)

Coordinate measuring machines are used to measure all the three axes required for the physical identification of an object. Such machines use a touch-trigger probe which exerts very low contact forces. They are capable of measuring extremely complex profiles. Measurements are shown on digital readouts. The air-bearing systems provide frictionless motion for the axes. Parts may be inspected on-line by connecting CMM to a computer. One of the major advantages of CMM is that CAD/CAM system can be used to generate the CNC program for machining the parts, and then the same database information can be used to generate the CMM inspection routine to ensure that the part conforms to the design specification.

22.13 Comparators

A comparator is used to compare known and unknown dimensions. The difference in two dimensions is usually small which can be magnified by the comparators. A comparator

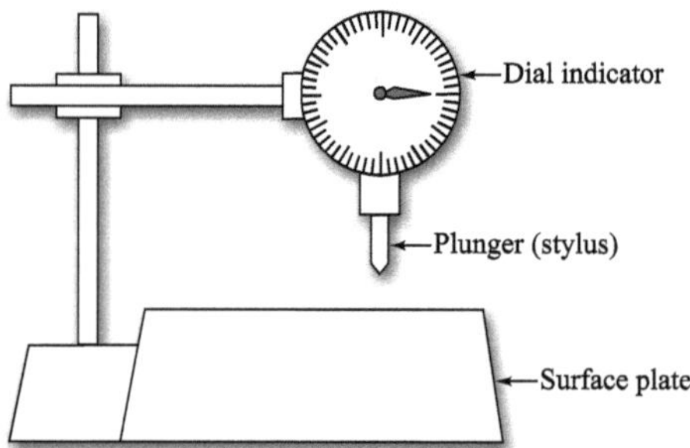

Fig. 22.19 A dial indicator

consists of three basic units: (a) a sensing device (plunger or stylus) to sense the input signal (b) a magnifying system to magnify the signal and (c) a display system to show the measurement. The comparators are generally used for linear measurements and are classified on the basis of their magnification units.

22.13.1 Mechanical Comparator

Mechanical comparators use levers for magnifying the small movement of the measuring stylus brought about due to the difference between the standard and the actual dimension being checked. The usual magnification lies between 300 and 5000. Sigma, Johansson Mikorkator and Venwick are some of the important mechanical comparators.

Dial indicator is one of the most commonly used mechanical comparators. It consists of a dial which is graduated into suitable number of divisions, a pointer and a plunger (Fig. 22.19). During the time of measurement, the plunger is pressed against the work to be tested and its magnified movement is displayed by the dial indicator. It is used to align machine tools, fixtures and work, to test and inspect the size and trueness of a finished work to an accuracy of one-hundredth of a mm.

22.13.2 Optical Comparator

The optical comparator is also known as optical projector or projecting comparator. It has a high degree of precision and suffers less wear during its usage than the mechanical type comparators.

The optical comparator consists of a light source, a condensing lens, a projection lens, a mirror and a screen. The object is placed on the worktable in front of the light source. The image of the object is produced by a beam of light. The shadow image of the object is projected at some enlarged scale on a screen, where it is compared to a master chart

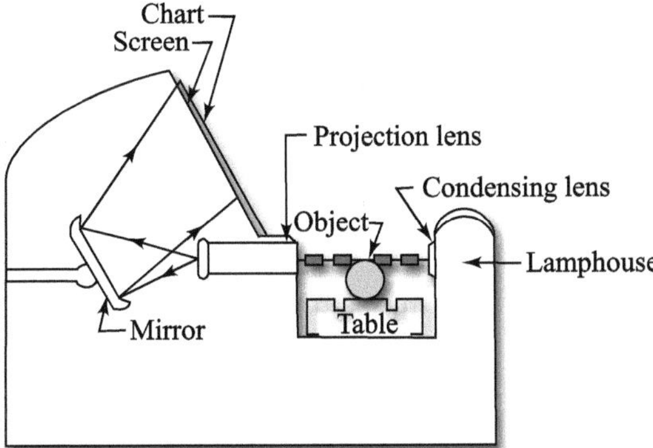

Fig. 22.20 An optical comparator

or drawing (Fig. 22.20). The light source is a tungsten lamp, filament lamp or high-pressure mercury or zenon arc lamp. The optical comparator displays a two-dimensional projection of a part, and it may be noted that all parts are not ideally suited to optical gauging. It is used to magnify parts of very small size and of complex configuration that requires accurate and enlarged profile. The magnification may vary from $\times 5$ to $\times 100$. The accuracy of the measurement is usually limited to 0.001 mm.

It is widely used in the inspection of small parts such as needles, saw teeth, gear teeth, taps and screw threads.

22.13.3 Pneumatic Comparator

A pneumatic comparator uses compressed air for magnifying the measurements. Very high magnifications (upto 10,000) are possible. Since there is no physical contact between the setting gauge and the object being measured, there is no loss of accuracy because of gauge wear. Additionally, pneumatic comparators are simple, relatively inexpensive, require minimum skill and can measure with high accuracy of about 0.013 mm and that too without affecting the finest or softest finish. They can check internal and external dimensions and multiple checking of several dimensions can be done simultaneously.

The operation of pneumatic comparator is based on a double-orifice arrangement (Fig. 22.21). If one of them is kept uniform, the pressure will vary according to the size of the other. The air pressure P_2 depends on the source pressure P_1 (constant) and is also dependent on the size of measuring orifice. As the gap is changed, pressure P_2 changes and this change can be used as a measure of the gap.

Initially air is supplied at constant pressure of about 150 kPa to the control orifice. If there is no gap, i.e., when the object closes the gap, the pressure in the chamber is equal to P_1. When the gap is increased, the pressure in the chamber falls to P_2.

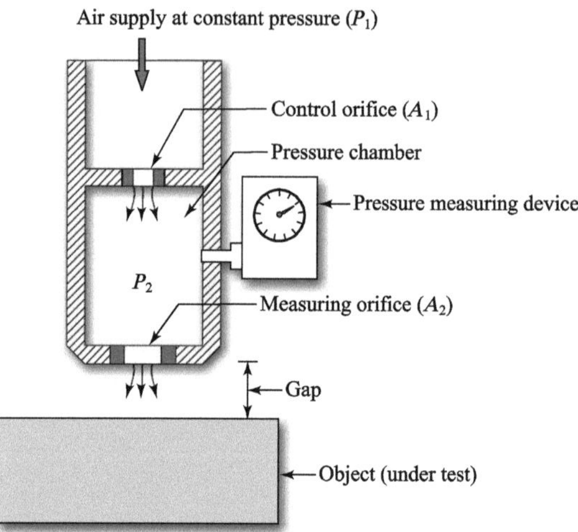

Fig. 22.21 A pneumatic comparator

The pneumatic sensitivity varies directly proportional to the source pressure and inversely proportional to the area of the control orifice. Also, there is an approximate linear relationship between pressure ratio (P_2/P_1) and area ratio (A_1/A_2).

22.13.4 Electrical Comparator

Electrical comparator uses electrical signal for measuring the small displacement of a workpiece. As compared to mechanical comparator, it has several advantages. Mechanical comparators are actuated by levers, gears, racks and pinions, and their accuracy is affected by wear and friction of these elements. Electrical comparators, on the other hand, have little or no moving parts, and therefore, they can maintain their accuracy over long periods. Higher magnification, ranging from 600 to 10,000 is possible. They can measure displacement as small as 0.0001 mm.

Electrical comparator consists of a transducer which converts mechanical displacement into a corresponding change in current and a metre or recorder to show the electrical change, calibrated to show in terms of displacement (Fig. 22.22). One of the important types of electrical comparator, called an electrolimit gauge uses wheatstone bridge for electrical magnification. The object to be checked is placed under the gauging spindle (stylus). Movement in the spindle disturbs the balance of the wheatstone bridge resulting in unbalanced current, thereby indicating the difference between the datum size and the object size by suitable calibration of the unbalanced current.

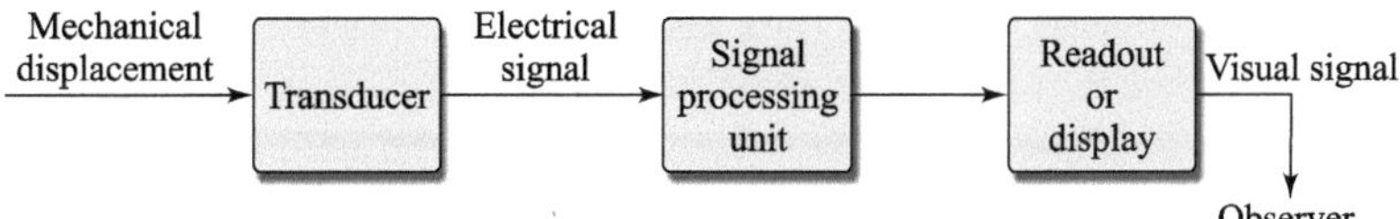

Fig. 22.22 An electrical comparator

22.13.5 Electronic Comparator

Electronic comparator can give magnifications of extremely high order. They do not contain any mechanical parts and are high precision instruments. Also, their sensitivity and speed reduce the time lag during measurements. Electronic comparators are particularly useful, when measurements of a moving feature are required. Measurements along a strip passing through a rolling mill are typical example of dynamic measurement. This comparator operates on a variable inductance transducer, commonly known as liner variable differential transformer (LVDT). The LVDT converts length displacement into a proportional electric voltage and it cannot be overloaded mechanically and is relatively insensitive to temperature. Electronic comparators have magnification ranging from ± 0.200 mm to ± 0.002 mm. Amplifier readouts are provided on either a sensitive meter or a direct digital readout display. The meter version has a precision-calibrated meter dial that is designed for easy readability without parallax error.

22.14 Gauges

Gauges are used for checking the size, shape and relative positions of various parts. Gauging is the name given to the process of measurement to assure the specified uniformity of size and contour. Basically gauging performs two functions:

- It controls the dimensions of a product within the prescribed limitations.
- It segregates or rejects products that are outside these limitations.

There are many gauging methods used to determine when a product conforms dimensionally with drawings, specifications or other prescribed requirements. The workpiece's geometry such as length, flatness, parallelism, squareness, concentricity and surface texture are the parameters which form the basis of measurement. A clear distinction between measuring instruments and gauges is not always observed.

Gauges are broadly classified into three groups:

- Fixed gauges
- Indicating gauges
- Combination gauges

Fixed gauges are replicas of the shapes of the parts to be measured. They are so named because they are tied to a particular operation and in most cases are built in the plant where the gauge is to be used. They are the most common and are used for both large- and small-scale production. They are divided into two categories: Standard or master gauges and limit gauges. Standard gauges are used as reference standards for setting, checking, adjusting or calibrating different types of measuring instruments. They can check if the part has been made to the design tolerance. Their dimensions may be the basic size or the median size of the tolerance zone.

Fixed-limit gauges are used to ensure that a product is within the prescribed limits of size and are intended for use as inspection gauges. Since there is a high and low limit on the product, two limit gauges are usually required, although both may be on the same device. GO gauge and NO GO gauges are twin gauges measuring the maximum and minimum limits of a workpiece. The GO gauge is smaller than NO GO gauge and slides into any hole whose smallest dimension is less than the diameter of the gauge. The NO GO gauge must not go into the hole.

Ring, plug and thread gauges are commonly used fixed gauges. Ring gauges are used to measure cylindrical surfaces such as shafts, tapers on shafts, and similar workpieces, as well as external threads. They are made in GO and NO GO types. Plug gauges are accurately ground cylinders, may be straight or tapered and of any cross section. They are used to check the uniformity of holes.

Indicating gauges are used for inspection purposes and are comparatively complex. The function of the gauge is to indicate coincidence, or the amount and direction of deviation from the basic size, when the occurring variations are within the measuring range of the instrument. The majority of indicating gauges compare the actual dimension of the workpiece with the dimension of a master setting gauge. For this reason, indicating gauging methods are often referred to as comparative gauging. Such variations in dimensions are shown on a dial or a graduated scale. The amplification of indicating gauge movement may be mechanical, optical, electrical or pneumatic.

Combination gauges are specially designed to measure or check more than one dimension of a workpiece at a given set-up.

22.15 Autocollimator

Autocollimator is an optical instrument used for the measurement of small angular displacement on a flat surface. It is also used to determine straightness, flatness, perpendicularity, parallelism, squareness and alignment. It resembles a telescope with a light beam that bounces back from the object to be measured. Most of the elements of their optical system are arranged in line and are mounted in a cylindrical housing, the barrel. This system is called a barrel-type autocollimator and is the most widely used one.

22.16 Automatic Gauges

Automatic gauges can perform the following functions:

- It can be used in the successive inspection of a number of independent dimensions and the automatic passing of only those workpieces which come within set tolerances; i.e., it serves as a basis of acceptance or rejection.
- It can be adapted to assigning the accepted parts to different size categories, such as would be needed for selective assembly with the corresponding categories of similarly inspected mating parts. This process of dimensional sorting is called segregation.
- A combination of the above two functions.

Automated inspection is based on on-line measurement systems that monitor the parts while being made and use these measurements to correct the process.

The use of accurate sensors and computer-controlled systems have integrated automatic measurement into manufacturing operations. Automated inspection is flexible and responsive to product design changes. Other advantages include reduced involvement of operator and increased productivity, and parts have higher-quality reliability and dimensional accuracy.

Automatic gauging systems are generally applied to precision machine tools only, e.g., grinding, honing, etc. They may be of pneumatic, electrical, mechanical or electronic type. Consider the example of a simple automatic gauge for checking the size of steel balls. During the test, the balls are sent rolling down an inclined raceway fitted with knife edges so set that the correct-sized ball reaches the proper point, where it drops into an already place container. Undersized balls drop through before this point, and oversized balls roll by and drop through at the end of the raceway.

22.17 Screw Thread Terminology

A screw thread is a helical groove on the surface of a cylinder. The profile of the screw thread, when vented in the axial cross section of the threaded part, is most commonly a symmetrical vee-shape, although screw threads with other cross-sectional shapes, and also some with nonsymmetrical profiles, are used for particular purposes. In the majority of screw thread types, the symmetrical vee-shape of both the groove and the ridge have a 60° included angle.

Nominal size, number of threads per inch, thread series symbol and thread class taken in order are used to designate a screw thread. The nominal size is the basic major diameter.

The basic screw thread terminology is shown in Fig. 22.23.

Pitch is the distance, measured parallel to the part axis, between corresponding points on adjacent thread flanks, e.g., the distance from crest to crest or root to root. Pitch is also

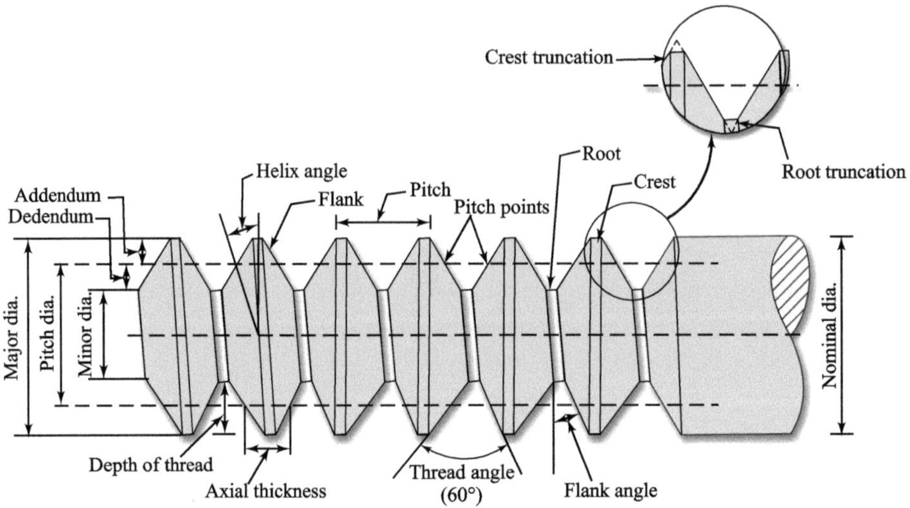

Fig. 22.23 Screw thread terminology

expressed as the number of threads per axial inch. For example, a screw having 16 single threads per inch is said to have a pitch of 1/16. But it is not the correct way of defining a pitch.

Lead is the distance moved by the screw axially in one complete revolution. The relationship used to connect pitch with lead is given as

$$\text{Lead} = n \times \text{pitch}$$

where $n =$ number of starts.

For single start threads, lead is equal to the pitch. In multistart threads, lead equals the number of starts multiplied by pitch.

Nominal diameter is the diameter of a cylinder on which threads are cut. The screw is specified by this diameter.

Major diameter, also known as crest diameter, is the largest diameter of the screw thread.

Minor diameter, also known as root diameter, is the smallest diameter of the screw thread.

Pitch diameter, also known as effective diameter, is the diameter of an imaginary cylinder passing through the width of the threads and is located where the widths of the thread ridge and the thread groove are equal.

Root is defined as the bottom of the groove between the two flanks of the screw thread.

Flank is defined as the surface which connects the crest with the root.

Depth of the thread is the distance between the crest and root of a screw thread measured perpendicular to its axis.

Angle of thread is the angle between the flanks, measured in an axial plane.

Helix angle is the angle made by the helix at any point with a plane perpendicular to the axis.

Internal thread is formed on the inside of a workpiece, e.g., on a nut or female screw gauge.

External thread is formed on the outside of a workpiece, e.g., on bolts or studs etc.

Axial thickness is the distance between the opposite faces of the same thread measured on the pitch cylinder in a direction parallel to the axis of screw thread.

Addendum is the radial distance between the major and pitch cylinders for external screw thread and is the radial distance between the minor and pitch cylinders for internal screw thread.

Dedendum is the radial distance between the minor and pitch cylinders for external screw thread and is the radial distance between the major and pitch cylinders for internal screw threads.

22.17.1 Truncation

A thread is sometimes truncated at the crest or at the root or at both. Crest truncation is the radial distance from the crest to the nearest apex of the fundamental triangle, and root truncation is the radial distance from the root to the nearest apex.

22.18 Class or Grade of a Screw Thread

Screw threads are standardized according to their cross-sectional form. They are classified into two systems.

- Unified screw thread standards
- ISO standards.

The unified standards are based on the American National thread system and are used in USA, Canada and UK to obtain interchangeable screw threads. The threads in this system are called inch threads.

The ISO standard, developed in 1969, is an universal standard and screw threads in this system are called metric threads.

Other types of threads include square, knuckle, acme and buttress.

22.19 Screw Thread Gauging and Measurement

Screw threads are important elements of mechanical design which are used for the purpose of transmission of power and motion, and to act as a fastener such as bolts, nuts, machine screws, studs and foundation bolts.

The objective of screw thread measurement is to determine whether the different parameters of the threads agree, within the tolerance limits, with the pertinent design dimensions. The processes and means of screw thread measurement usually differ for external and internal threads. The different thread features are measured basically by means of thread gauges of various designs that compare the thread produced against a standard thread.

Short Answer Questions

1. What is the difference between line measurement and end measurement? Name a few line-measuring and end-measuring instruments.

Ans. Line measurement refers to measuring of distance between two lines or edges, whereas end measurement refers to measuring of distance between two surfaces. Line-measuring instruments include vernier calliper and micrometer, and end-measuring instruments include slip gauges.

2. What is a comparator? Name a few important comparators.

Ans. Comparator is an inspection tool for comparing two measured dimensions. A few important comparators include mechanical, hydraulic, optical, pneumatic and electronic.

3. How does surface plate differ from try square? Of what material a surface plate is made of?

Ans. A surface plate is used to measure flatness of a given surface, whereas a try square is used to measure perpendicularity of two surfaces of a component. A surface plate is made of cast iron or graphite.

4. What is the difference between bevel protractor and autocollimator?

Ans. A bevel protractor is used to measure angles, whereas an autocollimator is used to measure small angular differences.

5. Of the two threads square and vee, which one is used for power transmission and why? Which one is used for locking purpose?

Ans. Square threads are used for power transmission as they offer smaller friction as compared to vee threads. Vee threads because of high friction are used for locking purpose.

Multiple Choice Questions

1. Match List I with List II and select the correct answer using the codes given below the lists:

List I	List II
A. Accuracy	1. Degree of repeatability
B. Calibration	2. Ability to measure small difference of measurement
C. Sensitivity	3. Degree to which the measured value agrees with the true value
D. Precision	4. Comparison of a measuring instrument with a known standard

Codes:	A	B	C	D
(a)	3	4	2	1
(b)	3	4	1	2
(c)	1	3	4	2
(d)	1	4	2	3

2. The standard temperature adopted for metrology is

(a) 65 °F (b) 68 °C
(c) 67 °F (d) 68 °F

3. Gage blocks are usually made of which of the following shapes?

(a) Triangular (b) Square
(c) Rectangular (d) Square and rectangular

4. Match List I with List II and select the correct answer using the codes given below the lists:

List I	List II
A. Sine bar	1. Measures linear dimensions

B. Micrometer	2. Measures flatness and parallelism
C. Profilometer	3. Measures angle
D. Coordinate measuring machine	4. Measures surface roughness

Codes:	A	B	C	D
(a)	3	1	2	4
(b)	3	1	4	2
(c)	3	2	4	1
(d)	1	3	4	2

5. Match List I with List II and select the correct answer using the codes given below the lists:

List I	List II
(Measuring devices)	(Measurement parameters)
A. Diffraction grating	1. Small angular deviations on long flat surfaces
B. Optical flat	2. On-line measurement of moving parts
C. Autocollimator	3. Measurement of gear pitch
D. Laser scan micrometer	4. Surface texture using interferometry
	5. Measurement of very small displacements

Codes:	A	B	C	D
(a)	5	4	2	1
(b)	3	5	1	2
(c)	3	5	4	1
(d)	5	4	1	2

6. Match List I with List II and select the correct answer using the codes given below the lists:

List I	List II
(Comparator)	(Related to)
A. Mechanical	1. Transducer
B. Optical	2. Stylus
C. Pneumatic	3. Projection lens
D. Electrical	4. Double-orifice arrangement

Codes:	A	B	C	D
(a)	2	3	1	4
(b)	2	3	4	1
(c)	4	3	2	1
(d)	3	4	2	1

7. A surface plate is used to measure

(a) Roundness
(b) Angle
(c) Flatness
(d) Coordinates

8. Dial indicator is a

(a) Pneumatic comparator
(b) Electrical comparator
(c) Optical comparator
(d) Mechanical comparator

9. An autocollimator is used to

(a) measure small angular displacement on a flat surface
(b) compare known and unknown dimensions
(c) measure the three axes required for the physical identification of an object
(d) measure roundness between centres.

10. Consider the following statements:

(a) In multistart threads, lead equals the number of starts multiplied by pitch
(b) The centre-line average (CLA) value gives the arithmetic average.
(c) Root diameter is the smallest diameter of the screw thread.
(d) Random errors obey normal distribution curve.
Of these statements:

(a) 1, 2 and 4 are true
(b) 2 and 4 are true
(c) 1, 2, 3 and 4 are true
(d) 1, 3 and 4 are true

Answers

1. (a) 2. (d) 3. (d) 4. (b) 5. (d) 6. (b)
7. (c) 8. (d) 9. (a) 10. (c)

Review Questions and Discussions

Q.1. What are unilateral and bilateral tolerancing?
Q.2. What is the major advantage of electronic gauge?
Q.3. How are roughness and waviness of a machined surface measured?
Q.4. What is a comparator? How does it differ from a gauging system?
Q.5. What is vernier acuity?
Q.6. Why is a sine bar not preferred for generating angles greater than 45°? How is it used for generating angles greater than 45°?
Q.7. What are slip gauges?
Q.8. What are the advantages of optical standards of length?

Manufacturing Tools and Workshop Applications

23

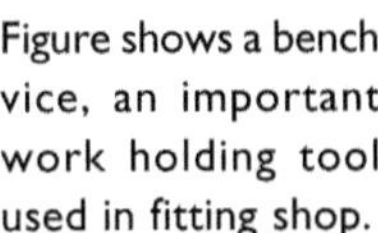
Figure shows a bench vice, an important work holding tool used in fitting shop.

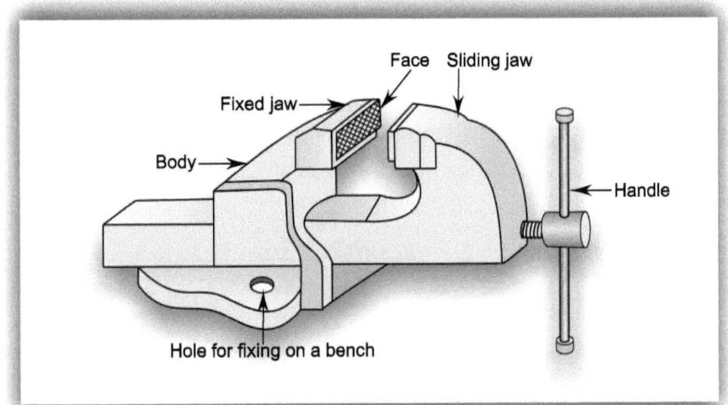

Learning Objectives

After reading this chapter, you will be able to answer some of the following questions:

- Why has a swage block numerous applications?
- Why is lathe called a basic machine tool?
- Why are lathe centres called live and dead?
- How is a mallet different from a hammer?

D. K. Singh, *Fundamentals of Manufacturing Engineering*,
https://doi.org/10.1007/978-981-99-8767-2_23

23.1 Introduction

Manufacturing tools are used to produce objects of desired shape and size from given metal stocks to their required dimensions. They are classified according to the shop, where they are used, e.g. fitting shop tools, welding shop tools, sheet metal shop tools, smithy tools, foundry shop tools, machine shop tools and others.

23.2 Fitting Shop and Fitting Shop Tools

Fitting shop is used to shape an object by cutting or removing materials from a given raw metal stock and securing proper fit. The important fitting shop tools are discussed below:

23.3 Files

Files are the most commonly used tools used in the fitting shop. They are used to remove undesired material from a workpiece in small quantity in order to give it smooth finish. The material removed is in the form of dust particles and not in the continuous or discontinuous chips. Files are made of high carbon steel to ensure its hardness during cutting operations. A file with its various parts is shown in Fig. 23.1.

It consists of a wooden handle to be fixed in the tang. The ferrule is a metal ring, which prevents the splitting of the handle. Cutting takes place during forward stroke by the face of file.

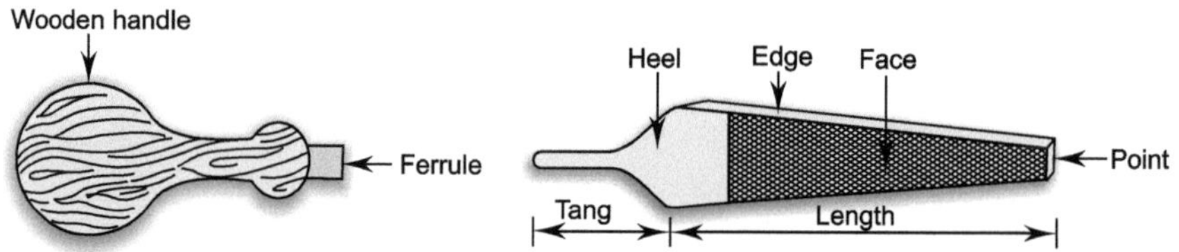

Fig. 23.1 A file and its parts

23.4 Classification of Files

Files are classified according to their size, cut and shapes.

(a) *Size* of a file indicates its length between point and heel (Fig. 23.1). For fine work, it may vary between 100 and 200 mm, and for rough work between 200 and 450 mm.

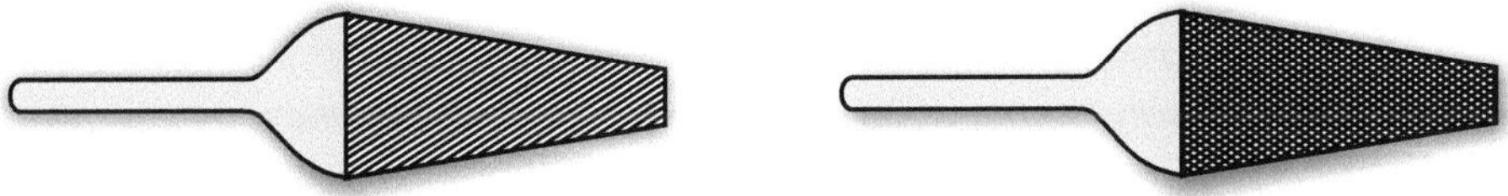

Fig. 23.2 Single and double cut files

Fig. 23.3 A flat file

(b) *Cut* implies single or double cut of a file (Fig. 23.2), which further divides it on the basis of coarseness into rough, bastard, smooth, dead smooth or super smooth category.
(c) *Shape* of a file indicates its cross section. Based on cross section, files are classified as flat, hand, square, round, pillar, triangular, half-round or knife-edge files.

23.5 Types of Files

23.5.1 Flat File

A flat file (Fig. 23.3) is parallel for about two-thirds of its length and then tapers in width and thickness. There are double cut on its both faces and single cut on edges. It is mostly used for filing flat surfaces and hence is so named. It has rectangular cross section.

23.5.2 Hand File

A hand file (Fig. 23.4) is parallel for its entire length but has tapered thickness. It is of rectangular cross section having double cut on faces, single cut on one edge whereas there is no cut on other edge, called safe edge. It is used in the filing of square corners of the work, without affecting its already filed adjacent surface.

Fig. 23.4 A hand file

Fig. 23.5 A square file

23.5.3 Square File

A square file (Fig. 23.5) has parallel width for about two-third of its length and then tapers towards point. It is of square cross section having double cut on all the sides (face and edge). It is used in the filing of square corners and slots, and hence is so named.

23.5.4 Triangular File

A triangular file (Fig. 23.6) has parallel width for its entire length or alternatively may have parallel width for one-half of its length and tapered for remaining length towards point. It has triangular cross section and double cut on all faces. It is used in the filing of square corners and sharpening of wood working saws.

23.5.5 Pillar File

A pillar file (Fig. 23.7) is similar to a hand file, but has reduced width and increased thickness. Its one or both edges may not have any cut. It is used in the filing of narrow works such as keyways, slots and grooves.

Fig. 23.6 A triangular file

Fig. 23.7 A pillar file

Fig. 23.8 A round file

Fig. 23.9 A needle file

23.5.6 Round File

A round file (Fig. 23.8) is uniformly round throughout its length or to one-half of its length and may be tapered for the remaining length. It is used in the filing of holes, round corners or round slots.

23.5.7 Needle File

A needle file (Fig. 23.9) is similar to a needle having very sharp point. It is used in the filing of very fine works such as pierced design in sheet metal.

23.5.8 Warding File

A warding file (Fig. 23.10) is similar to a flat file but has sharply tapered width towards point and has reduced thickness. It is used in the filing of narrow slots.

Fig. 23.10 A warding file

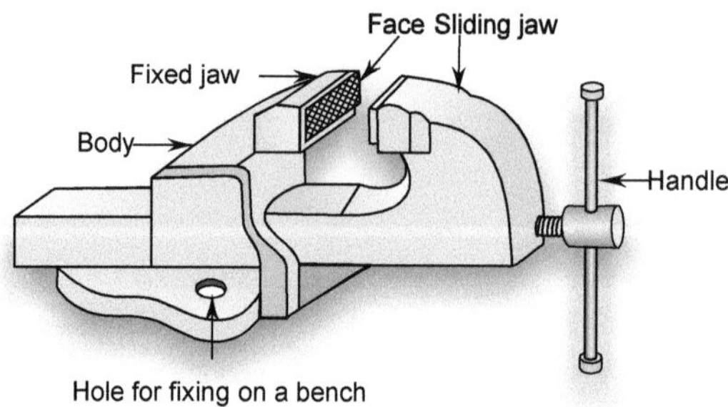

Fig. 23.11 A bench vice

23.6 Bench Vice

A bench vice (Fig. 23.11) is mounted on a bench. It is used for holding the job between its two jaws. Its one jaw is fixed and other movable. The workpiece, at the start of any operation on it, is kept close to the fixed jaw and the movable jaw is brought closer to support the workpiece.

23.7 Chisels

Chisels are used for cutting and chipping away pieces of metals. They are forged to shape from low-carbon steel and then hardened and tempered. The usual cross section of chisel may be rectangular, hexagonal or octagonal. Flat, cross-cut, half-round, diamond point or side chisels are important types of chisels. A flat chisel is shown in Fig. 23.12.

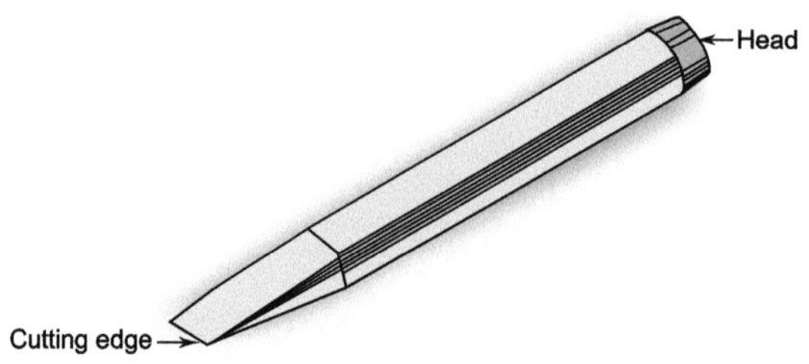

Fig. 23.12 A flat chisel

23.8 Hacksaw

A hacksaw (Fig. 23.13) is used for cutting (sawing) a workpiece into two parts. It consists of a frame, handle, prongs, tightening screws and nut and blade. Its blade is made of high carbon steel or high-speed steel.

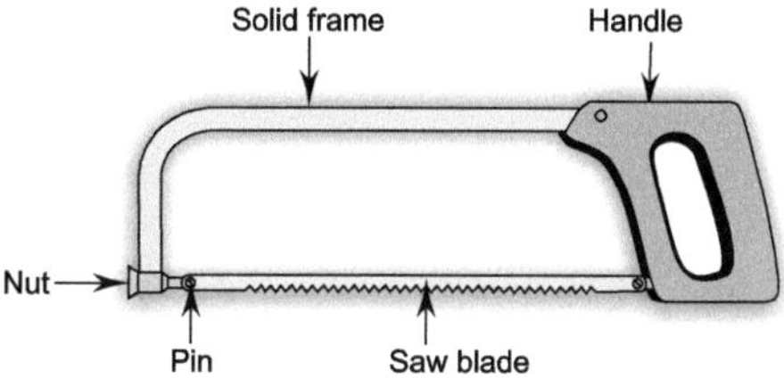

Fig. 23.13 A solid frame hacksaw

23.9 Scriber

A scriber (Fig. 23.14) is used to create marking (indented impression) on the workpiece to make cutting along the marked line. It looks like a needle having sharp point on one or both sides. It is made of hardened steel.

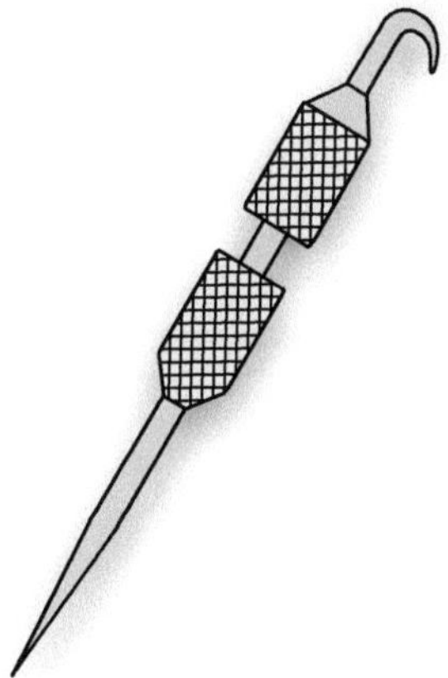

Fig. 23.14 A scriber

23.10 Punch

A punch (Fig. 23.15) is used for marking out work and locating centres prominently.

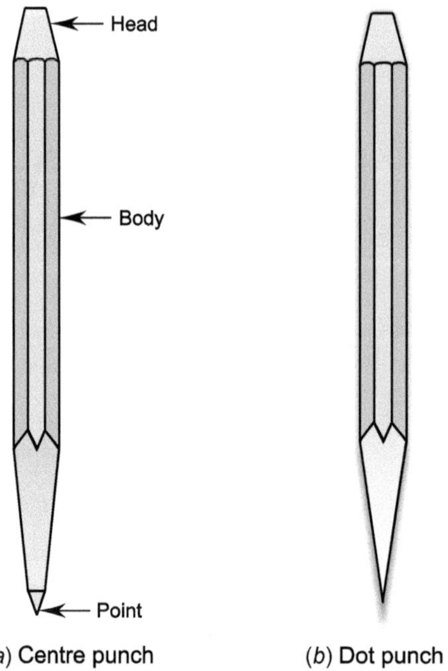

Fig. 23.15 Types of punches

23.11 Try Square

A try square (Fig. 23.16) is used to obtain an edge or surface exactly at right angles to an already trued surface and also for laying out work. It consists of a steel blade containing marked graduations and a stock.

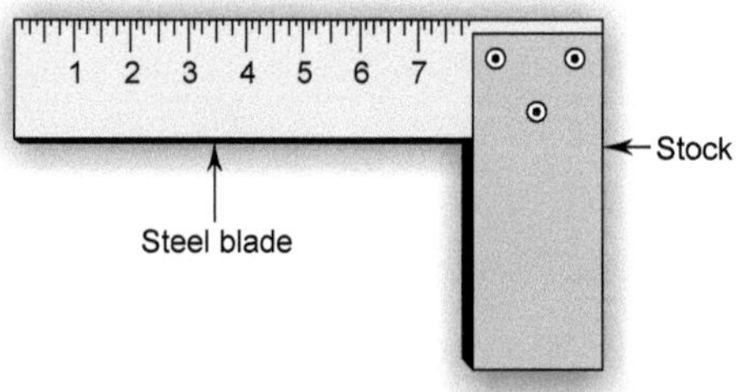

Fig. 23.16 A try square

23.12 Smithy Shop

Smithy shop is a place in the mechanical workshop, where workpieces after heating them in open fire or hearth are converted into the desired shapes through hammering using hand tools, also called Blacksmith's tools. It is used for small scale production with limited applications, and is essentially a manually operated process.

Forging is a special case of smithy, where medium or large size parts are produced using heavy blows, e.g. drop and hammer forging or steady pressure (pressure forging), while using closed heating furnaces. It is an improved version of smithy, where forging temperature and hammering actions are more accurately controlled using specialized machines. The better control of parameters helps to achieve uniform property in the part being shaped. Mechanical properties are significantly improved with improvement in grain flow. Strength and toughness are greatly increased by forging.

23.13 Forgeable Materials

Some materials undergo deformation more easily than others. Materials with higher ductility are most suitable for forging operations. Some important forgeable materials include alloys of aluminium, magnesium, copper and nickel, wrought iron, carbon and low alloy steels.

23.14 Forging Temperature

Forging is essentially a hot working operation. When working on a heated workpiece, deformation forces are greatly reduced and it becomes quite easier to give the required shape to the workpiece. This property is called forgeability.

Forging temperatures for some materials are given in Table 23.1.

Table 23.1 Forging temperature for some materials

Materials	Forging temperature (°C)	
	Starting	Finishing
Aluminium and magnesium alloys	500	350
Copper, brass and bronze	950	600
Wrought iron	1275	900
Mild steel (low-carbon steel)	1300	800
Medium-carbon steel	1250	750
High carbon steel	1150	825

23.15 Forging Tools

Some hand tools used in forging are discussed below.

23.15.1 Anvil

Anvil is perhaps the most important tool used in smithy shop. It is made of wrought iron or cast steel. It is used to support the workpiece during its shaping or to support the tool to be used on the workpiece, especially during swaging or fullering operations. The anvil is supported on a wooden block or a cast iron stand. Anvils are made in different sizes, weighing from 50 to 300 kg, and a 150 kg anvil is widely used. An anvil with its important parts is shown in Fig. 23.17.

The important parts of an anvil and their functions are discussed below:

Punching hole is also known as spud hole and has round shape. It is used to insert the ends of workpiece during its shaping.

The purpose of squared shape **hardy hole** is to hold the workpiece during its shaping using a hardie. Swaging tools (swages) or fullering tools (fullers) can also be inserted into these holes, while carrying out operations on the workpiece.

A **beak** is used to bend a workpiece to various diameters by placing it on the beak.
Working face is flat and hard and is used to carry out all the forging operations needed on a workpiece during its shaping.
Rounded edge is situated adjacent to working face and is used to give smooth curvature to a workpiece.

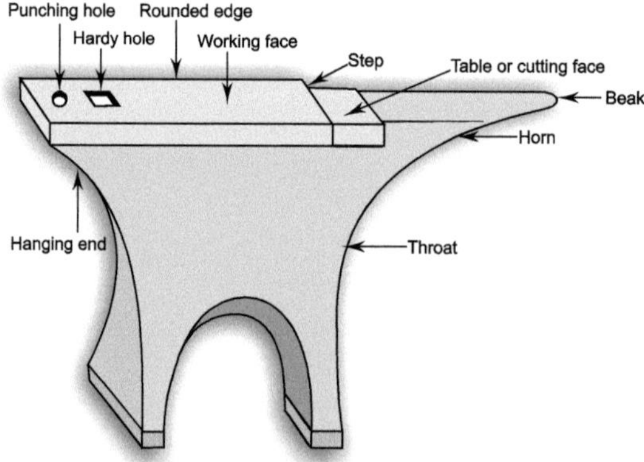

Fig. 23.17 An anvil

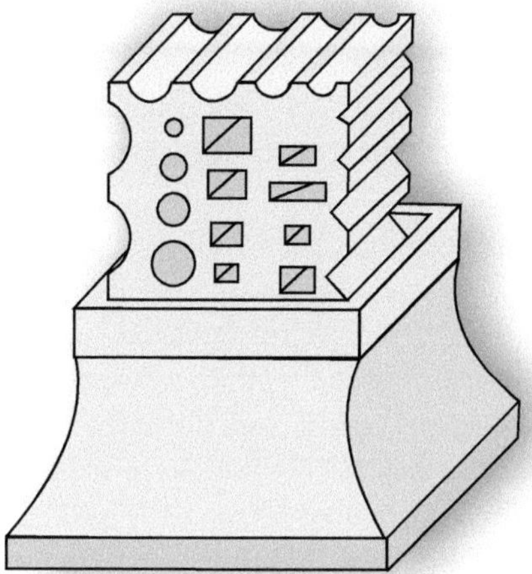

Fig. 23.18 A swage block

Hanging end is also called tail. It has rectangular cross section and is used to bend the workpiece at right angles.

23.15.2 Swage Block

Swage block (Fig. 23.18) is a solid metal block made of cast steel or forged steel. It has several curved surfaces on its face meant for performing operations such as bending, swaging, shaping and finishing on a workpiece and has several holes and grooves of various shapes and sizes on its side for inserting swaging and fullering tools to be used on the workpiece or the workpiece itself during its shaping. The swage block is usually kept on a stand at a certain height.

23.15.3 Tongs

Tongs are (Fig. 23.19) are used to hold the workpiece during its shaping. They are of various types and are usually made of low-carbon steel. Some of them are used to hold round bars, whereas others for holding plates.

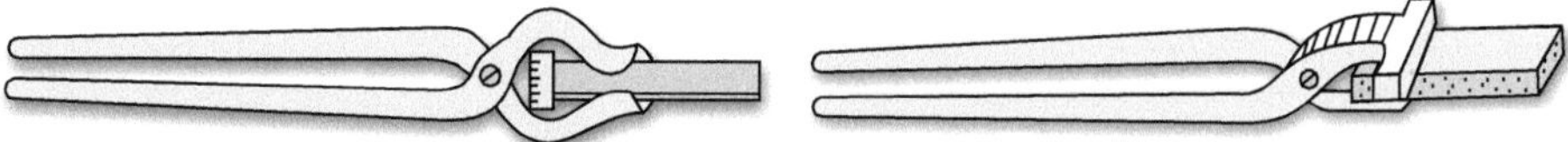

Fig. 23.19 Types of tongs

23.15.4 Hammers

Hammers are used to give blows on the heated workpiece to give it the required shape. They are made of cast steel or forged steel having their weights varying according to their applications. Lightweight hammers, also called hand hammers, are used for small and light components, whereas heavy weight hammers for thicker components requiring heavy blows. The hammers may be ball peen, straight peen, cross-peen or sledge type (Fig. 23.20).

Ball peen hammers (Fig. 23.20(*a*)) are most frequently used hand hammers. One end of the head is flat, called face, and is used for striking purpose. The other end of the head is half ball shaped and is called peen. It is used for riveting purpose. The head of the hammer is attached to a wooden handle. The weight of the hammer may vary between 1 and 1.5 kg. *Straight peen* and *cross-peen* hammers (Fig. 23.20(*b*) and (*c*) are defined on the basis of orientation of their peens. In former type, peen is straight, i.e. parallel to

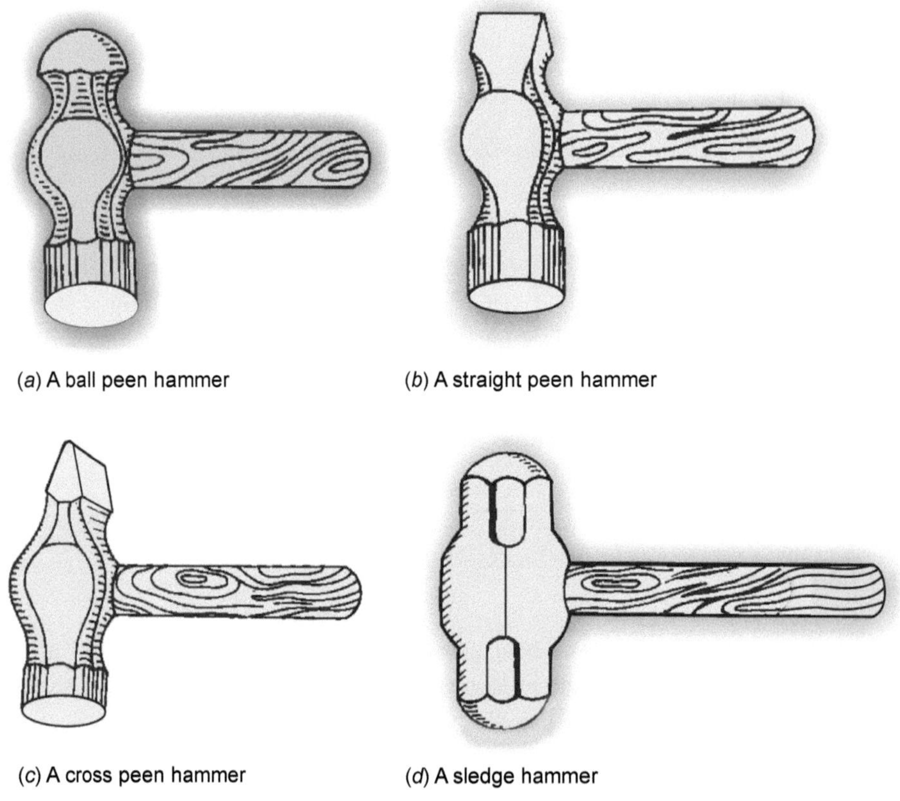

Fig. 23.20 Types of hammers

the axis of the handle; and in latter type, peen is across, i.e. perpendicular to the axis of handle.

Sledge hammers (Fig. 23.20(*d*)) are heavier in weight and are used for heavy blows applications. The basic difference between other hand hammers and sledge hammer is that, while using a hand hammer, a smith does not take help of others, whereas due to heavy weight of the sledge hammer, he has to take help of his helper, so that workpiece is held by one person and the hammer is operated by another fellow. The weight of sledge hammers can vary between 3 and 10 kg, depending upon quantum of blows required on the workpiece.

23.15.5 Swages

Swages (Fig. 23.21) are used for reducing or finishing the workpiece into round or hexagonal shape. There are usually two types of swages used in practice. The first type of swage has two separate parts: lower and upper. The upper part has a strip steel handle. The other type of swage, called spring swage, has two parts connected through a strip of steel having ability to produce spring effect. The workpiece is kept in the space between two parts during its shaping.

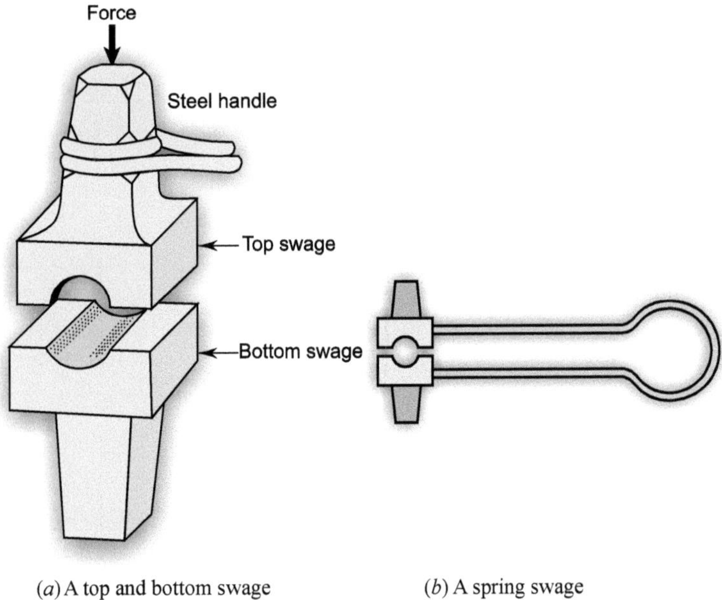

(*a*) A top and bottom swage (*b*) A spring swage

Fig. 23.21 Swages

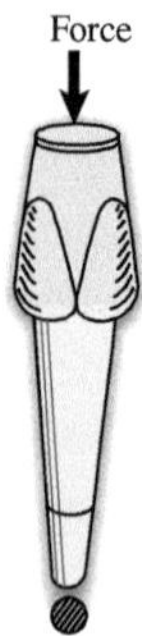

Fig. 23.22 A round punch

23.15.6 Punch

A punch is used to produce a hole in the heated workpiece. Punches are named on the basis of their cross sections, such as round punch, square punch and oval punch. A round punch is shown in Fig. 23.22. Punches are made of tool steel, hardened and tempered.

23.15.7 Drift

A drift (Fig. 23.23) is used to enlarge a hole, already produced by a punch to the required shape and size. Similar to a punch, a drift can be round, square and oval type.

Fig. 23.23 A drift

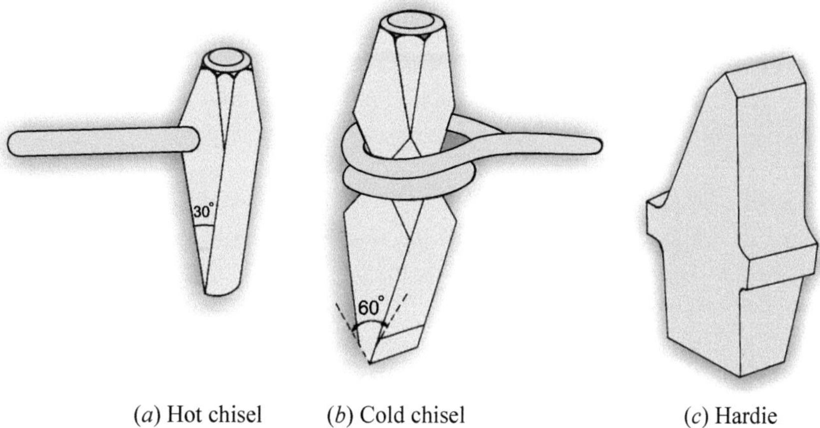

Fig. 23.24 Types of chisels

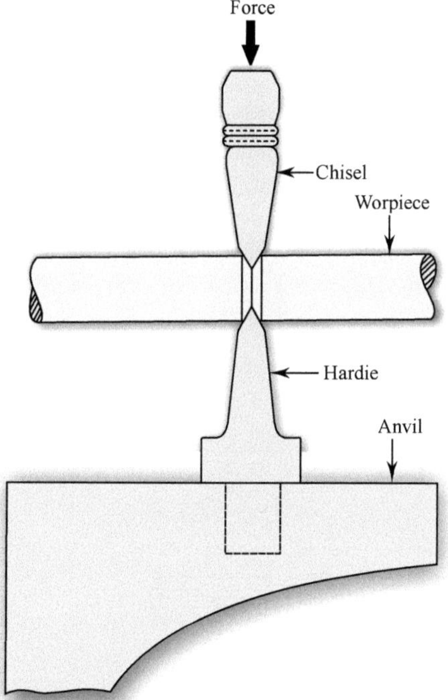

Fig. 23.25 Arrangement showing the use of a hardie and a chisel

23.15.8 Chisels

Chisels (Fig. 23.24) are used for cutting metals or for nicking prior to breaking. They may be hot or cold type, depending upon whether they are used for cutting hot or cold metal respectively. Cold chisels have higher cutting angle (60°) compared to hot chisels (30°), as the former is required to transmit more force on the workpiece. Chisels are generally used in pairs.

One of the chisels, is kept on top of the workpiece, while the other below it. The lower chisel, also called a hardie, is kept on the hardy hole of the anvil. The upper chisel is hammered, while the workpiece is rotated or turned after each blow to produce groove all around it. The grooving reduces the deformation force needed to be applied during cutting process. Use of a chisel and hardie is shown in Fig. 23.25.

23.15.9 Fullers

Fullers (Fig. 23.26) are used in pairs and are called top and bottom fullers. They are used to make groove or reduce thickness of the workpiece. The lower fuller with its square shank is fitted into the hardie hole of the anvil. The upper fuller is placed on the workpiece just above the bottom fuller. The force is applied on the workpiece through the upper fuller. Fullers are made of tool steel.

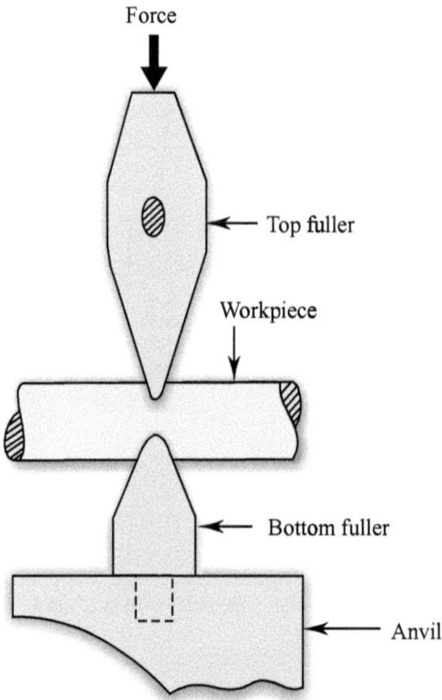

Fig. 23.26 Arrangement showing the use of top and bottom fullers

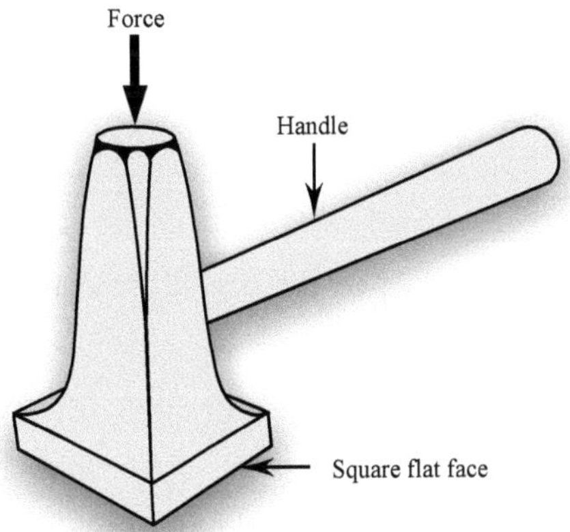

Fig. 23.27 A flatter

23.15.10 Flatter

A flatter (Fig. 23.27), also called a smoother, is used to flatten the uneven surfaces or to remove fullering marks left on workpiece while using fullers. It is made of tool steel and has a perfectly flat face of square or round shape.

23.15.11 Set Hammer

A set hammer (Fig. 23.28) can be regarded as a small size flatter, and is made of forged steel. Similar to a flatter, its face is also flat and of square shape but of reduced size. It is

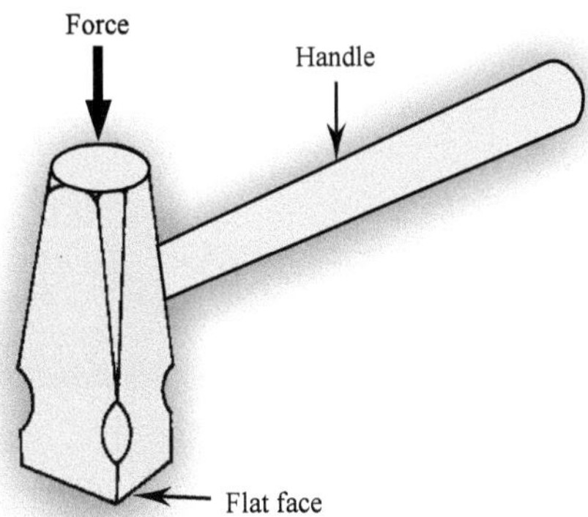

Fig. 23.28 A set hammer

used to produce good surface finish over a limited area, especially corners and inaccessible parts of the workpiece. It is called a set hammer, as it sets the work surface to a flat nature.

23.15.12 Calliper

A calliper is used to measure or compare the size of the work being forged.

23.15.13 Brass Scale

A brass scale, in place of a steel scale, is used to measure the dimensions of the work, because of its better heat resisting capacity.

23.15.14 Rake, Poker and Shovel

They are known as fire tools (Fig. 23.29). Rake is used to lift the heated work out of fire. Poker is a steel rod used to stir (poke) fire in the hearth. Shovel is used to put the coal in the hearth.

23.16 Sheet Metal Shop and Sheet Metal Shop Tools

In sheet metal shop, many useful products such as funnels, channels, trays, hoppers, canisters, boxes are made from metal sheets. The theory of development of surfaces finds extensive applications for deciding the requirements of sheet metal used for making various shaped containers. The major sheet metal operations include cutting, shearing and bending.

The most commonly used sheet metal tools are discussed below.

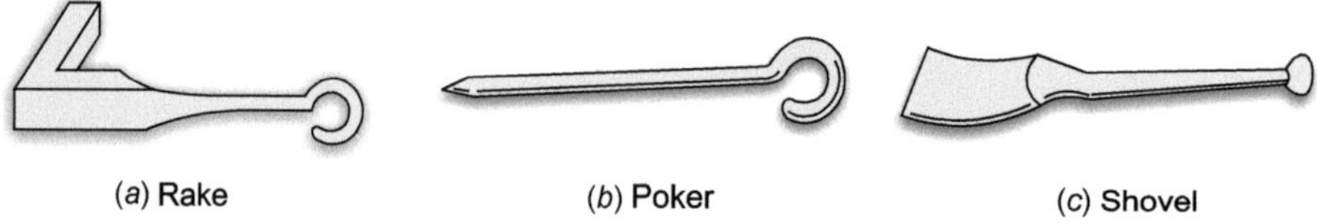

Fig. 23.29 Fire tools

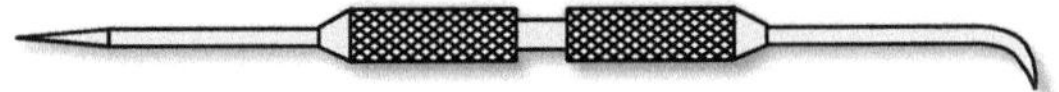

Fig. 23.30 A scriber

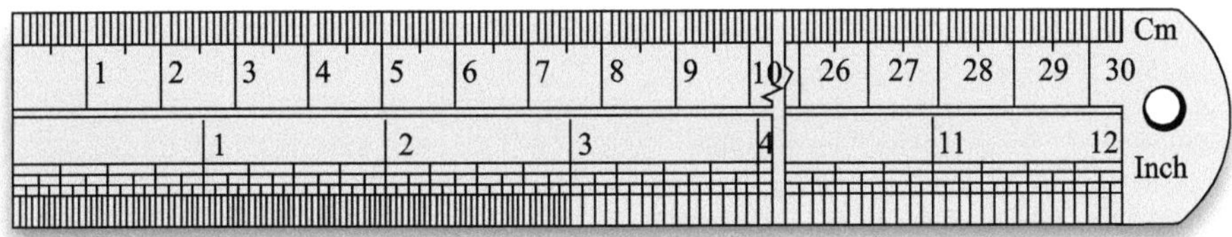

Fig. 23.31 A steel rule

23.16.1 Scriber

A scriber (Fig. 23.30) is used as a pencil to mark impression on the metal sheet so that the latter can be bent or cut at marked location. Its one or both ends have sharp points.

23.16.2 Rule

Rule is used for measuring and layout purpose. It can be a steel rule, folding rule or steel tape. A steel rule (Fig. 23.31) is used for small work and other rules for larger work. A circumference rule is used to measure circumference of a circle or cylinder.

23.16.3 Snips

Snips (Fig. 23.32) are similar to a scissor. They are used to cut sheet metal to the required shape and size. The two most commonly used snips are straight and curved or bent snips. The straight snips have straight edge and is used to cut along a straight line. The curved snips have curved blades for producing circular cuts.

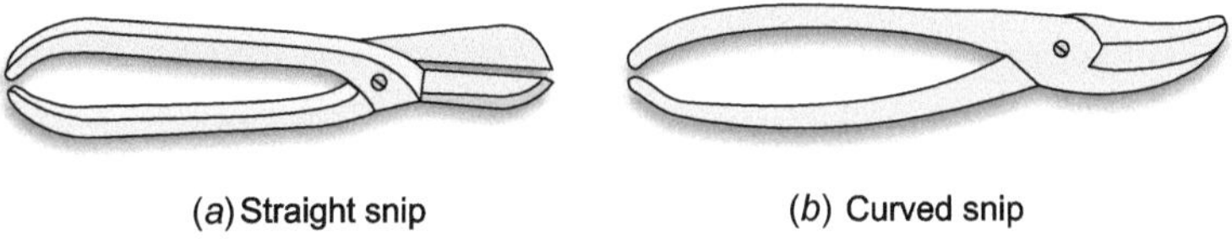

Fig. 23.32 Types of snips

23.16.4 Stakes

Stakes are used for carrying out sheet metal operations such as bending, seaming or other forming operations using a mallet or hammer. The various types of stakes and their specific uses are discussed in Table 23.2.

Table 23.2 Various types of stakes

Beak horn stake (Fig. 23.33)	It has a round tapered horn at one side and a square tapered horn on other side	It is the most general purpose anvil tool, used for riveting, straight bending or closed seam working
Hatchet stake (Fig. 23.34)	It has horizontal sharp straight edge	It is used to produce straight and sharp bends
Half moon stake (Fig. 23.35)	It has semi-circular shape	It is used to produce curved surfaces
Funnel stake (Fig. 23.36)	It has a tapered round working face	It is used to produce conical shapes
Creasing stake (Fig. 23.37)	It has a tapered square horn with grooved slots at one side and a round tapered horn on other side	The slots are used to produce wire rings and round horn for conical shapes

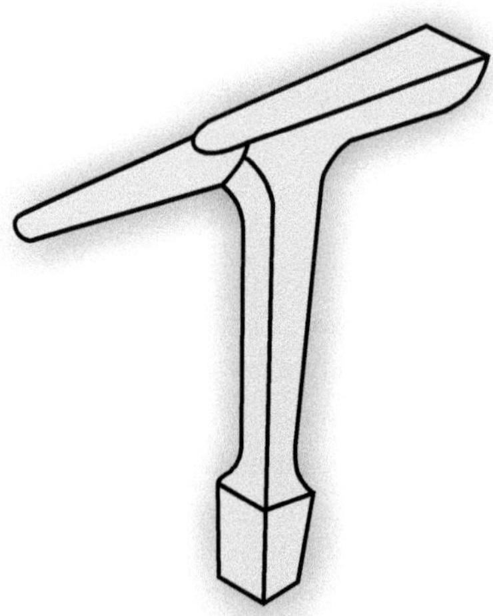

Fig. 23.33 A beak horn stake

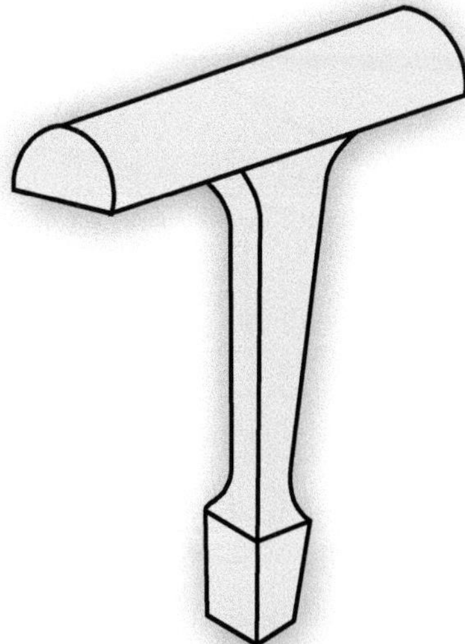

Fig. 23.34 A hatchet stake

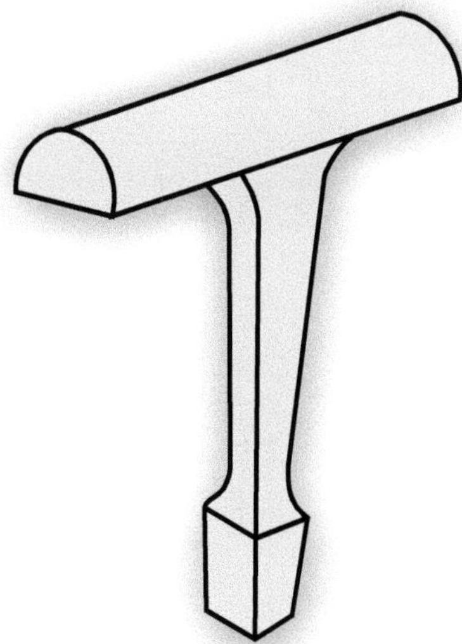

Fig. 23.35 A half moon stake

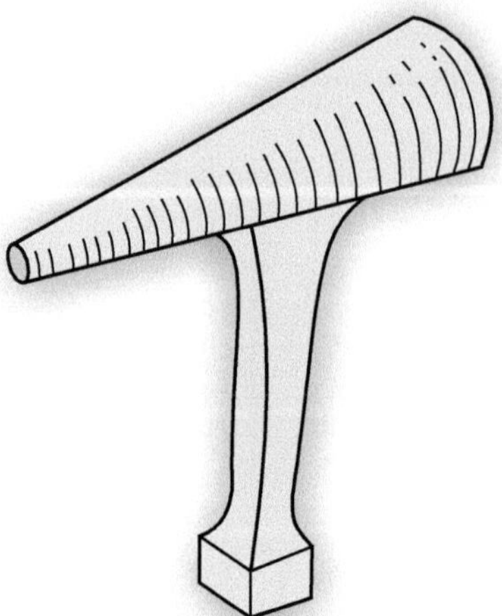

Fig. 23.36 A funnel stake

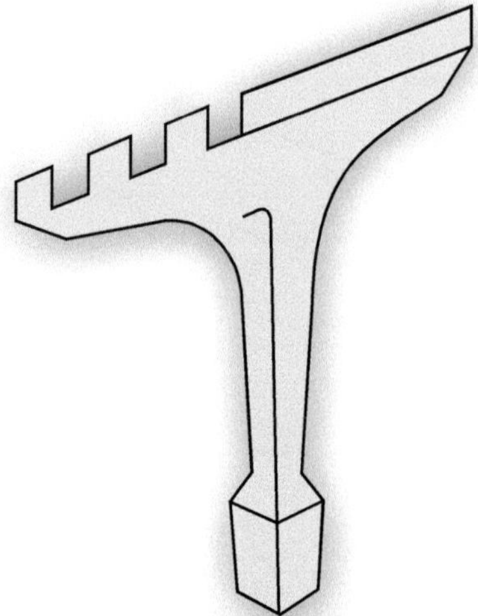

Fig. 23.37 A creasing stake

23.16.5 Mallets

Mallets (Fig. 23.38) are used to give soft and light blows on the sheet metal and are made of wood, hard rubber or metals such as copper and lead.

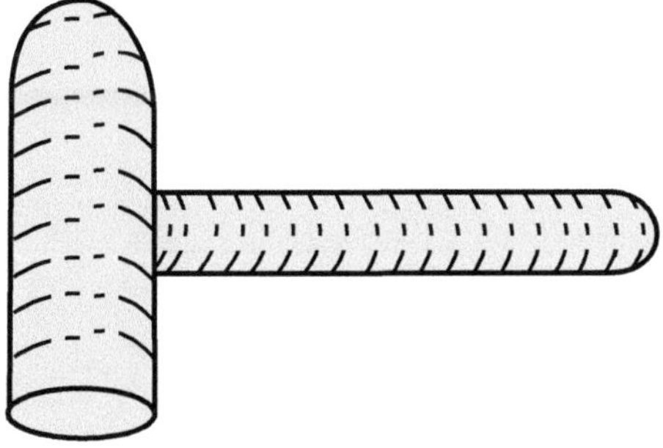

Fig. 23.38 A mallet

23.16.6 Hammers

Hammers are used for forming shapes by hollowing, raising, stretching or throwing off processes. Examples of few hammers include ball peen, raising, hollowing and riveting. The ball peen hammer shown in Fig. 23.20(*a*) is a general purpose hammer.

23.16.7 Chisels

Flat type cold chisels are the most widely used chisels in the sheet metal shop. They are used for cutting sheets and chipping operations. It has either hexagonal or octagonal shape and is made of high carbon steel. A flat chisel is shown in Fig. 23.12.

23.16.8 Punch

A punch is used for producing indentation marks or locating centres. The two important types of punch are pick and centre punch. In the former one, the punch is ground to an angle of 30°, whereas in the latter one, the punch is ground to 90° angle. Solid and hollow punches are used for producing holes in sheet metal; the former produces small size holes (2.5–10 mm) and the latter bigger size holes (10 mm or more). Punches are shown in Fig. 23.15.

23.16.9 Dividers and Trammel Points

Dividers and trammel points (Figs. 23.39 and 23.40) are used for drawing circles or arcs on the sheet metal. The basic difference between the two is that the former produces a

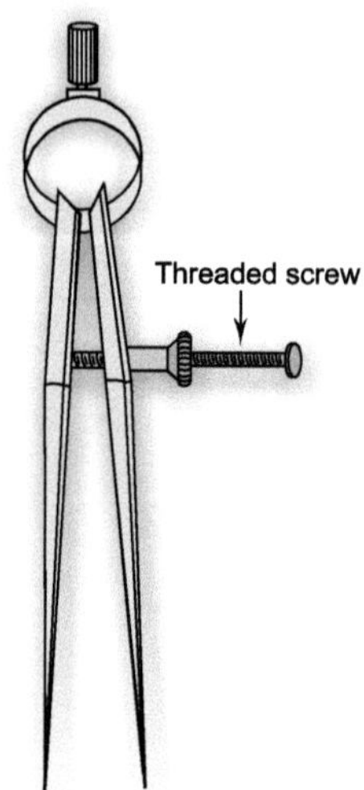

Fig. 23.39 A divider (spring type)

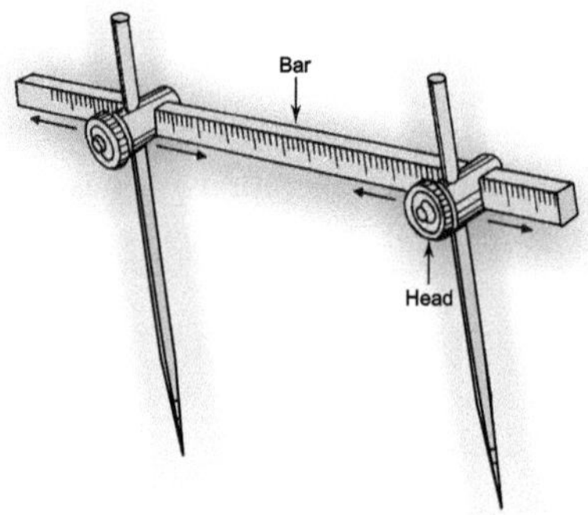

Fig. 23.40 A trammel point

small size circle, whereas the latter one is used for drawing large circles, which are not possible by the former one.

23.16.10 Swing Blade Protractor

It (Fig. 23.41) resembles with a semi-circular protractor commonly used for geometrical purposes, but has an arm. It is made of steel and is used for making or measuring angles.

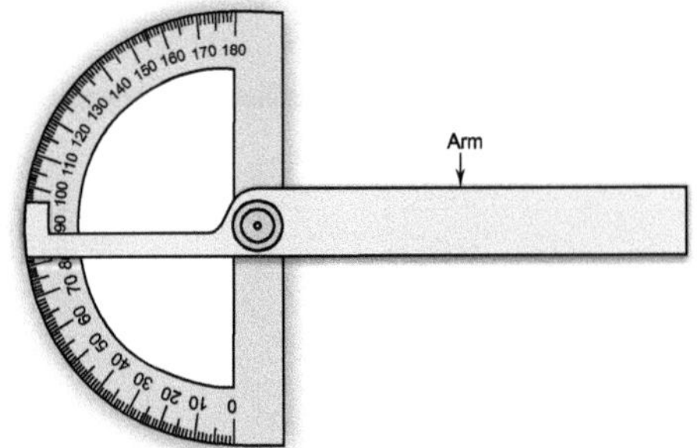

Fig. 23.41 A swing blade protractor

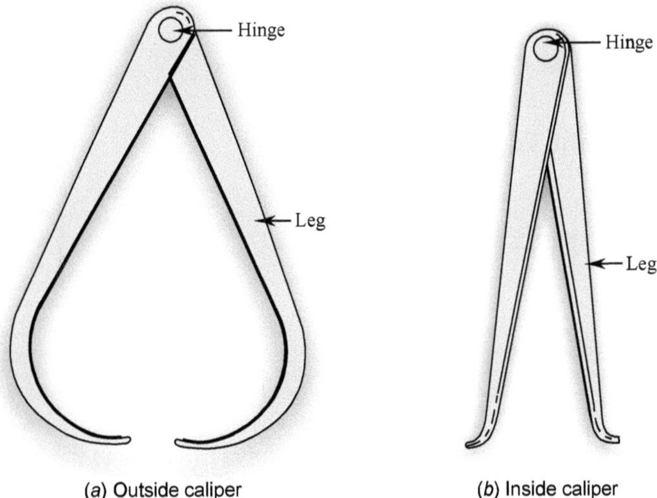

Fig. 23.42 Callipers

23.16.11 Callipers

Callipers are nonprecision linear measuring tools. Outside and inside callipers (Fig. 23.42) are used for measuring external and internal dimensions of an object respectively. A steel rule must be used along with a calliper to obtain direct reading. A vernier calliper is a precision tool. It has a main scale and a vernier scale. See Fig. 22.4.

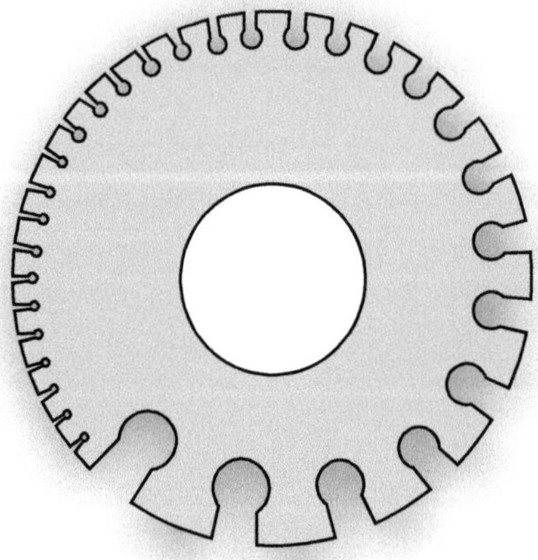

Fig. 23.43 A wire gauge

23.16.12 Straight Edge

It is a flat graduated bar of steel with a bevelled edge. It is used for scribing long straight lines. See Fig. 22.12.

23.16.13 Wire Gauge

A wire gauge (Fig. 23.43) is used to measure diameter of wire or thickness of sheet metal. The thickness of sheet metal is inversely proportional to its gauge number. Higher gauge number indicates a thinner sheet.

23.16.14 Micrometer

A micrometer is a precision instrument. Its main parts include frame, spindle, barrel, thimble and ratchet. See Fig. 22.7.

23.17 Important Sheet Metals

23.17.1 Galvanized Iron (GI)

The galvanized iron is basically mild steel coated with zinc, used to give it anti-rusting property. The GI sheets are most widely used in sheet metal shop and find applications in making buckets, cabinets, ducts, pan and other products.

23.17.2 Black Iron

The black iron is uncoated iron sheet and is black in colour. It corrodes very easily. To prevent it from corrosion, it may be painted. It is used in making tanks, stove pipes and pans.

23.17.3 Copper

Because of its good electrical conductivity, it finds extensive applications in electrical industry. It has also good anti-corrosion property, which makes it suitable for products like water pipes, roofing, gutters, etc.

23.17.4 Tin

Tin sheets are extensively used in food processing industry for making food containers, cans and pans, because of their nonpoisonous property.

23.17.5 Aluminium

It is a very light metal and has extensive applications in making household appliances, lighting fixtures, framing structures, refrigerator trays and in aeroplane industry.

23.18 Foundry Shop and Foundry Shop Tools

Foundry shop is used to shape an object from its molten state after solidification. Foundry shop tools are used in casting operations. The most important material used in foundry shop is sand, which finds applications in making moulds and mould cavity. Tools used in foundry shop along with their functions are discussed below.

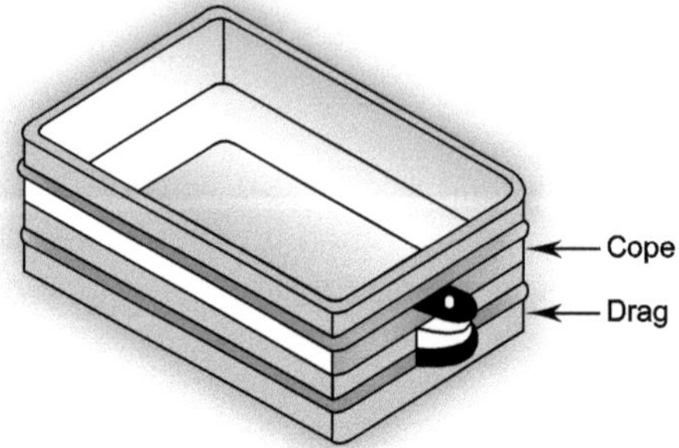

Fig. 23.44 A moulding flask

23.18.1 Moulding Box or Flask

Moulding box (Fig. 23.44) is a container used to make mould and mould cavity. It imparts rigidity and strength to the sand mass rammed in it during the preparation of mould. It is made in two or three parts; the upper part is called cope, the middle one cheek and the lower part drag. Cope and drag parts are separated by a parting line or plane. Two parts moulding flasks are most widely used, whereas three parts moulding flasks are used for complex castings.

23.18.2 Bellow

A bellow (Fig. 23.45) is used to blow loose sand particles from the pattern and mould cavity. It sucks air from the atmosphere, which it throws when directed against the undesired sand particles either on the pattern or in the mould cavity and thus making these components clear off extra sand particles.

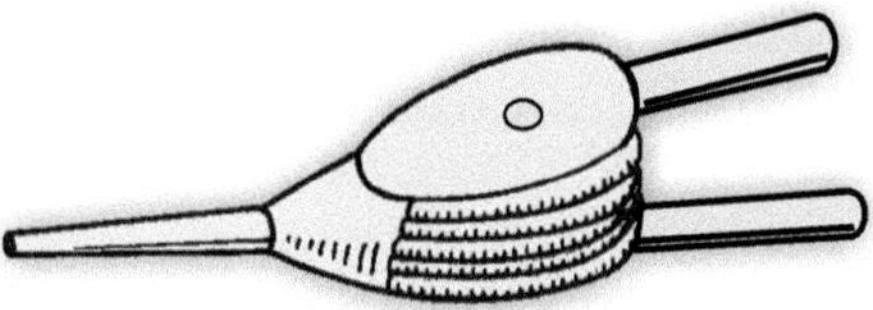

Fig. 23.45 A bellow

Fig. 23.46 A brush

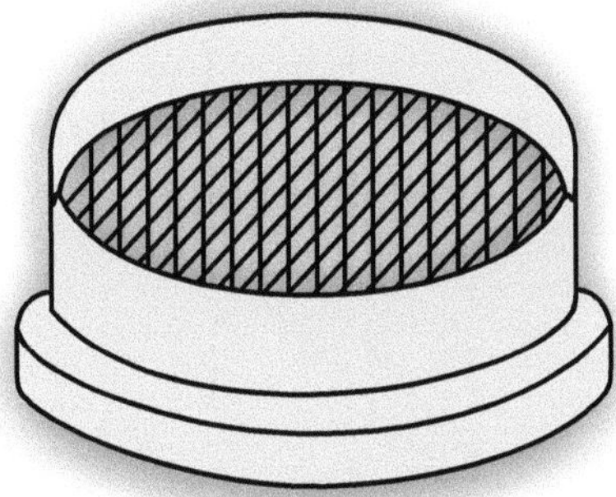

Fig. 23.47 A hand riddle

23.18.3 Brush

A brush (Fig. 23.46) is used to clean the dust or undesired sand particles or foreign materials from the pattern or parting sand at the time of making the moulds.

23.18.4 Riddle

A riddle (Fig. 23.47) is in the form of wire mesh fitted into a circular wooden frame. It is used to separate foreign materials from the moulding sand.

23.18.5 Rammer

Rammers (Fig. 23.48) are used for ramming the sand mass during making of mould. It may be peen, hand and floor type. Peen rammers are used for ramming sand in pockets and

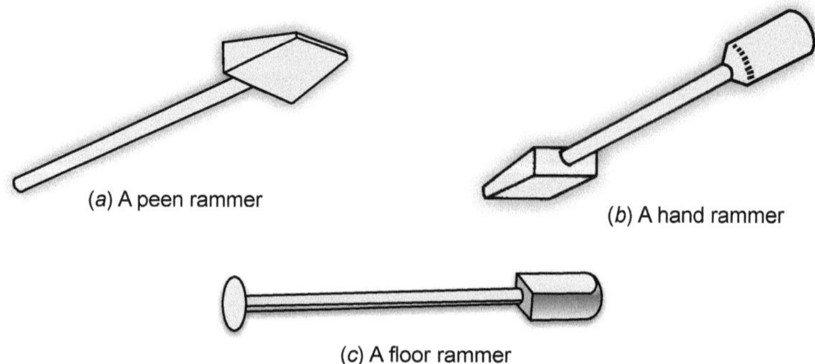

Fig. 23.48 Types of rammers

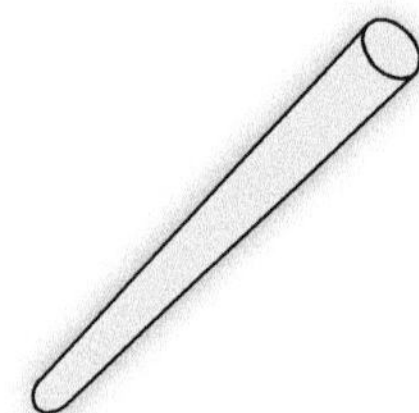

Fig. 23.49 A sprue pin

corners. Hand rammers are used in bench moulding and floor rammers in floor moulding for making large size moulds.

23.18.6 Sprue Pin

A sprue pin (Fig. 23.49) is a tapered wooden piece, in the shape of a sprue. It is used to create sprue impression by inserting it into the cope part of moulding sand and then taken out, thus creating cavity in the sand, called sprue.

23.18.7 Swab

A swab (Fig. 23.50) is a hemp fibre brush used for sprinkling water on sand mass around a pattern for better cohesion of sand particles.

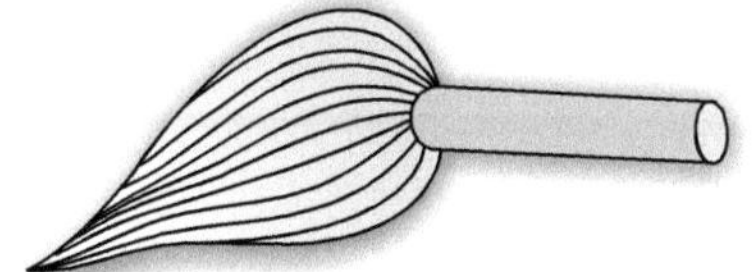

Fig. 23.50 A swab

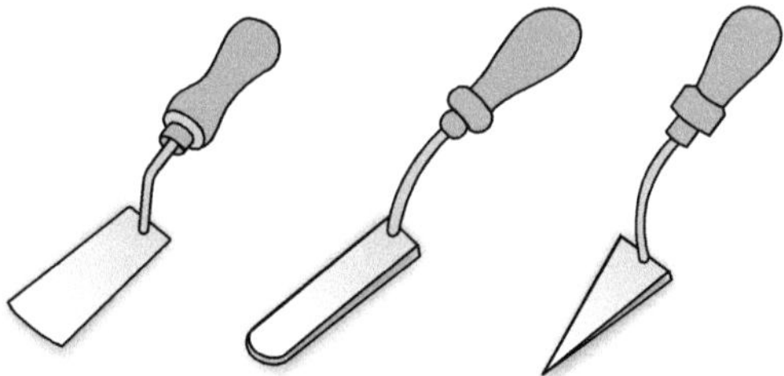

Fig. 23.51 Types of trowels

23.18.8 Trowel

A trowel (Fig. 23.51) is used for finishing or repairing a mould. It usually has a wooden handle for better grip by the hand.

23.18.9 Shovel

A shovel (Fig. 23.52) is either used to prepare the moulding sand or to bring it from the site of its preparation to the place of its use.

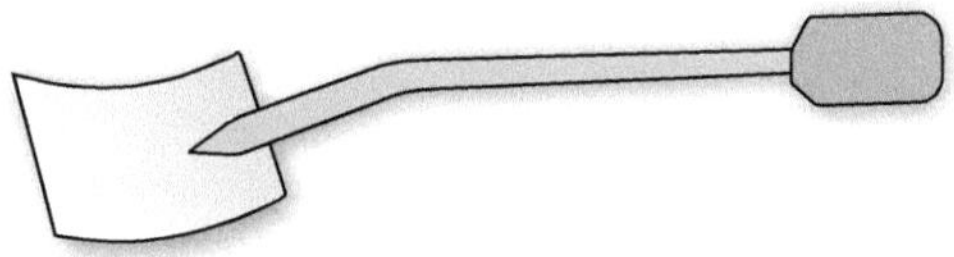

Fig. 23.52 A shovel

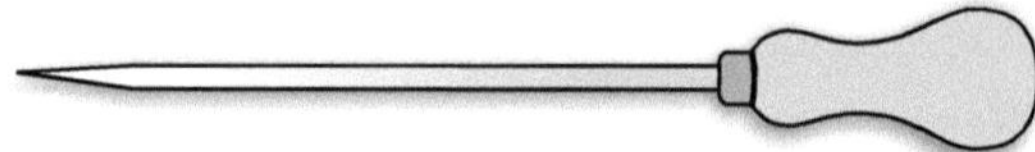

Fig. 23.53 A vent wire

Fig. 23.54 A heart and square tool

23.18.10 Vent Wire

A vent wire (Fig. 23.53) is in the form of a pointed metal tool, used to pierce holes in the cope part of rammed sand to facilitate escaping of gases from the moulding sand generated during pouring of molten metal.

23.18.11 Heart and Square

This tool (Fig. 23.54) is used for finishing the mould cavity.

23.18.12 Lifter

A lifter (Fig. 23.55) is also called a cleaner, as it is used to clean off the mould cavity of undesired sand particles by lifting them. It is used in the repair or finishing of sand mould cavity.

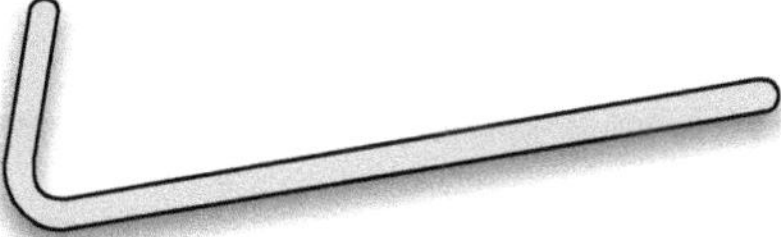

Fig. 23.55 A lifter

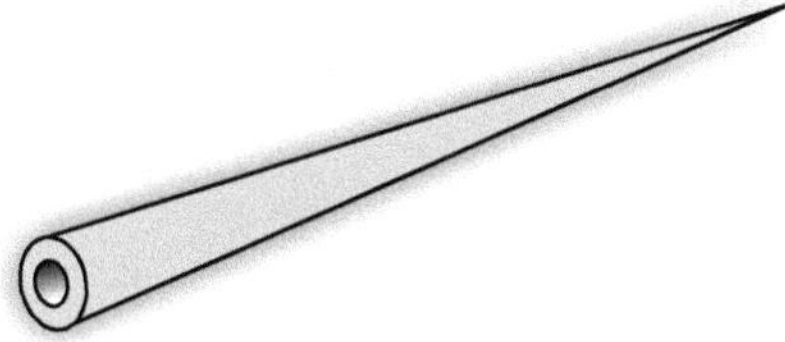

Fig. 23.56 A draw spike

23.18.13 Draw Spike

A draw spike (Fig. 23.56) is a sharp-pointed tapered metal rod used to take out the pattern (wooden) from the rammed sand mass by inserting it into the pattern once ramming is complete, to create the mould cavity.

23.18.14 Draw Screws and Rapping Plate

This assembly tool (Fig. 23.57) consisting of threaded draw screws and a rapping plate is used to take out a big size pattern from the mould at the time of creating mould cavity in the rammed sand mass. The rapping plate is clamped to the pattern with the help of draw screws, which are rapped or shaken during removal of pattern from the sand mass.

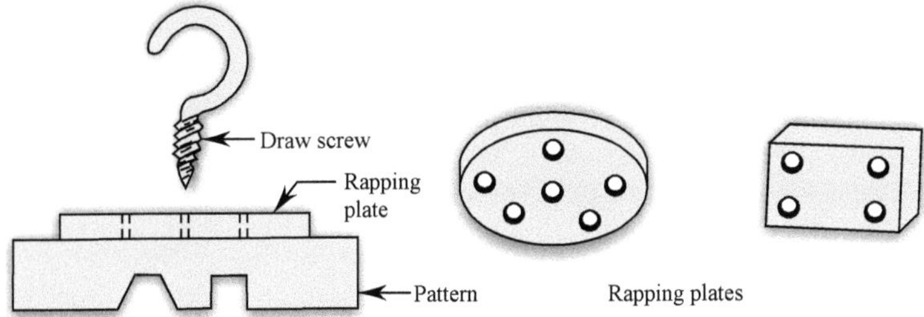

Fig. 23.57 Use of rapping plate and draw screw

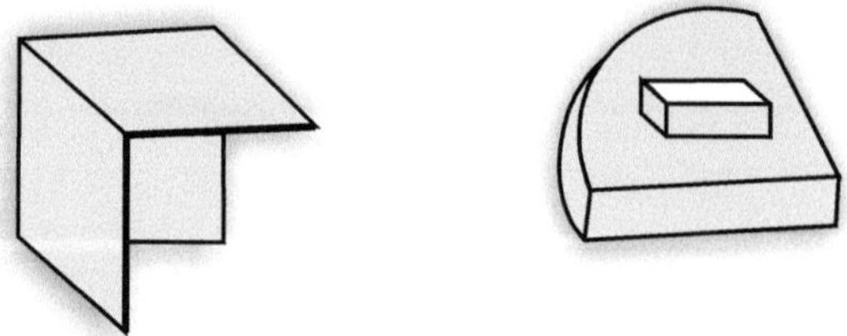

Fig. 23.58 Smoothers and corner slics

Fig. 23.59 A moulding sand muller

23.18.15 Smoothers and Corner Slics

These tools (Fig. 23.58) are used to repair or finish the corners and edges of the mould.

23.18.16 Moulding Sand Muller

A moulding sand muller (Fig. 23.59) is a mechanical mixer used to mix different ingredients such as sand, binders, water and additives in the moulding sand in order to make them uniformly distributed and thoroughly mixed.

23.18.17 Gaggers

Gaggers (Fig. 23.60) are iron rods, which are bent at one or both ends. It is used to reinforce the sand in the cope or to support the hanging bodies of sand.

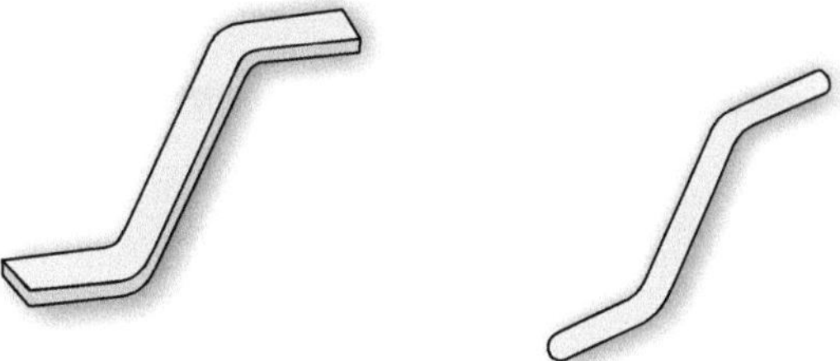

Fig. 23.60 Gaggers

23.18.18 Strike-off Bar

A strike-off bar (Fig. 23.61) is a wooden or steel bar of certain length with its cross-sectional rectangular or having one edge bevelled. It is used to remove excess sand from the top of the rammed sand mass in the moulding box to make its surface plane and smooth.

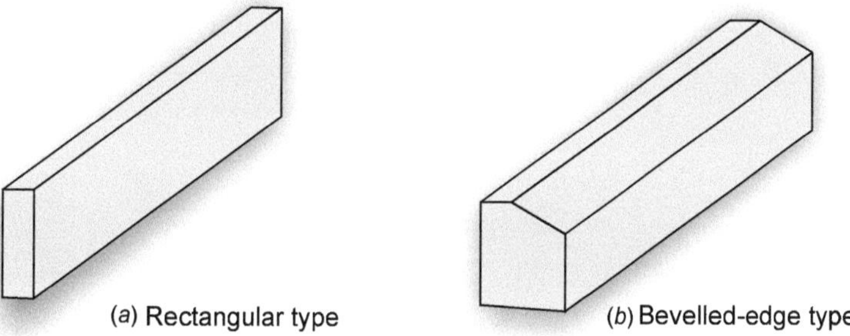

Fig. 23.61 Strike-off bars

23.19 Machine Shop and Machine Shop Tools

Machine shop is used to make different shaped objects by removing material from a given metal stock in bulk quantity. Machine shop tools are generally used to produce cylindrical or flat surfaces. Standard type machine tools are used for a wide range of operations, whereas special purpose machine tools are very specific and meant to perform very limited operations. A few standard machine tools include lathe, drilling, boring, shaping, planning and grinding machines. A few such tools are discussed below:

23.20 Lathe

Lathes (Fig. 23.62) are the most frequently used machines in industry. They are available in a wide range of sizes. It is the most basic and one of the oldest machine tools, which can be used to perform several operations such as turning, facing, drilling, boring, reaming threading, cutting, knurling, etc. Either the workpiece or tool can move with respect to each other during metal removal from the workpiece. Workpiece may be held between two centres of the lathe, more commonly known as live centre and dead centre. Also, workpiece may be held by a chuck. Typical products made on a lathe include table legs, bowls, crankshafts, camshafts, baseball bats, cue sticks and candle sticks. Today most commercial lathes are computer controlled, making them suitable for high precision works and allowing them for mass production.

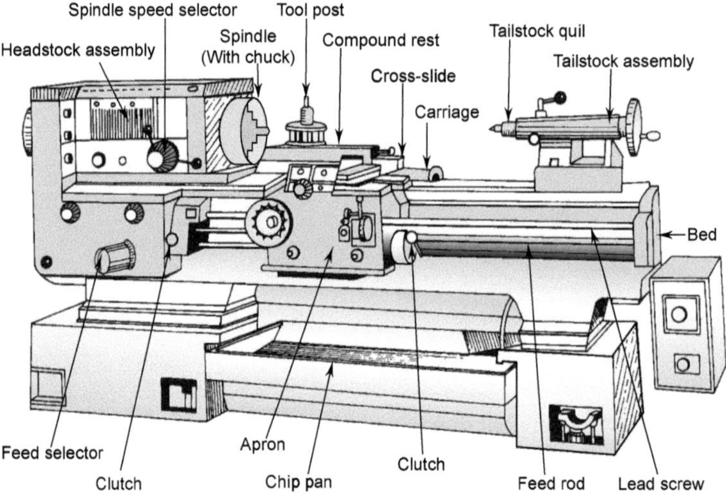

Fig. 23.62 Lathe and its important parts

23.21 Size of a Lathe

A lathe is specified by two parameters:

(a) **Maximum distance between lathe centres**, which represents the maximum length of the work that can be handled by the lathe.

(b) **Swing**, which represents the maximum diameter of the work that can be machined on the lathe. For example, a 200 mm × 600 mm lathe means a swing of 200 mm and a distance between centres of 600 mm. Sometimes, *'length of the bed'* is used in place of distance between centres to specify a lathe.

23.22 Types of Lathes

The main parts of a lathe are shown in Fig. 23.62 and are discussed below:

23.22.1 Engine Lathe

An engine lathe is the oldest and simplest lathe, used to be driven by the power from an engine. The workpiece is supported between the two centres of the lathe and hence this lathe is also called centre lathe. It is a basic turning machine.

23.22.2 Tool Room Lathe

Because of its extensive use in making tools and dies, it is so named. It is used for producing highly accurate and precision work such as test gauges, dies and small tools.

23.22.3 Speed Lathe

It is driven by a variable speed motor and hence the name speed lathe. Gear box, carriage and lead screw are normally missing in a speed lathe.

23.22.4 Bench Lathe

Because of its small size, this lathe can be mounted on a bench and hence it is so named. It is used for small and light work requiring precision.

23.22.5 Special Purpose Lathes

They are used to manufacture jobs, which cannot be conveniently produced by common lathes. They include duplicating lathe, crankshaft lathe, wheel lathe and gap lathe.

23.22.6 Turret Lathe

A turret lathe is used for mass production as a number of tools can operate on the job simultaneously. Other engine lathes permit one operation using only one tool at one time.

23.22.7 Capstan Lathe

A capstan lathe is so named, because it has a hexagon-shaped tool carrier, called capstan, which replaces the tailstock of the centre lathe. It is smaller in size as compared to turret lathe, but both of them are of semi-automatic type. Capstan lathes are used for small size workpiece, because of its short stroke of tool head, whereas turret lathes are suitable for long and heavy work, as their stroke length is larger.

23.23 Principal Parts of a Lathe

23.23.1 Bed

The bed forms the base of a lathe on which important parts such as headstock, tailstock, carriage are mounted. It should be rigid enough to withstand the vibrational impact during the operation and as such it is made of cast iron.

23.23.2 Headstock

The headstock, which lies usually on the left side, contains gear box for operating the spindle at various speeds during machining of the workpiece. The spindle is built into headstock on which a face plate or a chuck is fitted to hold and rotate the workpiece.

23.23.3 Tailstock

The tailstock is nonrotating part, which lies on the right side. It can only be moved on left or right side to adjust the length of the workpiece. The workpiece is held between headstock and tailstock to avoid its bending during machining. In other words, it is used to provide auxiliary support to the workpiece at its end. It can also be used to hold a drill or similar tool during making a hole in the workpiece.

23.23.4 Carriage

A carriage is used to carry the cutting tool and control its movements either in the longitudinal (along the axis of the workpiece) or transverse (perpendicular to the axis of the workpiece) direction. The former results in straight turning, whereas latter facing. The carriage can be moved on the left or right between headstock and tailstock by a handwheel or powered device. The important parts of the carriage include tool post, saddle, cross-slide, compound slide and apron.

23.23.5 Tool Post

A tool post is used to hold the cutting tool firmly. It is mounted on the compound slide, which can move the tool post on left or right side and can be clamped in position. Tool post can also be rotated at a certain angle to give cutting tool a position, best suited for working on the job. Tool post may be clamp, pillar, turret or quick-release type.

23.23.6 Saddle

A saddle is an H-shaped casting, which can be moved along the bedways, either manually or by a powered device. It carries cross-slide, compound slide and tool post.

23.23.7 Cross-Slide

The cross-slide is used to support compound slide. At the same time, it also allows movement of the cutting tool at right angle to the lathe centre line.

23.23.8 Compound Slide

A compound slide, also called compound rest, is mounted on the cross-slide. It is used to support tool post.

23.23.9 Apron

The apron houses the gears and controls for the carriage and feed mechanism.

23.24 Lathe Accessories and Attachments

Lathe accessories (Fig. 23.62) are used either to hold and support the work or to hold the tool. Important lathe accessories include centres, chuck, face plate, mandrel, steady rest, follower rest, lathe dog and drill holder.

On the other hand, the lathe attachments are used to increase the capabilities and scope of operations of the lathe. These attachments help to perform operations, which are not possible without their use. Grinding, milling and taper-turning attachments are common lathe attachments.

A few important lathe accessories are discussed below:

23.24.1 Lathe Centre

There are two lathe centres: live centre and dead centre. The live centre is fitted in the headstock and the dead centre in the tailstock. The live centre rotates, whereas the dead centre remains stationary. Lathe centres are used to support the work between them during its machining.

23.24.2 Chuck

A lathe chuck is attached to the headstock spindle and is used to hold the work firmly and rotating it simultaneously by means of adjustable jaws. There are many types of lathe chucks. Some of them are: three jaws chuck, four jaws chuck and collet chuck.

23.24.3 Face Plate

A face plate is a large flat disc that mounts to the spindle. It is used to hold the workpiece, when there is difficulty in holding the work in the chuck, because of its inconvenient shape and size. The various slots cut in the plate help to use nuts, bolts, clamps and angles during fixing the work in the face plate.

23.24.4 Mandrel

A mandrel is in the form of a shaft or bar, used to produce a hollow section as in the centre of a work.

23.24.5 Steady Rest

A steady rest, also called fixed steady, is used to hold long slender work, which is difficult to be supported in a chuck or on a face plate. Alternatively, when the workpiece needs certain operation at its end, e.g. making a hole, then steady rest can be used under such condition.

23.24.6 Follower Rest

A steady rest remains stationary, whereas a follower rest moves with the cutting tool, while supporting long thin workpiece to avoid it bending during cutting operation.

23.24.7 Lathe Dog

A lathe dog is also called a carrier, as it carries the workpiece while securing it firmly through its set-screw during turning operation.

23.25 Milling Machine

A milling machine tool can perform several operations of wide variety. It is used to produce complex shapes in work material. It can produce flat as well as internal or external curved surfaces. It has suitably replaced shapers and slotters because of its higher accuracy and higher production rates.

The cutting takes place using a multitooth rotating cutter, when the workpiece is feeded against it. There are mainly two types of milling machines: vertical and horizontal, based on the orientation of the spindle axis.

23.26 Size of a Milling Machine

A milling machine is specified by two ways:

(a) Length and breadth of the worktable
(b) Maximum length of longitudinal, cross and vertical travel of the worktable.

23.27 Milling Operations

Milling operations are tabulated in Table 23.3.

Table 23.3 Various milling operations

S. N	Milling operations	Description of operation
1	Slab (plain) milling	Produces plain or horizontal surfaces parallel to the axis of rotation of the plain milling cutter
2	Side milling	Produces a flat vertical surface on the side of a workpiece using a side milling cutter
3	End milling	Produces a flat surface (horizontal, vertical or at a certain angle w.r.t. the position of the table) using an end milling cutter
4	Face milling	Produces a flat surface either parallel to face of the milling cutter or perpendicular to its spindle axis
5	Slitting	Uses a slitting saw to cut a large stock workpiece into required number of pieces preferably by climb milling
6	Slotting	Used to produce a slot in the workpiece with the help of a slotting cutter
7	Form milling	Produces irregular contours such as convex, concave or any other shape with the help of a form cutter
8	Dovetailing	Used to produce a dovetail shape in the work using a dovetail cutter
9	Angular milling	Used to produce flat surfaces at certain angle but not at right angle
10	Straddle milling	Used to produce flat vertical surfaces on both sides of a workpiece by using two side milling cutters mounted on the same arbour. The distance between the two cutters is adjusted by suitable spacing collars
11	Gang milling	Used to machine several surfaces of a workpiece simultaneously in one pass by mounting several milling cutters on the arbour. It is widely used in repetitive work and is the extension of straddle milling
12	Profile milling	Used to reproduce an outline of a template or complex shape of a master die on workpiece using end milling cutter
13	Helical milling	Used to produce helical flutes or grooves around the periphery of a cylindrical or conical workpiece
14	Cam milling	Used to produce cams using a universal dividing head and a vertical milling attachment
15	Thread milling	Used to produce accurate threads on the work such dies, screws, worms etc. both externally or internally. As compared to thread cutting by a lathe, it offers higher production rate with better surface finish
16	Gear cutting milling	Used to cut teeth on a gear blank with the help of a dividing head and formed tooth cutters

23.28 Types of Milling Machines

Milling machines are classified on the basis of design, type of drive, feed and table movement, and work performed.

1. Column and Knee type milling machines
   - Hand miller
   - Plain milling machine
   - Universal milling machine
   - Vertical milling machine
2. Planer milling machine
3. Fixed-bed type milling machines
   - Simplex milling machine
   - Duplex milling machine
   - Triplex milling machine
4. Special types of milling machines
   - Rotary table milling machine
   - Planetary milling machine
   - Duplicator or profiling milling machine
   - Pantograph milling machine
5. CNC milling machines

23.28.1 Column and Knee Type Milling Machines

These milling machines consist of a column and a knee. The column is a vertical hollow structure containing spindle and driving mechanism. Knee is a projected part, which can move in a vertical way on the column. Such machines are used for small and medium works in tool rooms and prototype shops.

23.28.2 Hand Miller

A hand miller is also known as hand milling machine. It is the simplest type of milling machine to be operated by hand. It finds use in cutting grooves, short keyways and slotting.

23.28.3 Plain Milling Machine

A plane milling machine is similar to a hand miller, except that it is provided with a power feeding mechanism to control the movement of the table. These machines are more rigid and sturdy than a hand miller to handle large workpieces.

23.28.4 Universal Milling Machine

Because of its ability to orient the rotating cutter vertically or horizontally, it has increased flexibility, which enables it to produce variety of works, and hence it is so named. It can be used to produce spur, spiral, helical and level gears, twist drills, reamers, milling cutters, besides doing all conventional milling operations.

23.28.5 Vertical Milling Machine

In vertical milling machine, the spindle axis is vertically oriented. The end mills and face milling cutters are usually used in such machines. They are used to produce grooves, slots and flat surfaces and can be made to perform operations like drilling, boring reaming etc.

23.28.6 Planer Milling Machine

It resembles a planer, and hence is so named. It has a long table, which carries the work, and has only longitudinal movement. It is fed against the rotating cutter. It is used for the machining of large workpieces requiring bulk material removal.

23.28.7 Fixed-Bed Type Milling Machines

These are bigger milling machines to handle large production work. These are classified as simplex, duplex and triplex according to single, double and triple spindle heads, they are provided with. The table rests on a stationary bed, and hence these machines are so named. In duplex type, the spindle heads are mounted separately on both sides of the table, whereas in triplex type, the third spindle is mounted on a cross-rail.

23.28.8 Special Milling Machines

Special milling machines are used to perform specific operations on a workpiece. Its rotary type has a table, which can be rotated about a vertical axis. It is used for milling of flat surfaces. In planetary type, the cutter is given planetary motion during cutting and the workpiece remains stationary. It is used for producing internal and external threads on tapered surfaces, rear axle end holes and shell ends.

Duplicating, profiling and pantograph milling machines are used for die and mould cutting, engraving and profiling.

23.28.9 CNC Milling Machines

CNC milling machines are computer controlled vertical mills, and are also called machining centres. It has increased degree of freedom, which makes it suitable for engraving applications with increased precision. The most advanced CNC milling machine is a 5 axis machine, which has two more axes in addition to three normal axes (XYZ).

23.29 Principal Parts of a Milling Machine

The main parts of a column and knee type milling machine are discussed below:

Base refers to the lowest part of the machine, which supports the column, and is a cast part.

Column is vertical part of the machine, which contains the motor and gear mechanism to control the speed of the spindle and worktable.

Knee is supported on column face, and can be raised or lowered with the help of elevating screw attached to the base. It supports the saddle and the table.

Saddle is mounted on the top of the knee, and supports the table. Its accurately finished top surface guides the table during milling operation.

Table rests on top of the saddle and has T slots on its top surface for holding the workpiece firmly during milling operation.

Over-arm refers to the top part of the machine, which contains spindle and cutter. It can be adjusted horizontally to provide better position to the cutter during milling operation.

Spindle is a hallow shaft, supported by the column and over-arm. It is a rotating part, which contains milling cutters over it.

Arbour is an accurately machined shaft for holding spindle nose, and rotates along with spindle carrying milling cutters.

23.30 Shaper

A shaper, also called a shaping machine, is used to produce straight flat surfaces with the help of a single-point cutting tool, as used in a lathe. The tool has reciprocating motion and hence can move backward and forward across the stationary workpiece. The cutting takes place during the forward stroke of the tool and is governed by slotted link or whitworth link mechanism. Shapers are no longer used in modern manufacturing practice, as they involve time consuming operations. These machines have been suitably replaced by milling machines or grinding machines. Their use is reduced to repair of jobs or produce one or two pieces only. But compared to a planner, shaper has many positive points, namely occupying less space, consuming less power, less expensive and more quicker operations.

Shapers can produce a variety of surfaces, namely inclined, grooved, slotted, horizontal, vertical or stepped. Typical works include cutting of splines, gear teeth, keyways, dovetail shape etc.

23.30.1 Shaper Size

The shaper size is determined by the maximum worklength, that can be machined in one full stroke. Other parameters can also be used to complete the specifications of the shaper such as type of drive, type of speed reduction and motor power input.

23.30.2 Classification of Shapers

Shapers are mainly classified as standard, universal, horizontal, vertical, travelling head, hydraulic, geared and crank type. The first two shapers (standard and universal) are based on the design of worktable, the next three (horizontal, vertical and travelling head shapers) are based on the direction of travel of ram and the last three shapers, namely hydraulic, geared and crank type are classified on the basis of driving mechanism. Horizontal shapers are most commonly used. Vertical shaper is also known as a slotter.

23.30.3 Principal Parts of a Shaper

Base supports the whole machine structure and is bolted to the floor with the help of foundation bolts.

Column houses the driving mechanism components of the shaper, namely driving motor, variable speed gear box, levers, handles and other control parts.

Ram is reciprocating part of the shaper, and it moves on guide ways made on the top of the column. Its motion is governed by slotted link mechanism.

Crossrail, also called cross slide, support the table, and can be lowered or raised to give workpiece a suitable position by means of elevating screw.

Saddle is attached to the crossrail to support the **worktable** of the shaper, which is a rigid, box shaped structure, on which workpiece can be mounted. The table has T slots on its upper surface and on one of its side for clamping vice or fixture on it during machining operation.

Shaper head is also called tool head. It is fastened to the front end of the ram and is used to hold the cutting tool and impart it the necessary vertical and angular feed movements.

23.31 Planer

A planer, also called a planing machine, is similar to a shaper as far as the nature of job is concerned. Similar to a shaper, planer is also used to produce accurate flat surfaces and cutting slots, grooves and keyways. Surfaces produced may be horizontal, vertical or inclined. Both shaper and planer employ single-point cutting tools for machining and the tool moves back and forth operated by mechanical means such as rack and pinion gear or by a hydraulic cylinder. They differ in respect of motion of cutting tool and workpiece. In planer, the work is moved against a stationary cutting tool whereas in shaper, the cutting tool moves over a stationary workpiece. This gives more freedom to the size of the workpiece being used in the planer, and makes it suitable for machining of large size workpieces. On the other side, the shaper cannot be used for large workpieces, because of limitations on cutting stroke length and overhang of the ram.

Planers and shapers have now become obsolete and they have been replaced by milling machines. Planers are used in the machining of very large workpiece (4' × 8' size or more), which is either inconvenient to handle or too expensive while using other methods. They have the capacity to remove a large quantity of material from the workpiece in one pass with reasonable accuracy.

There are two important types of planer: double-housing and open-side. The double-housing planer is the most common type of planner, having vertical supports on both sides of its long bed. The horizontal cross-rail is mounted between vertical supports. The open-side planer has vertical support on only one side to which the cross-rail is mounted.

23.31.1 Planer Size

The size of a planer is specified by the size of the largest workpiece, that can be planed in one stroke. It includes the table size (length and width) and its distance from the rail.

23.31.2 Principal Parts of a Planer

As stated earlier, double-housing planer is the most common planer, hence its main parts are discussed below:

Bed forms the lower part of the planer on which worktable is mounted. It has the required grooves to slide the worktable. On both sides of the bed, there are two vertical supports, commonly known as columns or housings.
Worktable is mounted on the bed, and can move back and forth with the workpiece kept on it. The motion of the worktable is guided by a gear mechanism or a hydraulic cylinder. Its upper surface has T slots to facilitate clamping of the workpiece.
Column or housing provides vertical supports on both sides of the bed and are also used to support the cross-rail.
Crossrail is a horizontal structure, which can move up and down during positioning of the workpiece on the worktable. It carries the tool head.
Tool head, two in numbers, is mounted on the horizontal cross-rail. It carries a tool post, which in turn, holds the cutting tool.

23.32 Difference Between Shaper and Planer

- As stated earlier, in shaper, the workpiece remains stationary and the cutting tool moves back and forth on the workpiece, whereas in planer, it is just opposite; the cutting tool remains stationary and the workpiece moves back and forth beneath the cutting tool.
- Feed is provided to the cutting tool in a planer, whereas feed is provided to the workpiece in a shaper.
- Shapers are useful for small work, whereas planers for large work.
- A planer is a robust and heavy machine, having higher power consumption rate, which can only be justified in case of large quantity of similar jobs that can be produced in one go. For that, the planer is provided with several cutting tools, which can be used simultaneously during machining on the workpiece. This is in sharp contrast to a single tool being used at a time in a shaper during machining. Hence, shaper is a small machine tool to be used for small work only.
- The worktable is driven either by gears or hydraulic means in a planer. On the other side, the shaper ram can also be driven in this way, but in most of the cases, quick return link mechanism is used.

Short Answer Questions

1. What are files and where are they used?

Ans. Files are cutting tools mostly used in the fitting shop to remove undesired material from a workpiece in small quantity to give it smooth finish.

2. How does a flat file differ from a hand file?

Ans. A flat file is parallel for about two-third of its length and then tapers in width and thickness, and has double cut on its both faces and single cut on edges. On the other hand, a hand file is parallel for its entire length but has tapered thickness, and has double cut on its both faces, single cut on one edge and there is no cut on other edge, called safe edge.

3. Name a few fitting tools.

Ans. A few examples of fitting tools include files, chisels, scriber, punch and try square.

4. What is the difference among ball peen, straight peen and cross-peen hammers?

Ans. Ball peen hammer has a half ball shaped head called peen, whereas straight peen hammer has a straight peen which is parallel to the axis of the handle of the hammer. The peen of cross-peen hammer is across, that is, perpendicular to the axis of the handle of the hammer. All hammers have flat face which is used for hammering.

5. Name a few sheet metal tools.

Ans. A few examples of sheet metal tools include scriber, rule, snip, punch, stake, mallet and wire gauge.

6. What is wire gauge? For what purpose is it used?

Ans. A wire gauge is a measuring tool used in the sheet metal shop. It is used to measure diameter of a wire or thickness of a metal sheet in terms of gauge number whose higher value indicates a thinner sheet.

7. What is muller? For what purpose is it used?

Ans. Muller is a foundry shop tool. It is used to thoroughly mix the ingredients of the moulding sand such as sand, water, binders or additives so as to make them uniformly distributed.

8. How is a lathe specified?

Ans. A lathe can be specified either by the maximum distance between the lathe centres (headstock and tailstock) which also represents the maximum length of the workpiece to be placed between two centres or by the swing which represents the maximum diameter of the work that can be machined.

Multiple Choice Questions

1. Consider the following statements:

   1. It uses a large workpiece.
   2. Workpiece is stationary, while tool has reciprocating motion.
   3. Large number of jobs can be machined in one go.

   Which of the above statements is true in case of a shaper?

   (a) 1 and 2 (b) 2 alone
   (c) 2 and 3 (d) 1, 2 and 3

2. Consider the following statements:

   1. It uses a large workpiece.
   2. Workpiece is stationary, while tool has reciprocating motion.
   3. Large number of jobs can be machined in one go.
   4. Open-side planer has a vertical support on only one side.

   Which of the above statements is true in case of a planer?

   (a) 1 and 2 (b) 2 alone
   (c) 3 and 4 (d) 1, 3 and 4

3. Planers and shapers have limited applications in modern manufacturing practices. They have been suitably replaced by

   (a) lathe machines (b) milling machines
   (c) boring machines (d) drilling machines

4. Consider the following statements:

   1. Double housing is one of the types of a planer.
   2. A vertical cross-rail is used in a double-housing planer.
   3. An open-side planer has one vertical support.
   4. A planer is a bulk material removal machine tool.

   Which of the above statements is true in case of a shaper?

(a) 1 alone (b) 1 and 3
(c) 1, 3 and 4 (d) 1 and 2

5. Consider the following statements:

   1. A vertical shaper is sometimes referred to as a slotter.
   2. Vertical shapers are generally fitted with a rotary table to enable curved surfaces to be machined.
   3. Both forward and backward strokes are used in cutting operations.
   4. Backward strokes are idle strokes.

   Of these statements:

(a) 1 and 2 are true (b) 1, 2 and 4 are true
(c) 2 and 3 are true (d) 1 and 3 are true

6. Consider the following statements:

   1. In shaper, the tool is stationary, while the workpiece moves to and fro.
   2. Feed is provided to the cutting tool in a planer.
   3. A planer is a robust and heavy machine as compared to a shaper.
   4. A planer is suitable for large workpiece.

   Of these statements:

(a) 1 and 2 are true (b) 1 and 3 are true
(c) 2, 3 and 4 are true (d) 3 and 4 are true

7. Match List I with List II and select the correct answer using the codes given below the lists:

List I	List II
A. Double-housing planer	1. Shaper
B. Vertical shaper	2. Planer
C. Large workpiece	3. Two vertical supports
D. Moving tool	4. Slotter

Codes:	A	B	C	D
(a)	3	4	1	2
(b)	3	4	2	1
(c)	4	3	2	1
(d)	4	3	1	2

8. Match List I with List II and select the correct answer using the codes given below the lists:

List I	List II
A. Stationary workpiece	1. Planer
B. Stationary cutting tool	2. Single-point cutting tool
C. Lathe	3. Shaper
D. Milling machines	4. Multiple tooth cutters

Codes:	A	B	C	D
(a)	3	1	4	2
(b)	1	3	2	4
(c)	3	1	2	4
(d)	1	3	4	2

9. Match List I with List II and select the correct answer using the codes given below the lists:

List I	List II
A. Forward stroke	1. Idle stroke
B. Backward stroke	2. Lathe
C. Slotted link mechanism	3. Cutting stroke
D. Turning	4. Shaper

Codes:	A	B	C	D
(a)	1	3	4	2
(b)	3	1	4	2
(c)	3	1	2	4
(d)	1	3	2	4

10. Match List I with List II and select the correct answer using the codes given below the lists:

List I (Types of lathe)	List II (Description)
A. Engine lathe	1. Mass production
B. Turret lathe	2. Basic turning machine
C. Capstan lathe	3. Dies
D. Tool room lathe	4. Hexagonal-shaped tool carrier

Codes:	A	B	C	D
(a)	2	1	3	4
(b)	2	1	4	3
(c)	1	2	3	4
(d)	3	1	4	2

11. Match List I with List II and select the correct answer using the codes given below the lists:

List I	List II
A. Conical surface	1. Internal turning
B. Cylindrical surface	2. Taper turning
C. Boring	3. Straight turning
D. Flat surface	4. Facing

Codes:	A	B	C	D
(a)	2	4	1	3
(b)	1	3	2	4
(c)	2	3	1	4
(d)	2	3	4	1

12. Match List I with List II and select the correct answer using the codes given below the lists:

List I (Lathe components)	List II (Description)
A. Bed	1. H-shaped casting
B. Head stock	2. Tool post
C. Carriage	3. Hollow spindle
D. Compound rest	4. Base of a lathe

Codes:	A	B	C	D
(a)	4	3	2	1
(b)	4	3	1	2
(c)	1	3	2	4
(d)	2	1	4	3

13. Consider the following statements about a lathe:

    1. A headstock contains a hollow spindle and a motion control unit consisting of transmission gears.
    2. The tool post is mounted on a compound rest.
    3. The bed of a lathe is made of nodular cast iron.
    4. A leadscrew is used for threads cutting.

    Of these statements:

    (a) 1 and 2 are true (b) 2 and 3 are true
    (c) 1, 2, 3 and 4 are true (d) 1, 2 and 3 are true

14. Match List I with List II and select the correct answer using the codes given below the lists:

List I	List II
A. Swing	1. Uses six tools mounting
B. Engine lathe	2. Maximum diameter of a workpiece
C. Turret lathe	3. Maximum distance between centres
D. Chuck	4. Most frequently used
	5. Work holding device

Codes:	A	B	C	D
(a)	2	1	4	5
(b)	3	1	4	2
(c)	3	4	1	5
(d)	2	4	1	5

15. Consider the following statements:

    1. Lathe cutting tools are made of high-speed steel (HSS).
    2. In a turret lathe, workpiece is held in collet.
    3. Faceplates are used to support irregularly shaped workpiece.
    4. Inserts are disposable cutting tools.

    Of these statements:

    (a) 1 and 3 are true (b) 2 and 4 are true
    (c) 2, 3 and 4 are true (d) 1, 2, 3 and 4 are true

16. Consider the following statements about capstan and turret lathes:

1. They are modified forms of a centre lathe.
2. They have no tailstock.
3. They are used for mass production.
4. They are very much suitable for one or few jobs.

Of these statements:

(a) 2 and 3 are true (b) 1, 2 and 3 are true
(c) 1, 2 and 4 are true (d) 2 and 4 are true

17. Match List I with List II and select the correct answer using the codes given below the lists:

List I	List II
A. Engine lathe	1. Production lathe
B. Capstan lathe	2. Main part of a lathe
C. Chuck	3. Versatile machine
D. Tool post	4. Lathe accessory

Codes:	A	B	C	D
(a)	3	1	2	4
(b)	3	1	4	2
(c)	1	3	4	2
(d)	1	3	2	4

Answers

1. (b)	2. (d)	3. (b)	4. (c)	5. (b)	6. (c)
7. (b)	8. (c)	9. (b)	10. (b)	11. (c)	12. (b)
13. (c)	14. (d)	15. (d)	16. (b)	17. (b)	

Review Questions and Discussions

Q.1. How are files classified?
Q.2. Differentiate between a flat file and a hand file.
Q.3. Forging gives superior mechanical properties. How ?

Q.4. Differentiate between a ball peen hammer and a cross-peen hammer.
Q.5. What is the use of a swage block?
Q.6. Casting is an economical operation. How?
Q.7. Why is lathe called a basic machine tool?
Q.8. Why is boring termed as internal turning?
Q.9. Why are lathe centres called live and dead?
Q.10. What is the function of a tool post?
Q.11. How does headstock differ from tailstock?
Q.12. Why is centre lathe termed a versatile machine?
Q.13. Differentiate between capstan and turret lathes.
Q.14. Why are capstan and turret lathes called production lathes?
Q.15. Why are capstan and turret lathes used for limited number of jobs?
Q.16. Why is universal milling machine so named?
Q.17. Name a few special types of milling machines.

Bibliography

Alting, L. (1994). *Manufacturing engineering processes.* Marcel Dekker, Inc.

Amstead, B. H., et al. (1987). *Manufacturing processes.* Wiley (2nd ed.).

Armarego, E. J. A., & Brown, R. H. (1969). *Machining of metals.* Prentice Hall.

Avitzur, B. (1983). *Handbook of metal forming processes.* Wiley Interscience.

Black, S. C., Chites, V., & Lissman, A. J. (1966). *Principles of engineering manufacture* (3rd ed.). Arnold.

Campbell, J. S. (1999). *Principles of manufacturing materials and processes.* Tata McGraw Hill.

Cary, H. B. (1989). *Modern welding technology* (2nd ed.). Prentice Hall.

Chapman, W. A. J., & Martin, S. J. (1975). *Workshop technology* (Vol. 1, 2 and 3). Arnold Publisher (India) Pvt. Ltd.

Clark, D. S., & Varney, W. R. (1962). *Physical metallurgy for engineers* (2nd ed.). Affiliated East-West Press Pvt. Ltd.

DeGarmo, E. P., et al. (1997). *Materials and processes in manufacturing* (8th Ed.). Prentice Hall of India Pvt. Ltd.

Deiter, G. E. (1976). *Mechanical metallurgy.* McGraw Hill.

Donaldson, C., et al. (1976). *Tool design* (3rd ed.). Tata McGraw Hill Publishing Company Ltd.

Doyle, L. E., et al. (1985). *Manufacturing processes and materials for engineers* (3rd ed.). Prentice Hall.

Farago, F. T., & Curtis, M. A. (1994). *Handbook of dimensional measurement* (3rd ed.). Industrial Press Inc.

Flinn, R. A. (1963). *Fundamentals of metal casting.* Addison-Wesley.

Flinn, R. A., & Trojan, P. K. (1998). *Engineering materials and their applications* (2nd ed.). Jaico Publishing House.

German, R. M. (1994). *Powder metallurgy science.* Metal Powder Industries Federation.

Ghosh, A., & Mallik, A. K. (1985). *Manufacturing science* (1st ed.). Affiliated East-West Press Pvt. Ltd.

Groover, M. P. (2012). *Introduction to manufacturing processes.* Wiley.

Gupta, K. M. (1997). *Materials science and engineering* (1st ed.). Umesh Publications.

Habicht, F. H. (1963). *Modern machine tools.* D. Van Nostrand Co.

Heine, R., & Rosenthal, P. (1955). *Principles of metal casting* (International Student Edition). McGraw Hill Book Company, Inc.

Kalpakjian, S. (1995). *Manufacturing engineering and technology* (3rd ed.). Addison-Wesley Publishing Company, Inc.

Koenigsberger, F. (1964). *Design principles of metal cutting machine tools.* McMillian Co.

Lincoln, J. F. (1989). *Modern welding technology* (2nd ed.). Prentice Hall.

D. K. Singh, *Fundamentals of Manufacturing Engineering*,
https://doi.org/10.1007/978-981-99-8767-2

Lindberg, R. A. (1995). *Processes and materials of manufacturing* (4th ed.). Prentice Hall of India Pvt. Ltd.
Little, R. L. (1973). *Welding and welding technology*. McGraw Hill.
Niebel, B. W., et al. (1989). *Modern manufacturing process engineering*. McGraw Hill.
Olive, C. T. (1989). *Machine tool technology and manufacturing processes*. Galgotia Publications Pvt. Ltd.
Ostwald, P. F., & Jairo, M. (1998). *Manufacturing processes and systems*. Wiley.
Parmar, R. S. (1997a). *Welding engineering and technology* (1st ed.). Khanna Publishers.
Parmar, R. S. (1997b). *Welding processes and technology* (2nd ed.). Khanna Publishers.
Schey, J. A. (2012). *Introduction to manufacturing processes*. McGraw Hill.
Sharma, P. C. (1996). *A text book of production technology*. S. Chand and Company Ltd.
Shaw, M. C. (2006). *Metal cutting principles* (2nd ed.). Oxford University Press, Inc.
Springborn, R. K. (1967). *Nontraditional machining processes*. American Society of Tool and Manufacturing Engineers.
Tanner, J.P. (1991). *Manufacturing engineering*: an introduction to the basic functions (2nd ed.). Marcel Dekker, Inc.
Timing, R.L., & Wilkinson, S.P. (2000). *Manufacturing technology* (2nd ed.). Pearson.
Van Vlack, L. H. (1987). *Elements of materials science and engineering* (5th ed.). Addison-Wesley Publishing Company.
Walters, T.F. (1996). *Fundamentals of manufacturing for engineers*. UCL Press
Welker, E. J. (1984). *Non-traditional machining processes*. S.M.E.
Wilson, F. W. (1968). *Machining with carbides and oxides*. McGraw Hill.

Subject Index

D. K. Singh, *Fundamentals of Manufacturing Engineering*,
https://doi.org/10.1007/978-981-99-8767-2

D

MIX
Papier aus verantwortungsvollen Quellen
Paper from responsible sources
FSC® C105338

If you have any concerns about our products,
you can contact us on
ProductSafety@springernature.com

In case Publisher is established outside the EU,
the EU authorized representative is:
Springer Nature Customer Service Center GmbH
Europaplatz 3, 69115 Heidelberg, Germany

Printed by Libri Plureos GmbH
in Hamburg, Germany